Student Solutions Ma

MW00333349

Applied Calculus

SEVENTH EDITION

Geoffrey C. Berresford
Long Island University

Andrew M. Rockett
Long Island University

Prepared by

Christine Verity

CENGAGE
Learning

Australia • Brazil • Mexico • Singapore • United Kingdom • United States

ISBN: 978-1-305-10795-3

Cengage Learning
20 Channel Center Street
Boston, MA 02210
USA

Cengage Learning is a leading provider of customized learning solutions with office locations around the globe, including Singapore, the United Kingdom, Australia, Mexico, Brazil, and Japan. Locate your local office at: **www.cengage.com/global**.

Cengage Learning products are represented in Canada by Nelson Education, Ltd.

To learn more about Cengage Learning Solutions, visit **www.cengage.com**.

Purchase any of our products at your local college store or at our preferred online store **www.cengagebrain.com**.

Printed in the United States of America
Print Number: 01 Print Year: 2014

Contents

Diagnostic Test

DIAGNOSTIC TEST

1. The inequality symbols < and > *point* to the smaller quantity, so $\frac{1}{2} < -3$ is *false* because the positive number $\frac{1}{2}$ is *not* smaller than the negative number –3. Using the real number line to visualize the real numbers, smaller values are to the left and larger values are to the right:

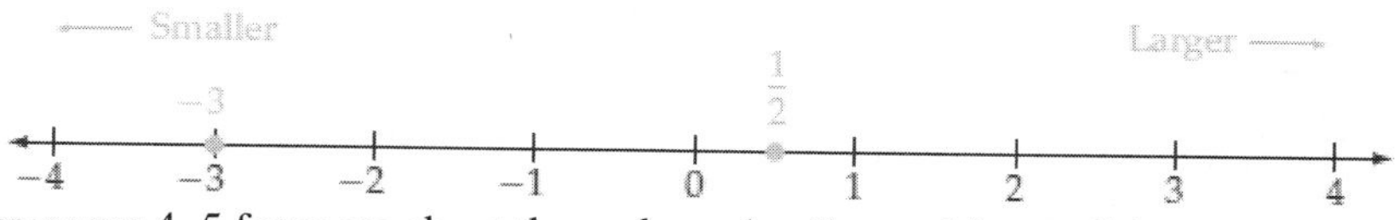

See pages 4–5 for more about the real number line and inequalities.

Be Careful Inequalities indicate the relative *left/right position* of values on the number line, while the absolute value is the *distance* from the number to zero. So $\left|\frac{1}{2}\right| < |-3|$ because $\frac{1}{2}$ is *closer to zero* than is –3. See pages 52–53 for more about the absolute value function.

2. The numbers in the set $\{x \mid -4 < x \le 5\}$ form an interval from –4 to 5 excluding the left endpoint –4 but including the right endpoint 5. We say that the interval is *open* on the left and *closed* on the right. Interval notation uses parentheses "(" and ")" to indicate open endpoints, and square brackets "[" and "]" to indicate closed endpoints. So this interval is written (–4, 5}. See pages 5–6 for more about intervals.

3. The slope of the line through $(x_1, y_1) = (6, -7)$ and $(x_2, y_2) = (9, 8)$ is

$$m = \frac{\Delta y}{\Delta x} = \frac{y_2 - y_1}{x_2 - x_1} = \frac{8-(-7)}{9-6} = \frac{15}{3} = 5$$

See page 7 for more about slope, and pages B5–B6 later in this Appendix for more about order of operations and evaluating expressions.

4. The line $y = 3x + 4$ is written in slope-intercept form $y = mx + b$, so the slope is $m = 3$. Since

$$m = \frac{\Delta y}{\Delta x}$$

we have

$$3 = \frac{\Delta y}{2} \quad \text{giving} \quad \Delta y = 3 \cdot 2 = 6$$

See page 7 for more about slope, and page 9 for more about the slope-intercept form of the equation for a line.

5. The line $y = 2x - 1$ has slope is $m = 2$ and y-intercept $(0, b) = (0, -1)$. Since the slope is positive, the line slopes upward, eliminating (d). Since $b = -1$ is negative, the y-intercept is below the x-axis, eliminating (c). Since the slope is the amount the line rises when x increases by 1, starting from $(0, -1)$ in either of (a) or (b) and moving *right* 1 and *up* 2, we see (a) is correct while (b) is not.

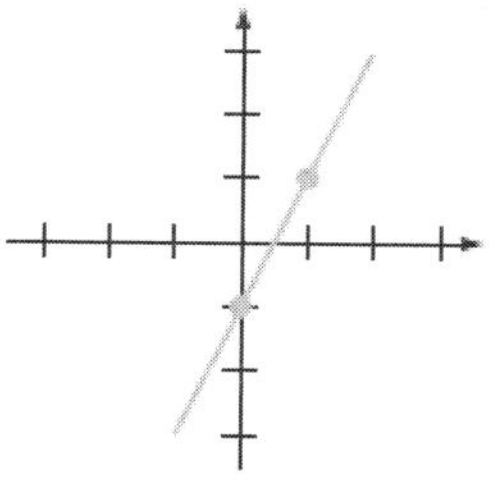

See pages 7 and 9 for more about slopes, y-intercepts, slope-intercept equations, and graphs of lines.

6. The statement is *true* because

$$\left(\frac{\sqrt{x}}{y}\right)^{-2} = \left(\frac{y}{\sqrt{x}}\right)^{2} = \frac{y^2}{(\sqrt{x})^2} = \frac{y^2}{x}$$

where in the first step we used the fact that a negative exponent means the *reciprocal*. See pages 22–23 for more about exponents and roots.

7. Since $9x^2 - 6x - 1$ does not have a simple factorization, we solve the equation $9x^2 - 6x - 1 = 0$ by the Quadratic Formula

$$x = \frac{-b \pm \sqrt{b^2 - 4ac}}{2a}$$

Using $a = 9$, $b = -6$, and $c = -1$, we obtain

$$x = \frac{-(-6) \pm \sqrt{(-6)^2 - 4(9)(-1)}}{2(9)} = \frac{6 \pm \sqrt{72}}{18} = \frac{6 \pm 6\sqrt{2}}{18} = \frac{1 \pm \sqrt{2}}{3}$$

See page 39 for more about quadratic functions and the Quadratic Formula, and pages B5 and B7 later in this Appendix for more about order of operations and evaluating expressions.

8. Multiplying out to remove the parentheses,

$$x(8 - x) - (3x + 7) = 8x - x^2 - 3x - 7$$

Reordering and combining *like* terms (that is, terms with the same power of x) gives

$$-x^2 + (8x - 3x) - 7 = -x^2 + 5x - 7$$

See pages 48–49 for more about working with polynomials, and page B5 later in this Appendix for more about the distributive law.

9. The domain of the rational function $f(x) = \dfrac{x^2 - 3x + 2}{x^3 + x^2 - 6x}$ is the set of numbers for which the denominator $x^3 + x^2 - 6x$ is not zero, so we factor:

$$x^3 + x^2 - 6x = x(x^2 + x - 6) = x(x + 3)(x - 2)$$

Since the denominator is zero when x is 0, –3, or 2, the domain is

$$\{x \mid x \neq -3,\ x \neq 0,\ x \neq 2\}$$

See pages 49–50 for more about rational functions, and pages B11–B12 later in this Appendix for more about factoring.

Be Careful When finding the domain of a rational function, do not begin by "simplifying" the function. If you first factor the numerator as $(x-1)(x-2)$ and cancel the $(x-2)$'s, you would overlook the fact that $x = 2$ makes the original denominator zero. In fact, "cancelling" the $(x-2)$'s requires assuming that $x \neq 2$, or else you are dividing by zero, which is not allowed.

10. For $f(x) = x^2 - 5x$, the difference quotient is

$$\frac{f(x+h) - f(x)}{h} = \frac{(x+h)^2 - 5(x+h) - (x^2 - 5x)}{h}$$

$$= \frac{x^2 + 2hx + h^2 - 5x - 5h - x^2 + 5x}{h}$$

$$= \frac{2hx + h^2 - 5h}{h} = 2x + h - 5$$

See pages 57–58 for more about difference quotients, and pages B8–B11 later in this Appendix for more about expanding and simplifying expressions.

Chapter 1: Functions

EXERCISES 1.1

1. $\{x \mid 0 \le x < 6\}$

3. $\{x \mid x \le 2\}$

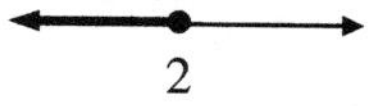

5. **a.** Since $\Delta x = 3$ and $m = 5$, then Δy, the change in y, is

$$\Delta y = 3 \bullet m = 3 \bullet 5 = 15$$

b. Since $\Delta x = -2$ and $m = 5$, then Δy, the change in y, is

$$\Delta y = -2 \bullet m = -2 \bullet 5 = -10$$

7. For (2, 3) and (4, –1), the slope is

$$\frac{-1-3}{4-2} = \frac{-4}{2} = -2$$

9. For (–4, 0) and (2, 2), the slope is

$$\frac{2-0}{2-(-4)} = \frac{2}{2+4} = \frac{2}{6} = \frac{1}{3}$$

11. For (0, –1) and (4, –1), the slope is

$$\frac{-1-(-1)}{4-0} = \frac{-1+1}{4} = \frac{0}{4} = 0$$

13. For (2, –1) and (2, 5), the slope is

$$\frac{5-(-1)}{2-2} = \frac{5+1}{0} \qquad \text{undefined}$$

15. Since $y = 3x - 4$ is in slope-intercept form, $m = 3$ and the y-intercept is (0, –4). Using the slope $m = 3$, we see that the point 1 unit to the right and 3 units up is also on the line.

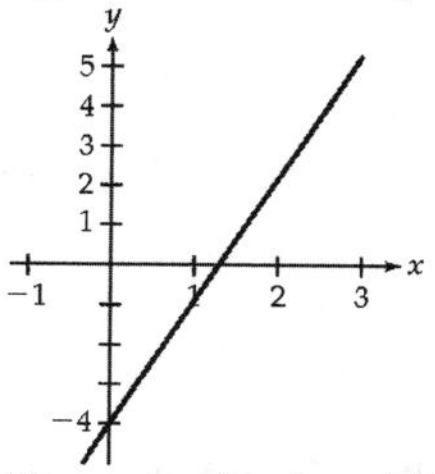

17. Since $y = -\frac{1}{2}x$ is in slope-intercept form, $m = -\frac{1}{2}$ and the y-intercept is (0, 0). Using $m = -\frac{1}{2}$, we see that the point 2 units to the right and 1 unit down is also on the line.

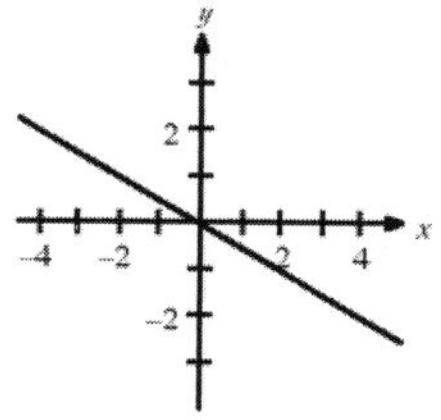

19. The equation $y = 4$ is the equation of the horizontal line through all points with y-coordinate 4. Thus, $m = 0$ and the y-intercept is (0, 4).

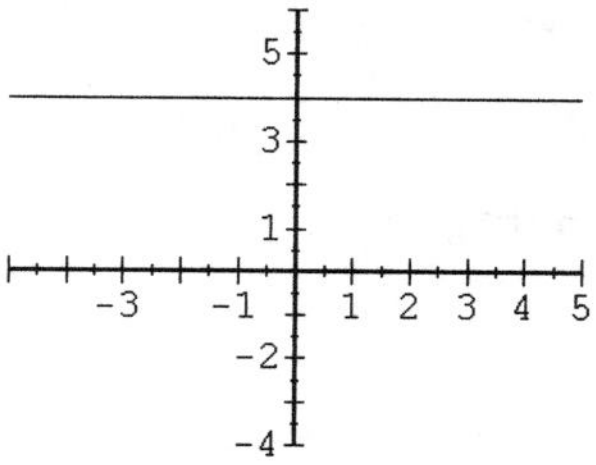

21. The equation $x = 4$ is the equation of the vertical line through all points with x-coordinate 4. Thus, m is not defined and there is no y-intercept.

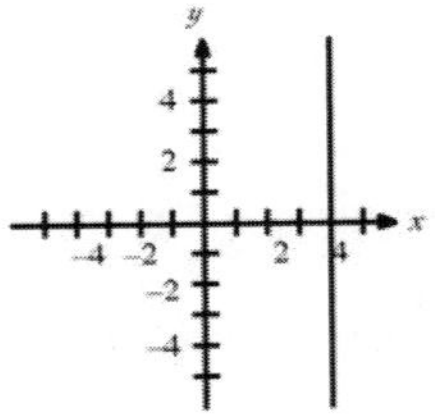

23. First, solve for y:

$$2x - 3y = 12$$
$$-3y = -2x + 12$$
$$y = \frac{2}{3}x - 4$$

Therefore, $m = \frac{2}{3}$ and the y-intercept is (0, –4).

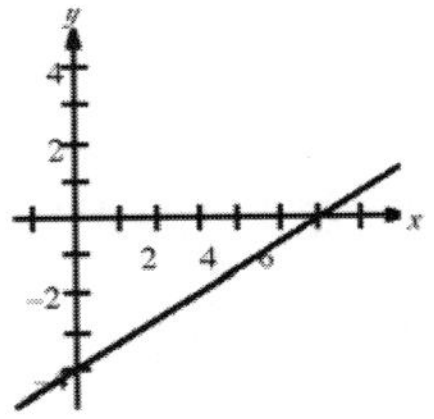

25. First, solve for y:

$$x + y = 0$$
$$y = -x$$

Therefore, $m = -1$ and the y-intercept is (0, 0).

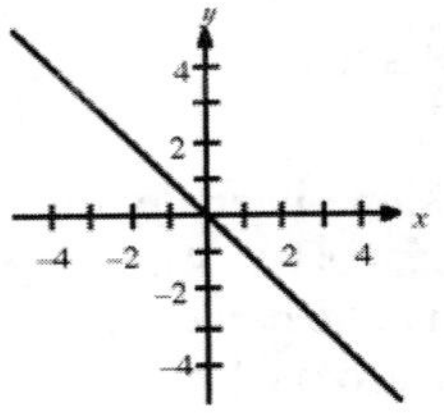

27. First, solve for y:

$$x - y = 0$$
$$-y = -x$$
$$y = x$$

Therefore, $m = 1$ and the y-intercept is (0, 0).

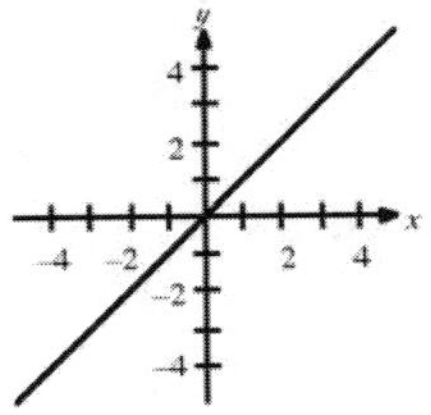

29. First, put the equation in slope-intercept form:

$$y = \frac{x+2}{3}$$
$$y = \frac{1}{3}x + \frac{2}{3}$$

Therefore, $m = \frac{1}{3}$ and the y-intercept is $\left(0, \frac{2}{3}\right)$.

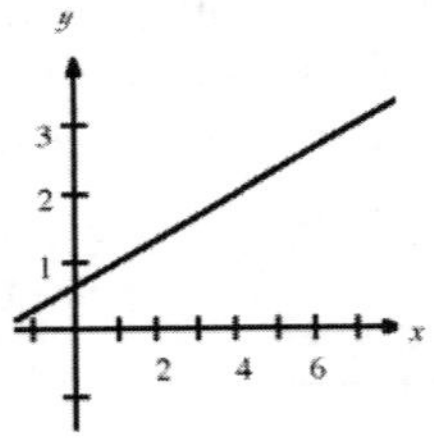

31. First, solve for y:

$$\frac{2x}{3} - y = 1$$
$$-y = -\frac{2}{3}x + 1$$
$$y = \frac{2}{3}x - 1$$

Therefore, $m = \frac{2}{3}$ and the y-intercept is (0, –1).

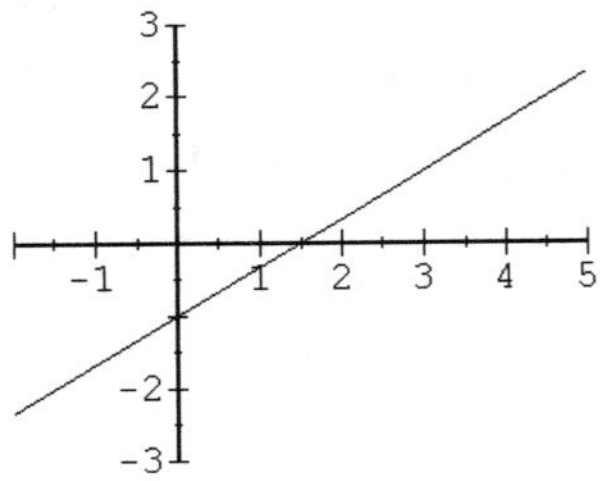

33. $y = -2.25x + 3$

35.

$$y - (-2) = 5[x - (-1)]$$
$$y + 2 = 5x + 5$$
$$y = 5x + 3$$

37. $y = -4$

39. $x = 1.5$

41. First, find the slope.

$m = \frac{-1-3}{7-5} = \frac{-4}{2} = -2$

Then use the point-slope formula with this slope and the point (5, 3).

$$y-3=-2(x-5)$$
$$y-3=-2x+10$$
$$y=-2x+13$$

43. First, find the slope.

$m = \frac{-1-(-1)}{5-1} = \frac{-1+1}{4} = 0$

Then use the point-slope formula with this slope and the point (1,–1).

$$y-(-1)=0(x-1)$$
$$y+1=0$$
$$y=-1$$

45. **a.** First find the slope of the line $4y-3x=5$. Write the equation in slope-intercept form.

$$y=\frac{3}{4}x+\frac{5}{4}$$

The slope of the parallel line is $m=\frac{3}{4}$.

Next, use the point-slope form with the point (12, 2):

$$y-y_1=m(x-x_1)$$
$$y-2=\frac{3}{4}(x-12)$$
$$y=\frac{3}{4}x-7$$

b. The slope of the line perpendicular to $y=\frac{3}{4}x+\frac{5}{4}$. is $m=-\frac{4}{3}$.

Next, use the point-slope form with the point (12, 2):

$$y-y_1=m(x-x_1)$$
$$y-2=-\frac{4}{3}(x-12)$$
$$y=-\frac{4}{3}x+18$$

47. The y-intercept of the line is (0, 1), and $\Delta y=-2$ for $\Delta x = 1$. Thus, $m=\frac{\Delta y}{\Delta x}=\frac{-2}{1}=-2$. Now, use the slope-intercept form of the line:
$y=-2x+1$.

49. The y-intercept is (0, –2), and $\Delta y = 3$ for $\Delta x = 2$. Thus, $m=\frac{\Delta y}{\Delta x}=\frac{3}{2}$. Now, use the slope-intercept form of the line: $y=\frac{3}{2}x-2$

51. First, consider the line through the points (0, 5) and (5, 0). The slope of this line is $m=\frac{0-5}{5-0}=\frac{-5}{5}=-1$. Since (0, 5) is the y-intercept of this line, use the slope-intercept form of the line: $y=-1x+5$ or $y=-x+5$.

Now consider the line through the points (5, 0) and (0, –5). The slope of this line is $m=\frac{-5-0}{0-5}=\frac{-5}{-5}=1$. Since (0,–5) is the y-intercept of the line, use the slope-intercept form of the line: $y=1x-5$ or $y=x-5$

Next, consider the line through the points (0, –5) and (–5, 0). The slope of this line is $m=\frac{0-(-5)}{-5-0}=\frac{5}{-5}=-1$. Since (0, –5) is the y-intercept, use the slope-intercept form of the line: $y=-1x-5$ or $y=-x-5$

Finally, consider the line through the points (–5, 0) and (0, 5). The slope of this line is $m=\frac{5-0}{0-(-5)}=\frac{5}{5}=1$. Since (0, 5) is the y-intercept, use the slope-intercept form of the line: $y=1x+5$ or $y=x+5$

53. If the point (x_1, y_1) is the y-intercept $(0, b)$, then substituting into the point-slope form of the line gives

$$y-y_1=m(x-x_1)$$
$$y-b=m(x-0)$$
$$y-b=mx$$
$$y=mx+b$$

55. **a.**

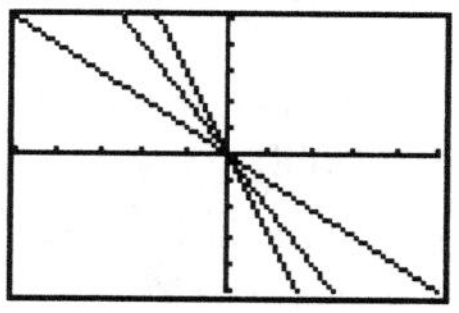

on [–5, 5] by [–5, 5]

b.

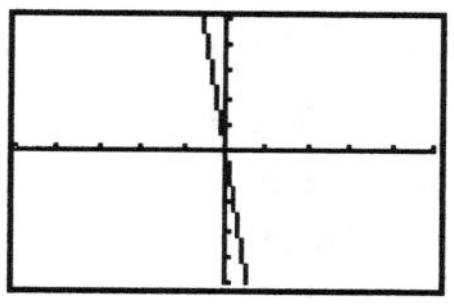

on [–5, 5] by [–5, 5]

57. Low demand: [0, 8);
average demand: [8, 20);
high demand: [20, 40);
critical demand: [40, ∞)

59. **a.** The value of x corresponding to the year 2020 is $x = 2020 - 1900 = 120$. Substituting $x = 120$ into the equation for the regression line gives

$$y = -0.356x + 257.44$$
$$y = -0.356(120) + 257.44 = 214.72 \text{ seconds}$$

Since 3 minutes = 180 seconds, 214.72 = 3 minutes 34.72 seconds. Thus, the world record in the year 2020 will be 3 minutes 34.72 seconds.

b. To find the year when the record will be 3 minutes 30 seconds, first convert 3 minutes 30 seconds to seconds: 3 minutes 30 seconds = 3 minutes • $\frac{60 \text{ sec}}{1 \text{ min}}$ + 30 seconds = 210 seconds.

Now substitute $y = 210$ seconds into the equation for the regression line and solve for x.

$$y = -0.356x + 257.44$$
$$210 = -0.356x + 257.44$$
$$0.356x = 257.44 - 210$$
$$0.356x = 47.44$$
$$x = \frac{47.44}{0.356} \approx 133.26$$

Since x represents the number of years after 1900, the year corresponding to this value of x is $1900 + 133.26 = 2033.26 \approx 2033$. The world record will be 3 minutes 30 seconds in 2033.

61. **a.** To find the linear equation, first find the slope of the line containing these points.

$$m = \frac{146-70}{3-1} = \frac{76}{2} = 38$$

Next, use the point-slope form with the point (1, 70):

$$y - y_1 = m(x - x_1)$$
$$y - 70 = 38(x-1)$$
$$y = 38x + 32$$

b. Sales are increasing by 38 million units per year.

c. The sales at the end of 2020 is $y = 38(10) + 32 = 412$ million units.

63. **a.** First, find the slope of the line containing the points.

$$m = \frac{212-32}{100-0} = \frac{180}{100} = \frac{9}{5}$$

Next, use the point-slope form with the point (0, 32):

$$y - y_1 = m(x - x_1)$$
$$y - 32 = \frac{9}{5}(x-0)$$
$$y = \frac{9}{5}x + 32$$

b. Substitute 20 into the equation.

$$y = \frac{9}{5}x + 32$$
$$y = \frac{9}{5}(20) + 32 = 36 + 32 = 68^\circ \text{F}$$

65. **a.** Price = \$50,000; useful lifetime = 20 years; scrap value = \$6000

$$V = 50{,}000 - \left(\frac{50{,}000-6000}{20}\right)t \quad 0 \le t \le 20$$
$$= 50{,}000 - 2200t \quad 0 \le t \le 20$$

b. Substitute $t = 5$ into the equation.

$$V = 50{,}000 - 2200t$$
$$= 50{,}000 - 2200(5)$$
$$= 50{,}000 - 11{,}000 = \$39{,}000$$

c.

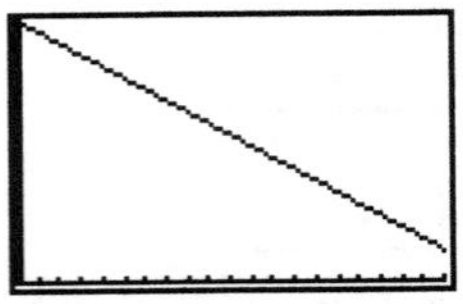

on [0, 20] by [0, 50,000]

67. **a.** Substitute $w = 10$, $r = 5$, $C = 1000$ into the equation.

$10L + 5K = 1000$

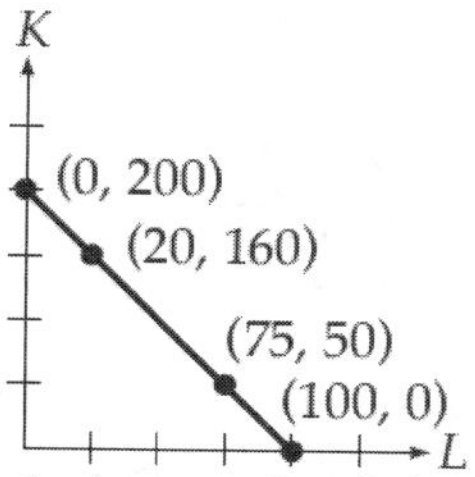

b. Substitute each pair into the equation.

For (100, 0), $10(100) + 5(0) = 1000$

For (75, 50), $10(75) + 5(50) = 1000$

For (20, 160), $10(20) + 5(160) = 1000$

For (0, 200), $10(0) + 5(200) = 1000$

Every pair gives 1000.

69. **a.** Median Marriage Age for Men and Women

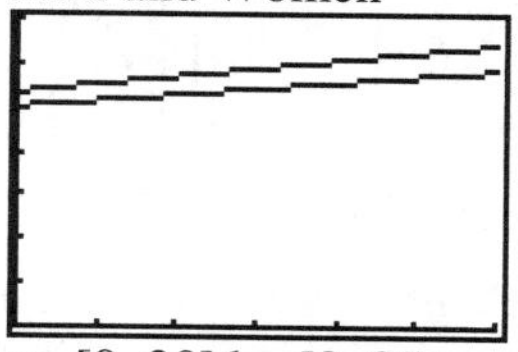

on [0, 30] by [0, 35]

b. The x-value corresponding to the year 2020 is $x = 2020 - 2000 = 20$. The following screens are a result of the CALCULATE command with $x = 20$.

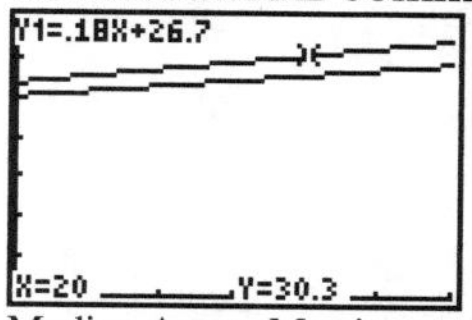

Median Age at Marriage for Men in 2020

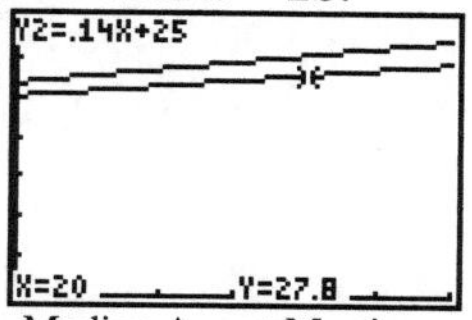

Median Age at Marriage for Women in 2020.

So, the median marriage age for men in 2020 will be 30.3 years and for women it will be 27.8 years.

c. The x-value corresponding to the year 2030 is $x = 2030 - 2000 = 30$. The following screens are a result of the CALCULATE command with $x = 30$.

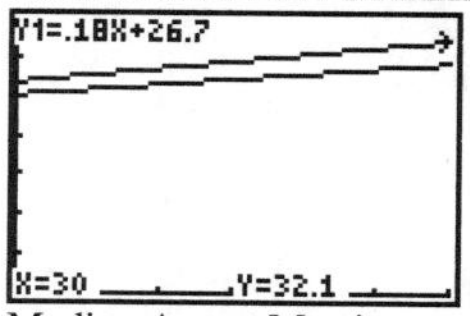

Median Age at Marriage for Men in 2030

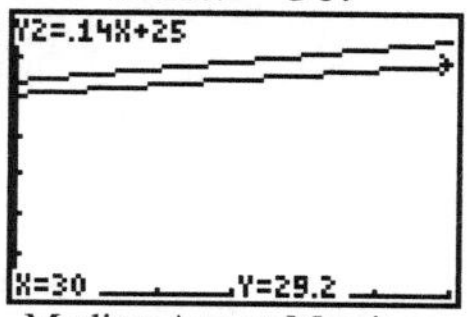

Median Age at Marriage for Women in 2030.

So, the median marriage age for men in 2030 will be 32.1 years and for women it will be 29.2 years.

71. **a.**

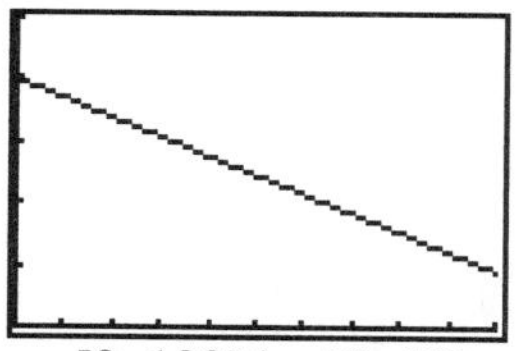

on [0, 100] by [0, 50]

b. To find the probability that a person with a family income of \$40,000 is a smoker, substitute 40 into the equation

$y = -0.31x + 40$

$y = -0.31(40) + 40 = 27.6$ or 28%.

c. The probability that a person with a family income of \$70,000 is a smoker is $y = -0.31(70) + 40 = 18.3$ or 18%.

73. **a.**

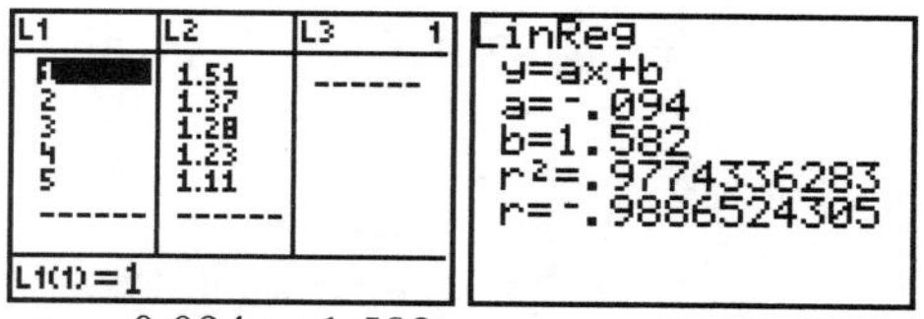

$y = -0.094x + 1.582$

b. Cigarette consumption is declining by about 94 cigarettes (from 0.094 thousand, so about 5 packs) per person per year.

c. $y = -0.094(13) + 1.582 = 0.36$ thousand (360 cigarettes)

75. **a.**

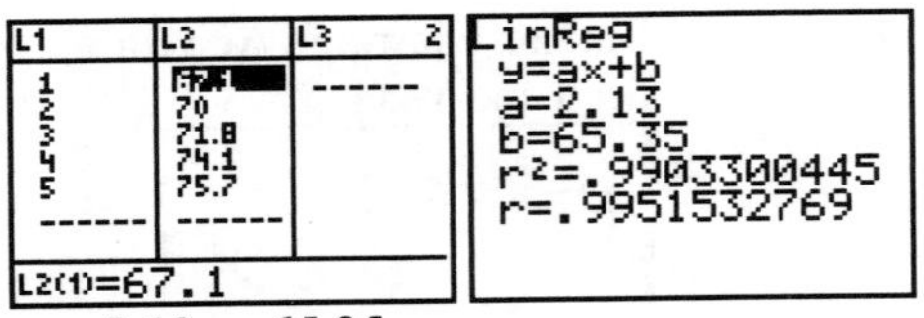

$y = 2.13x + 65.35$

b. The male life expectancy is increasing by 2.13 years per decade, which is 0.213 years (or about 2.6 months per year).

c. $y = 2.13(6.5) + 65.35 \approx 79.2$ years

77. **a.**

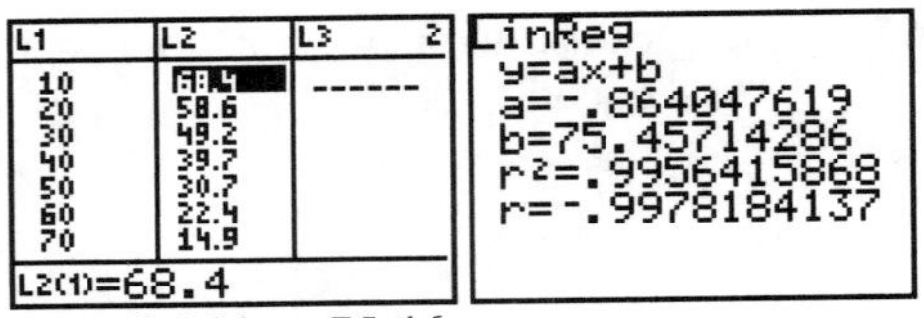

$y = -0.864x + 75.46$

b. Future longevity decreases by 0.864 (or about 10.44 months) per year.

c. $y = -0.864(25) + 75.46 \approx 53.9$ years

d. It would not make sense to use the regression line to predict future longevity at age 90 because the line predicts –2.3 years of life remaining.

79. False: Infinity is not a number.

81. $m = \dfrac{y_2 - y_1}{x_2 - x_1}$ for any two points (x_1, y_1) and (x_1, y_1) on the line *or* the slope is the amount that the line rises when x increases by 1.

83. False: The slope of a vertical line is undefined.

85. True: The slope is $\dfrac{-a}{b}$ and the y-intercept is $\dfrac{c}{b}$.

87. False: It should be $\dfrac{y_2 - y_1}{x_2 - x_1}$.

89. Drawing a picture of a right triangle.

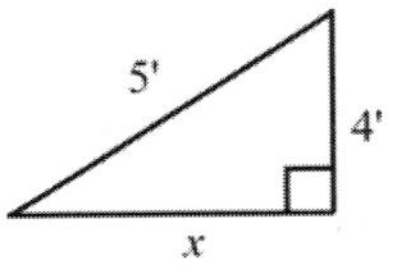

$$x^2 + 4^2 = 5^2$$
$$x^2 + 16 = 25$$
$$x^2 = 9$$
$$x = 3$$

The slope is $m = \dfrac{4}{3}$ or $-\dfrac{4}{3}$ if the ladder slopes downward.

91. To find the x-intercept, substitute $y = 0$ into the equation and solve for x:

$$y = mx + b$$
$$0 = mx + b$$
$$-mx = b$$
$$x = -\frac{b}{m}$$

If $m \neq 0$, then a single x-intercept exists. So $a = -\dfrac{b}{m}$. Thus, the x-intercept is $\left(-\dfrac{b}{m}, 0\right)$.

93. Consider $R > 1$ and $0 < x < K$

$x < K$ means that $K - x > 0$ and $0 < \dfrac{x}{K} < 1$.

Since $K - x > 0$, then

$$K - x + Rx > Rx$$
$$K + (R-1)x > Rx$$
$$K\left[1 + \left(\frac{R-1}{K}\right)x\right] > Rx$$

Therefore, $K > \dfrac{Rx}{1 + \left(\frac{R-1}{K}\right)x} = y$

Additionally, since $0 < \dfrac{x}{K} < 1$,

$$1 + (R-1)\left(\frac{x}{K}\right) < 1 + (R-1)\cdot 1$$

So, $y = \dfrac{Rx}{1 + \left(\frac{R-1}{K}\right)x} > \dfrac{Rx}{1 + (R-1)} = \dfrac{Rx}{R} = x$

We have $x < y < K$.

EXERCISES 1.2

1. $\left(2^2\cdot 2\right)^2=\left(2^2\cdot 2^1\right)^2=\left(2^3\right)^2=2^6=64$

3. $2^{-4}=\frac{1}{2^4}=\frac{1}{16}$

5. $\left(\frac{1}{2}\right)^{-3}=\left(2^{-1}\right)^{-3}=2^3=8$

7. $\left(\frac{5}{8}\right)^{-1}=\frac{8}{5}$

9. $4^{-2}\cdot 2^{-1}=\left(2^2\right)^{-2}\cdot 2^{-1}$
$=2^{-4}\cdot 2^{-1}=2^{-5}=\frac{1}{2^5}=\frac{1}{32}$

11. $\left(\frac{3}{2}\right)^{-3}=\left(\frac{2}{3}\right)^3=\frac{2^3}{3^3}=\frac{8}{27}$

13. $\left(\frac{1}{3}\right)^{-2}-\left(\frac{1}{2}\right)^{-3}=\left(\frac{3}{1}\right)^2-\left(\frac{2}{1}\right)^3=\frac{3^2}{1^2}-\frac{2^3}{1^3}$
$=9-8=1$

15. $\left[\left(\frac{2}{3}\right)^{-2}\right]^{-1}=\left[\left(\frac{3}{2}\right)^{2}\right]^{-1}=\left(\frac{3}{2}\right)^{-2}=\left(\frac{2}{3}\right)^{2}=\frac{2^2}{3^2}=\frac{4}{9}$

17. $25^{1/2}=\sqrt{25}=5$

19. $25^{3/2}=\left(\sqrt{25}\right)^3=5^3=125$

21. $16^{3/4}=\left(\sqrt[4]{64}\right)^3=2^3=8$

23. $(-8)^{2/3}=\left(\sqrt[3]{-8}\right)^2=(-2)^2=4$

25. $(-8)^{5/3}=\left(\sqrt[3]{-8}\right)^5=(-2)^5=-32$

27. $\left(\frac{25}{36}\right)^{3/2}=\left(\sqrt{\frac{25}{36}}\right)^3=\left(\frac{5}{6}\right)^3=\frac{5^3}{6^3}=\frac{125}{216}$

29. $\left(\frac{27}{125}\right)^{2/3}=\left(\sqrt[3]{\frac{27}{125}}\right)^2=\left(\frac{3}{5}\right)^2=\frac{9}{25}$

31. $\left(\frac{1}{32}\right)^{2/5}=\left(\sqrt[5]{\frac{1}{32}}\right)^2=\left(\frac{1}{2}\right)^2=\frac{1}{4}$

33. $4^{-1/2}=\frac{1}{4^{1/2}}=\frac{1}{\sqrt{4}}=\frac{1}{2}$

35. $4^{-3/2}=\frac{1}{4^{3/2}}=\frac{1}{\left(\sqrt{4}\right)^3}=\frac{1}{2^3}=\frac{1}{8}$

37. $8^{-2/3}=\frac{1}{8^{2/3}}=\frac{1}{\left(\sqrt[3]{8}\right)^2}=\frac{1}{2^2}=\frac{1}{4}$

39. $(-8)^{-1/3}=\frac{1}{(-8)^{1/3}}=\frac{1}{\left(\sqrt[3]{-8}\right)}=\frac{1}{-2}=-\frac{1}{2}$

41. $(-8)^{-2/3}=\frac{1}{(-8)^{2/3}}=\frac{1}{\left(\sqrt[3]{(-8)^2}\right)}=\frac{1}{(-2)^2}=\frac{1}{4}$

43. $\left(\frac{25}{16}\right)^{-1/2}=\left(\frac{16}{25}\right)^{1/2}=\left(\sqrt{\frac{16}{25}}\right)=\frac{4}{5}$

45. $\left(\frac{25}{16}\right)^{-3/2}=\left(\frac{16}{25}\right)^{3/2}=\left(\sqrt{\frac{16}{25}}\right)^3=\left(\frac{4}{5}\right)^3=\frac{64}{125}$

47. $\left(-\frac{1}{27}\right)^{-5/3}=(-27)^{5/3}=\left(\sqrt[3]{-27}\right)^5=(-3)^5=-243$

49. $7^{0.39}\approx 2.14$

51. $8^{2.7}\approx 274.37$

53. $\left[(0.1)^{0.1}\right]^{0.1}\approx 0.977$

55. $\left(1-\frac{1}{1000}\right)^{-1000}\approx 2.720$

57. $\frac{4}{x^5}=4x^{-5}$

59. $\frac{4}{\sqrt[3]{8x^3}}=\frac{4}{2x^{4/3}}=2x^{-4/3}$

61. $\frac{24}{\left(2\sqrt{x}\right)^3}=\frac{24}{8x^{3/2}}=3x^{-3/2}$

63. $\sqrt{\frac{9}{x^4}}=\frac{3}{x^2}=3x^{-2}$

65. $\frac{5x^2}{\sqrt{x}}=\frac{5x^2}{x^{1/2}}=5x^{2-1/2}=5x^{3/2}$

67. $\frac{12\sqrt[3]{x^2}}{3x^2}=\frac{12x^{2/3}}{3x^2}=\frac{12}{3}x^{2/3-2}=4x^{-4/3}$

69. $\frac{\sqrt{36x}}{2x}=\frac{6\sqrt{x}}{2x}=\frac{6x^{1/2}}{2x}=\frac{6}{2}x^{1/2-1}=3x^{-1/2}$

71. $\left(x^3\cdot x^2\right)^2=\left(x^5\right)^2=x^{10}$

73. $\left[z^2\left(z\cdot z^2\right)^2 z\right]^3=\left[z^2\left(z^3\right)^2 z\right]^3=\left(z^2\cdot z^6\cdot z\right)^3$
$=\left(z^9\right)^3=z^{27}$

75. $\left[\left(x^2\right)^2\right]^2=\left(x^4\right)^2=x^8$

77. $(3x^2y^5z)^3 = 3^3x^{2\cdot3}y^{5\cdot3}z^3 = 27x^6y^{15}z^3$

79. $\dfrac{(ww^2)^3}{w^3w} = \dfrac{(w^3)^3}{w^4} = \dfrac{w^9}{w^4} = w^5$

81. $\dfrac{(5xy^4)^2}{25x^3y^3} = \dfrac{25x^2y^8}{25x^3y^3} = \dfrac{y^5}{x}$

83. $\dfrac{(9xy^3z)^2}{3(xyz)^2} = \dfrac{81x^2y^6z^2}{3x^2y^2z^2} = 27y^4$

85. $\dfrac{(2u^2vw^3)^2}{4(uw^2)^2} = \dfrac{4u^4v^2w^6}{4u^2w^4} = u^2v^2w^2$

87 Average body thickness
$= 0.4(\text{hip-to-shoulder length})^{3/2}$
$= 0.4(16)^{3/2}$
$= 0.4\left(\sqrt{16}\right)^3$
≈ 25.6 ft

89. $C' = x^{0.6}C$
$= 4^{0.6}C \approx 2.3C$
To quadruple the capacity costs about 2.3 times as much.

91. **a.** Given the unemployment rate of 2 percent, the inflation rate is
$y = 9.638(2)^{-1.394} - 0.900$
≈ 2.77 percent.

b. Given the unemployment rate of 5 percent, the inflation rate is
$y = 9.638(5)^{-1.394} - 0.900$
≈ 0.12 percent.

93. $(\text{Heart rate}) = 250(\text{weight})^{-1/4}$
$= 250(16)^{-1/4}$
$= 125$ beats per minute

95. (Time to build the 50th Boeing 707)
$= 150(50)^{-0.322}$
≈ 42.6 thousand work-hours

It took approximately 42,600 work-hours to build the 50th Boeing 707.

97. Increase in energy $= 32^{B-A}$
$= 32^{7.8-6.7}$
$= 32^{1.1} \approx 45$
The 1906 San Francisco earthquake had about 45 times more energy released than the 1994 Northridge earthquake.

99. $K = 3000(225)^{-1/2} = 200$

101. $S = \dfrac{60}{11}x^{0.5}$
$= \dfrac{60}{11}(3281)^{0.5} \approx 312$ mph

103.

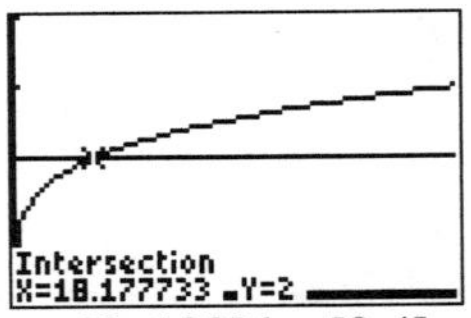

on [0, 100] by [0,4]
$x \approx 18.2$. Therefore, the land area must be increased by a factor of more than 18 to double the number of species.

105. $y = 9.4x^{0.37}$
$y = 9.4(150)^{0.37} = 60$ miles per hour
The speed of a car that left 150-foot skid marks was 60 miles per hour.

107. **a.**

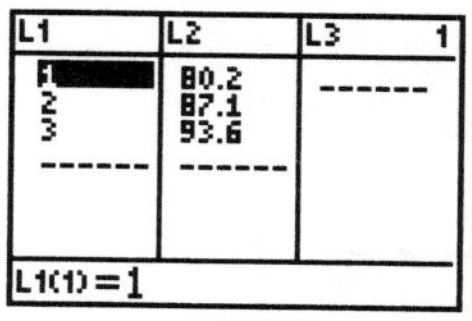

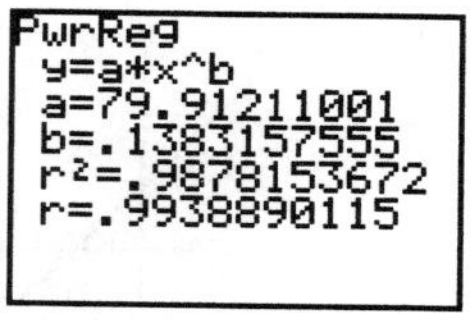

$y = 79.9x^{0.138}$ (rounded)

b. For year 2020, $x = 10$.

$y = 79.9(10)^{0.138} \approx \$110$ billion

109. **a.**

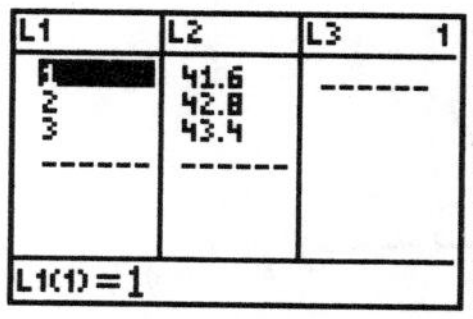

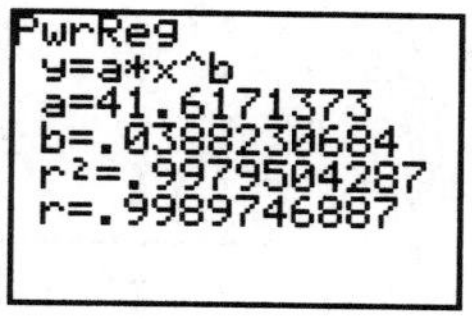

$y = 41.6x^{0.0388}$ (rounded)

b. For year 2020, $x = 11$.

$y = 41.6(11)^{0.0388} \approx \$45.7$ million

111. 3, since $\sqrt{9}$ means the principal square foot. (To get ± 3 you would have to write $\pm\sqrt{9}$.)

113. False: $\frac{2^6}{2^6} = \frac{64}{4} = 16$, while $2^{6/2} = 2^3 = 8$.

(The correct statement is $\frac{x^m}{x^n} = x^{m-n}$.)

115. $x^{1/2} = \sqrt{x}$, so x must be nonnegative for the expression to be defined.

117. $x^{-1} = \frac{1}{x}$, so all values of x except 0, because you cannot divide by 0.

EXERCISES 1.3

1. Yes

3. No

5. No

7. No

9. Domain = $\{x \mid x \le 0 \text{ or } x \ge 1\}$; Range = $\{y \mid y \ge -1\}$

11. **a.** $f(x) = \sqrt{x-1}$

$f(10) = \sqrt{10-1} = \sqrt{9} = 3$

b. Domain = $\{x \mid x \ge 1\}$ since $f(x) = \sqrt{x-1}$ is defined for all values of $x \ge 1$.

c. Range = $\{y \mid y \ge 0\}$

13. **a.** $h(z) = \frac{1}{z+4}$

$h(-5) = \frac{1}{-5+4} = -1$

b. Domain $= \{z \mid z \ne -4\}$ since $h(z) = \frac{1}{z+4}$ is defined for all values of z except $z = -4$.

c. Range = $\{y \mid y \ne 0\}$

15. **a.** $h(x) = x^{1/4}$

$h(81) = 81^{1/4} = \sqrt[4]{81} = 3$

b. Domain = $\{x \mid x \ge 0\}$ since $h(x) = x^{1/4}$ is defined only for nonnegative values of x.

c. Range = $\{y \mid y \ge 0\}$

17. **a.** $f(x) = x^{2/3}$

$f(-8) = (-8)^{2/3} = \left(\sqrt[3]{-8}\right)^2 = (-2)^2 = 4$

b. Domain = $\mathbb{R}$

c. Range = $\{y \mid y \ge 0\}$

19. **a.** $f(x) = \sqrt{4 - x^2}$

$f(0) = \sqrt{4 - 0^2} = \sqrt{4} = 2$

b. $f(x) = \sqrt{4 - x^2}$ is defined for values of x such that $4 - x^2 \ge 0$. Thus,

$4 - x^2 \ge 0$

$-x^2 \ge -4$

$x^2 \le 4$

$-2 \le x \le 2$

Domain = $\{x \mid -2 \le x \le 2\}$

c. Range = $\{y \mid 0 \le y \le 2\}$

21. **a.** $f(x) = \sqrt{-x}$

$f(-25) = \sqrt{-(-25)} = \sqrt{25} = 5$

b. $f(x) = \sqrt{-x}$ is defined only for values of x such that $-x \ge 0$. Thus $x \le 0$.

Domain = $\{x \mid x \le 0\}$

c. Range = $\{y \mid y \ge 0\}$

23.

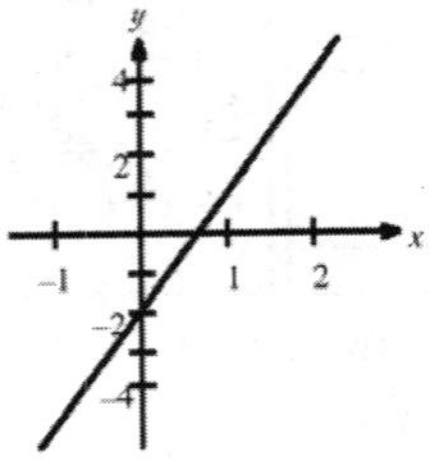

25.

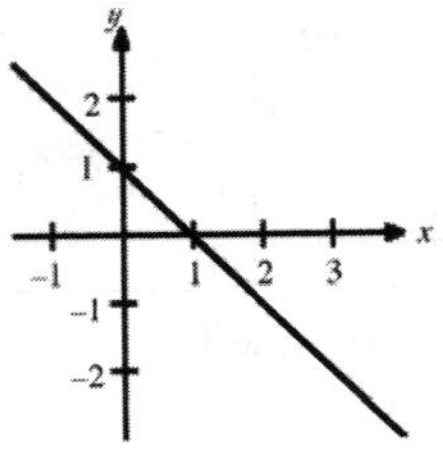

27.

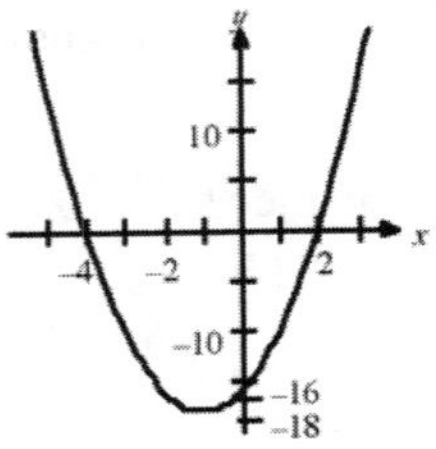

29.

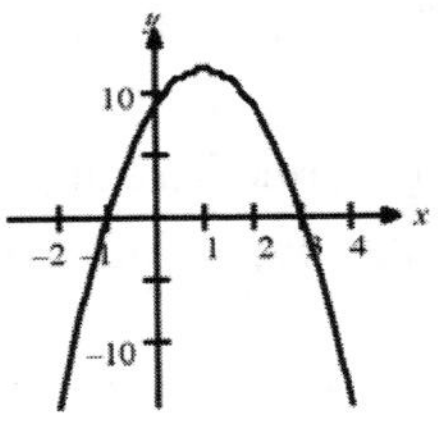

31. **a.** $x = \frac{-b}{2a} = \frac{-(-40)}{2(1)} = \frac{40}{2} = 20$

To find the y-coordinate, evaluate f at $x = 20$.

$f(20) = (20)^2 - 40(20) + 500 = 100$

The vertex is (20, 100).

b.

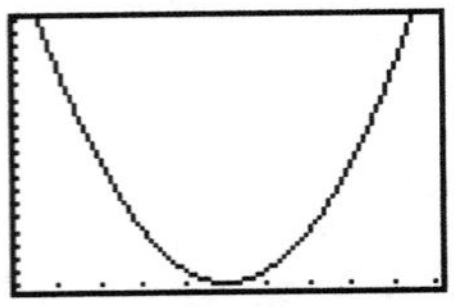

on [15, 25] by [100, 120]

33. **a.** $x = \frac{-b}{2a} = \frac{-(-80)}{2(-1)} = \frac{80}{-2} = -40$

To find the y-coordinate, evaluate f at $x = -40$.

$f(-40) = -(-40)^2 - 80(-40) - 1800$
$= -200$

The vertex is (–40, –200).

b.

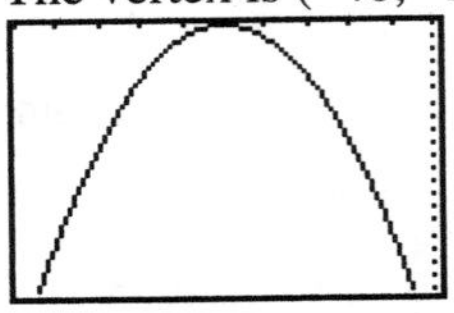

on [–45, –35] by [–220, –200]

35.
$x^2 - 6x - 7 = 0$
$(x-7)(x+1) = 0$
Equals 0 at $x = 7$; Equals 0 at $x = -1$
$x = 7, \quad x = -1$

37.
$x^2 + 2x = 15$
$x^2 + 2x - 15 = 0$
$(x+5)(x-3) = 0$
Equals 0 at $x = -5$; Equals 0 at $x = 3$
$x = -5, \quad x = 3$

39.
$2x^2 + 40 = 18x$
$2x^2 - 18x + 40 = 0$
$x^2 - 9x + 20 = 0$
$(x-4)(x-5) = 0$
Equals 0 at $x = 4$; Equals 0 at $x = 5$
$x = 4, \quad x = 5$

41.
$5x^2 - 50x = 0$
$x^2 - 10x = 0$
$x(x-10) = 0$
Equals 0 at $x=0$; Equals 0 at $x=10$
$x=0, \quad x=10$

43.
$2x^2 - 50 = 0$
$x^2 - 25 = 0$
$(x-5)(x+5) = 0$
Equals 0 at $x = 5$; Equals 0 at $x = -5$
$x = 5, \quad x = -5$

45.
$4x^2 + 24x + 40 = 4$
$4x^2 + 24x + 36 = 0$
$x^2 + 6x + 9 = 0$
$(x+3)^2 = 0$
Equals 0 at $x = -3$

47.
$$-4x^2+12x=8$$
$$-4x^2+12x-8=0$$
$$x^2-3x+2=0$$
$$(x-2)(x-1)=0$$
Equals 0 at $x=2$ Equals 0 at $x=1$

$x=2,\quad x=1$

49.
$$2x^2-12x+20=0$$
$$x^2-6x+10=0$$
Use the quadratic formula with $a=1$, $b=-6$, and $c=10$.
$$x=\frac{-(-6)\pm\sqrt{(-6)^2-4(1)(10)}}{2(1)}$$
$$=\frac{6\pm\sqrt{36-40}}{2}$$
$$=\frac{6\pm\sqrt{-4}}{2}\quad \text{Undefined}$$
$2x^2-12x+20=0$ has no real solutions.

51.
$$3x^2+12=0$$
$$x^2+4=0$$
$$x^2=-4$$
$$x=\pm\sqrt{-4}\quad \text{Undefined}$$
$3x^2+12=0$ has no real solutions.

53.

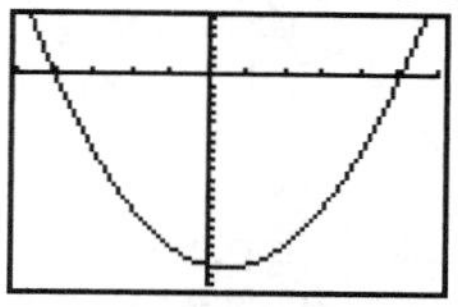

on [–5, 6] by [–22, 6]

$x=-4, x=5$

55.

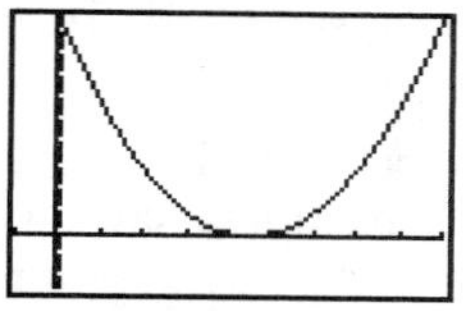

on [–1, 9] by [–10, 40]

$x=4, x=5$

57.

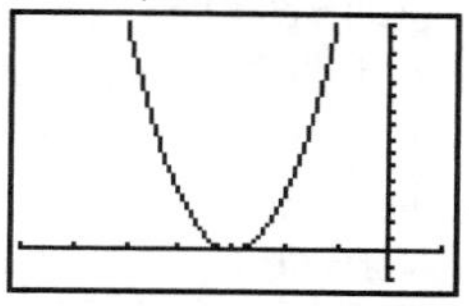

on [–7, 1] by [–2, 16]

$x=-3$

59.

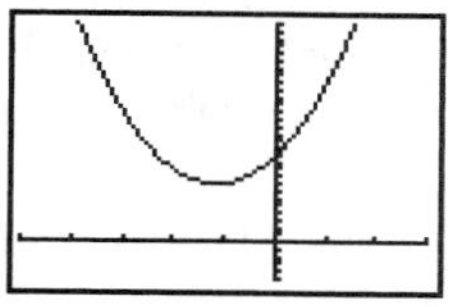

on [–5, 3] by [–5, 30]

No real solutions

61.

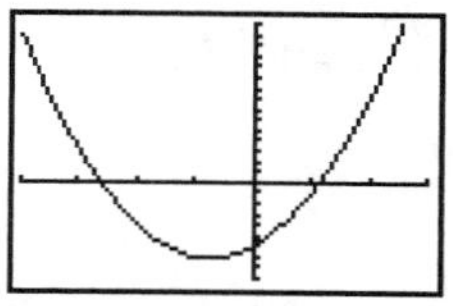

on [–4, 3] by [–9, 15]

$x=-2.64, x=1.14$

63.

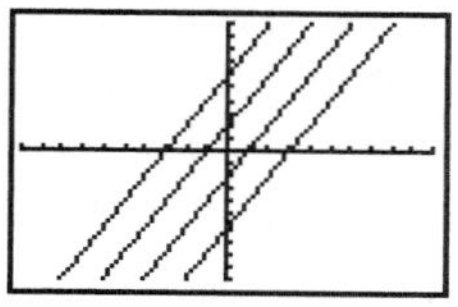

on [–10, 10] by [–10, 10]

a. Their slopes are all 2, but they have different y-intercepts.

b. The line 2 units below the line of the equation $y=2x-6$ must have y-intercept –8. Thus, the equation of this line is $y=2x-8$.

65. Let x = the number of board feet of wood. Then $C(x)=4x+20$

67. Let x = the number of hours of overtime. Then $P(x)=15x+500$

69.

a.
$$p(d)=0.45d+15$$
$$p(6)=0.45(6)+15$$
$$=17.7 \text{ pounds per square inch}$$

b.
$$p(d)=0.45d+15$$
$$p(35{,}000)=0.45(35{,}000)+15$$
$$=15{,}765 \text{ pounds per square inch}$$

71. $D(v) = 0.055v^2 + 1.1v$
$D(40) = 0.055(40)^2 + 1.1(40) = 132$ ft

73. **a.** $N(t) = 200 + 50t^2$
$N(2) = 200 + 50(2)^2$
$= 400$ cells

b. $N(t) = 200 + 50t^2$
$N(10) = 200 + 50(10)^2$
$= 5200$ cells

75. $v(x) = \frac{60}{11}\sqrt{x}$
$v(1776) = \frac{60}{11}\sqrt{1776} \approx 230$ mph

77.

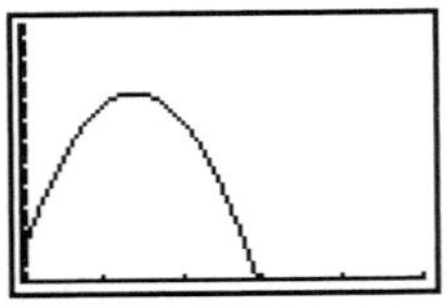

on [0, 5] by [0, 50]

The object hits the ground in about 2.92 seconds

79. **a.** To find the break-even points, set $C(x)$ equal to $R(x)$ and solve the resulting equation.

$$C(x) = R(x)$$
$$180x + 16,000 = -2x^2 + 660x$$
$$2x^2 - 480x + 16,000 = 0$$

Use the quadratic formula with $a = 2$, $b = -480$ and $c = 16,000$.

$$x = \frac{480 \pm \sqrt{(-480)^2 - 4(2)(16,000)}}{2(2)}$$
$$x = \frac{480 \pm \sqrt{102,400}}{4} = \frac{480 \pm 320}{4}$$
$$x = \frac{800}{4} \text{ or } \frac{160}{4}$$
$$x = 200 \text{ or } 40$$

The company will break even when it makes either 40 devices or 200 devices.

b. To find the number of devices that maximizes profit, first find the profit function, $P(x) = R(x) - C(x)$.

$$P(x) = (-2x^2 + 660x) - (180x + 16,000)$$
$$= -2x^2 + 480x - 16,000$$

Since this is a parabola that opens downward, the maximum profit is found at the vertex.

$$x = \frac{-480}{2(-2)} = \frac{-480}{-4} = 120$$

Thus, profit is maximized when 120 devices are produced per week. The maximum profit is found by evaluating $P(120)$.

$$P(120) = -2(120)^2 + 480(120) - 16,000$$
$$= \$12,800$$

Therefore, the maximum profit is \$12,800.

81. **a.** To find the break-even points, set $C(x)$ equal to $R(x)$ and solve the resulting equation.

$$C(x) = R(x)$$
$$100x + 3200 = -2x^2 + 300x$$
$$2x^2 - 200x + 3200 = 0$$

Use the quadratic formula with $a = 2$, $b = -200$ and $c = 3200$.

$$x = \frac{200 \pm \sqrt{(-1020)^2 - 4(2)(3200)}}{2(2)}$$
$$x = \frac{200 \pm \sqrt{14,400}}{4} = \frac{200 \pm 120}{4}$$
$$x = \frac{320}{4} \text{ or } \frac{80}{4}$$
$$x = 80 \text{ or } 20$$

The store will break even when it sells either 20 exercise machines or 80 exercise machines.

b. To find the number of exercise machines that maximizes profit, first find the profit function, $P(x) = R(x) - C(x)$.

$$P(x) = (-2x^2 + 300x) - (100x + 3200)$$
$$= -2x^2 + 200x - 3200$$

Since this is a parabola that opens downward, the maximum profit is found at the vertex.

$$x = \frac{-200}{2(-2)} = \frac{-200}{-4} = 50$$

Thus, profit is maximized when 50 exercise machines are sold per day. The maximum profit is found by evaluating $P(50)$.

$$P(50) = -2(50)^2 + 200(50) - 3200$$
$$= \$1800$$

Therefore, the maximum profit is \$1800.

83. $(w+a)(v+b)=c$

$$v+b=\frac{c}{w+a}$$

$$v=\frac{c}{w+a}-b$$

85. **a.**

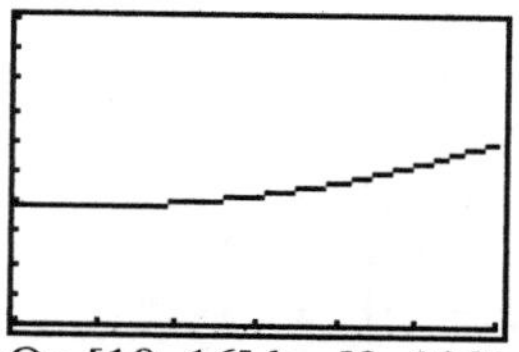

On [10, 16] by [0, 100].

b.

$y=0.831x^2-18.1x+137.3$

$y=0.831(12)^2-18.1(12)+137.3$

$y=39.764$

The probability that a high school graduate smoker will quit is 40%.

c.

$y=0.831x^2-18.1x+137.3$

$y=0.831(16)^2-18.1(16)+137.3$

$y=60.436$

The probability that a college graduate smoker will quit is 60%.

87. **a.** $(100-x)x=100x-x^2$ or $-x^2+100$

$f(x)=100x-x^2$ or $f(x)=-x^2+100x$

b. Since this function represents a parabola opening downward (because $a=-1$), it is maximized at its vertex, which is found using the vertex formula, $x=\frac{-b}{2a}$, with $a=-1$ and $b=100$.

$$x=\frac{-100}{-2}=50$$

She should charge \$50 to maximize her revenue.

89. **a.**

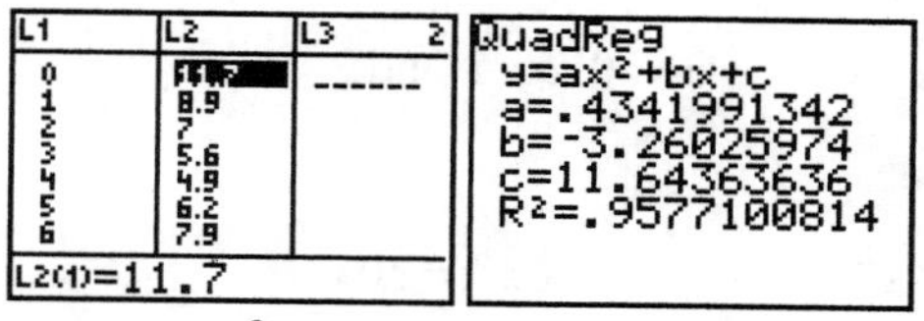

$y=0.434x^2-3.26x+11.6$

b. For year 2016, $x=8.6$.

$y=0.434(8.6)^2-3.26(8.6)+11.6\approx 15.7\%$

91. A function can have more than one x-intercept. Many parabolas cross the x-axis twice. A function cannot have more than one y-intercept because that would violate the vertical line test.

93. Because the function is linear and 5 is halfway between 4 and 6, $f(5)=9$ (halfway between 7 and 11).

95. m is blargs per prendle and $\frac{\Delta y}{\Delta x}$, so x is in prendles and y is in blargs.

97. No, that would violate the vertical line test. Note: A *parabola* is a geometric shape and so *may* open sideways, but a quadratic function, being a *function*, must pass the vertical line test.

EXERCISES 1.4

1. Domain = $\{x \mid x<-4 \text{ or } x>0\}$
Range = $\{y \mid y<-2 \text{ or } y>0\}$

3. **a.**

$$f(x)=\frac{1}{x+4}$$

$$f(-3)=\frac{1}{-3+4}=1$$

b. Domain = $\{x \mid x\neq -4\}$

c. Range = $\{y \mid y\neq 0\}$

5. **a.** $f(x)=\dfrac{x^2}{x-1}$

$f(-1)=\dfrac{(-1)^2}{-1-1}=-\dfrac{1}{2}$

b. Domain = $\{x \mid x \neq 1\}$

c. Range = $\{y \mid y \leq 0 \text{ or } y \geq 4\}$

7. **a.** $f(x)=\dfrac{12}{x(x+4)}$

$f(2)=\dfrac{12}{2(2+4)}=1$

b. Domain = $\{x \mid x \neq 0, x \neq -4\}$

c. Range = $\{y \mid y \leq -3 \text{ or } y > 0\}$

9. **a.** $g(x)=|x+2|$

$g(-5)=|(-5)+2|=|-3|=3$

b. Domain = $\mathbb{R}$

c. Range = $\{y \mid y \geq 0\}$

11.
$$x^5+2x^4-3x^3=0$$
$$x^3(x^2+2x-3)=0$$
$$x^3(x+3)(x-1)=0$$

Equals 0 at $x=0$ Equals 0 at $x=-3$ Equals 0 at $x=1$

$x=0, x=-3, \text{and } x=1$

13.
$$5x^3-20x=0$$
$$5x(x^2-4)=0$$
$$5x(x-2)(x+2)=0$$

Equals 0 at $x=0$ Equals 0 at $x=2$ Equals 0 at $x=-2$

$x=0, x=2, \text{and } x=-2$

15.
$$2x^3+18x=12x^2$$
$$2x^3-12x^2+18x=0$$
$$2x(x^2-6x+9)=0$$
$$2x(x-3)^3=0$$

Equals 0 at $x=0$ Equals 0 at $x=3$

$x=0$ and $x=3$

17.
$$6x^5=30x^4$$
$$6x^5-30x^4=0$$
$$6x^4(x-5)=0$$

Equals 0 at $x=0$ Equals 0 at $x=5$

$x=0$ and $x=5$

19.
$$3x^{5/2}-6x^{3/2}=9x^{1/2}$$
$$3x^{5/2}-6x^{3/2}-9x^{1/2}=0$$
$$3x^{1/2}(x^2-2x-3)=0$$
$$3x^{1/2}(x-3)(x+1)=0$$

Equals 0 at $x=0$ Equals 0 at $x=3$ Equals 0 at ~~$x=-1$~~

$x=0$, $x=3$ and ~~$x=-1$~~

Valid solutions are $x=0$ and $x=3$.

21.

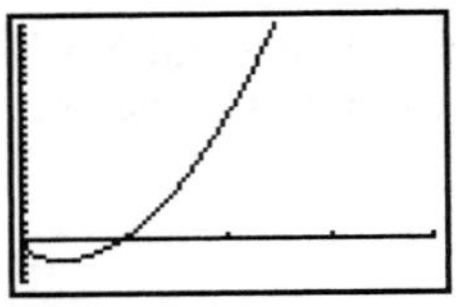

on [0, 4] by [–5,25]

$x=0$ and $x=1$

23.

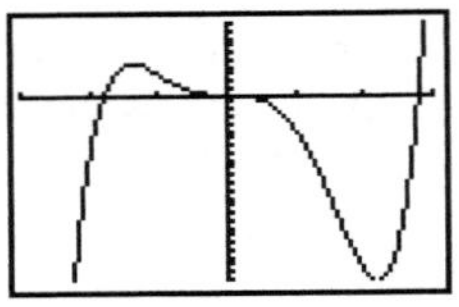

on [–3, 3] by [–25, 10]

$x \approx -1.79$, $x=0$, and $x \approx 2.79$

25.

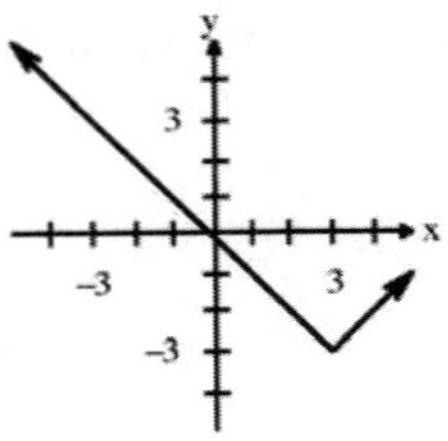

27.

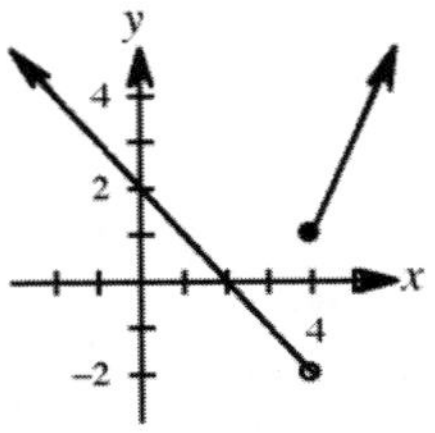

29.

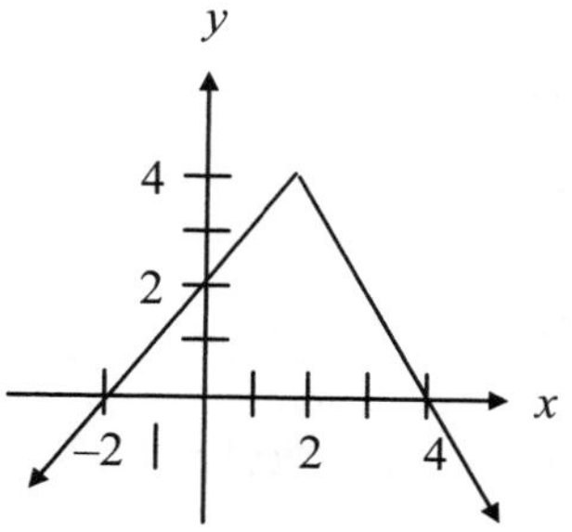

31.

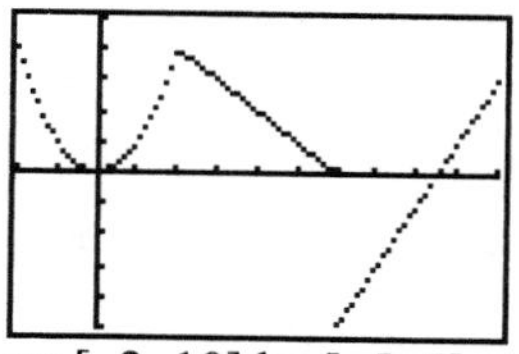

on [–2, 10] by [–5, 5]

33. Polynomial

35. Piecewise linear function

37. Polynomial

39. Rational function

41. Piecewise linear function

43. Polynomial

45. None of these

47. **a.** y_4

b. y_1

c.

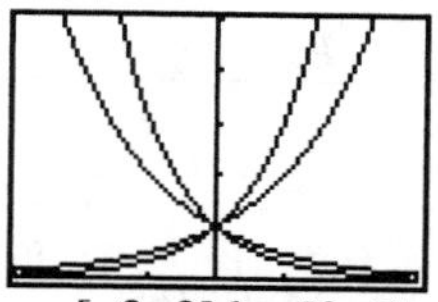

on [–3, 3] by [0, 5]

d. (0, 1) because $a^0 = 1$ for any constant $a \neq 0$.

49. **a.** $f(g(x)) = [g(x)]^5 = (7x-1)^5$

b. $$g(f(x)) = 7[f(x)] - 1$$
$$= 7(x^5) - 1$$
$$= 7x^5 - 1$$

51. **a.** $f(g(x)) = \frac{1}{g(x)} = \frac{1}{x^2+1}$

b. $g(f(x)) = [f(x)]^2 + 1 = \frac{1}{x^2} + 1$

53. **a.** $$f(g(x)) = [g(x)]^3 - [g(x)]^2$$
$$= (\sqrt{x}-1)^3 - (\sqrt{x}-1)^2$$

b. $g(f(x)) = \sqrt{f(x)} - 1 = \sqrt{x^3 - x^2} - 1$

55. **a.** $$f(g(x)) = \frac{[g(x)]^3 - 1}{[g(x)]^3 + 1} = \frac{(x^2-x)^3 - 1}{(x^2-x)^3 + 1}$$

b. $$g(f(x)) = [f(x)]^2 - f(x)$$
$$= \left(\frac{x^3-1}{x^3+1}\right)^2 - \frac{x^3-1}{x^3+1}$$

57. **a.** $$f(g(x)) = 2[g(x)] - 6$$
$$= 2\left(\frac{x}{2}+3\right) - 6$$
$$= x + 6 - 6$$

b. $$g(f(x)) = \frac{f(x)}{2} + 3$$
$$= \frac{2x-6}{2} + 3 = x - 3 + 3$$
$$= x$$

59. $$f(x+h) = 5(x+h)^2 = 5(x^2 + 2hx + h^2)$$
$$= 5x^2 + 10hx + 5h^2$$

61. $$f(x+h) = 2(x+h)^2 - 5(x+h) + 1$$
$$= 2x^2 + 4xh + 2h^2 - 5x - 5h + 1$$

63. $f(x) = 5x^2$

$$\frac{f(x+h)-f(x)}{h} = \frac{5(x+h)^2 - 5x^2}{h} = \frac{5(x^2+2xh+h^2)-5x^2}{h} = \frac{5x^2+10xh+5h^2-5x^2}{h} = \frac{10xh+5h^2}{h} = \frac{h(10x+5h)}{h} = 10x+5h \text{ or } 5(2x+h)$$

65. $f(x) = 2x^2 - 5x + 1$

$$\frac{f(x+h)-f(x)}{h} = \frac{2(x+h)^2-5(x+h)+1-(2x^2-5x+1)}{h} = \frac{2(x^2+2xh+h^2)-5(x+h)+1-(2x^2-5x+1)}{h} = \frac{2x^2+4xh+2h^2-5x-5h+1-2x^2+5x-1}{h} = \frac{4xh+2h^2-5h}{h} = \frac{h(4x+2h-5)}{h} = 4x+2h-5$$

67. $f(x) = 7x^2 - 3x + 2$

$$\frac{f(x+h)-f(x)}{h} = \frac{7(x+h)^2-3(x+h)+2-(7x^2-3x+2)}{h} = \frac{7(x^2+2xh+h^2)-3(x+h)+2-(7x^2-3x+2)}{h} = \frac{7x^2+14xh+7h^2-3x-3h+2-7x^2+3x-2}{h} = \frac{14xh+7h^2-3h}{h} = \frac{h(14x+7h-3)}{h} = 14x+7h-3$$

69. $f(x) = x^3$

$$\frac{f(x+h)-f(x)}{h} = \frac{(x+h)^3 - x^3}{h} = \frac{x^3+3x^2h+3xh^2+h^3-x^3}{h} = \frac{3x^2h+3xh^2+h^3}{h} = \frac{h(3x^2+3xh+h^2)}{h} = 3x^2+3xh+h^2$$

71. $f(x) = \frac{2}{x}$

$$\frac{f(x+h)-f(x)}{h} = \frac{\frac{2}{x+h}-\frac{2}{x}}{h} = \frac{\frac{2}{x+h}-\frac{2}{x}}{h}\cdot\frac{x(x+h)}{x(x+h)} = \frac{2x-2(x+h)}{hx(x+h)} = \frac{2x-2x-2h}{hx(x+h)} = \frac{-2h}{hx(x+h)} = \frac{-2}{x(x+h)}$$

73. $f(x) = \frac{1}{x^2}$

$$\frac{f(x+h)-f(x)}{h} = \frac{\frac{1}{(x+h)^2}-\frac{1}{x^2}}{h} = \frac{\frac{1}{(x+h)^2}-\frac{1}{x^2}}{h}\cdot\frac{x^2(x+h)^2}{x^2(x+h)^2} = \frac{x^2-(x+h)^2}{hx^2(x+h)^2} = \frac{x^2-x^2-2xh-h^2}{hx^2(x+h)^2} = \frac{-2xh-h^2}{hx^2(x+h)^2} = \frac{h(-2x-h)}{hx^2(x+h)^2} = \frac{-2x-h}{x^2(x+h)^2} \text{ or } \frac{-2x-h}{x^2(x^2+2xh+h^2)} \text{ or } \frac{-2x-h}{x^4+2x^3h+x^2h^2}$$

75. **a.** 2.70481
b. 2.71815
c. 2.71828
d. Yes, 2.71828

77. The graph of $y = (x+3)^3 + 6$ is the same shape as the graph of $y = x^3$ but it is shifted left 3 units and up 6 units.
Check:

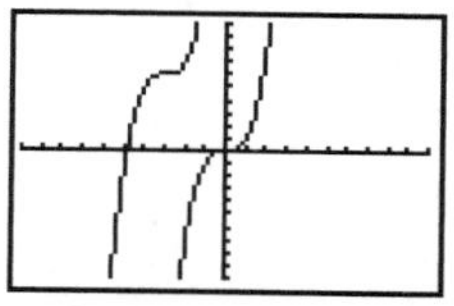

on [–10, 10] by [–10, 10]

79. $P(x) = 522(1.0053)^x$
$P(50) = 522(1.0053)^{50} \approx 680$ million people in 1750

81. **a.** For $x = 3000$, use $f(x) = 0.10x$.
$f(x) = 0.10x$
$f(3000) = 0.10(3000) = \$300$

b. For $x = 5000$,
use $f(x) = 0.10x$.
$f(x) = 0.10x$
$f(5000) = 0.10(5000) = \$500$

c. For $x = 10{,}000$,
use $f(x) = 500 + 0.30(x - 5000)$.
$f(x) = 500 + 0.30(x - 5000)$
$f(10{,}000) = 500 + 0.30(10{,}000 - 5000)$
$= 500 + 0.30(5000)$
$= \$2000$

d.

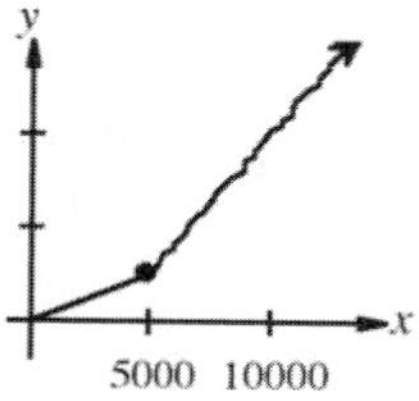

83. **a.** For $x = \frac{2}{3}$, use $f(x) = 10.5x$.
$f(x) = 10.5x$
$f\left(\frac{2}{3}\right) = 10.5\left(\frac{2}{3}\right)$
$= 7$ years

For $x = \frac{4}{3}$, use $f(x) = 10.5x$.
$f(x) = 10.5x$
$f\left(\frac{4}{3}\right) = 10.5\left(\frac{4}{3}\right)$
$= 14$ years

For $x = 4$, use $f(x) = 21 + 4(x-2)$.
$f(x) = 21 + 4(x-2)$
$f(4) = 21 + 4(4-2)$
$= 21 + 4(2)$
$= 29$ years

For $x = 10$, use $f(x) = 21 + 4(x-2)$.
$f(x) = 21 + 4(x-2)$
$f(10) = 21 + 4(10-2)$
$= 21 + 4(8)$
$= 53$ years

b.

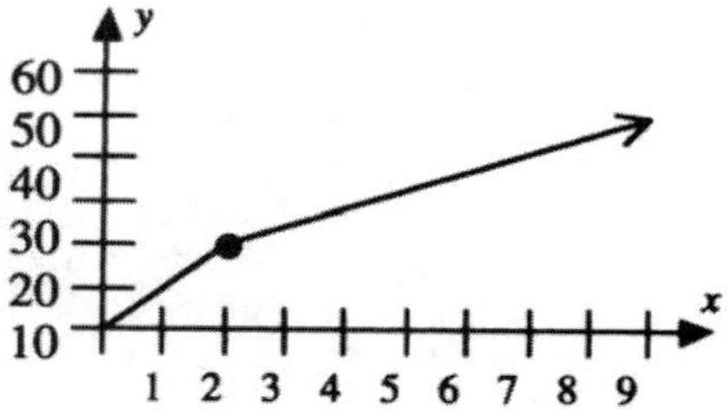

85. Substitute $K = 24 \cdot L^{-1}$ into $3L + 8K = 48$.

$$3L + 8(24 \cdot L^{-1}) = 48$$
$$3L + \frac{192}{L} = 48$$
$$3L^2 + 192 = 48L$$
$$3L^2 - 48L + 192 = 0$$
$$3(L^2 - 16L + 64) = 0$$
$$3(L-8)^2 = 0$$

So, $L = 8$.

And $K = 24 \cdot 8^{-1} = \frac{24}{8} = 3$

The intersection point is (8, 3).

87. First find the composition $R(v(t))$.

$$R(v(t)) = 2[v(t)]^{0.3}$$
$$= 2(60+3t)^{0.3}$$

Then find $R(v(10))$.

$$R(v(10)) = 2(60+3(10))^{0.3} = 2(90)^{0.3}$$
$$\approx \$7.714 \text{ million}$$

89. **a.** $f(x) = 4^{10} = 1{,}048{,}576$ cells ≈ 1 million cells

b. $f(15) = 4^{15} = 1{,}073{,}741{,}824$ cells
No, the mouse will not survive beyond day 15.

91. **a.**

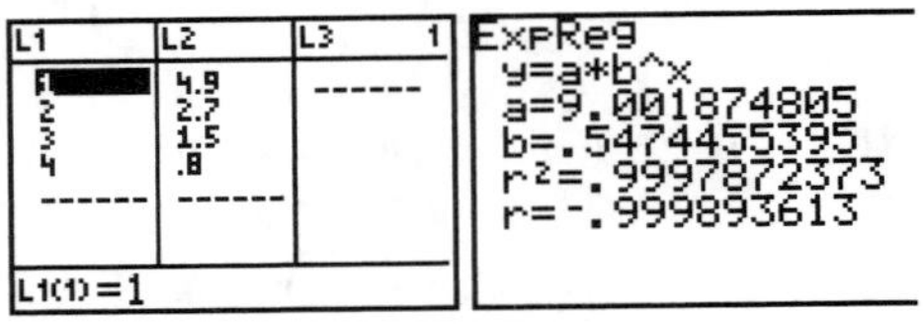

$y = 9 \cdot 0.547^x$ (rounded)

b. For year 2020, $x = 6$.
$y = 9 \cdot 0.547^6 \approx 0.24$ weeks
or less than 2 days

93. One will have "missing points" at the excluded x-values.

95. A slope of 1 is a tax of 100%. That means, all dollars taxed are paid as the tax.

97.
$$f(f(x)) = (f(x)) + a$$
$$= (x+a) + a$$
$$= x + 2a$$

99. $f(x+10)$ is shifted to the left by 10 units.

101. False: $f(x+h) = (x+h)^2 = x^2 + 2xh + h^2$, not $x^2 + h^2$.

103. **a.**

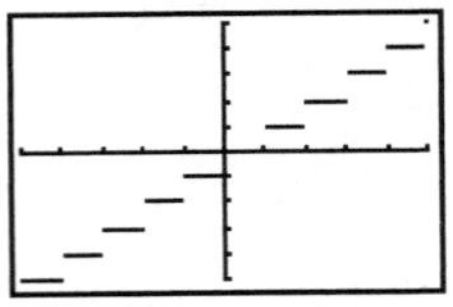

on [–5, 5] by [–5, 5]
Note that each line segment in this graph includes its left end-point, but excludes its right endpoint. So it should be drawn like •—o.

b. Domain = $\mathbb{R}$; range = the set of integers

105. **a.**
$$f(g(x)) = a[g(x)] + b = a(cx+d) + b$$
$$= acx + ad + b$$

b. Yes

REVIEW EXERCISES AND CHAPTER TEST FOR CHAPTER 1

1. $\{x \mid 2 < x \leq 5\}$

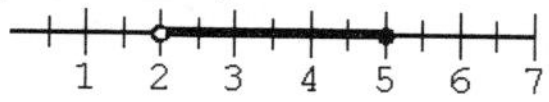

2. $\{x \mid -2 \leq x < 0\}$

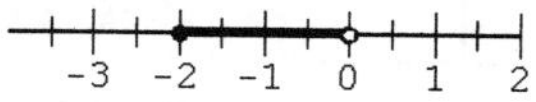

3. $\{x \mid x \geq 100\}$

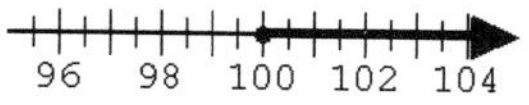

4. $\{x \mid x \leq 6\}$

5. Hurricane: $[74, \infty)$; storm: $[55, 74)$; gale: $[38, 55)$; small craft warning: $[21, 38)$

6.
a. $(0, \infty)$
b. $(-\infty, 0)$
c. $[0, \infty)$
d. $(-\infty, 0]$

7.
$$y-(-3)=2(x-1)$$
$$y+3=2x-2$$
$$y=2x-5$$

8.
$$y-6=-3[x-(-1)]$$
$$y-6=-3x-3$$
$$y=-3x+3$$

9. Since the vertical line passes through the point with x-coordinate 2, the equation of the line is $x = 2$.

10. Since the horizontal line passes through the point with y-coordinate 3, the equation of the line is
$y = 3$.

11. First, calculate the slope from the two points.
$$m=\frac{-3-3}{2-(-1)}=\frac{-6}{3}=-2$$
Now use the point-slope formula with this slope and the point $(-1, 3)$.
$$y-3=-2[x-(-1)]$$
$$y-3=-2x-2$$
$$y=-2x+1$$

12. First find the slope of the line $x+2y=8$.
Write the equation in slope-intercept form.
$$y=-\frac{1}{2}x+4.$$
The slope of the perpendicular line is $m=2$.
Next, use the point-slope form with the point $(6, -1)$:
$$y-y_1=m(x-x_1)$$
$$y+1=2(x-6)$$
$$y=2x-13$$

13. Since the y-intercept is $(0, -1)$, $b = -1$. To find the slope, use the slope formula with the points $(0, -1)$ and $(1, 1)$.
$$m=\frac{1-(-1)}{1-0}=2$$
Thus the equation of the line is $y = 2x - 1$.

14. Since the y-intercept is $(0, 1)$, $b = 1$. To find the slope, use the slope formula with the points $(0, 1)$ and $(2, 0)$.
$$m=\frac{0-1}{2-0}=-\frac{1}{2}$$
The equation of the line is $y=-\frac{1}{2}x+1$

15.
a. Use the straight-line depreciation formula with price = 25,000, useful lifetime = 8, and scrap value = 1000.
$$\text{Value}=\text{price}-\left(\frac{\text{price}-\text{scrap value}}{\text{useful lifetime}}\right)t$$
$$=25{,}000-\left(\frac{25{,}000-1000}{8}\right)t$$
$$=25{,}000-\left(\frac{24{,}000}{8}\right)t$$
$$=25{,}000-3000t$$
b. Value after 4 years $= 25{,}000-3000(4)$
$= 25{,}000-12{,}000$
$= \$13{,}000$

16.
a. Use the straight-line depreciation formula with price = 78,000, useful lifetime = 15, and scrap value = 3000.
$$\text{Value}=\text{price}-\left(\frac{\text{price}-\text{scrap value}}{\text{useful lifetime}}\right)t$$
$$=78{,}000-\left(\frac{78{,}000-3000}{15}\right)t$$
$$=78{,}000-\left(\frac{75{,}000}{15}\right)t$$
$$=78{,}000-5000t$$
b. Value after 8 years $= 78{,}000-5000(8)$
$= 78{,}000-40{,}000$
$= \$38{,}000$

17. **a.**

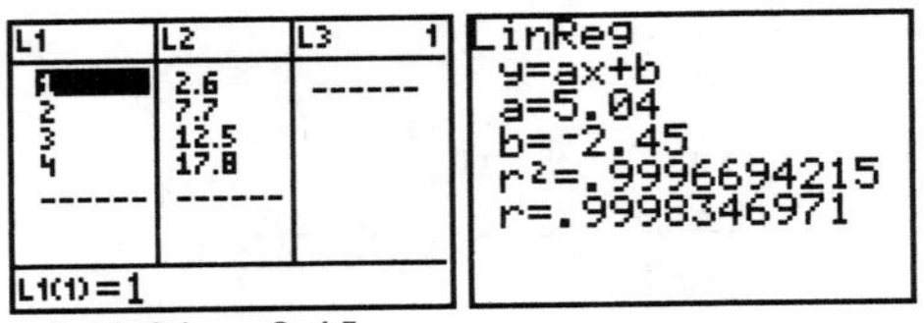

$y = 5.04x - 2.45$

b. The number is increasing by about 5000 3D screens per year.

c. For the year 2020, $x = 12$
$y = 5.04(12) - 2.45 = 58.03$
or about 58 thousand

18. $\left(\frac{6}{1}\right)^2 = \frac{6^2}{1^2} = 36$

19. $\left(\frac{4}{3}\right)^{-1} = \frac{3}{4}$

20. $64^{1/2} = \sqrt{64} = 8$

21. $1000^{1/3} = \sqrt[3]{1000} = 10$

22. $(81)^{-3/4} = \left(\frac{1}{81}\right)^{3/4} = \left(\sqrt[4]{\frac{1}{81}}\right)^3 = \left(\frac{1}{3}\right)^3 = \frac{1}{27}$

23. $100^{-3/2} = \left(\frac{1}{100}\right)^{3/2} = \left(\sqrt{\frac{1}{100}}\right)^3 = \left(\frac{1}{10}\right)^3 = \frac{1}{1000}$

24. $\left(-\frac{8}{27}\right)^{-2/3} = \left(-\frac{27}{8}\right)^{2/3} = \left(\sqrt[3]{-\frac{27}{8}}\right)^2$
$= \left(-\frac{3}{2}\right)^2 = \frac{9}{4}$

25. $\left(\frac{9}{16}\right)^{-3/2} = \left(\frac{16}{9}\right)^{3/2} = \left(\sqrt{\frac{16}{9}}\right)^3 = \left(\frac{4}{3}\right)^3 = \frac{64}{27}$

26. 13.97

27. 112.32

28. **a.** $y = 0.86x^{0.47}$
$y = 0.86(4000)^{0.47} = 42.4$
The weight for the top cold-blooded meat-eating animals in Hawaii is 42.4 lbs.

b. $y = 0.86x^{0.47}$
$y = 0.86(9{,}400{,}000)^{0.47} = 1628.8$
The weight for the top cold-blooded meat-eating animals in North America is 1628.8 lbs.

29. **a.** $y = 1.7x^{0.52}$
$y = 1.7(4000)^{0.52} = 126.9$
The weight for the top warm-blooded plant-eating animals in Hawaii is 126.9 lbs.

b. $y = 1.7x^{0.52}$
$y = 1.7(9{,}400{,}000)^{0.52} = 7185.8$
The weight for the top warm-blooded plant-eating animals in Hawaii is 7185.8 lbs.

30. **a.**

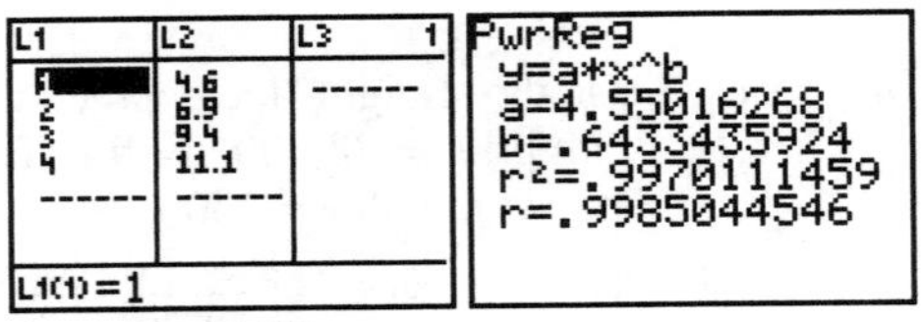

$y = 4.55x^{0.643}$ (rounded)

b. For year 2020, $x = 12$.
$y = 4.55(12)^{0.643} \approx \$22.5$ billion

31. **a.** $f(11) = \sqrt{11-7} = \sqrt{4} = 2$

b. Domain = $\{x \mid x \geq 7\}$ because $\sqrt{x-7}$ is defined only for all values of $x \geq 7$.

c. Range = $\{y \mid y \geq 0\}$

32. **a.** $g(-1) = \frac{1}{-1+3} = \frac{1}{2}$

b. Domain = $\{t \mid t \neq -3\}$

c. Range = $\{y \mid y \neq 0\}$

33. **a.** $h(16) = 16^{-3/4} = \left(\frac{1}{16}\right)^{3/4} = \left(\sqrt[4]{\frac{1}{16}}\right)^3$

$= \left(\frac{1}{2}\right)^3 = \frac{1}{8}$

b. Domain = $\{w \mid w > 0\}$ because the fourth root is defined only for nonnegative numbers and division by 0 is not defined.

c. Range = $\{y \mid y > 0\}$

34. **a.** $w(8) = 8^{-4/3} = \left(\frac{1}{8}\right)^{4/3} = \left(\sqrt[3]{\frac{1}{8}}\right)^4$

$= \left(\frac{1}{2}\right)^4 = \frac{1}{16}$

b. Domain = $\{z \mid z \neq 0\}$ because division by 0 is not defined.

c. Range = $\{y \mid y > 0\}$

35. Yes

36. No

37.

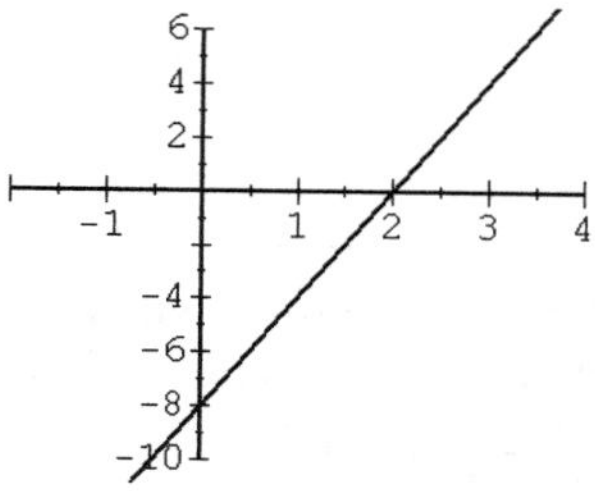

38.

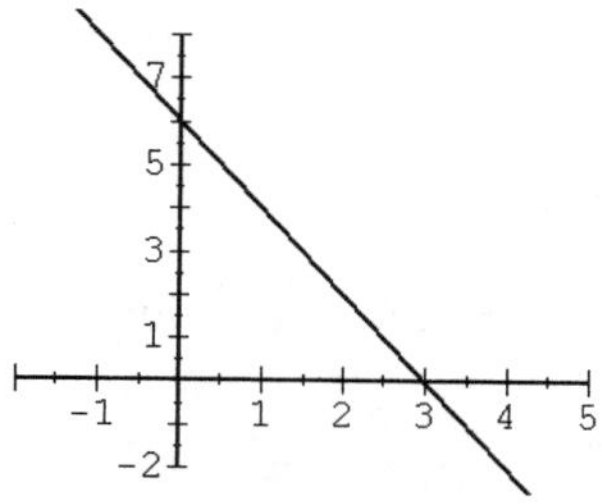

39.

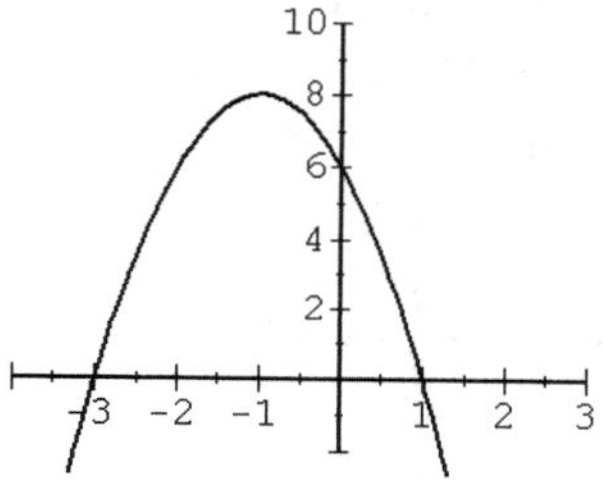

40.

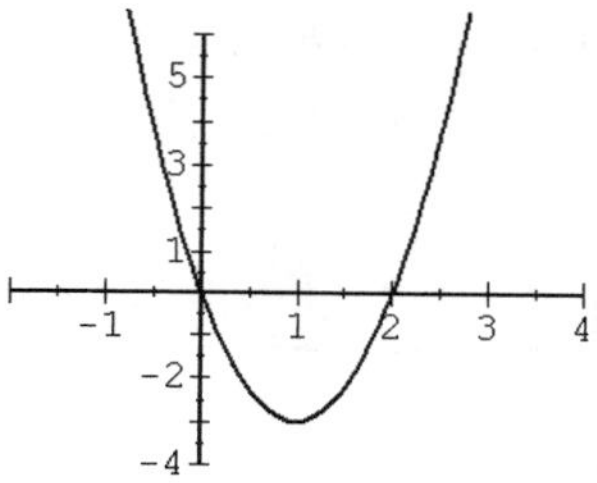

41. **a.**

$3x^2 + 9x = 0$

$3x(x+3) = 0$

Equals 0 at $x = 0$ Equals 0 at $x = -3$

$x = 0$ and $x = -3$

b. Use the quadratic formula with $a = 3$, $b = 9$, and $c = 0$

$\frac{-9 \pm \sqrt{9^2 - 4(3)(0)}}{2(3)} = \frac{-9 \pm \sqrt{81}}{6}$

$= \frac{-9 \pm 9}{6}$

$= 0, -3$

$x = 0$ and $x = -3$

42. **a.**

$2x^2 - 8x - 10 = 0$

$2(x^2 - 4x - 5) = 0$

$(x-5)(x+1) = 0$

Equals 0 at $x = 5$ Equals 0 at $x = -1$

$x = 5$ and $x = -1$

b. Use the quadratic formula with $a = 2$, $b = -8$, and $c = -10$

$\frac{-(-8) \pm \sqrt{(-8)^2 - 4(2)(-10)}}{2(2)}$

$= \frac{8 \pm \sqrt{64 + 80}}{4}$

$= \frac{8 \pm 12}{4}$

$= 5, -1$

$x = 5$ and $x = -1$

43. **a.**
$$3x^2+3x+5=11$$
$$3x^2+3x-6=0$$
$$3(x^2+x-2)=0$$
$$(x+2)(x-1)=0$$
Equals 0 at $x=-2$ Equals 0 at $x=1$

$x=-2$ and $x=1$

b.
$$\frac{-3\pm\sqrt{3^2-4(3)(-6)}}{2(3)}=\frac{-3\pm\sqrt{9+72}}{6}=\frac{-3\pm\sqrt{81}}{6}=\frac{-3\pm 9}{6}=-2,1$$
$x=-2$ and $x=1$

44. **a.**
$$4x^2-2=2$$
$$4x^2=4$$
$$x^2=1$$
$$x=\pm 1$$
$x=1$ and $x=-1$

b.
$$\frac{-0\pm\sqrt{0^2-4(4)(-4)}}{2(4)}=\frac{\pm\sqrt{64}}{8}=\frac{\pm 8}{8}=\pm 1$$
$x=1$ and $x=-1$

45. **a.** Use the vertex formula with $a=1$ and $b=-10$.
$$x=\frac{-b}{2a}=\frac{-(-10)}{2(1)}=\frac{10}{2}=5$$
To find y, evaluate $f(5)$.
$$f(5)=(5)^2-10(5)-25=-50$$
The vertex is $(5,-50)$.

b.
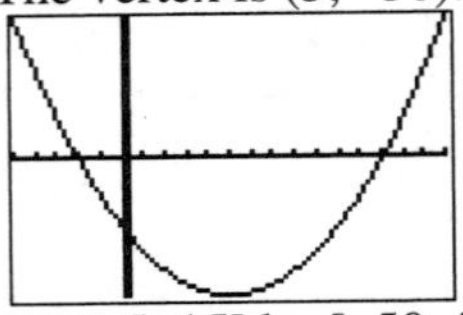
on [–5, 15] by [–50, 50]

46. **a.** Use the vertex formula with $a=1$ and $b=14$.
$$x=\frac{-b}{2a}=\frac{-14}{2(1)}=-7$$
To find y, evaluate $f(-7)$.
$$f(-7)=(-7)^2+14(-7)-15=-64$$
The vertex is $(-7,-64)$

b.
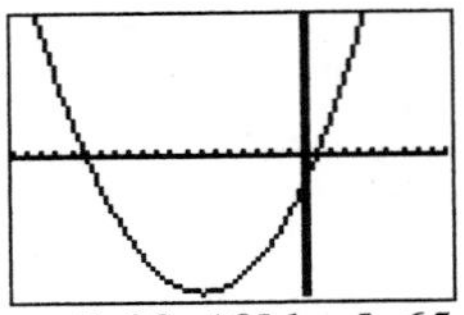
on [–20, 10] by [–65, 65]

47. Let x = number of miles per day.
$C(x)=0.12x+45$

48. Use the interest formula with $P=10{,}000$ and $r=0.08$.
$I(t)=10{,}000(0.08)t=800t$

49. Let x = the altitude in feet.
$$T(x)=70-\frac{x}{300}$$

50. Let t = the number of years after 2010.
$$C(t)=0.45t+20.3$$
$$25=0.45t+20.3$$
$t\approx 10.4$ years after 2010; in the year 2020

51. **a.** To find the break even points, solve the equation $C(x)=R(x)$ for x.
$$C(x)=R(x)$$
$$80x+1950=-2x^2+240x$$
$$2x^2-160x+1950=0$$
$$x^2-80x+975=0$$
$$(x-65)(x-15)=0$$
Equals 0 at $x=65$ Equals 0 at $x=15$

$x=65$ and $x=15$

The store breaks even at 15 receivers and at 65 receivers.

52. **a.** To find the break even points, solve the equation $C(x)=R(x)$ for x.
$$C(x)=R(x)$$
$$220x+202{,}500=-3x^2+2020x$$
$$3x^2-1800x+202{,}500=0$$
$$x^2-600x+67{,}500=0$$
$$(x-450)(x-150)=0$$
Equals 0 at $x=450$ Equals 0 at $x=150$

$x=450$ and $x=150$

The outlet breaks even at 150 units and 450 units.

51. b. To find the number of receivers that maximizes profit, first find the profit function, $P(x) = R(x) - C(x)$.

$$P(x) = (-2x^2 + 240x) - (80x + 1950)$$
$$= -2x^2 + 160x - 1950$$

Since this is a parabola that opens downward, the maximum profit is found at the vertex.

$$x = \frac{-b}{2a} = \frac{-160}{2(-2)} = \frac{-160}{-4} = 40$$

Thus, profit is maximized when 40 receivers are installed per week. The maximum profit is found by evaluating $P(40)$.

$$P(40) = -2(40)^2 + 160(40) - 1950$$
$$= \$1250$$

Therefore, the maximum profit is $1250.

52. b. To find the number of units that maximizes profit, first find the profit function, $P(x) = R(x) - C(x)$.

$$P(x) = (-3x^2 + 2020x) - (220x + 202{,}500)$$
$$= -3x^2 + 1800x - 202{,}500$$

Since this is a parabola that opens downward, the maximum profit is found at the vertex.

$$x = \frac{-b}{2a} = \frac{-1800}{2(-3)} = \frac{-1800}{-6} = 300$$

Thus, profit is maximized when 300 units are installed per month. The maximum profit is found by evaluating $P(300)$.

$$P(300) = -3(300)^2 + 1800(300) - 202{,}500$$
$$= \$67{,}500$$

Therefore, the maximum profit is $67,500.

53. a.

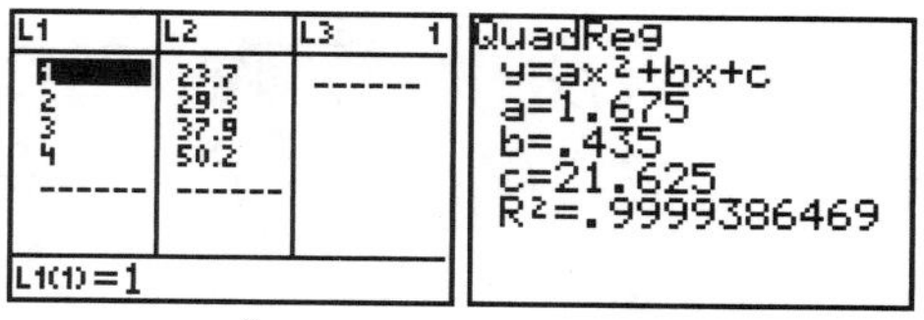

$y = 1.675x^2 + 0.435x + 21.625$ (rounded)

b. For year 2020, $x = 12$.

$$y = 1.675(12)^2 + 0.435(12) + 21.6$$
$$\approx \$268 \text{ billion}$$

54. a. $f(-1) = \dfrac{3}{(-1)(-1-2)} = \dfrac{3}{3} = 1$

b. Domain = $\{x \mid x \neq 0, x \neq 2\}$

c. Range = $\{y \mid y > 0 \text{ or } y \leq -3\}$

55. a. $f(-8) = \dfrac{16}{(-8)(-8+4)} = \dfrac{16}{32} = \dfrac{1}{2}$

b. Domain = $\{x \mid x \neq 0, x \neq -4\}$

c. Range = $\{y \mid y > 0 \text{ or } y \leq -4\}$

56. a. $g(-4) = |-4+2| - 2 = 2 - 2 = 0$

b. Domain = $\Re$

c. Range = $\{y \mid y \geq 0\}$

57. a. $g(-5) = (-5) - |-5| = -5 - 5 = -10$

b. Domain = $\Re$

c. Range = $\{y \mid y \geq 0\}$

58.

$$5x^4 + 10x^3 = 15x^2$$
$$5x^4 + 10x^3 - 15x^2 = 0$$
$$5x^2(x^2 + 2x - 3) = 0$$
$$x^2(x+3)(x-1) = 0$$

Equals 0 at $x = 0$ Equals 0 at $x = -3$ Equals 0 at $x = 1$

$x = 0$, $x = -3$, and $x = 1$

59.

$$4x^5 + 8x^4 = 32x^3$$
$$4x^5 + 8x^4 - 32x^3 = 0$$
$$4x^3(x^2 + 2x - 8) = 0$$
$$x^3(x+4)(x-2) = 0$$

Equals 0 at $x = 0$ Equals 0 at $x = -4$ Equals 0 at $x = 2$

$x = 0$, $x = -4$, and $x = 2$

60.

$$2x^{5/2} - 8x^{3/2} = 10x^{1/2}$$
$$2x^{5/2} - 8x^{3/2} - 10x^{1/2} = 0$$
$$2x^{1/2}(x^2 - 4x - 5) = 0$$
$$x^{1/2}(x-5)(x+1) = 0$$

Equals 0 at $x = 0$ Equals 0 at $x = 5$ Equals 0 at ~~$x = -1$~~

$x = 0$, $x = 5$ and ~~$x = -1$~~

Only $x = 0$ and $x = 5$ are solutions.

61.

$$3x^{5/2} + 3x^{3/2} = 18x^{1/2}$$
$$3x^{5/2} + 3x^{3/2} - 18x^{1/2} = 0$$
$$3x^{1/2}(x^2 + x - 6) = 0$$
$$x^{1/2}(x+3)(x-2) = 0$$

Equals 0 at $x = 0$ Equals 0 at ~~$x = -3$~~ Equals 0 at $x = 2$

$x = 0$, ~~$x = -3$~~ and $x = 2$

Only $x = 0$ and $x = 2$ are solutions.

62.

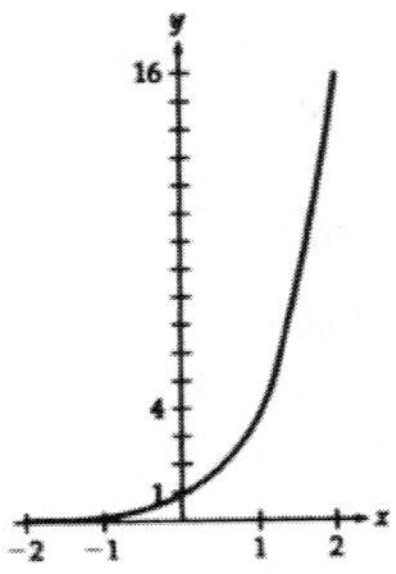

63.

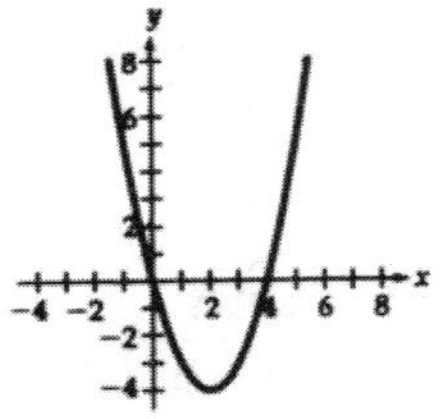

64.

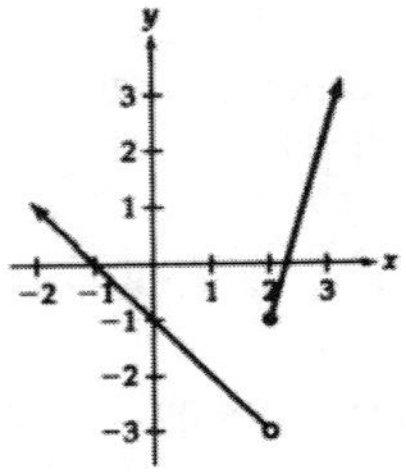

65.

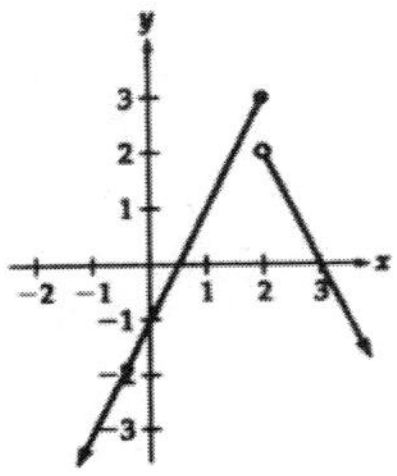

66. a.
$$f(g(x)) = [g(x)]^2 + 1 = \left(\frac{1}{x}\right)^2 + 1 = \frac{1}{x^2} + 1$$

b.
$$g(f(x)) = \frac{1}{f(x)} = \frac{1}{x^2+1}$$

67. a.
$$f(g(x)) = \sqrt{g(x)} = \sqrt{5x-4}$$

b.
$$g(f(x)) = 5[f(x)] - 4 = 5\sqrt{x} - 4$$

68. a.
$$f(g(x)) = \frac{g(x)+1}{g(x)-1} = \frac{x^3+1}{x^3-1}$$

b.
$$g(f(x)) = [f(x)]^3 = \left(\frac{x+1}{x-1}\right)^3$$

69. a.
$$f(g(x)) = |\,g(x)\,| = |\,x+2\,|$$

b.
$$g(f(x)) = [f(x)] + 2 = |\,x\,| + 2$$

70.
$$f(x) = 2x^2 - 3x + 1$$
$$\frac{f(x+h) - f(x)}{h}$$
$$= \frac{2(x+h)^2 - 3(x+h) + 1 - (2x^2 - 3x + 1)}{h}$$
$$= \frac{2(x^2 + 2xh + h^2) - 3(x+h) + 1 - (2x^2 - 3x + 1)}{h}$$
$$= \frac{2x^2 + 4xh + 2h^2 - 3x - 3h + 1 - 2x^2 + 3x - 1}{h}$$
$$= \frac{4xh + 2h^2 - 3h}{h}$$
$$= \frac{h(4x + 2h - 3)}{h}$$
$$= 4x + 2h - 3$$

71.
$$f(x) = \frac{5}{x}$$
$$\frac{f(x+h) - f(x)}{h} = \frac{\frac{5}{x+h} - \frac{5}{x}}{h}$$
$$= \frac{\frac{5}{x+h} - \frac{5}{x}}{h} \cdot \frac{x(x+h)}{x(x+h)}$$
$$= \frac{5x - 5(x+h)}{hx(x+h)}$$
$$= \frac{5x - 5x - 5h}{hx(x+h)}$$
$$= \frac{-5h}{hx(x+h)}$$
$$= \frac{-5}{x(x+h)}$$

72. The advertising budget A as a function of t is the composition of $A(p)$ and $p(t)$.

$$A(p(t)) = 2[p(t)]^{0.15} = 2(18+2t)^{0.15}$$
$$A(4) = 2[18+2(4)]^{0.15} = 2(26)^{0.15}$$
$$\approx \$3.26 \text{ million}$$

73. **a.**

$$x^4 - 2x^3 - 3x^2 = 0$$
$$x^2(x^2 - 2x - 3) = 0$$
$$x^2(x-3)(x+1) = 0$$

Equals 0 at $x = 0$ Equals 0 at $x = 3$ Equals 0 at $x = -1$

$x = 0,$ $x = 3$ and $x = -1$

b.

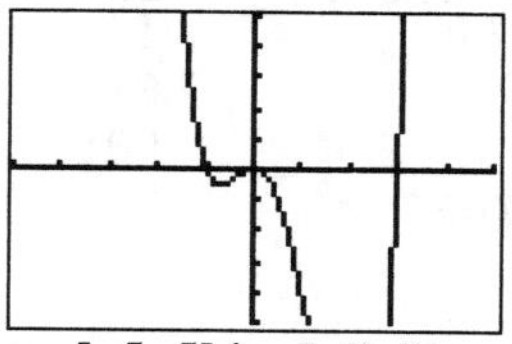

on [–5, 5] by [–5, 5]

74. **a.**

$$x^3 + 2x^2 - 3x = 0$$
$$x(x^2 + 2x - 3) = 0$$
$$x(x+3)(x-1) = 0$$

Equals 0 at $x = 0$ Equals 0 at $x = -3$ Equals 0 at $x = 1$

$x = 0,$ $x = -3,$ and $x = 1$

b.

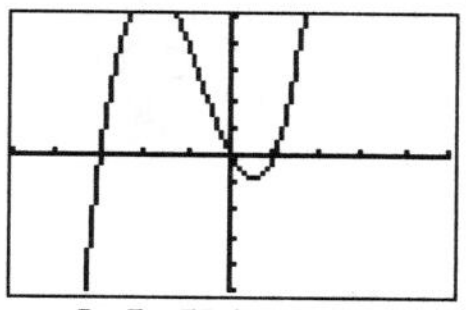

on [–5, 5] by [–5, 5]

75. **a.**

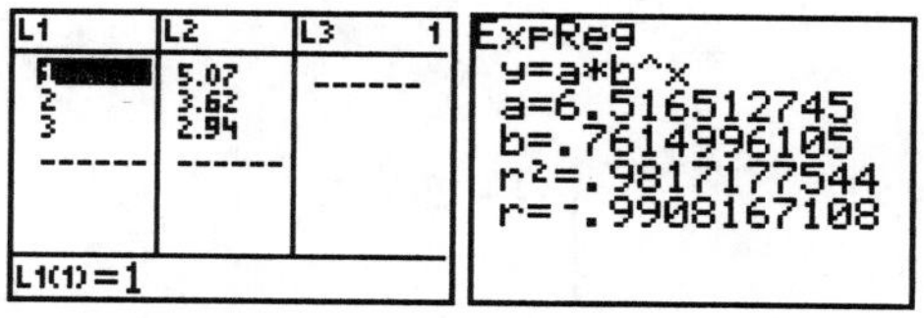

$y = 6.52 \cdot 0.761^x$

b. For year 2020, $x = 4$.

$y = 6.52 \cdot 0.761^4 \approx 2.2$

2.2 crimes per 100,000

Chapter 2: Derivatives and Their Uses

EXERCISES 2.1

1.

x	$5x-7$
1.9	2.500
1.99	2.950
1.999	2.995

x	$5x-7$
2.1	3.500
2.01	3.050
2.001	3.005

a. $\lim_{x\to 2^-}(5x-7)=3$

b. $\lim_{x\to 2^+}(5x-7)=3$

c. $\lim_{x\to 2}(5x-7)=3$

3.

x	$\frac{x^3-1}{x-1}$
0.900	2.710
0.990	2.970
0.999	2.997

x	$\frac{x^3-1}{x-1}$
1.1	3.310
1.01	3.030
1.001	3.003

a. $\lim_{x\to 1^-}\frac{x^3-1}{x-1}=3$

b. $\lim_{x\to 1^+}\frac{x^3-1}{x-1}=3$

c. $\lim_{x\to 1}\frac{x^3-1}{x-1}=3$

5.

x	$(1+2x)^{1/x}$
–0.1	9.313
–0.01	7.540
–0.001	7.404

x	$(1+2x)^{1/x}$
0.1	6.192
0.01	7.245
0.001	7.374

$\lim_{x\to 0}(1+2x)^{1/x}\approx 7.4$

7.

x	$\frac{\frac{1}{x}-\frac{1}{2}}{x-2}$
1.9	–0.263
1.99	–0.251
1.999	–0.250

x	$\frac{\frac{1}{x}-\frac{1}{2}}{x-2}$
2.1	-0.238
2.01	-0.249
2.001	-0.250

$\lim_{x\to 2}\frac{\frac{1}{x}-\frac{1}{2}}{x-2}=-0.25$

9.

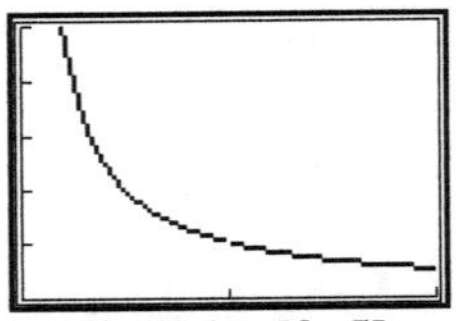

on [0, 2] by [0, 5]

$\lim_{x\to 1}\frac{\frac{1}{x}-1}{1-x}=1$

11.

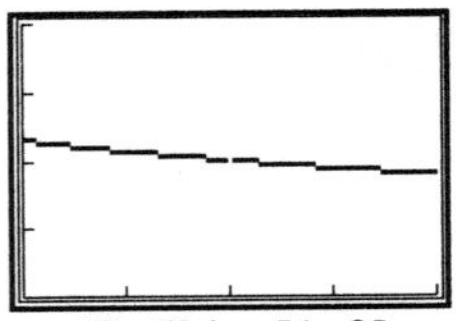

on [3, 5] by [1, 3]

$\lim_{x\to 4}\frac{x^{1.5}-4x^{0.5}}{x^{1.5}-2x}=2$

13. $\lim_{x\to 3} 4x^2-10x+2=4(3)^2-10(3)+2=8$

15. $\lim_{x\to 5}\frac{3x^2-5x}{7x-10}=\frac{3(5)^2-5(5)}{7(5)-10}=2$

17. $\lim_{x\to 3}\sqrt{2}=\sqrt{2}$ because the limit of a constant is just the constant.

19. $\lim_{t\to 25}\left[(t+5)t^{-1/2}\right]=(25+5)(25)^{-1/2}=6$

21. $\lim_{h\to 0}(5x^3+2x^2h-xh^2)=5x^3+2x^2\cdot 0-x(0)^2=5x^3$

23. $\lim_{x\to 2}\frac{x^2-4}{x-2}=\lim_{x\to 2}\frac{(x+2)(x-2)}{x-2}$
$=\lim_{x\to 2}(x+2)=2+2=4$

25. $\lim_{x\to -3}\frac{x+3}{x^2+8x+15}=\lim_{x\to -3}\frac{x+3}{(x+3)(x+5)}$
$=\lim_{x\to -3}\frac{1}{x+5}=\frac{1}{-3+5}=\frac{1}{2}$

27. $\lim_{x\to -1}\frac{3x^3-3x^2-6x}{x^2+x}=\lim_{x\to -1}\frac{3x(x^2-x-2)}{x(x+1)}$
$=\lim_{x\to -1}\frac{3x(x-2)(x+1)}{x(x+1)}$
$=\lim_{x\to -1}3(x-2)=3(-1-2)=-9$

29. $\lim_{h\to 0} \frac{2xh-3h^2}{h} = \lim_{h\to 0}(2x-3h) = 2x-3(0) = 2x$

31. $\lim_{h\to 0} \frac{4x^2h+xh^2-h^3}{h} = \lim_{h\to 0}(4x^2+xh-h^2)$
$= 4x^2 + x(0) - (0)^2 = 4x^2$

33. **a.** $\lim_{x\to 2^-} f(x) = 1$
b. $\lim_{x\to 2^+} f(x) = 3$
c. $\lim_{x\to 2} f(x)$ does not exist.

35. **a.** $\lim_{x\to 2^-} f(x) = -1$
b. $\lim_{x\to 2^+} f(x) = -1$
c. $\lim_{x\to 2} f(x) = -1$

37. **a.** $\lim_{x\to 4^-} f(x) = \lim_{x\to 4^-}(3-x) = 3-(4) = -1$
b. $\lim_{x\to 4^+} f(x) = \lim_{x\to 4^+}(10-2x) = 10-2(4) = 2$
c. $\lim_{x\to 4} f(x)$ does not exist.

39. **a.** $\lim_{x\to 4^-} f(x) = \lim_{x\to 4^-}(2-x) = 2-4 = -2$
b. $\lim_{x\to 4^+} f(x) = \lim_{x\to 4^+}(x-6) = 4-6 = -2$
c. $\lim_{x\to 4} f(x) = -2$

41. **a.** $\lim_{x\to 0^-} f(x) = \lim_{x\to 0^-}(-x) = -0 = 0$
b. $\lim_{x\to 0^+} f(x) = \lim_{x\to 0^+}(x) = 0$
c. $\lim_{x\to 0} f(x) = 0$

43. **a.**

x	$\frac{\lvert x\rvert}{x}$
−0.1	−1
−0.01	−1
−0.001	−1

$\lim_{x\to 0} f(x) = -1$

b.

x	$\frac{\lvert x\rvert}{x}$
0.1	1
0.01	1
0.001	1

$\lim_{x\to 0} f(x) = 1$

c. $\lim_{x\to 0} f(x)$ does not exist.

45. $\lim_{x\to -\infty} f(x) = 0$; $\lim_{x\to 3^-} f(x) = \infty$ and $\lim_{x\to 3^+} f(x) = -\infty$,
so $\lim_{x\to 3} f(x)$ does not exist and $\lim_{x\to\infty} f(x) = 0$.

47. $\lim_{x\to -\infty} f(x) = 0$; $\lim_{x\to 0^-} f(x) = \infty$ and $\lim_{x\to 0^+} f(x) = \infty$,
so $\lim_{x\to 0} f(x) = \infty$ and $\lim_{x\to\infty} f(x) = 0$.

49. $\lim_{x\to -\infty} f(x) = 2$; $\lim_{x\to 1^-} f(x) = -\infty$ and $\lim_{x\to 1^+} f(x) = \infty$,
so $\lim_{x\to 1} f(x)$ does not exist and $\lim_{x\to\infty} f(x) = 2$.

51. $\lim_{x\to -\infty} f(x) = 1$; $\lim_{x\to -2^-} f(x) = \infty$ and $\lim_{x\to -2^+} f(x) = \infty$,
so $\lim_{x\to -2} f(x) = \infty$ and $\lim_{x\to\infty} f(x) = 1$.

53. Continuous

55. Discontinuous at c because $\lim_{x\to c} f(x) \neq f(c)$.

57. Discontinuous at c because $f(c)$ is not defined.

59. Discontinuous at c because $\lim_{x\to c} f(x)$ does not exist.

61. **a.**

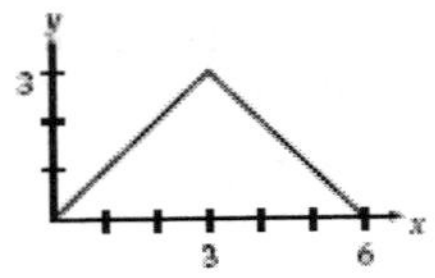

b. $\lim_{x\to 3^-} f(x)=3;\ \lim_{x\to 3^+} f(x)=3$

c. Continuous

63. **a.**

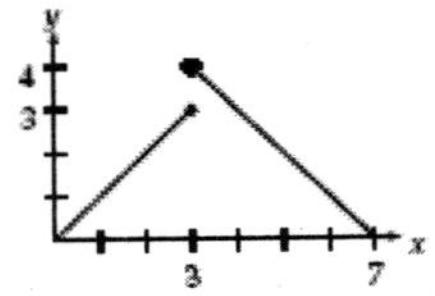

b. $\lim_{x\to 3^-} f(x)=3;\ \lim_{x\to 3^+} f(x)=4$

c. Discontinuous because $\lim_{x\to 3} f(x)$ does not exist.

65. Continuous

67. Discontinuous at $x = 1$

69. $f(x)=\dfrac{12}{5x^3-5x}$ is discontinuous at values of x for which the denominator is zero. Thus, consider

$$5x^3 - 5x = 0$$

$$5x(x^2 - 1) = 0$$

$5x$ equals zero at $x = 0$ and $x^2 - 1$ equals zero at $x = \pm 1$.
Thus, the function is discontinuous at $x = 0$, $x = -1$, and $x = 1$.

71. From Exercise 37, we know $\lim_{x\to 4} f(x)$ does not exist. Therefore, the function is discontinuous at $x = 4$.

73. From Exercise 39, we know $\lim_{x\to 4} f(x) = -2 = f(4)$. Therefore, the function is continuous.

75. From Exercise 41, we know $\lim_{x\to 0} f(x) = 0 = f(0)$. Therefore, the function is continuous.

77. From the graph, we can see that $\lim_{x\to 6} f(x)$ does not exist because the left and right limits do not agree. $f(x)$ is discontinuous at $x = 6$.

79. Since the function $\dfrac{(x-1)(x+2)}{x-1}$ is not defined at $x = 1$ and the function $x + 2$ equals 3 at $x = 1$, the functions are not equal.

81.

x	$\left(1+\frac{x}{10}\right)^{1/x}$
−0.1	1.11
−0.01	1.11
−0.001	1.11
−0.0001	1.11

x	$\left(1+\frac{x}{10}\right)^{1/x}$
0.1	1.10
0.01	1.11
0.001	1.11
0.0001	1.11

$\lim_{x\to 0}\left(1+\frac{x}{10}\right)^{1/x} \approx \$1.11$

83. $\lim_{x\to 0}\dfrac{100}{1+.001x^2}=\dfrac{100}{1+.001(0)^2}=\dfrac{100}{1}$
$=100$

85. As x approaches c, the function is approaching $\lim_{x\to c} f(x)$ even if the value of the function at c is different, so the limit is where the function is "going".

87. False: The value of the function at 2 has nothing to do with the limit as x approaches 2.

89. False: Both one-sided limits would have to exist and agree to guarantee that the limit exists.

91. False: On the left side of the limit exists and equals 2 (as we saw in Example 4), but on the right side of the denominator of the fraction is zero. Therefore one side of the equation is defined and the other is not.

93. True: The third requirement for continuity at $x = 2$ is that the limit and the value at 2 must agree, so if one is 7 the other must be 7.

95. $\lim_{x\to 0^+} f(x)$ does not exist; $\lim_{x\to 0^-} f(x)$ does not exist; $\lim_{x\to 0} f(x)$ does not exist

EXERCISES 2.2

1. The slope is positive at P_1.
The slope is negative at P_2.
The slope is zero at P_3.

3. The slope is positive at P_1.
The slope is negative at P_2.
The slope is zero at P_3.

5. The tangent line at P_1 contains the points (0, 2) and (1, 5). The slope of this line is

$$m = \frac{5-2}{1-0} = 3.$$

The slope of the curve at P_1 is 3.
The tangent line at P_2 contains the points (3, 5) and (5, 4). The slope of this line is

$$m = \frac{4-5}{5-3} = -\frac{1}{2}$$

The slope of the curve at P_2 is $-\frac{1}{2}$.

7. Your graph should look roughly like the following:

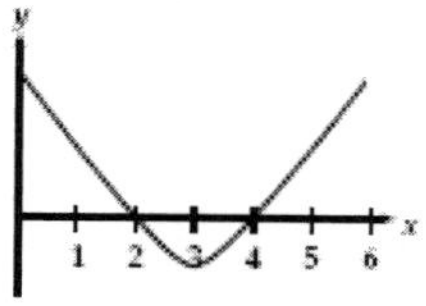

9.
a. $\frac{f(3)-f(1)}{2} = \frac{12-2}{2} = 5$
b. $\frac{f(2)-f(1)}{1} = \frac{6-2}{1} = 4$
c. $\frac{f(1.5)-f(1)}{.5} = \frac{3.75-2}{.5} = 3.5$
d. $\frac{f(1.1)-f(1)}{.1} = \frac{2.31-2}{.1} = 3.1$
e. $\frac{f(1.01)-f(1)}{.01} = \frac{2.0301-2}{.01} = 3.01$
f. Answers seem to be approaching 3.

11.
a. $\frac{f(4)-f(2)}{2} = \frac{34-8}{2} = 13$
b. $\frac{f(3)-f(2)}{1} = \frac{19-8}{1} = 11$
c. $\frac{f(2.5)-f(2)}{.5} = \frac{13-8}{.5} = 10$
d. $\frac{f(2.1)-f(2)}{.1} = \frac{8.92-8}{.1} = 9.2$
e. $\frac{f(2.01)-f(1)}{.01} = \frac{8.0902-8}{.01} = 9.02$
f. Answers seen to be approaching 9.

13.
a. $\frac{f(5)-f(3)}{2} = \frac{26-16}{2} = 5$
b. $\frac{f(4)-f(3)}{1} = \frac{21-16}{1} = 5$
c. $\frac{f(3.5)-f(3)}{.5} = \frac{18.5-16}{.5} = 5$
d. $\frac{f(3.1)-f(3)}{.1} = \frac{16.5-16}{.1} = 5$
e. $\frac{f(3.01)-f(3)}{.01} = \frac{16.05-16}{.01} = 5$
f. Answers seem to be approaching 5.

15.
a. $\frac{f(6)-f(4)}{2} = \frac{2.4495-2}{2} = 0.2247$
b. $\frac{f(5)-f(4)}{1} = \frac{2.236-2}{1} = 0.2361$
c. $\frac{f(4.5)-f(4)}{.5} = \frac{2.121-2}{.5} = 0.2426$
d. $\frac{f(4.1)-f(4)}{.1} = \frac{2.025-2}{.1} = 0.2485$
e. $\frac{f(4.01)-f(4)}{.01} = \frac{2.0025-2}{.01} = 0.2498$
f. Answers seem to approach 0.25.

17. $\lim\limits_{h\to 0}\dfrac{f(x+h)-f(x)}{h}$
$= \lim\limits_{h\to 0}\dfrac{(x+h)^2+(x+h)-(x^2+x)}{h}$
$= \lim\limits_{h\to 0}\dfrac{x^2+2xh+h^2+x+h-x^2-x}{h}$
$= \lim\limits_{h\to 0}\dfrac{2xh+h^2+h}{h}$
$= \lim\limits_{h\to 0} 2x+h+1 = 2x+1$
Evaluating at $x=1$ gives $2(1)+1=3$, which matches the answer from Exercise 9.

19. $\lim\limits_{h\to 0}\dfrac{f(x+h)-f(x)}{h}$
$= \lim\limits_{h\to 0}\dfrac{5(x+h)+1-(5x+1)}{h}$
$= \lim\limits_{h\to 0}\dfrac{5x+5h+1-5x-1}{h}$
$= \lim\limits_{h\to 0}\dfrac{5h}{h}$
$= \lim\limits_{h\to 0} 5 = 5$
Evaluating at $x=3$ gives 5, which matches the answer from Exercise 13.

21. $\lim\limits_{h\to 0}\dfrac{f(x+h)-f(x)}{h}$
$= \lim\limits_{h\to 0}\dfrac{2(x+h)^2+(x+h)-2-(2x^2+x-2)}{h}$
$= \lim\limits_{h\to 0}\dfrac{2x^2+4xh+2h^2+x+h-2-2x^2-x+2}{h}$
$= \lim\limits_{h\to 0}\dfrac{4xh+2h^2+h}{h}$
$= \lim\limits_{h\to 0} 4x+2h+1 = 4x+1$
The slope of the tangent line is at $x=2$ is $4(2)+1=9$, which matches the answer from Exercise 11.

23. $\lim\limits_{h\to 0}\dfrac{f(x+h)-f(x)}{h} = \lim\limits_{h\to 0}\dfrac{\sqrt{x+h}-\sqrt{x}}{h}$
$= \lim\limits_{h\to 0}\left(\dfrac{\sqrt{x+h}-\sqrt{x}}{h}\right)\left(\dfrac{\sqrt{x+h}+\sqrt{x}}{\sqrt{x+h}+\sqrt{x}}\right)$
$= \lim\limits_{h\to 0}\dfrac{(x+h)-x}{h(\sqrt{x+h}+\sqrt{x})}$
$= \lim\limits_{h\to 0}\dfrac{1}{\sqrt{x+h}+\sqrt{x}} = \dfrac{1}{\sqrt{x+0}+\sqrt{x}} = \dfrac{1}{2\sqrt{x}}$
The slope of the tangent line at $x=4$ is $\dfrac{1}{2\sqrt{4}} = 0.25$, which matches the answer from Exercise 15.

25. $f'(x) = \lim\limits_{h\to 0}\dfrac{f(x+h)-f(x)}{h} = \lim\limits_{h\to 0}\dfrac{(x+h)^2-3(x+h)+5-(x^2-3x+5)}{h}$
$= \lim\limits_{h\to 0}\dfrac{x^2+2xh+h^2-3x-3h+5-x^2+3x-5}{h}$
$= \lim\limits_{h\to 0}(2x+h-3)$
$= 2x-3$

27. $f'(x) = \lim\limits_{h\to 0}\dfrac{f(x+h)-f(x)}{h}$
$= \lim\limits_{h\to 0}\dfrac{1-(x+h)^2-(1-x^2)}{h}$
$= \lim\limits_{h\to 0}\dfrac{1-x^2-2xh-h^2-1+x^2}{h}$
$= \lim\limits_{h\to 0}(-2x-h)$
$= -2x$

29. $f'(x) = \lim\limits_{h\to 0}\dfrac{f(x+h)-f(x)}{h}$
$= \lim\limits_{h\to 0}\dfrac{9(x+h)-2-(9x-2)}{h}$
$= \lim\limits_{h\to 0}\dfrac{9x+9h-2-9x+2}{h} = 9$

31. $f'(x) = \lim\limits_{h\to 0}\dfrac{f(x+h)-f(x)}{h}$
$= \lim\limits_{h\to 0}\dfrac{\frac{x+h}{2}-\frac{x}{2}}{h} = \lim\limits_{h\to 0}\dfrac{\frac{h}{2}}{h}$
$= \lim\limits_{h\to 0}\dfrac{h}{2h} = \dfrac{1}{2}$

33. $f'(x) = \lim\limits_{h\to 0}\dfrac{f(x+h)-f(x)}{h} = \lim\limits_{h\to 0}\dfrac{4-4}{h} = 0$

35. $$f'(x)=\lim_{h\to0}\frac{f(x+h)-f(x)}{h}=\lim_{h\to0}\frac{a(x+h)^2+b(x+h)+c-(ax^2+bx+c)}{h}$$
$$=\lim_{h\to0}\frac{ax^2+2axh+ah^2+bx+bh+c-ax^2-bx-c}{h}$$
$$=\lim_{h\to0}(2ax+ah+b)=2ax+b$$

37. $$f'(x)=\lim_{h\to0}\frac{f(x+h)-f(x)}{h}$$
$$=\lim_{h\to0}\frac{(x+h)^5-x^5}{h}$$
$$=\lim_{h\to0}\frac{x^5+5x^4h+10x^3h^2+5xh^4+h^5-x^5}{h}$$
$$=\lim_{h\to0}5x^4+10x^3h+5xh^3+h^4=5x^4$$

39. $$f'(x)=\lim_{h\to0}\frac{f(x+h)-f(x)}{h}$$
$$=\lim_{h\to0}\frac{\frac{2}{x+h}-\frac{2}{x}}{h}$$
$$=\lim_{h\to0}\frac{\frac{2x}{x(x+h)}-\frac{2(x+h)}{x(x+h)}}{h}$$
$$=\lim_{h\to0}\frac{2x-2x-2h}{x(x+h)}\cdot\frac{1}{h}$$
$$=\lim_{h\to0}-\frac{2h}{x(x+h)}\cdot\frac{1}{h}$$
$$=\lim_{h\to0}-\frac{2}{x(x+h)}$$
$$=-\frac{2}{x^2}$$

41. $$f'(x)=\lim_{h\to0}\frac{f(x+h)-f(x)}{h}=\lim_{h\to0}\frac{\sqrt{x+h}-\sqrt{x}}{h}=\lim_{h\to0}\frac{\sqrt{x+h}-\sqrt{x}}{h}\cdot\frac{\sqrt{x+h}+\sqrt{x}}{\sqrt{x+h}+\sqrt{x}}$$
$$=\lim_{h\to0}\frac{x+h-x}{h\left(\sqrt{x+h}+\sqrt{x}\right)}=\lim_{h\to0}\frac{h}{h\left(\sqrt{x+h}+\sqrt{x}\right)}$$
$$=\lim_{h\to0}\frac{1}{\sqrt{x+h}+\sqrt{x}}=\frac{1}{2\sqrt{x}}$$

43. $$f'(x)=\lim_{h\to0}\frac{f(x+h)-f(x)}{h}=\lim_{h\to0}\frac{(x+h)^3+(x+h)^2-(x^3+x^2)}{h}$$
$$=\lim_{h\to0}\frac{(x^3+3x^2h+3xh^2+h^3)+(x^2+2xh+h^2)-x^3-x^2}{h}$$
$$=\lim_{h\to0}\frac{3x^2h+3xh^2+h^3+2xh+h^2}{h}=\lim_{h\to0}3x^2+3xh+h^2+2x+h$$
$$=3x^2+2x$$

45. **a.** The slope of the tangent line at $x=2$ is $f'(2)=2(2)-3=1$. To find the point of the curve at $x=2$, we calculate $y=f(2)=2^2-3(2)+5=3$. Using the point-slope form with the point (2, 3), we have

$y-3=1(x-2)$

$y-3=x-2$

$y=x+1$

b.

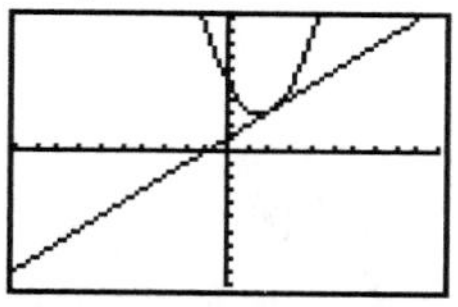

on viewing window
[–10, 10] by [–10, 10]

47. **a.**

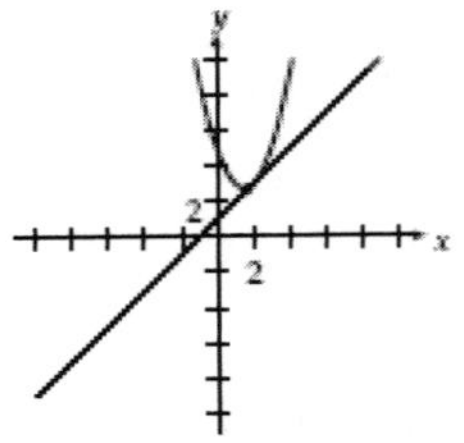

b.

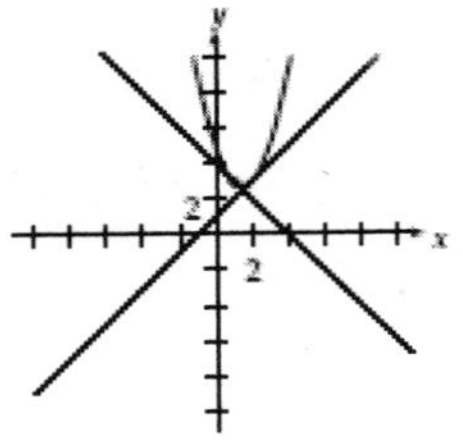

49. **a.** $f'(x) = \lim_{h\to 0} \frac{f(x+h)-f(x)}{h} = \lim_{h\to 0} \frac{3(x+h)-4-(3x-4)}{h} = \lim_{h\to 0} \frac{3x+3h-4-3x+4}{h} = \lim_{h\to 0} 3 = 3$

b. The graph of $f(x) = 3x - 4$ is a straight line with slope 3.

51. **a.** $f'(x) = \lim_{h\to 0} \frac{f(x+h)-f(x)}{h} = \lim_{h\to 0} \frac{5-5}{h} = \lim_{h\to 0} 0 = 0$

b. The graph of $f(x) = 5$ is a straight line with slope 0.

53. **a.** $f'(x) = \lim_{h\to 0} \frac{f(x+h)-f(x)}{h} = \lim_{h\to 0} \frac{m(x+h)+b-(mx+b)}{h} = \lim_{h\to 0} \frac{mx+mh+b-mx-b}{h} = \lim_{h\to 0} \frac{mh}{h}$
$= \lim_{h\to 0} m = m$

b. The graph of $f(x) = mx + b$ is a straight line with slope m.

55. **a.** $f'(x) = \lim_{h\to 0} \frac{f(x+h)-f(x)}{h} = \lim_{h\to 0} \frac{(x+h)^2-8(x+h)+110-(x^2-8x+110)}{h}$
$= \lim_{h\to 0} \frac{x^2+2hx+h^2-8x-8h+110-x^2+8x-110}{h}$
$= \lim_{h\to 0} 2x+h-8 = 2x-8$

b. $f'(2) = 2(2)-8 = -4$. The temperature is decreasing at a rate of 4 degrees per minute after 2 minutes.

c. $f'(5) = 2(5)-8 = 2$. The temperature is increasing at a rate of 2 degrees per minute after 5 minutes.

57. **a.** $f'(x) = \lim_{h\to 0} \frac{f(x+h)-f(x)}{h} = \lim_{h\to 0} \frac{2(x+h)^2-(x+h)-(2x^2-x)}{h}$
$= \lim_{h\to 0} \frac{2x^2+4hx+2h^2-x-h-2x^2+x}{h} = \lim_{h\to 0} 4x+2h-1 = 4x-1$

b. $f'(5) = 4(5)-1 = 19$. When 5 words have been memorized, the memorization time is increasing at a rate of 19 seconds per word.

59. **a.** $f'(x) = \lim_{h\to 0} \frac{f(x+h)-f(x)}{h} = \lim_{h\to 0} \frac{\frac{1}{2}(x+h)^2-3.7(x+h)+12-\left(\frac{1}{2}x^2-3.7x+12\right)}{h}$
$= \lim_{h\to 0} \frac{\frac{1}{2}x^2+xh+\frac{1}{2}h^2-3.7x-3.7h+12-\left(\frac{1}{2}x^2-3.7x+12\right)}{h}$
$= \lim_{h\to 0} \frac{\frac{1}{2}x^2+xh+\frac{1}{2}h^2-3.7x-3.7h+12-\frac{1}{2}x^2+3.7x-12}{h}$
$= \lim_{h\to 0} \left(x+\frac{1}{2}h-3.7\right) = x-3.7$

b. $f'(1) = 1-3.7 = -2.7$. In 1940, the percentage of immigrants was decreasing by 2.7 percentage points per decade (which is about 0.27 of a percentage point decrease per year).

c. $f'(8) = 8-3.7 = 4.3$. Increasing by 4.3 percentage points per decade (so about 0.43 of a percentage point per year).

61. The average rate of change requires two x-values and is the change in the function - values divided by the change in the x-values. The instantaneous rate of change is at a single x-value, and is found from the formula $f(x)=\lim_{x\to\infty}\frac{f(x+h)-f(x)}{h}$.

63. Substitution $h=0$ would make the denominator zero, and we can't divide by zero. That's why we need to do some algebra on the difference quotient, to cancel out the terms that are zero so that afterwards we can evaluate by direct substitution.

65. The units of x are blargs and the units of f are prendles because the derivative $f'(x)$ is equivalent to the slope of $f(x)$, which is the change in f over the change in x.

67. The patient's health is deteriorating during the first day (temperature is rising above normal). The patient's health is improving during the second day (temperature is falling).

EXERCISES 2.3

1. $f'(x)=\frac{d}{dx}(x^4)=4x^{4-1}=4x^3$

3. $f'(x)=\frac{d}{dx}(x^{500})=500x^{500-1}=500x^{499}$

5. $f'(x)=\frac{d}{dx}(x^{1/2})=\frac{1}{2}x^{1/2-1}=\frac{1}{2}x^{-1/2}$

7. $g'(x)=\frac{d}{dx}\left(\frac{1}{4}x^4\right)=\frac{1}{2}\cdot 4x^{4-1}=2x^3$

9. $g'(w)=\frac{d}{dw}(6w^{1/3})=6\cdot\frac{1}{3}w^{1/3-1}=2w^{-2/3}$

11. $h'(x)=\frac{d}{dx}(3x^{-2})=3(-2)x^{-2-1}=-6x^{-3}$

13.
$$\begin{aligned}f'(x)&=\frac{d}{dx}(4x^2-3x+2)\\&=4\cdot 2x^{2-1}-3\cdot 1x^{1-1}+0\\&=8x-3\end{aligned}$$

15.
$$\begin{aligned}f'(x)&=\frac{d}{dx}\left(\frac{1}{x^{1/2}}\right)=\frac{d}{dx}(x^{-1/2})\\&=-\frac{1}{2}x^{-3/2}=-\frac{1}{2x^{3/2}}\end{aligned}$$

17.
$$\begin{aligned}f'(x)&=\frac{d}{dx}\left(\frac{6}{\sqrt[3]{x}}\right)=\frac{d}{dx}(6x^{-1/3})\\&=6\left(-\frac{1}{3}\right)x^{-4/3}=-\frac{2}{x^{4/3}}\end{aligned}$$

19. $f'(r)=\frac{d}{dr}(\pi r^2)=\pi(2r)=2\pi r$

21.
$$\begin{aligned}f'(x)&=\frac{d}{dx}\left(\frac{1}{6}x^3+\frac{1}{2}x^2+x+1\right)\\&=\frac{1}{6}(3x^2)+\frac{1}{2}(2x)+1\\&=\frac{1}{2}x^2+x+1\end{aligned}$$

23.
$$\begin{aligned}g'(x)&=\frac{d}{dx}(x^{1/2}-x^{-1})=\frac{1}{2}x^{1/2-1}-(-1)x^{-1-1}\\&=\frac{1}{2}x^{-1/2}+x^{-2}\end{aligned}$$

25.
$$\begin{aligned}h'(x)&=\frac{d}{dx}(6x^{2/3}-12x^{-1/3})=6\cdot\frac{3}{2}x^{2/3-1}-12\left(-\frac{1}{3}\right)x^{-1/3-1}\\&=4x^{-1/3}+4x^{-4/3}\end{aligned}$$

27.
$$\begin{aligned}f'(x)&=\frac{d}{dx}\left(10x^{-1/2}-\frac{9}{5}x^{5/3}+17\right)=10\left(-\frac{1}{2}\right)x^{-1/2-1}-\frac{9}{5}\left(\frac{5}{3}\right)x^{5/3-1}+0\\&=-5x^{-3/2}-3x^{2/3}\end{aligned}$$

29. $f'(x)=\frac{d}{dx}\left(\frac{x^2+x^3}{x}\right)=\frac{d}{dx}(x+x^2)=1+2x$

31. $f'(x)=\frac{d}{dx}(x^5)=5x^4$; $f'(-2)=5(-2)^4=80$

33. $f'(x) = \frac{d}{dx}(6x^{2/3} - 48x^{-1/3}) = 6\left(\frac{2}{3}\right)x^{-1/3} - 48\left(-\frac{1}{3}\right)x^{-4/3}$

$= 4x^{-1/3} + 16x^{-4/3}$

$f'(8) = 4(8)^{-1/3} + 16(8)^{-4/3} = \frac{4}{\sqrt[3]{8}} + \frac{16}{\sqrt[3]{8^4}} = \frac{4}{2} + \frac{16}{\sqrt[3]{8^3} \cdot \sqrt[3]{8}} = 2 + \frac{16}{8 \cdot 2} = 3$

35. $\frac{df}{dx} = \frac{d}{dx}(x^3) = 3x^2; \left.\frac{df}{dx}\right|_{x=-3} = 3(-3)^2 = 27$

37. $\frac{df}{dx} = \frac{d}{dx}(16x^{-1/2} + 8x^{1/2}) = 16\left(-\frac{1}{2}\right)x^{-3/2} + 8\left(\frac{1}{2}\right)x^{-1/2}$

$= 8x^{-3/2} + 4x^{-1/2}$

$\left.\frac{df}{dx}\right|_{x=4} = -8(4)^{-3/2} + 4(4)^{-1/2} = -\frac{8}{\sqrt{4^3}} + \frac{4}{\sqrt{4}} = -\frac{8}{8} + \frac{4}{2} = 1$

39. **a.** $f'(x) = 2 \cdot x^1 - 2 + 0 = 2x + 2$

$f'(3) = 2(3) - 2 = 4$

The slope of the tangent line is 4.

$y - 5 = 4(x - 3)$

$y - 5 = 4x - 12$

$y = 4x - 7$

The equation for the tangent line is $y = 4x - 7$.

b.

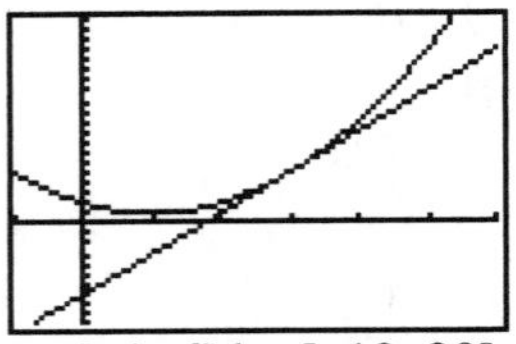

on [–1, 6] by [–10, 20]

41. **a.** $f'(x) = 3 \cdot x^2 - 3(2)x^1 + 2 + 0 = 3x^2 - 6x + 2$

$f'(2) = 3(2)^2 - 6(2) + 2 = 2$

The slope of the tangent line is 2.

$y + 2 = 2(x - 2)$

$y + 2 = 2x - 4$

$y = 2x - 6$

The equation for the tangent line is $y = 2x - 6$.

b.

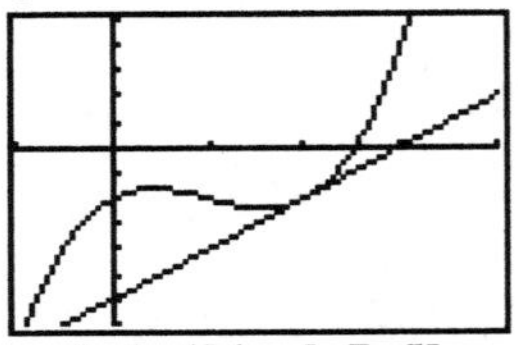

on [–1, 4] by [–7, 5]

43. For $y_1 = 5$ and viewing rectangle [–10, 10] by [–10, 10], your graph should look roughly like the following:

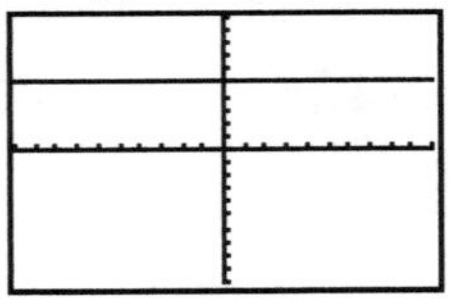

45. $y' = 9.638(-1.394)x^{-2.394} = -13.435372x^{-2.394}$

a. $y'(2) = -13.435372(2)^{-2.394} \approx -2.6$

b. $y'(5) = -13.435372(5)^{-2.394} \approx -0.29$

Interpretation: Near full employment, inflation greatly decreases with small increases in unemployment but with higher unemployment, inflation hardly decreases with increases in unemployment.

47. **a.** $C'(x) = \frac{16}{\sqrt[3]{x}}$

$C'(8) = \frac{16}{\sqrt[3]{8}} = \frac{16}{2} = 8$

When 8 items are purchased, the cost of the last item is about \$8.

b. $C'(64) = \frac{16}{\sqrt[3]{64}} = \frac{16}{4} = 4$

When 8 items are purchased, the cost of the last item is about \$4.

49. $C(64)-C(63)=24(64)^{2/3}-24(63)^{2/3}\approx 4.01$
The answer is close to \$4.

51. **a.** The rate of change of the population in x years is the derivative of the population function
$P'(x)=-x^2+5x-3$
To find the rate of change of the population 20 years from now, evaluate $P'(x)$ for $x=2$.
$P'(2)=-2^2+5(2)-3=3$
In 2030, this population group will be increasing by 3 million per decade.

b. To find the rate of change of the population 0 years from now, evaluate $P'(x)$ for $x=0$.
$P'(0)=-0^2+5(0)-3=-3$
In 2010, this population group will be decreasing by 3 million per decade.

53. The rate of change of the pool of potential customers is the derivative of the function
$N(x)=400,000-\dfrac{200,000}{x}$.
$N'(x)=\dfrac{d}{dx}(400,000-200,000x^{-1})$
$N'(x)=0-(-1)200,000x^{-2}=\dfrac{200,000}{x^2}$
To find the rate of change of the pool of potential customers when the ad has run for 5 days, evaluate $N'(x)$ for $x=5$.
$N'(5)=\dfrac{200,000}{5^2}=8000$
The pool of potential customers is increasing by about 8000 people per additional day.

55. The rate of change of 3D movies is the derivative of the function $f(x)=\frac{1}{2}x^2+11x+8$.
$f'(x)=x+11$
The rate of change of number of 3D movies in 10 years is found by evaluating $f'(x)$ at $x=10$.
$f'(10)=10+11=21$
In 2020, the number of 3D movies will be increasing by 21 per year.

57. The instantaneous rate of change of the number of phrases students can memorize is the derivative of the function $p(t)=24\sqrt{t}$.
$p'(t)=\dfrac{d}{dt}(24t^{1/2})$
$=\dfrac{1}{2}(24t^{-1/2})=12t^{-1/2}$
$p'(4)=12(4)^{-1/2}=\dfrac{12}{\sqrt{4}}=6$
The number of phrases students can memorize after 4 hours is increasing by 6.

59. **a.** $U(x)=100\sqrt{x}=100x^{1/2}$
$MU(x)=U'(x)=\frac{1}{2}(100)x^{1/2-1}=50x^{-1/2}$

b. $MU(1)=U'(1)=50(1)^{-1/2}=50$
The marginal utility of the first dollar is 50.

c. $MU(1,000,000)=U'(1,000,000)=50(10^6)^{-1/2}=50(10)^{-3}=\dfrac{50}{1000}=0.05$
The marginal utility of the millionth dollar is 0.05.

61. **a.** $f(12) = 0.831(12)^2 - 18.1(12) + 137.3 = 39.764$

A smoker who is a high school graduate has a 39.8% chance of quitting.

$f'(x) = 0.831(2)x - 18.1 = 1.662x - 18.1$

$f'(12) = 1.662(12) - 18.1 = 1.844$

When a smoker has a high school diploma, the chance of quitting is increasing at the rate of 1.8% per year of education.

b. $f(16) = 0.831(16)^2 - 18.1(16) + 137.3 = 60.436$

A smoker who is a college graduate has a 60.4% chance of quitting.

$f'(16) = 1.662(16) - 18.1 = 8.492$

When a smoker has a college degree, the chance of quitting is increasing at the rate of 8.5% per year of education.

63. **a.** $f(5) = 650(5)^2 + 3000(5) + 12{,}000$
$= \$43{,}250$

b. $f'(x) = 1300x + 3000$

$f'(5) = 1300(5) + 3000 = \$9500$

In 2020-2021, private college tuition will be increasing by \$9500 every 5 years.

c. \$1900

65. $f(x) = 2$ will have a graph that is a horizontal line at height 2, and the slope (the derivative) of a horizontal line is zero. A function that stays constant ill have a rate of change of zero, so its derivative (instantaneous rate of change) will be zero.

67. If *f* has a particular rate of change, then the rate of change of $2 \cdot f(x)$ will be twice as large, and the rate of change of $c \cdot f(x)$ will be $c \cdot f'(x)$, which is just the constant multiple rule.

69. Since $-f$ slopes down by the same amount that *f* slopes up, the slope of $-f$ should be the negative of the slope of *f*. The constant multiple rule with $c = -1$ also says that the slope of $-f$ will be the negative of the slope of *f*.

71. Evaluating first would give a constant and the derivative of a constant is zero, so evaluating and the differentiating would always give zero, regardless of the function and number. This supports the idea that we should always differentiate and then evaluate to obtain anything meaningful.

73. Each additional year of education increases life expectancy by 1.7 years.

75. **a.** If $D(c) = 0$, then $f(c) - L(c) = 0$ so $f(c) = L(c)$.

b. Since the line $L(x)$ must pass through the point $(c, L(c))$, part (a) shows that this point is $(c, f(c))$. Then the point-slope form $y - y_1 = m(x - x_1)$ with $x_1 = c$ and $y_1 = f(c)$ becomes

$$y - f(c) = m(x - c)$$

or $y = m(x - c) + f(c)$, giving the line

$$L(x) = m(x - c) + f(c)$$

c. $D(x) = f(x) - L(x) = f(x) - (m(x - c) + f(c)) = f(x) - m(x - c) - f(c)$

d. Differentiating the last expression for $D(x)$ above gives

$$D'(x) = f'(x) - m$$

which at $x = c$ becomes $D'(c) = f'(c) - m$, and setting this equal to zero gives $m = f'(c)$.

e. $L(x) = f'(c)(x - c) + f(c)$ so $y = L(x)$ is the same as $y - f(c) = f'(c)(x - c)$, which is the point-slope form of the equation of the tangent line to the curve $y = f(x)$ at $x = c$.

f. The tangent line has zero error at the point and errors near the point that differ as little as possible from zero. So the tangent line is the best linear approximation to the function since it has the smallest possible differences between a line and the curve at and near the point.

EXERCISES 2.4

1. **a.** Using the product rule:
$$\frac{d}{dx}(x^4 \cdot x^6) = 4x^3 \cdot x^6 + x^4(6x^5) = 4x^9 + 6x^9 = 10x^9$$
b. Using the power rule:
$$\frac{d}{dx}(x^4 \cdot x^6) = \frac{d}{dx}(x^{10}) = 10x^9$$

3. **a.** Using the product rule:
$$\frac{d}{dx}[x^4(x^5+1)] = 4x^3(x^5+1) + x^4(5x^4) = 4x^8 + 4x^3 + 5x^8 = 9x^8 + 4x^3$$
b. Using the power rule:
$$\frac{d}{dx}[x^4(x^5+1)] = \frac{d}{dx}(x^9 + x^4) = 9x^8 + 4x^3$$

5. $f'(x) = 2x(x^3+1) + x^2(3x^2) = 2x^4 + 2x + 3x^4 = 5x^4 + 2x$

7. $f'(x) = 1(5x^2 - 1) + x(10x) = 5x^2 - 1 + 10x^2 = 15x^2 - 1$

9. $f'(x) = \frac{1}{2}x^{-1/2}(6x+2) + x^{1/2}(6) = 3x^{1/2} + x^{-1/2} + 6x^{1/2} = 9x^{1/2} + x^{-1/2} = 9\sqrt{x} + \frac{1}{\sqrt{x}}$

11. $f'(x) = (2x)(x^2 - 1) + (x^2 + 1)(2x) = 2x^3 - 2x + 2x^3 + 2x = 4x^3$

13. $f'(x) = (2x+1)(3x+1) + (x^2 + x)(3) = 6x^2 + 5x + 1 + 3x^2 + 3x = 9x^2 + 8x + 1$

15. $f'(x) = 2x(x^2 + 3x - 1) + x^2(2x+3) = 2x^3 + 6x^2 - 2x + 2x^3 + 3x^2 = 4x^3 + 9x^2 - 2x$

17. $f'(x) = (4x)(1-x) + (2x^2+1)(1) = 4x - 4x^2 - 2x^2 - 1 = -6x^2 + 4x - 1$

19. $f'(x) = \left(\frac{1}{2}x^{-1/2}\right)x^{1/2} + 1 + (x^{1/2} - 1)\left(\frac{1}{2}x^{-1/2}\right) = \frac{1}{2} + \frac{1}{2}x^{-1/2} + \frac{1}{2} - \frac{1}{2}x^{-1/2} = 1$

21. $f'(t) = 8t^{1/3}(3t^{2/3} + 1) + 6t^{4/3}(2t^{-1/3}) = 24t + 8t^{1/3} + 12t = 36t + 8t^{1/3}$

23.
$$\begin{aligned} f'(z) &= (4z^3 + 2z)(z^3 - z) + (z^4 + z^2 + 1)(3z^2 - 1) \\ &= 4z^6 - 2z^4 - 2z^2 + 3z^6 + 3z^4 + 3z^2 - z^4 - z^2 - 1 \\ &= 7z^6 - 1 \end{aligned}$$

25.
$$\begin{aligned} f'(x) &= \left(2 + 4\cdot\tfrac{1}{2}z^{-1/2}\right)\left(2z^{1/2} + 1\right) + \left(2z + 4z^{1/2} - 1\right)\left(2\cdot\tfrac{1}{2}z^{-1/2}\right) \\ &= 4z^{1/2} + 2 + 4 + 2z^{-1/2} + 2z^{1/2} + 4 - z^{-1/2} \\ &= 6z^{1/2} + z^{-1/2} + 10 \\ &= 6\sqrt{z} + \frac{1}{\sqrt{z}} + 10 \end{aligned}$$

27. **a.** Using the quotient rule:
$$\frac{d}{dx}\left(\frac{x^8}{x^2}\right) = \frac{x^2(8x^7) - 2x(x^8)}{(x^2)^2} = \frac{8x^9 - 2x^9}{x^4} = 6x^5$$
b. Using the power rule:
$$\frac{d}{dx}\left(\frac{x^8}{x^2}\right) = \frac{d}{dx}(x^6) = 6x^5$$

29. **a.** Using the quotient rule:

$$\frac{d}{dx}\left(\frac{1}{x^3}\right) = \frac{x^3(0)-3x^2(1)}{(x^3)^2} = -\frac{3x^2}{x^6} = -\frac{3}{x^4}$$

b. Using the power rule:

$$\frac{d}{dx}\left(\frac{1}{x^3}\right) = \frac{d}{dx}(x^{-3}) = -3x^{-4} = -\frac{3}{x^4}$$

31. $$f'(x) = \frac{x^3(4x^3) - 3x^2(x^4+1)}{(x^3)^2} = \frac{4x^6 - 3x^6 - 3x^2}{x^6} = \frac{x^6 - 3x^2}{x^6} = 1 - \frac{3}{x^4}$$

33. $$f'(x) = \frac{(x-1)(1)-(1)(x+1)}{(x-1)^2} = \frac{x-1-x-1}{(x-1)^2} = -\frac{2}{(x-1)^2}$$

35. $$f'(x) = \frac{(2+x)(3)-(1)(3x+1)}{(2+x)^2} = \frac{6+3x-3x-1}{(2+x)^2} = \frac{5}{(2+x)^2}$$

37. $$f'(t) = \frac{(t^2+1)(2t)-(2t)(t^2-1)}{(t^2+1)^2} = \frac{2t^3+2t-2t^3+2t}{(t^2+1)^2} = \frac{4t}{(t^2+1)^2}$$

39. $$f'(s) = \frac{(s+1)(3s^2)-(1)(s^3-1)}{(s+1)^2} = \frac{3s^3+3s^2-s^3+1}{(s+1)^2} = \frac{2s^3+3s^2+1}{(s+1)^2}$$

41. $$f'(x) = \frac{(x+1)(2x-2)-(1)(x^2-2x+3)}{(x+1)^2} = \frac{2x^2-2-x^2+2x-3}{(x+1)^2} = \frac{x^2+2x-5}{(x+1)^2}$$

43. $$f'(x) = \frac{(x^2+1)(4x^3+2x)-(2x)(x^4+x^2+1)}{(x^2+1)^2} = \frac{4x^3(x^2+1)+2x(x^2+1)-2x(x^4)-2x(x^2+1)}{(x^2+1)^2}$$
$$= \frac{4x^5+4x^3-2x^5}{(x^2+1)^2} = \frac{2x^5+4x^3}{(x^2+1)^2}$$

45. $$f'(t) = \frac{(t^2+t-3)(2t+2)-(t^2+2t-1)(2t+1)}{(t^2+t-3)^2} = \frac{2t^3+4t^2-4t-6-2t^3-5t^2+1}{(t^2+t-3)^2} = -\frac{t^2+4t+5}{(t^2+t-3)^2}$$

47.

Rewrite	Differentiate	Rewrite
$y = 3x^{-1}$	$\frac{dy}{dx} = -3x^{-2}$	$\frac{dy}{dx} = -\frac{3}{x^2}$

49.

Rewrite	Differentiate	Rewrite
$y = \frac{3}{8}x^4$	$\frac{dy}{dx} = \frac{3}{2}x^3$	$\frac{dy}{dx} = \frac{3x^3}{2}$

51.

Rewrite	Differentiate	Rewrite
$y = \frac{1}{3}x^2 - \frac{5}{3}x$	$\frac{dy}{dx} = \frac{2}{3}x - \frac{5}{3}$	$\frac{dy}{dx} = \frac{2x-5}{3}$

53. $$\frac{d}{dx}\left[(x^3+2)\frac{x^2+1}{x+1}\right] = \frac{d}{dx}(x^3+2)\cdot\frac{x^2+1}{x+1} + (x^3+2)\cdot\frac{d}{dx}\left(\frac{x^2+1}{x+1}\right)$$
$$3x^2\left(\frac{x^2+1}{x+1}\right) + (x^3+2)\frac{(x+1)(2x)-(1)(x^2+1)}{(x+1)^2}$$
$$= 3x^2\left(\frac{x^2+1}{x+1}\right) + (x^3+2)\frac{x^2+2x-1}{(x+1)^2}$$

55. $$\frac{d}{dx}\frac{(x^2+3)(x^3+1)}{x^2+2}=\frac{(x^2+2)\frac{d}{dx}\left[(x^2+3)(x^3+1)\right]-\left[\frac{d}{dx}(x^2+2)\right]\left[(x^2+3)(x^3+1)\right]}{(x^2+2)^2}$$

$$=\frac{(x^2+2)\left\{\left[\frac{d}{dx}(x^2+3)\right](x^3+1)+(x^2+3)\frac{d}{dx}(x^3+1)\right\}-2x\left[(x^2+3)(x^3+1)\right]}{(x^2+2)^2}$$

$$=\frac{(x^2+2)\left[(2x)(x^3+1)+(x^2+3)(3x^2)\right]-2x(x^2+3)(x^3+1)}{(x^2+2)^2}$$

$$=\frac{(x^2+2)(2x^4+2x+3x^4+9x^2)-(2x)(x^5+3x^3+x^2+3)}{(x^2+2)^2}$$

$$=\frac{2x^6+2x^3+3x^6+9x^4+4x^4+4x+6x^4+18x^2-2x^6-6x^4-2x^3-6x}{(x^2+2)^2}$$

$$=\frac{3x^6+13x^4+18x^2-2x}{(x^2+2)^2}$$

57. $$\frac{d}{dx}\left(\frac{\sqrt{x}-1}{\sqrt{x}+1}\right)=\frac{d}{dx}\left(\frac{x^{1/2}-1}{x^{1/2}+1}\right)=\frac{(x^{1/2}+1)(\frac{1}{2}x^{-1/2})-\frac{1}{2}x^{-1/2}(x^{1/2}-1)}{(x^{1/2}+1)^2}$$

$$=\frac{\frac{1}{2}+\frac{1}{2}x^{-1/2}-\frac{1}{2}+\frac{1}{2}x^{-1/2}}{(x^{1/2}+1)^2}=\frac{x^{-1/2}}{(x^{1/2}+1)^2}=\frac{1}{\sqrt{x}(\sqrt{x}+1)^2}$$

59. $$\frac{d}{dx}\left[\frac{R(x)}{x}\right]=\frac{x\cdot R'(x)-1\cdot R(x)}{x^2}=\frac{xR'(x)-R(x)}{x^2}$$

61. **a.** The instantaneous rate of change of cost with respect to purity is the derivative of the cost function $C(x)=\frac{100}{100-x}$ on $50 \bullet x<100$.

$$C'(x)=\frac{(100-x)\cdot 0-(-1)(100)}{(100-x)^2}=\frac{100}{(100-x)^2} \quad \text{on } 50 \bullet x<100$$

b. To find the rate of change for a purity of 95%, evaluate $C'(x)$ at $x=95$.

$$C'(95)=\frac{100}{(100-95)^2}=\frac{100}{5^2}=4$$

The cost is increasing by 4 cents per additional percent of purity.

c. To find the rate of change for a purity of 98%, evaluate $C'(x)$ at $x=98$.

$$C'(98)=\frac{100}{(100-98)^2}=\frac{100}{2^2}=25$$

The cost is increasing by 25 cents per additional percent of purity.

63. **a.**

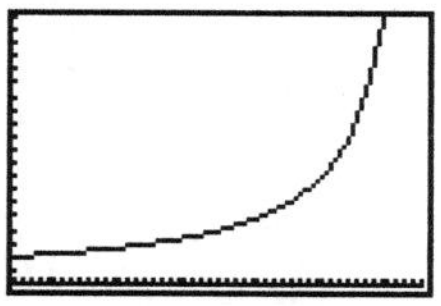

on [50, 100] by [0, 20]

b. Rate of change of cost is 4 for $x=95$; rate of change of cost is 25 for $x=98$.

65. **a.** $AC(x)=\frac{C(x)}{x}=\frac{8x+45}{x}$

b. The marginal average cost function $MAC(x)$ is the derivative of the average cost function $AC(x)=\frac{C(x)}{x}=\frac{8x+45}{x}$, or $8+\frac{45}{x}$.

$$MAC(x)=\frac{d}{dx}\left(\frac{8x+45}{x}\right)=\frac{x(8)-1(8x+45)}{x^2}=\frac{8x-8x-45}{x^2}=\frac{-45}{x^2}$$

c. $MAC(30)=\frac{-45}{30^2}=-\frac{1}{20}=-0.05$

The average cost is decreasing at the rate of 5 cents per additional clock.

67. To find the rate of change of temperature, find $T'(x)$.

$$T'(x)=3x^2(4-x^2)+x^3(-2x)$$
$$=12x^2-3x^4-2x^4=12x^2-5x^4$$

For $x=1$, $T'(1)=12(1)^2-5(1)^4=12-5=7$.

After 1 hour, the person's temperature is increasing by 7 degrees per hour.

69. **a.**

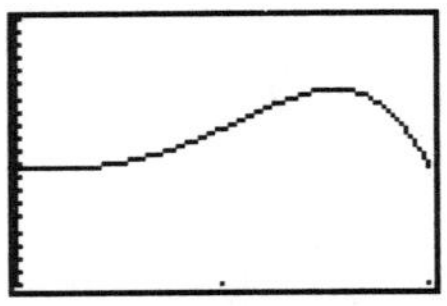

on [0, 2] by [90, 110]

b. The rate of change at $x = 1$ is 7.

c. The maximum temperature is about 104.5 degrees.

71. **a.** $y_3(8)=\frac{-73.1(8)^2+1270(8)+16,280}{2.3(8)+314}\approx 65.468$

The per capita national debt in 2020 would be $65,468.

$y_3(13)=\frac{-73.1(13)^2+1270(13)+16,280}{2.3(13)+314}\approx 59.425$

The per capita national debt in 2025 would be $59,425.

b. $y_3'(8)=-0.151$

In 2020 the per capita national debt will be shrinking by $151 per year.

$y_3'(13)=-2.231$

In 2025 the per capita national debt will be shrinking by $2231 per year.

73. $I(x)=0.45(x-1.7)(x^2-12.5x+43)$

$$I'(x)=0.45\left[1(x^2-12.5x+43)+(x-1.7)(2x-12.5)\right]$$
$$I'(5)=0.45\left[(5)^2-12.5(5)+43+(5-1.7)(2(5)-12.5)\right]$$
$$=-1.2375$$
$$I'(6)=0.45\left[(6)^2-12.5(6)+43+(6-1.7)(2(6)-12.5)\right]$$
$$=0.8325$$

Interest rates were declining by about 1.24 percent per year in 2010 and rising by 0.83 percent per year in 2011.

75. $E(x)=\dfrac{261x}{x+8.84}+647$

$E'(x)=\dfrac{(x+8.84)(261)-(1)(261x)}{(x+8.84)^2}=\dfrac{2307.24}{(x+8.84)^2}$

$E'(4)=\dfrac{2307.24}{(4+8.84)^2}\approx 13.99$

$E'(8)=\dfrac{2307.24}{(8+8.84)^2}\approx 8.14$

Median weekly earnings were rising by \$13.99 per year in 2009 and by \$8.14 per year in 2013.

77. **a.** $\dfrac{d}{dx}\left(\dfrac{1}{x^2}\right)=\dfrac{d}{dx}(x^{-2})=-2x^{-3}=-\dfrac{2}{x^3}$

$\left.\dfrac{d}{dx}\right|_{x=0}=-\dfrac{2}{0^3}$ Undefined

b. Answers will vary.

79. False: the product rule gives the correct right-hand side.

81. True:
$\dfrac{d}{dx}(x\cdot f)=\dfrac{d}{dx}x\cdot f+x\cdot f'=f+x\cdot f'$

83. $\dfrac{d}{dx}(f\cdot g)=\dfrac{(f\cdot g)f'}{f}+\dfrac{(f\cdot g)g'}{g}=g\cdot f'+f\cdot g'$

The right-hand side multiplies out to $g\cdot f'+f\cdot g'$ which agrees with the product rule.

85. False: This would be the same as saying that the derivative (instantaneous rate of change) of a product is a product of the derivatives. The product rule gives the correct way of finding the derivative of a product.

87.
$$\begin{aligned}\frac{d}{dx}(f\cdot g\cdot h)&=\frac{d}{dx}[f\cdot(g\cdot h)]=\frac{df}{dx}\cdot(g\cdot h)+f\cdot\frac{d}{dx}(g\cdot h)\\&=\frac{df}{dx}\cdot(g\cdot h)+f\left(\frac{dg}{dx}\cdot h+g\cdot\frac{dh}{dx}\right)\\&=\frac{df}{dx}\cdot g\cdot h+f\cdot\frac{dg}{dx}\cdot h+f\cdot g\cdot\frac{dh}{dx}\end{aligned}$$

89.
$$\begin{aligned}\frac{d}{dx}[f(x)]^2&=\frac{d}{dx}[f(x)\cdot f(x)]\\&=\left[\frac{d}{dx}f(x)\right]f(x)+f(x)\left[\frac{d}{dx}f(x)\right]\\&=f'(x)\cdot f(x)+f(x)\cdot f'(x)=2f(x)\cdot f'(x)\end{aligned}$$

91.
$$\frac{dy}{dx}\left(\frac{Rx}{1+\left(\frac{R-1}{K}\right)x}\right)=\frac{\left(1+\left(\frac{R-1}{K}\right)x\right)R-Rx\left(\frac{R-1}{K}\right)}{\left(1+\left(\frac{R-1}{K}\right)x\right)^2}=\frac{R+Rx\left(\frac{R-1}{K}\right)-Rx\left(\frac{R-1}{K}\right)}{\left(1+\left(\frac{R-1}{K}\right)x\right)^2}=\frac{R}{\left(1+\left(\frac{R-1}{K}\right)x\right)^2}$$

To show $y'>0$, show that both the numerator and denominator are greater than zero.

$R>1>0 \Rightarrow$ numerator is greater than zero.

$R>1 \Rightarrow R-1>0.$

$R-1>0;\ K>0 \Rightarrow \dfrac{R-1}{K}>0$

$\dfrac{R-1}{K}>0,\ x>0 \Rightarrow \left(1+\left(\dfrac{R-1}{K}\right)x\right)>0 \Rightarrow$ denominator is greater than zero.

$\therefore y'>0.$

This means the density of the offspring is always increasing faster than the density of the parents.

EXERCISES 2.5

1. **a.** $f'(x) = 4x^3 - 3(2)x^2 - 2(3)x + 5 - 0$
$= 4x^3 - 6x^2 - 6x + 5$

b. $f''(x) = 3(4)x^2 - 2(6)x - 6 + 0 = 12x^2 - 12x - 6$

c. $f'''(x) = 2(12)x - 12 - 0 = 24x - 12$

d. $f^{(4)}(x) = 24$

3. **a.** $f'(x) = 1 + x + \frac{1}{2}x^2 + \frac{1}{6}x^3 + \frac{1}{24}x^4$

b.

c. $f'''(x) = 1 + x + \frac{1}{2}x^2$

d. $f^{(4)}(x) = 1 + x$

5. $f(x) = \sqrt{x^5} = x^{5/2}$

a. $f'(x) = \frac{5}{2}x^{3/2}$

b. $f''(x) = \frac{3}{2}\left(\frac{5}{2}\right)x^{1/2} = \frac{15}{4}\sqrt{x}$

c. $f'''(x) = \frac{d}{dx}\left(\frac{15}{4}x^{1/2}\right) = \frac{1}{2}\left(\frac{15}{4}\right)x^{-1/2}$
$= \frac{15}{8}x^{-1/2} = \frac{15}{8\sqrt{x}}$

d. $f^{(4)}(x) = \frac{d}{dx}\left(\frac{15}{8}x^{-1/2}\right) = -\frac{1}{2}\left(\frac{15}{8}\right)x^{-3/2}$
$= -\frac{15}{16}x^{-3/2}$

7. $f(x) = \frac{x-1}{x} = 1 - \frac{1}{x}$

a. $f'(x) = \frac{1}{x^2}$

$f''(x) = \frac{d}{dx}\left(\frac{1}{x^2}\right) = \frac{d}{dx}\left(x^{-2}\right) = -2x^{-3}$
$= -\frac{2}{x^3}$

b. $f''(3) = -\frac{2}{3^3} = -\frac{2}{27}$

9. $f(x) = \frac{x+1}{2x} = \frac{1}{2} + \frac{1}{2x}$

a. $f'(x) = -\frac{1}{2x^2}$

$f''(x) = \frac{d}{dx}\left(-\frac{1}{2x^2}\right) = \frac{d}{dx}\left(-\frac{1}{2}x^{-2}\right) = \frac{1}{x^3}$

b. $f''(3) = \frac{1}{3^3} = \frac{1}{27}$

11. $f(x) = \frac{1}{6x^2}$

a. $f'(x) = -2\left(\frac{1}{6}\right)x^{-3} = -\frac{1}{3x^3}$

$f''(x) = \frac{d}{dx}\left(-\frac{1}{3}x^{-3}\right) = x^{-4} = \frac{1}{x^4}$

b. $f''(3) = \frac{1}{3^4} = \frac{1}{81}$

13. $f(x) = (x^2 - 2)(x^2 + 3) = x^4 + x^2 - 6$
$f'(x) = 4x^3 + 2x$
$f''(x) = 12x^2 + 2$

15. $f(x) = \frac{27}{\sqrt[3]{x}} = 27x^{-1/3}$
$f'(x) = \left(-\frac{1}{3}\right)(27x^{-4/3}) = -9x^{-4/3}$
$f''(x) = \left(-\frac{4}{3}\right)(-9x^{-7/3}) = 12x^{-7/3}$

17. $f(x) = \frac{x}{x-1}$

$$f'(x) = \frac{(x-1)\frac{dx}{dx} - \left[\frac{d}{dx}(x-1)\right](x)}{(x-1)^2} = \frac{(x-1)(1) - (1)(x)}{(x-1)^2} = -\frac{1}{(x-1)^2} = -\frac{1}{x^2 - 2x + 1}$$

$$f''(x) = \frac{(x^2 - 2x + 1)\frac{d}{dx}(-1) - \left[\frac{d}{dx}(x^2 - 2x + 1)\right](-1)}{[(x-1)^2]^2} = \frac{0 - (2x-2)(-1)}{(x-1)^4} = \frac{2x-2}{(x-1)^4} = \frac{2(x-1)}{(x-1)^4} = \frac{2}{(x-1)^3}$$

19. $\frac{d}{dr}(\pi r^2) = 2\pi r$

$\frac{d^2}{dr^2}(\pi r^2) = \frac{d}{dr}(2\pi r) = 2\pi$

21. $\frac{d}{dx}x^{10} = 10x^9$

$\frac{d^2}{dx^2}x^{10} = \frac{d}{dx}(10x^9) = 90x^8$

$\left.\frac{d^2}{dx^2}x^{10}\right|_{x=-1} = 90(-1)^8 = 90$

23. From Exercise 21, we know

$$\frac{d^2}{dx^2}x^{10} = 90x^8$$

$$\frac{d^3}{dx^3}x^{10} = \frac{d}{dx}(90x^8) = 720x^7$$

Thus, $\left.\frac{d^3}{dx^3}x^{10}\right|_{x=-1} = 720(-1)^7 = -720$

25. $$\frac{d}{dx}\sqrt{x^3} = \frac{d}{dx}x^{3/2} = \frac{3}{2}x^{1/2}$$

$$\frac{d^2}{dx^2}\sqrt{x^3} = \frac{d}{dx}\left(\frac{3}{2}x^{1/2}\right) = \frac{3}{4}x^{-1/2}$$

$$\left.\frac{d^2}{dx^2}\sqrt{x^3}\right|_{x=1/16} = \frac{3}{4}\left(\frac{1}{16}\right)^{-1/2} = \frac{3}{4}\sqrt{16} = 3$$

27. $$\frac{d}{dx}[(x^2 - x + 1)(x^3 - 1)] = (2x-1)(x^3-1) + (x^2 - x + 1)(3x^2)$$

$$= 2x^4 - x^3 - 2x + 1 + 3x^4 - 3x^3 + 3x^2$$

$$= 5x^4 - 4x^3 + 3x^2 - 2x + 1$$

$$\frac{d^2}{dx^2}[(x^2 - x + 1)(x^3 - 1)] = \frac{d}{dx}(5x^4 - 4x^3 + 3x^2 - 2x + 1)$$

$$= 20x^3 - 12x^2 + 6x - 2$$

29. $$\frac{d}{dx}\left(\frac{x}{x^2+1}\right) = \frac{(x^2+1)\left(\frac{dx}{dx}\right) - x\frac{d}{dx}(x^2+1)}{(x^2+1)^2} = \frac{x^2+1-x(2x)}{(x^2+1)^2} = \frac{1-x^2}{(x^2+1)^2} = \frac{1-x^2}{x^4+2x^2+1}$$

$$\frac{d^2}{dx^2}\left(\frac{x}{x^2+1}\right) = \frac{d}{dx}\left(\frac{1-x^2}{x^4+2x^2+1}\right) = \frac{(x^4+2x^2+1)\frac{d}{dx}(1-x^2)-(1-x^2)\frac{d}{dx}(x^4+2x^2+1)}{[(x^2+1)^2]^2}$$

$$= \frac{(x^4+2x^2+1)(-2x)-(1-x^2)(4x^3+4x)}{(x^2+1)^4} = \frac{(x^2+1)[(x^2+1)(-2x)-(1-x^2)(4x)]}{(x^2+1)^4}$$

$$= \frac{-2x^3-2x-4x+4x^3}{(x^2+1)^3} = \frac{2x(x^2-3)}{(x^2+1)^3}$$

31. $$\frac{d}{dx}\left(\frac{2x-1}{2x+1}\right) = \frac{(2x+1)\cdot 2-2(2x-1)}{(2x+1)^2} = \frac{4x+2-4x+2}{4x^2+4x+1} = \frac{4}{4x^2+4x+1}$$

$$\frac{d^2}{dx^2}\left(\frac{2x-1}{2x+1}\right) = \frac{(4x^2+4x+1)\cdot 0-(8x+4)(4)}{(4x^2+4x+1)^2} = -\frac{32x+16}{[(2x+1)^2]^2} = -\frac{32x+16}{(2x+1)^4}$$

$$= -\frac{16(2x+1)}{(2x+1)^4} = -\frac{16}{(2x+1)^3}$$

33. To find velocity, we differentiate the distance function $s(t) = 18t^2 - 2t^3$.

$v(t) = s'(t) = 36t - 6t^2$

- **a.** For $t = 3$, $v(3) = s'(3) = 36(3) - 6(3)^2 = 108 - 54 = 54$ miles per hour
- **b.** For $t = 7$, $v(7) = s'(7) = 36(7) - 6(7)^2 = 252 - 294 = -42$ miles per hour
- **c.** To find the acceleration, we differentiate the velocity function $s'(t) = 36t - 6t^2$.
  $a(t) = s''(t) = 36 - 12t$.
  For $t = 1$, $a(1) = s''(1) = 36 - 12(1) = 24$ mi / hr^2

35. $v(t) = h'(t) = 3t^2 + 2(0.5)t = 3t^2 + t$

$v(10) = 3(10)^2 + 10 = 310$ feet per second

$a(t) = v'(t) = \frac{d}{dt}(3t^2 + t) = 6t + 1$

$a(10) = 6(10) + 1 = 61$ ft / sec^2

37. **a.** To find when the steel ball will reach the ground, we need to determine what value of t produces $s(t)=0$. Thus, set $s(t)=0$ and solve the equation.

$$0=s(t)=2717-16t^2$$
$$0=2717-16t^2$$
$$16t^2=2717$$
$$t^2=169.8125$$
$$t\approx 13.0312 \text{ sec}$$

The steel ball will reach the ground after about 13.03 seconds.

b. $v(t)=s'(t)=-32t$
$v(13.03)=s'(13.03)\approx -417$ feet per second

c. $a(t)=v'(t)=s''(t)$
$=-32$ feet per second per second

39. **a.** $v(t)=s'(t)=-32t+1280$

b. When $s(t)$ is a maximum, $s'(t)=0$.

$$-32t+1280=0$$
$$32t=1280$$
$$t=40 \text{ seconds}$$

c. At $t=40$,

$$s(40)=-16(40)^2+1280(40)$$
$$=-25,600+51,200$$
$$=25,600 \text{ feet}$$

41. $D'(t)=\frac{4}{3}(9t^{1/3})=12t^{1/3}$

$D'(8)=12(8)^{1/3}=12(2)=24$

The national debt is increasing by 24 billion dollars per year after 8 years.

$$D''(t)=\frac{d}{dt}(12t^{1/3})=\frac{1}{3}(12t^{-2/3})=4t^{-2/3}$$
$$D''(8)=4(8)^{-2/3}=4\left(\frac{1}{8}\right)^{2/3}=4\left(\frac{1}{4}\right)=1$$

The rate of growth of the national debt is increasing by 1 billion dollars per year each year after 8 years.

43. $L(10)=9.3$

By 2100, sea levels may have risen by 93 cm (about 3 feet).

$L'(t)=0.06x^4-0.14x+8$
$L'(10)=12.6$

In 2100 sea levels will be rising by about 12.6 centimeters per decade, or about 1.26 cm (about half an inch) per year.

$L''(x)=0.12x-0.14$
$L''(10)=1.06$

The rise in the sea level will be speeding up (by about 1 cm per decade per decade).

45. **a.** For $x = 15$, approximately 21.589°; for $x = 30$, approximately 17.597°.

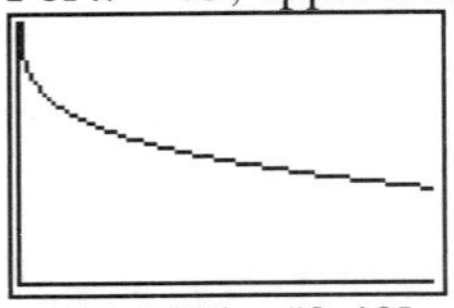

on [0,50] by [0,40]

b. Each 1-mph increase in wind speed lowers the wind-chill index. As wind speed increases, the rate with which the wind-chill index decreases slows.

c. For $x = 15$, $y' = -0.363$. For a wind speed of 15 mph, each additional mile per hour decreases the wind-chill index by about 0.363. For $x = 30$, $y' = -0.203$. For a wind speed of 30 mph, each additional mile per hour decreases the wind-chill index by about 0.203°.

47. **a.** The first derivative is negative because the temperature is dropping.

b. The second derivative is negative because the temperature is dropping increasingly rapidly making the change in temperature speed up in a negative direction.

49. **a.** The first derivative is negative, because the stock market is declining.

b. The second derivative is positive because the change in the stock market is slowing down in a negative direction.

51. True: For example $\frac{d^4}{dx^4}x^3=0$.

53. **a.** iii (showing a stop, then a slower velocity)
b. i (showing a stop, then a negative velocity)
c. ii (showing stops and starts and then a higher velocity)

55. $\frac{d^{100}}{dx^{100}}(x^{99}-4x^{98}+3x^{50}+6)=0$

57. $$\frac{d^2}{dx^2}(f\cdot g)=\frac{d}{dx}\left(\frac{df}{dx}\cdot g+f\cdot\frac{dg}{dx}\right)\quad\left[\text{or } \frac{d}{dx}(f'\cdot g+f\cdot g')\right]$$
$$=\frac{d}{dx}\left(\frac{df}{dx}\cdot g\right)+\frac{d}{dx}\left(f\cdot\frac{dg}{dx}\right)\quad\left[\text{or } \frac{d}{dx}(f'\cdot g+\frac{d}{dx}(f\cdot g')\right]$$
$$\frac{d^2f}{dx^2}\cdot g+\frac{df}{dx}\cdot\frac{dg}{dx}+\frac{df}{dx}\cdot\frac{dg}{dx}+f\cdot\frac{d^2g}{dx^2}\quad[\text{or } f''\cdot g+f'\cdot g'+f'\cdot g'+f\cdot g'']$$
$$=f''\cdot g+2f'\cdot g'+f\cdot g''$$

EXERCISES 2.6

1. $f(g(x))=\sqrt{x^2-3x+1}$
The outside function is $\sqrt{x}$ and the inside function is x^2-3x+1.
Thus, we take $\begin{cases} f(x)=\sqrt{x} \\ g(x)=x^2-3x+1 \end{cases}$

3. $f(g(x))=(x^2-x)^{-3}$
The outside function is x^{-3} and the inside function is x^2-x.
Thus, we take $\begin{cases} f(x)=x^{-3} \\ g(x)=x^2-x \end{cases}$

5. $f(g(x))=\frac{x^3+1}{x^3-1}$
$f(x)=\frac{x+1}{x-1}$
$g(x)=x^3$

7. $f(g(x))=\left(\frac{x+1}{x-1}\right)^4$
$f(x)=x^4$
$g(x)=\frac{x+1}{x-1}$

9. $f(g(x))=\sqrt{x^2-9}+5$
$f(x)=\sqrt{x}+5$
$g(x)=x^2-9$

11. $f(x)=(x^2+1)^3$
$f'(x)=3(x^2+1)^2(2x)=6x(x^2+1)^2$

13. $g(x)=(2x^2-7x+3)^4$
$g'(x)=4(2x^2-7x+3)^3(4x-7)$

15. $h(z)=(3z^2-5z+2)^4$
$h'(z)=4(3z^2-5z+2)^3(6z-5)$

17. $f(x)=\sqrt{x^4-5x+1}=(x^4-5x+1)^{1/2}$
$f'(x)=\frac{1}{2}(x^4-5x+1)^{-1/2}(4x^3-5)$

19. $w(z)=\sqrt[3]{9z-1}=(9z-1)^{1/3}$
$w'(z)=\frac{1}{3}(9z-1)^{-2/3}(9)=3(9z-1)^{-2/3}$

21. $y=(4-x^2)^4$
$y'=4(4-x^2)^3(-2x)=-8x(4-x^2)^3$

23. $y=\left(\frac{1}{w^3-1}\right)^4=[(w^3-1)^{-1}]^4=(w^3-1)^{-4}$
$y'=-4(w^3-1)^{-5}(3w^2)=-12w^2(w^3-1)^{-5}$

25. $y=x^4+(1-x)^4$
$y'=4x^3+4(1-x)^3(-1)=4x^3-4(1-x)^3$

27. $f(x)=\frac{1}{\sqrt{3x^2-5x+1}}=(3x^2-5x+1)^{-1/2}$
$f'(x)=-\frac{1}{2}(3x^2-5x+1)^{-3/2}(6x-5)$

29. $f(x)=\dfrac{1}{\sqrt[3]{(9x+1)^2}}=\dfrac{1}{(9x+1)^{2/3}}=(9x+1)^{-2/3}$

$f'(x)=-\dfrac{2}{3}(9x+1)^{-5/3}(9)=-6(9x+1)^{-5/3}$

31. $f(x)=\dfrac{1}{\sqrt[3]{(2x^2-3x+1)^2}}=(2x^2-3x+1)^{-2/3}$

$f'(x)=-\dfrac{2}{3}(2x^2-3x+1)^{-5/3}(4x-3)$

33. $f(x)=[(x^2+1)^3+x]^3$

$f'(x)=3[(x^2+1)^3+x]^2[3(x^2+1)^2(2x)+1]$

$=3[(x^2+1)^3+x]^2[6x(x^2+1)^2+1]$

35. $f(x)=3x^2(2x+1)^5$

$f'(x)=6x(2x+1)^5+3x^2[5(2x+1)^4(2)]$

$=6x(2x+1)^5+30x^2(2x+1)^4$

37. $f(x)=(2x+1)^3(2x-1)^4$

$f'(x)=3(2x+1)^2(2)(2x-1)^4+(2x+1)^3[4(2x-1)^3(2)]$

$=6(2x+1)^2(2x-1)^4+8(2x+1)^3(2x-1)^3$

39. $g(z)=2z(3z^2-z+1)^4$

$g'(z)=2(3z^2-z+1)^4+2z(4)(3z^2-z+1)^3(6z-1)$

41. $f(x)=\left(\dfrac{x+1}{x-1}\right)^3$

$f'(x)=3\left(\dfrac{x+1}{x-1}\right)^2\left[\dfrac{(x-1)(1)-(1)(x+1)}{(x-1)^2}\right]$

$=3\left(\dfrac{x+1}{x-1}\right)^2\left[\dfrac{x-1-x-1}{(x-1)^2}\right]=3\left(\dfrac{x+1}{x-1}\right)^2\left[\dfrac{-2}{(x-1)^2}\right]$

$=-6\dfrac{(x+1)^2}{(x-1)^4}$

43. $f(x)=x^2\sqrt{1+x^2}=x^2(1+x^2)^{1/2}$

$f'(x)=2x(1+x^2)^{1/2}+x^2\left[\dfrac{1}{2}(1+x^2)^{-1/2}(2x)\right]$

$=2x(1+x^2)^{1/2}+x^3(1+x^2)^{-1/2}$

45. $f(x)=\sqrt{1+\sqrt{x}}=(1+x^{1/2})^{1/2}$

$f'(x)=\frac{1}{2}(1+x^{1/2})^{-1/2}\left(\frac{1}{2}x^{-1/2}\right)$

$=\frac{1}{4}x^{-1/2}(1+x^{1/2})^{-1/2}$

47. $f(x)=(2x+1)^4$

$f(-1)=(2(-1)+1)^4=1$

The point is (–1, 1).

$f'(x)=4(2x+1)^3(2)=8(2x+1)^3$

$f'(-1)=8(2(-1)+1)^3=-8$

The tangent line has slope –8 at (–1, 1).

$y-1=-8(x+1)$

$y=-8x-7$

49. $f(x)=\left[(x-1)^2-x\right]^2$

$f(2)=\left[(2-1)^2-2\right]^2=1$

The point is (2, 1).

$f'(x)=2\left[(x-1)^2-x\right][2(x-1)(1)-1]$

$=2\left[(x-1)^2-x\right](2x-3)$

$f'(2)=2\left[(2-1)^2-2\right](2(2)-3)=-2$

The tangent line has slope –2 at (2, 1).

$y-1=-2(x-2)$

$y=-2x+5$

51. **a.** $\dfrac{d}{dx}[(x^2+1)^2]=2(x^2+1)(2x)=4x^3+4x$

b. $\dfrac{d}{dx}[(x^2+1)^2]=\dfrac{d}{dx}(x^4+2x^2+1)=4x^3+4x$

53. **a.** $\frac{d}{dx}\left(\frac{1}{3x+1}\right)=\frac{(3x+1)0-3(1)}{(3x+1)^2}=-\frac{3}{(3x+1)^2}$

b. $\frac{d}{dx}[(3x+1)^{-1}]=-(3x+1)^{-2}(3)$

$$=-\frac{3}{(3x+1)^2}$$

55. $f(x)=(x^2+1)^{10}$

$$f'(x)=10(x^2+1)^9(2x)=20x(x^2+1)^9$$
$$f''(x)=20(x^2+1)^9+20x[9(x^2+1)^8(2x)]$$
$$=20(x^2+1)^9+40x^2[9(x^2+1)^8]$$
$$=20(x^2+1)^9+360x^2(x^2+1)^8$$

57. The marginal cost function is the derivative of the function $C(x)=\sqrt{4x^2+900}=(4x^2+900)^{1/2}$.

$$MC(x)=\tfrac{1}{2}\left(4x^2+900\right)^{-1/2}(8x)$$
$$=4x\left(4x^2+900\right)^{-1/2}$$
$$MC(20)=4(20)\left[4(20)^2+900\right]^{-1/2}$$
$$=80[4(400)+900]^{-1/2}$$
$$=80(2500)^{-1/2}=\frac{80}{50}=1.60$$

59. $x=27$

61. $S(i)=17.5(i-1)^{0.53}$

$$S'(i)=9.275(i-1)^{-0.47}$$
$$S'(25)=9.275(25-1)^{-0.47}\approx 2.08$$

At an income of $25,000 social status increases by about 2.08 units per additional $1000 of income.

63. $R(x)=4x\sqrt{11+0.5x}=4x(11+0.5x)^{1/2}$

$$R'(x)=4x\left(\frac{1}{2}\right)(11+0.5x)^{-1/2}(0.5)+4(11+0.5x)^{1/2}$$
$$=\frac{x}{\sqrt{11+0.5x}}+4\sqrt{11+0.5x}$$

The sensitivity to a dose of 50 mg is

$$R'(50)=\frac{x}{\sqrt{11+0.5(50)}}+4\sqrt{11+0.5(50)}$$
$$=\frac{50}{\sqrt{11+25}}+4\sqrt{11+25}=\frac{50}{\sqrt{36}}+4\sqrt{36}$$
$$=\frac{50}{6}+24=\frac{25+72}{3}=\frac{97}{3}=32\frac{1}{3}$$

65.

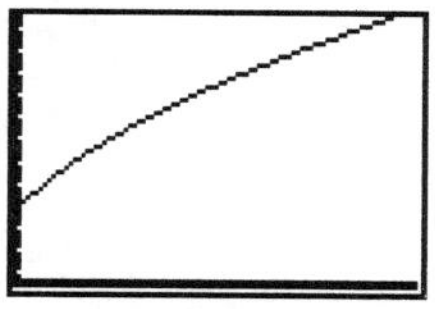

on [0, 140] by [0, 50]
$x\approx 26$ mg

67. $P(t)=0.02(12+2t)^{3/2}+1$

$$P'(t)=0.03(12+2t)^{1/2}(2)$$
$$=0.06(12+2t)^{1/2}$$
$$P'(2)=0.06[12+2(2)]^{1/2}=0.24$$

69. **b.**

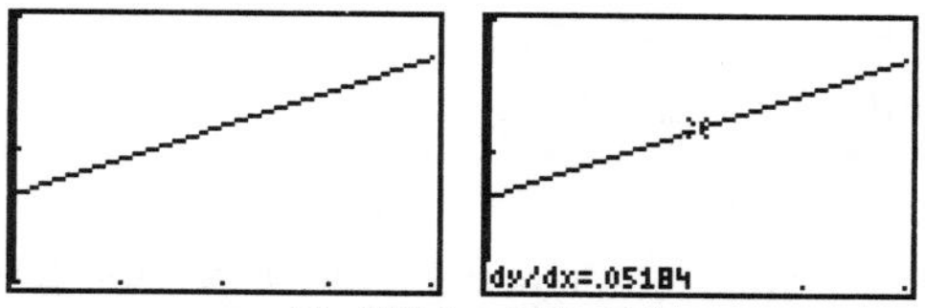

on [0, 20] by [60, 62]
slope = 0.05184

c. $\frac{1.8}{x}=0.05184$

$x\approx 34.7$, so about 35 years

71. False: There should not be a prime of the first g on the right-hand side.

73. The generalized power rule is a special case of the chain rule, when the outer function is just a power of the variable.

75. True: $\frac{d}{dx} f(5x) = \frac{d}{dx} 5x \cdot f'(5x) = 5 \cdot f'(5x)$

77. False: The outer function, $\sqrt{x}$, was not differentiated. The correct right-hand side is $\frac{1}{2}[g(x)]^{-1/2} \cdot g'(x)$.

79. No: Since instantaneous rates of change are derivatives, this would be saying that $\frac{d}{dx}[f(x)]^2 = [f'(x)]^2$, where $f(x)$ is the length of a side. The chain rule gives the correct derivative of $[f(x)]^2$.

81. $\frac{d}{dx} L(g(x)) = L'(g(x))g'(x)$

But since $L'(x) = \frac{1}{x}$, $L'(g(x)) = \frac{1}{g(x)}$.

Thus, $\frac{d}{dx} L(g(x)) = \frac{g'(x)}{g(x)}$.

83. Let $G(x) = g(h(x))$, then the chain rule gives $G'(x) = g'(h(x)) \cdot h'(x)$. We use the chain rule to find $\frac{d}{dx} f(G(x))$ and substitute the expressions for $G(x)$ and $G'(x)$.

$$\begin{aligned}\frac{d}{dx} f(G(x)) &= f'(G(x)) \cdot G'(x)\\ &= f'(g(h(x)) \cdot g'(h(x)) \cdot h'(x)\end{aligned}$$

85. Solve $f(x+h) - f(x) = F(h) \cdot h$ for $F(h)$ to find $F(h) = \frac{f(x+h) - f(x)}{h}$ and use the continuity of F at 0 to obtain

$$F(0) = \lim_{h \to 0} F(h) = \lim_{h \to 0} \frac{f(x+h) - f(x)}{h},$$

which is the definition of the derivative.

EXERCISES 2.7

1. The derivative does not exist at the corner points $x = -2, 0, 2$.

3. The derivative does not exist at the discontinuous points $x = -3, 3$.

5. For positive h, $\lim_{h \to 0} \frac{f(x+h) - f(x)}{h} = \lim_{h \to 0} \frac{|2x + 2h| - |2x|}{h}$. For $x = 0$, this becomes $\lim_{h \to 0} \frac{|0 + 2h| - |0|}{h} = \frac{|2h|}{h} = 2$ because h is positive. For h negative, $\lim_{h \to 0} \frac{f(x+h) - f(x)}{h} = \lim_{h \to 0} \frac{|2x + 2h| - |2x|}{h}$. Since $x = 0$, we get $\lim_{h \to 0} \frac{|0 + 2h| - |0|}{h} = \frac{|2h|}{h} = -2$. Thus, the derivative does not exist.

7. For $x = 0$, $\lim_{h \to 0} \frac{f(x+h) - f(x)}{h} = \lim_{h \to 0} \frac{(0+h)^{2/5} - 0}{h} = \lim_{h \to 0} \frac{h^{2/5}}{h} = \lim_{h \to 0} \frac{1}{h^{3/5}}$ which does not exist. Thus, the derivative does not exist at $x = 0$.

9. If you get a numerical answer, it is wrong because the function is undefined at $x = 0$. Thus, the derivative at $x = 0$ does not exist.

11. **a.** For $x = 0$,

$$\begin{aligned}\lim_{h \to 0} \frac{f(x+h) - f(x)}{h} &= \lim_{h \to 0} \frac{\sqrt[5]{0+h} - \sqrt[5]{0}}{h}\\ &= \lim_{h \to 0} \frac{\sqrt[5]{h}}{h}\end{aligned}$$

b.

h	$\frac{\sqrt[5]{h}}{h}$
0.1	6.3
0.001	251.2
0.00001	10,000

c. No, the limit does not exist. No, the derivative does not exist at $x = 0$.

d.

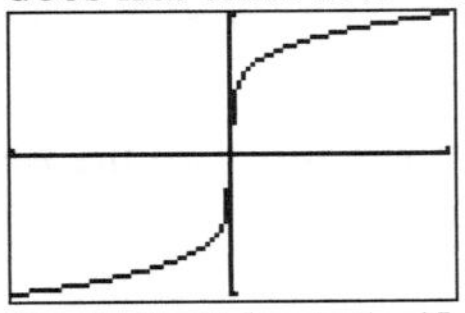

On [–1, 1] by [–1, 1]

13. At a corner point, a proposed tangent line can tip back and forth and so there is no well-defined slope.

15. At a discontinuity, the values of the function take a sudden jump, and so a (steady) rate of change cannot be defined.

17. True: For a function to be differentiable, it cannot jump or break.

REVIEW EXERCISES AND CHAPTER TEST FOR CHAPTER 2

1.

x	$4x+2$
1.9	9.6
1.99	9.96
1.999	9.996

x	$4x+2$
2.1	10.4
2.01	10.04
2.001	10.004

a. $\lim_{x\to 2^-} 4x+2=10$

b. $\lim_{x\to 2^+} 4x+2=10$

c. $\lim_{x\to 2} 4x+2=10$

2.

x	$\frac{\sqrt{x+1}-1}{x}$
–0.1	0.513
–0.01	0.501
–0.0001	0.500

x	$\frac{\sqrt{x+1}-1}{x}$
0.1	0.488
0.01	0.499
0.001	0.500

a. $\lim_{x\to 0^-} \frac{\sqrt{x+1}-1}{x}=0.5$

b. $\lim_{x\to 0^+} \frac{\sqrt{x+1}-1}{x}=0.5$

c. $\lim_{x\to 0} \frac{\sqrt{x+1}-1}{x}=0.5$

3. **a.** $\lim_{x\to 5^-} f(x)=\lim_{x\to 5^-}(2x-7)$
$=2(5)-7)=3$

b. $\lim_{x\to 5^+} f(x)=\lim_{x\to 5+}(3-x)$
$=3-5=-2$

c. $\lim_{x\to 5} f(x)$ does not exist.

4. **a.** $\lim_{x\to 5^-} f(x)=\lim_{x\to 5^-}(4-x)$
$=4-5=-1$

b. $\lim_{x\to 5^+} f(x)=\lim_{x\to 5^+}(2x-11)$
$=2(5)-11=-1$

c. $\lim_{x\to 5} f(x)=-1$

5. $\lim_{x\to 4}\sqrt{x^2+x+5}=\sqrt{4^2+4+5}=\sqrt{16+9}$
$=\sqrt{25}=5$

6. $\lim_{x\to 0}\pi=\pi$

7. $\lim_{s\to 16}\left(\frac{1}{2}s-s^{1/2}\right)=\frac{1}{2}(16)-16^{1/2}=8-4=2$

8. $\lim_{r\to 8}\frac{r}{r^2-30\sqrt[3]{r}}=\frac{8}{8^2-30\sqrt[3]{8}}=\frac{8}{64-30(2)}=2$

9. $\lim_{x\to 1}\frac{x^2-x}{x^2-1}=\lim_{x\to 1}\frac{x(x-1)}{(x+1)(x-1)}$
$=\lim_{x\to 1}\frac{x}{x+1}=\frac{1}{2}$

10. $\lim_{x\to -1}\frac{3x^3-3x}{2x^2+2x}=\lim_{x\to -1}\frac{3x(x^2-1)}{2x(x+1)}$
$=\lim_{x\to -1}\frac{3x(x-1)(x+1)}{2x(x+1)}$
$=\lim_{x\to -1}\frac{3(x-1)}{2}=\frac{3(-1-1)}{2}=-3$

11. $\lim_{h\to 0}\frac{2x^2h-xh^2}{h}=\lim_{h\to 0}\frac{xh(2x-h)}{h}$
$=\lim_{h\to 0}x(2x-h)$
$=x(2x-0)=2x^2$

12. $\lim_{h\to 0}\frac{6xh^2-x^2h}{h}=\lim_{h\to 0}\frac{xh(6h-x)}{h}$
$=\lim_{h\to 0}x(6h-x)$
$=x(0-x)=-x^2$

13. $\lim_{x\to -\infty} f(x)=0$; $\lim_{x\to -2^-} f(x)=\infty$ and $\lim_{x\to -2^+} f(x)=\infty$,
so $\lim_{x\to -2} f(x)=\infty$ and $\lim_{x\to \infty} f(x)=0$.

14. $\lim_{x\to -\infty} f(x)=3$; $\lim_{x\to 2^-} f(x)=-\infty$ and $\lim_{x\to 2^+} f(x)=\infty$,
so $\lim_{x\to 2} f(x)$ does not exist and $\lim_{x\to \infty} f(x)=3$.

15. Continuous

16. Continuous

17. Discontinuous at $x = -1$

18. Continuous

19. The function is discontinuous at values of x for which the denominator is zero. Thus, we consider $x^2 + x = 0$ and solve.

$$x^2 + x = 0$$
$$x(x+1) = 0$$
$$x = 0, -1$$

Discontinuous at $x = 0, -1$

20. Discontinuous at $x = -3, 3$

21. From Exercise 3, we know $\lim_{x\to 5} f(x)$ does not exist. Therefore, the function is discontinuous at $x = 5$.

22. From Exercise 4, we know $\lim_{x\to 5} f(x) = -1 = f(5)$. Therefore, the function is continuous.

23. $$\lim_{h\to 0}\frac{f(x+h)-f(x)}{h} = \lim_{h\to 0}\frac{2(x+h)^2+3(x+h)-1-(2x^2+3x-1)}{h} = \lim_{h\to 0}\frac{2(x^2+2xh+h^2)+3x+3h-1-2x^2-3x+1}{h}$$
$$= \lim_{h\to 0}\frac{2x^2+4xh+2h^2+3x-3x+3h-1+1-2x^2}{h} = \lim_{h\to 0}\frac{4xh+2h^2+3h}{h} = \lim_{h\to 0} 4x+2h+3 = 4x+3$$

24. $$\lim_{h\to 0}\frac{f(x+h)-f(x)}{h} = \lim_{h\to 0}\frac{3(x+h)^2+2(x+h)-3-(3x^2+2x-3)}{h} = \lim_{h\to 0}\frac{3x^2+6xh+3h^2+2x+2h-3-3x^2-2x+3}{h}$$
$$= \lim_{h\to 0}\frac{6xh+3h^2+2h}{h} = \lim_{h\to 0} 6x+3h+2 = 6x+2$$

25. $$\lim_{h\to 0}\frac{f(x+h)-f(x)}{h} = \lim_{h\to 0}\frac{\frac{3}{x+h}-\frac{3}{x}}{h} = \lim_{h\to 0}\frac{\frac{3x-3(x+h)}{x(x+h)}}{h} = \lim_{h\to 0}\frac{3x-3x-3h}{x(x+h)}\cdot\frac{1}{h} = \lim_{h\to 0}\frac{-3h}{x(x+h)h} = \lim_{h\to 0} -\frac{3}{x(x+h)} = -\frac{3}{x^2}$$

26. $$\lim_{h\to 0}\frac{f(x+h)-f(x)}{h} = \lim_{h\to 0}\frac{4\sqrt{x+h}-4\sqrt{x}}{h} = \lim_{h\to 0} 4\frac{\sqrt{x+h}-\sqrt{x}}{h}\cdot\frac{\sqrt{x+h}+\sqrt{x}}{\sqrt{x+h}+\sqrt{x}} = \lim_{h\to 0} 4\frac{x+h-x}{h\left(\sqrt{x+h}+\sqrt{x}\right)}$$
$$= \lim_{h\to 0} 4\frac{h}{h\left(\sqrt{x+h}+\sqrt{x}\right)} = \lim_{h\to 0}\frac{4}{\sqrt{x+h}+\sqrt{x}} = \frac{4}{2\sqrt{x}} = \frac{2}{\sqrt{x}}$$

27. $$f(x) = 6\sqrt[3]{x^5} - \frac{4}{\sqrt{x}} + 1 = 6x^{5/3} - 4x^{-1/2} + 1$$
$$f'(x) = \frac{5}{3}\left(6x^{2/3}\right) - \left(-\frac{1}{2}\right)\left(4x^{-3/2}\right) + 0$$
$$= 10x^{2/3} + 2x^{-3/2}$$

28. $$f(x) = 4\sqrt{x^5} - \frac{6}{\sqrt[3]{x}} + 1 = 4x^{5/2} - 6x^{-1/3} + 1$$
$$f'(x) = \frac{5}{2}\left(4x^{3/2}\right) - \left(-\frac{1}{3}\right)\left(6x^{-4/3}\right) + 0$$
$$= 10x^{3/2} + 2x^{-4/3}$$

29. $$f(x) = \frac{1}{x^2} = x^{-2}$$
$$f'(x) = -2x^{-3}$$
$$f'\left(\frac{1}{2}\right) = -2\left(\frac{1}{2}\right)^{-3} = -2(2)^3 = -16$$

30. $$f(x) = \frac{1}{x} = x^{-1}$$
$$f'(x) = -x^{-2}$$
$$f'\left(\frac{1}{3}\right) = -\left(\frac{1}{3}\right)^{-2} = -(3)^2 = -9$$

31. $f(x) = 12\sqrt[3]{x} = 12x^{1/3}$

$f'(x) = \frac{1}{3}(12x^{-2/3}) = 4x^{-2/3}$

$f'(8) = 4(8)^{-2/3} = 4\left(\frac{1}{8}\right)^{2/3} = 4\left(\frac{1}{8^{2/3}}\right)$

$= 4\left(\frac{1}{\sqrt[3]{8^2}}\right) = 4\left(\frac{1}{\sqrt[3]{64}}\right) = 1$

32. $f(x) = 6\sqrt[3]{x} = 6x^{1/3}$

$f'(x) = \frac{1}{3}(6x^{-2/3}) = 2x^{-2/3}$

$f'(-8) = 2(-8)^{-2/3} = 2\left(-\frac{1}{8}\right)^{2/3} = 2\left(\frac{1}{\sqrt[3]{(-8)^2}}\right) = \frac{1}{2}$

33. $y = \frac{1}{x} + x^2 = x^{-1} + x^2$

$y(1) = \frac{1}{1} + 1^2 = 2$

The point is (1, 2).

$y' = -1x^{-2} + 2x = -\frac{1}{x^2} + 2x$

$y'(1) = -\frac{1}{(1)^2} + 2(1) = 1$

The tangent line has slope 1 at (1, 2).

$y - 2 = 1(x - 1)$

$y = x + 1$

34. **a.** $C'(x) = 20x^{-1/6}$

$C'(1) = \frac{20}{\sqrt[6]{1}} = 20$

Costs are increasing by about \$20 per additional license.

b. $C'(64) = \frac{20}{\sqrt[6]{64}} = \frac{20}{2} = 10$

Costs are increasing by about \$10 per additional license.

35. $f(x) = 150x^{-0.322}$

$f'(x) = -0.322(150x^{-1.322}) = -48.3x^{-1.322}$

$f'(10) = -48.3(10)^{-1.322} \approx -2.3$

After 10 planes, construction time is decreasing by about 2300 hours for each additional plane built.

36. **a.** $A = \pi r^2$

$A' = 2\pi r$

b. As the radius increases, the area grows by "a circumference."

37. **a.** $V = \frac{4}{3}\pi r^3$

$A' = 3\left(\frac{4}{3}\pi r^2\right) = 4\pi r^2$

b. As the radius increases, the volume grows by "a surface area."

38. $f(x) = 2x(5x^3 + 3)$

$f'(x) = 2(5x^3 + 3) + 2x(15x^2)$

$= 2(5x^3 + 3) + 30x^3 = 40x^3 + 6$

39. $f(x) = x^2(3x^3 - 1)$

$f'(x) = 2x(3x^3 - 1) + x^2(9x^2)$

$= 2x(3x^3 - 1) + 9x^4 = 15x^4 - 2x$

40. $f(x) = (x^2 + 5)(x^2 - 5)$

$f'(x) = 2x(x^2 - 5) + (x^2 + 5)(2x)$

$= 2x^3 - 10x + 2x^3 + 10x = 4x^3$

41. $f(x) = (x^2 + 3)(x^2 - 3)$

$f'(x) = 2x(x^2 - 3) + (x^2 + 3)(2x)$

$= 2x^3 - 6x + 2x^3 + 6x = 4x^3$

42. $y = (x^4 + x^2 + 1)(x^5 - x^3 + x)$

$y' = (4x^3 + 2x)(x^5 - x^3 + x) + (x^4 + x^2 + 1)(5x^4 - 3x^2 + 1)$

$= 4x^8 - 4x^6 + 4x^4 + 2x^6 - 2x^4 + 2x^2 + 5x^8 - 3x^6 + x^4 + 5x^6 - 3x^4 + x^2 + 5x^4 - 3x^2 + 1$

$= 9x^8 + 5x^4 + 1$

43. $y = (x^5 + x^3 + x)(x^4 - x^2 + 1)$

$$y' = (5x^4 + 3x^2 + 1)(x^4 - x^2 + 1) + (x^5 + x^3 + x)(4x^3 - 2x)$$
$$= 5x^8 - 5x^6 + 5x^4 + 3x^6 - 3x^4 + 3x^2 + x^4 - x^2 + 1 + 4x^8 + 4x^6 + 4x^4 - 2x^6 - 2x^4 - 2x^2$$
$$= 5x^8 - 2x^6 + 3x^4 + 2x^2 + 1 + 4x^8 + 2x^6 + 2x^4 - 2x^2$$
$$9x^8 + 5x^4 + 1$$

44. $y = \frac{x-1}{x+1}$

$$y' = \frac{(x+1)(1) - (1)(x-1)}{(x+1)^2} = \frac{x+1-x+1}{(x+1)^2} = \frac{2}{(x+1)^2}$$

45. $y = \frac{x+1}{x-1}$

$$y' = \frac{(x-1)(1) - (1)(x+1)}{(x-1)^2} = \frac{x-1-x-1}{(x-1)^2} = -\frac{2}{(x-1)^2}$$

46. $y = \frac{x^5+1}{x^5-1}$

$$y' = \frac{(x^5-1)(5x^4) - (5x^4)(x^5+1)}{(x^5-1)^2}$$
$$= \frac{5x^9 - 5x^4 - 5x^9 - 5x^4}{(x^5-1)^2} = -\frac{10x^4}{(x^5-1)^2}$$

47. $y = \frac{x^6-1}{x^6+1}$

$$y' = \frac{(x^6+1)(6x^5) - (6x^5)(x^6-1)}{(x^6+1)^2}$$
$$= \frac{6x^{11} + 6x^5 - 6x^{11} + 6x^5}{(x^6+1)^2} = \frac{12x^5}{(x^6+1)^2}$$

48. **a.** $f(x) = \frac{2x+1}{x}$

$$f'(x) = \frac{x(2) - (1)(2x+1)}{x^2}$$
$$= \frac{2x-2x-1}{x^2} = -\frac{1}{x^2}$$

b. $f(x) = (2x+1)(x^{-1})$

$$f'(x) = 2(x^{-1}) + (2x+1)(-1)x^{-2}$$
$$= \frac{2}{x} - \frac{2x+1}{x^2} = \frac{2x}{x^2} - \frac{2x+1}{x^2} = -\frac{1}{x^2}$$

c. Dividing $2x+1$ by x, we get

$$f(x) = 2 + \frac{1}{x} = 2 + x^{-1}$$
$$f'(x) = 0 + (-1)x^{-2} = -\frac{1}{x^2}$$

49. $y = \frac{x^3-1}{x^4+1}$

$$y(-1) = \frac{(-1)^3 - 1}{(-1)^4 + 1} = -1$$

The point is $(-1, -1)$.

$$y' = \frac{(x^4+1)(3x^2) - (4x^3)(x^3-1)}{(x^4+1)^2}$$
$$= \frac{-x^6 + 4x^3 + 3x^2}{(x^4+1)^2}$$
$$y'(-1) = \frac{-(-1)^6 + 4(-1)^3 + 3(-1)^2}{((-1)^4+1)^2} = \frac{-1-4+3}{4}$$
$$= -\frac{1}{2}$$

The tangent line has slope $-\frac{1}{2}$ at $(-1, -1)$.

$$y + 1 = -\frac{1}{2}(x+1)$$
$$y = -\frac{1}{2}x - \frac{3}{2}$$

50. $S(x)=\frac{2250}{x+9}=2250(x+9)^{-1}$

$S'(x)=-2250(x+9)^{-2}=-\frac{2250}{(x+9)^2}$

$S'(6)=-\frac{2250}{(6+9)^2}=-10$

At \$6 per flash drive, the number of drives sold is decreasing by 10 per dollar increase in price.

51. **a.** To find the average profit function, divide $P(x)=6x-200$ by x.

$AP(x)=\frac{6x-200}{x}$

b. To find the marginal average profit, take the derivative of $AP(x)$.

$MAP(x)=\frac{x(6)-(1)(6x-200)}{x^2}=\frac{200}{x^2}$

c. $MAP(10)=\frac{200}{(10)^2}=\frac{200}{100}=2$

Average profit is increasing by about \$2 for each additional unit.

52. **a.** $AC(x)=\frac{C(x)}{x}=\frac{7.5x+50}{x}=7.5+\frac{50}{x}$

b. $MAC(x)=\frac{d}{dx}\left(7.5+\frac{50}{x}\right)=\frac{-50}{x^2}$

c. $MAC(50)=\frac{-50}{50^2}=-\frac{1}{50}=-0.02$

Average cost is decreasing by about \$0.02 per additional mouse.

53. $f(x)=12\sqrt{x^3}-9\sqrt[3]{x}=12x^{3/2}-9x^{1/3}$

$f'(x)=\frac{3}{2}(12x^{1/2})-\frac{1}{3}(9x^{-2/3})=18x^{1/2}-3x^{-2/3}$

$f''(x)=\frac{1}{2}(18x^{-1/2})-\frac{2}{3}(-3x^{-5/3})$

$=9x^{-1/2}+2x^{-5/3}$

54. $f(x)=18\sqrt[3]{x^2}-4\sqrt{x^3}=18x^{2/3}-4x^{3/2}$

$f'(x)=12x^{-1/3}-6x^{1/2}$

$f''(x)=-4x^{-4/3}-3x^{-1/2}$

55. $f(x)=\frac{1}{3x^2}=\frac{1}{3}x^{-2}$

$f'(x)=-2\left(\frac{1}{3}x^{-3}\right)=-\frac{2}{3}x^{-3}$

$f''(x)=-3\left(-\frac{2}{3}x^{-4}\right)=2x^{-4}$

56. $f(x)=\frac{1}{2x^3}=\frac{1}{2}x^{-3}$

$f'(x)=-\frac{3}{2}x^{-4}$

$f''(x)=6x^{-5}$

57. $f(x)=\frac{2}{x^3}=2x^{-3}$

$f'(x)=-3(2x^{-4})=-6x^{-4}=-\frac{6}{x^4}$

$f''(x)=-4(-6x^{-5})=24x^{-5}=\frac{24}{x^5}$

$f''(-1)=\frac{24}{(-1)^5}=-24$

58. $f(x)=\frac{3}{x^4}=3x^{-4}$

$f'(x)=-12x^{-5}$

$f''(x)=60x^{-6}=\frac{60}{x^6}$

$f''(-1)=\frac{60}{(-1)^6}=60$

59. $\frac{d}{dx}x^6=6x^5$

$\frac{d^2}{dx^2}x^6=\frac{d}{dx}6x^5=30x^4$

$\left.\frac{d^2}{dx^2}\right|_{x=-2}x^6=30(-2)^4=480$

60. $\frac{d}{dx}x^{-2}=-2x^{-3}$

$\frac{d^2}{dx^2}x^{-2}=\frac{d}{dx}(-2x^{-3})=6x^{-4}=\frac{6}{x^4}$

$\left.\frac{d^2}{dx^2}\right|_{x=-2}x^{-2}=\frac{6}{(-2)^4}=\frac{6}{16}=\frac{3}{8}$

61. $\frac{d}{dx}\sqrt{x^5}=\frac{d}{dx}x^{5/2}=\frac{5}{2}x^{3/2}$

$\frac{d^2}{dx^2}\sqrt{x^5}=\frac{d}{dx}\left(\frac{5}{2}x^{3/2}\right)=\frac{15}{4}x^{1/2}$

$\left.\frac{d^2}{dx^2}\right|_{x=16}\sqrt{x^5}=\frac{15}{4}(16)^{1/2}=15$

62. $\frac{d}{dx}\sqrt{x^7} = \frac{d}{dx}x^{7/2} = \frac{7}{2}x^{5/2}$

$\frac{d^2}{dx^2}\sqrt{x^7} = \frac{d}{dx}\left(\frac{7}{2}x^{5/2}\right) = \frac{35}{4}x^{3/2}$

$\left.\frac{d^2}{dx^2}\right|_{x=4}\sqrt{x^7} = \frac{35}{4}(4)^{3/2} = 70$

63. $P(t) = 0.25t^3 - 3t^2 + 5t + 200$

$P'(t) = 3(0.25t^2) - 2(3t) + 1(5) + 0 = 0.75t^2 - 6t + 5$

$P''(t) = 2(0.75t) - 6(1) + 0 = 1.5t - 6$

$P(10) = 0.25(10)^3 - 3(10)^2 + 5(10) + 200 = 250 - 300 + 50 + 200 = 200$

In 10 years, the population will be 200,000.

$P'(10) = 0.75(10)^2 - 6(10) + 5 = 75 - 60 + 5 = 20$

In 10 years, the population will be increasing by about 20,000 per year.

$P''(10) = 1.5(10) - 6 = 9$

In 10 years, the rate of growth of the increase will be 9000 per year each year.

64. $s(t) = 8t^{5/2}$

$v(t) = s'(t) = \frac{5}{2}(8t^{3/2}) = 20t^{3/2}$

$a(t) = v'(t) = \frac{3}{2}(20t^{1/2}) = 30t^{1/2}$

$v(25) = 20(25)^{3/2} = 2500$ ft / sec

$a(25) = 30(25)^{1/2} = 150$ ft / sec^2

65. **a.** When the height is a maximum, the velocity is zero. Thus, to find the maximum height, set $v(t) = 0$ and solve. First, we find $v(t) = s'(t)$.

$s(t) = -16t^2 + 148t + 5$

$v(t) = s'(t) = -32t + 148$

Now set $v(t) = 0$ and solve.

$v(t) = -32t + 148 = 0$

$-32t = -148$

$t = 4.625$

To find the height when $t = 4.625$, evaluate $s(4.625)$.

$s(4.625) = -16(4.625)^2 + 148(4.625) + 5$

$= 347.25$ feet

b.

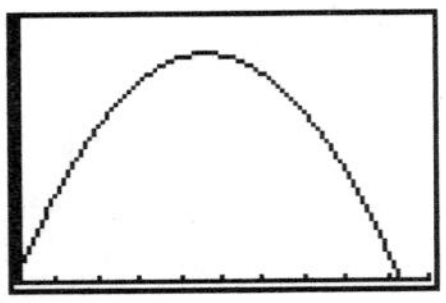

66. $h(z) = (4z^2 - 3z + 1)^3$

$h'(z) = 3(4z^2 - 3z + 1)^2(8z - 3)$

67. $h(z) = (3z^2 - 5z - 1)^4$

$h'(z) = 4(3z^2 - 5z - 1)^3(6z - 5)$

68. $g(x) = (100 - x)^5$

$g'(x) = 5(100 - x)^4(-1) = -5(100 - x)^4$

69. $g(x) = (1000 - x)^4$

$g'(x) = 4(1000 - x)^3(-1) = -4(1000 - x)^3$

70. $f(x) = \sqrt{x^2 - x + 2} = (x^2 - x + 2)^{1/2}$

$f'(x) = \frac{1}{2}(x^2 - x + 2)^{-1/2}(2x - 1)$

71. $f(x) = \sqrt{x^2 - 5x - 1} = (x^2 - 5x - 1)^{1/2}$

$f'(x) = \frac{1}{2}(x^2 - 5x - 1)^{-1/2}(2x - 5)$

72. $w(z) = \sqrt[3]{6z-1} = (6z-1)^{1/3}$

$w'(z) = \frac{1}{3}(6z-1)^{-2/3}(6) = 2(6z-1)^{-2/3}$

73. $w(z) = \sqrt[3]{3z+1} = (3z+1)^{1/3}$

$w'(z) = \frac{1}{3}(3z+1)^{-2/3}(3) = (3z+1)^{-2/3}$

74. $h(x) = \dfrac{1}{\sqrt[5]{(5x+1)^2}} = (5x+1)^{-2/5}$

$h'(x) = -\frac{2}{5}(5x+1)^{-7/5}(5) = -2(5x+1)^{-7/5}$

75. $h(x) = \dfrac{1}{\sqrt[5]{(10x+1)^3}} = (10x+1)^{-3/5}$

$h'(x) = -\frac{3}{5}(10x+1)^{-8/5}(10) = -6(10x+1)^{-8/5}$

76. $g(x) = x^2(2x-1)^4$

$g'(x) = 2x(2x-1)^4 + x^2[4(2x-1)^3(2)]$

$= 2x(2x-1)^4 + 8x^2(2x-1)^3$

77. $g(x) = 5x(x^3-2)^4$

$g'(x) = 5(x^3-2)^4 + 5x[4(x^3-2)^3(3x^2)]$

$= 5(x^3-2)^4 + 60x^3(x^3-2)^3$

78. $y = x^3\sqrt[3]{x^3+1} = x^3(x^3+1)^{1/3}$

$y' = 3x^2(x^3+1)^{1/3} + x^3\left[\frac{1}{3}(x^3+1)^{-2/3}(3x^2)\right]$

$= 3x^2(x^3+1)^{1/3} + x^5(x^3+1)^{-2/3}$

79. $y = x^4\sqrt{x^2+1} = x^4(x^2+1)^{1/2}$

$y' = 4x^3(x^2+1)^{1/2} + x^4\left[\frac{1}{2}(x^2+1)^{-1/2}(2x)\right]$

$= 4x^3(x^2+1)^{1/2} + x^5(x^2+1)^{-1/2}$

80. $f(x) = [(2x^2+1)^4 + x^4]^3$

$f'(x) = 3[(2x^2+1)^4 + x^4]^2[4(2x^2+1)^3(4x) + 4x^3]$

$= 3[(2x^2+1)^4 + x^4]^2[16x(2x^2+1)^3 + 4x^3]$

81. $f(x) = [(3x^2-1)^3 + x^3]^2$

$f'(x) = 2[(3x^2-1)^3 + x^3][3(3x^2-1)^2(6x) + 3x^2]$

$= 2[(3x^2-1)^3 + x^3][18x(3x^2-1)^2 + 3x^2]$

82. $f(x) = \sqrt{(x^2+1)^4 - x^4} = [(x^2+1)^4 - x^4]^{1/2}$

$f'(x) = \frac{1}{2}[(x^2+1)^4 - x^4]^{-1/2}[4(x^2+1)^3(2x) - 4x^3]$

$= \frac{1}{2}[(x^2+1)^4 - x^4]^{-1/2}[8x(x^2+1)^3 - 4x^3]$

83. $f(x) = \sqrt{(x^3+1)^2 + x^2} = [(x^3+1)^2 + x^2]^{1/2}$

$f'(x) = \frac{1}{2}[(x^3+1)^2 + x^2]^{-1/2}[2(x^3+1)(3x^2) + 2x]$

$= \frac{1}{2}[(x^3+1)^2 + x^2]^{-1/2}[6x^2(x^3+1) + 2x]$

$= [(x^3+1)^2 + x^2]^{-1/2}[3x^2(x^3+1) + x]$

84. $f(x) = (3x+1)^4(4x+1)^3$

$f'(x) = 4(3x+1)^3(3)(4x+1)^3 + (3x+1)^4(3)(4x+1)^2(4)$

$= 12(3x+1)^3(4x+1)^3 + 12(3x+1)^4(4x+1)^2$

$= 12(3x+1)^3(4x+1)^2[(4x+1) + (3x+1)]$

$= 12(3x+1)^3(4x+1)^2(7x+2)$

85. $f(x) = (x^2+1)^3(x^2-1)^4$

$f'(x) = 3(x^2+1)^2(2x)(x^2-1)^4 + (x^2+1)^3[4(x^2-1)^3(2x)]$

$= 6x(x^2+1)^2(x^2-1)^4 + 8x(x^2+1)^3(x^2-1)^3$

86. $f(x) = \left(\dfrac{x+5}{x}\right)^4$

$f'(x) = 4\left(\dfrac{x+5}{x}\right)^3\left[\dfrac{x(1)-(1)(x+5)}{x^2}\right]$

$= 4\left(\dfrac{x+5}{x}\right)^3\left(\dfrac{x-x-5}{x^2}\right) = -\dfrac{20}{x^2}\dfrac{(x+5)^3}{x^3}$

$= -\dfrac{20(x+5)^3}{x^5}$

87. $f(x) = \left(\dfrac{x+4}{x}\right)^5$

$f'(x) = 5\left(\dfrac{x+4}{x}\right)^4\left[\dfrac{x(1)-(1)(x+4)}{x^2}\right]$

$= 5\left(\dfrac{x+4}{x}\right)^4\left(\dfrac{x-x-4}{x^2}\right) = -\dfrac{20}{x^2}\left(\dfrac{x+4}{x}\right)^4$

$= -\dfrac{20(x+4)^4}{x^6}$

88.

$$y=\sqrt[3]{x^2+4}=\left(x^2+4\right)^{1/3}$$

$$y(2)=\sqrt[3]{(2)^2+4}=2$$

The point is (2, 2).

$$y'=\frac{1}{3}\left(x^2+4\right)^{-2/3}(2x)=\frac{2x}{3\left(x^2+4\right)^{2/3}}$$

$$y'(2)=\frac{2(2)}{3\left((2)^2+4\right)^{2/3}}=\frac{1}{3}$$

The tangent line has slope $\frac{1}{3}$ at (2, 2).

$$y-2=\frac{1}{3}(x-2)$$
$$y=\frac{1}{3}x+\frac{4}{3}$$

89.

$$h(w)=(2w^2-4)^5$$
$$h'(w)=5(2w^2-4)^4(4w)$$
$$h''(w)=20(2w^2-4)^3(4w)(4w)+5(2w^2-4)^4(4)$$
$$=20(16w^2)(2w^2-4)^3+20(2w^2-4)^4$$
$$=320w^2(2w^2-4)^3+20(2w^2-4)^4$$

90.

$$h(w)=(3w^2+1)^4$$
$$h'(w)=4(3w^2+1)^3(6w)$$
$$h''(w)=12(3w^2+1)^2(6w)(6w)+4(3w^2+1)^3(6)$$
$$=432w^2(3w^2+1)^2+24(3w^2+1)^3$$

91.

$$g(z)=z^3(z+1)^3$$
$$g'(z)=3z^2(z+1)^3+z^3[3(z+1)^2]=3z^2(z+1)^3+3z^3(z+1)^2$$
$$g''(z)=6z(z+1)^3+3z^2[3(z+1)^2]+9z^2(z+1)^2+3z^3[2(z+1)]$$
$$=6z(z+1)^3+18z^2(z+1)^2+6z^3(z+1)$$

92.

$$g(z)=z^4(z+1)^4$$
$$g'(z)=4z^3(z+1)^4+4z^4(z+1)^3$$
$$g''(z)=12z^2(z+1)^4+4z^3[4(z+1)^3]+16z^3(z+1)^3+4z^4[3(z+1)^2]$$
$$=12z^2(z+1)^4+32z^3(z+1)^3+12z^4(z+1)^2$$

93. **a.** $\frac{d}{dx}(x^3-1)^2=2(x^3-1)(3x^2)=6x^2(x^3-1)=6x^5-6x^2$

b. $\frac{d}{dx}(x^3-1)^2=\frac{d}{dx}(x^6-2x^3+1)=6x^5-6x^2$

94. **a.** $\frac{d}{dx}\left(\frac{1}{x^3+1}\right)=\frac{(x^3+1)(0)-(3x^2)(1)}{(x^3+1)^2}=-\frac{3x^2}{(x^3+1)^2}$

b. $\frac{d}{dx}\left(\frac{1}{x^3+1}\right)=\frac{d}{dx}(x^3+1)^{-1}=-1(x^3+1)^{-2}(3x^2)=-\frac{3x^2}{(x^3+1)^2}$

95.

$$P(x)=\sqrt{x^3-3x+34}=(x^3-3x+34)^{1/2}$$
$$P'(x)=\frac{1}{2}(x^3-3x+34)^{-1/2}(3x^2-3)=\frac{3x^2-3}{2\sqrt{x^3-3x+34}}$$
$$P'(5)=\frac{3(5)^2-3}{2\sqrt{(5)^3-3(5)+34}}=\frac{72}{2\sqrt{144}}=\frac{72}{24}=3$$

When 5 tons is produced, profit is increasing at about \$3000 for each additional ton.

96. $V(r) = 500(1 + 0.01r)^3$

$V'(r) = 1500(1 + 0.01r)^2(0.01) = 15(1 + 0.01r)^2$

$V'(8) = 15[1 + 0.01(8)]^2 \approx 17.50$

For 8 percent interest, the value increases by about \$17.50 for each additional percent interest.

97. 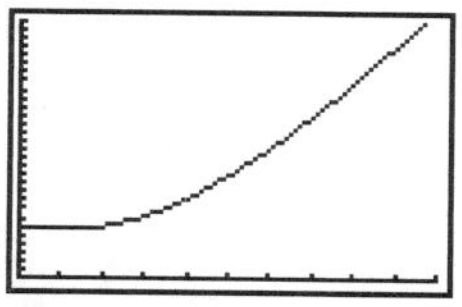

on [0, 10] by [0, 30]

a. $P(5) - P(4) \approx 12 - 9.27 = 2.73$

$P(6) - P(5) \approx 15.23 - 12 = 3.23$

Both values are near 3.

b. $x \approx 7.6$

98. 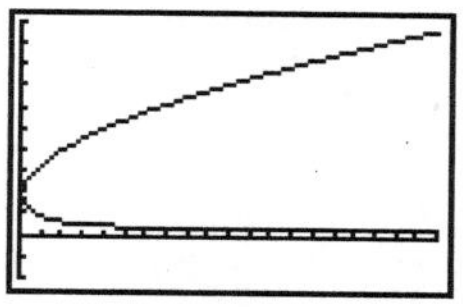

on [0, 20] by [–2, 10]

$x \approx 16$

99. $R(x) = 0.25(0.01x + 1)^4$

$R'(x) = 0.25[4(0.01x + 1)^3(0.01)]$

$= 0.01(0.01x + 1)^3$

$R'(100) = 0.01[0.01(100) + 1)]^3 = 0.08$

100. $N(x) = 1000\sqrt{100 - x} = 1000(100 - x)^{1/2}$

$N'(x) = 500(100 - x)^{-1/2}(-1) = -\dfrac{500}{\sqrt{100 - x}}$

$N'(96) = -\dfrac{500}{\sqrt{100 - 96}} = -250$

At age 96, the number of survivors is decreasing by 250 people per year.

101. The derivative does not exist at corner points $x = -3$ and $x = 3$ and at the discontinuous point $x = 1$.

102. The derivative does not exist at the corner point $x = 2$ and at the discontinuous point $x = -2$.

103. The derivative does not exist at the corner point $x = 3.5$ and the discontinuous point $x = 0$.

104. The derivative does not exist at the corner points $x = 0$ and $x = 3$.

105. For positive h,

$$\lim_{h \to 0} \frac{f(x+h) - f(x)}{h} = \lim_{h \to 0} \frac{|5x + 5h| - |5x|}{h} = \lim_{h \to 0} \frac{|5(0) + 5h| - |5 \cdot 0|}{h} \quad \text{for} \quad x = 0$$

$$= \lim_{h \to 0} \frac{5h}{h} = 5$$

For negative h,

$$\lim_{h \to 0} \frac{f(x+h) - f(x)}{h} = \lim_{h \to 0} \frac{|5x + 5h| - |5x|}{h} = \lim_{h \to 0} \frac{|5(0) + 5h| - |5(0)|}{h} \quad \text{for} \quad x = 0$$

$$= \lim_{h \to 0} -\frac{5h}{h} = -5$$

Thus, the limit does not exist, and so the derivative does not exist.

106.

$$\lim_{h \to 0} \frac{f(x+h) - f(x)}{h} = \lim_{h \to 0} \frac{(x+h)^{3/5} - x^{3/5}}{h} = \lim_{h \to 0} \frac{h^{3/5}}{h} \quad \text{for} \quad x = 0$$

$$= \lim_{h \to 0} \frac{1}{h^{2/5}}$$

which does not exist. Thus, the derivative does not exist.

Chapter 3: Further Applications of Derivatives

EXERCISES 3.1

1. **a.** The derivative is positive on $(-\infty, -2)$ and $(0, \infty)$.

b. The derivative is negative on $(-2, 0)$.

3. All the numbers except 3 are critical numbers.

5. $f(x) = x^3 - 48x$

$$f'(x) = 3x^2 - 48 = 0$$
$$x^2 = \frac{48}{3} = 16$$
$$x = \pm 4$$

The critical numbers are $x = 4$ and -4.

7. $f(x) = x^3 - 6x^2 - 15x + 30$

$$f'(x) = 3x^2 - 12x - 15$$
$$3x^2 - 12x - 15 = 0$$
$$3(x^2 - 4x - 5) = 0$$
$$3(x-5)(x+1) = 0$$
$$x = 5, \ -1$$

The critical numbers are $x = 5$ and -1.

9. $f(x) = x^4 + 4x^3 - 8x^2 + 1$

$$f'(x) = 4x^3 + 12x^2 - 16x = 0$$
$$4x(x^2 + 3x - 4) = 0$$
$$4x(x+4)(x-1) = 0$$
$$x = 0, -4, 1$$

The critical numbers are $x = 0$, -4 and 1.

11. $f(x) = (2x-6)^4$

$$f'(x) = 4(2x-6)^3(2)$$
$$f'(x) = 4(2x-6)^3(2)$$
$$= 8(2x-6)^3 = 0$$
$$2x - 6 = 0$$
$$x = 3$$

The critical number is $x = 3$.

13. $f(x) = 3x + 5$

$$f'(x) = 3$$

Since $f'(x) = 3$, there are no critical numbers.

15. $f(x) = x^3 + x^2 - x + 4$

$$f'(x) = 3x^2 + 2x - 1$$
$$3x^2 + 2x - 1 = 0$$
$$(3x-1)(x+1) = 0$$
$$x = \tfrac{1}{3}, \ -1$$

The critical numbers are $x = \frac{1}{3}$ and -1.

17. $f(x) = x^3 - 3x^2 - 9x + 10$

$$f'(x) = 3x^2 - 6x - 9$$
$$3x^2 - 6x - 9 = 0$$
$$3(x^2 - 2x - 3) = 0$$
$$3(x+1)(x-3) = 0$$
$$x = -1, \ 3$$

The critical numbers are $x = -1$ and 3.

$f' > 0$	$f' = 0$	$f' < 0$	$f' = 0$	$f' > 0$
	$x = -1$		$x = 3$	
↗	rel max $(-1, 15)$	↘	rel min $(3, -17)$	↗

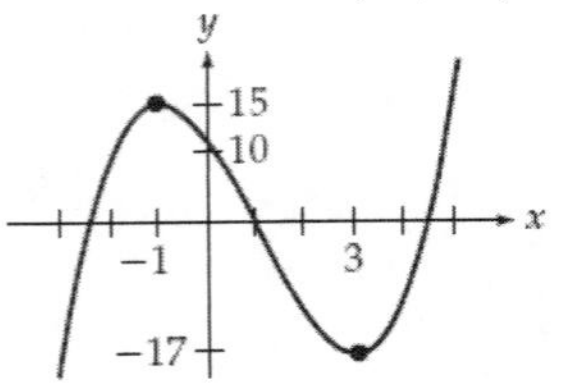

Increase: $(-\infty, -1)$ and $(3, \infty)$
Decrease: $(-1, 3)$

19. $f(x) = x^4 + 4x^3 - 8x^2 + 64$

$$f'(x) = 4x^3 + 12x^2 - 16x = 0$$
$$4x(x^2 + 3x - 4) = 0$$
$$4x(x+4)(x-1) = 0$$
$$x = 0, -4, 1$$

The critical numbers are $x = 0, -4$, and 1.

$f' < 0$	$f' = 0$	$f' > 0$	$f' = 0$	$f' < 0$	$f' = 0$	$f' > 0$
	$x = -4$		$x = 0$		$x = 1$	
╲	rel min (−4, −64)	╱	rel max (0, 64)	╲	rel min (1, 61)	╱

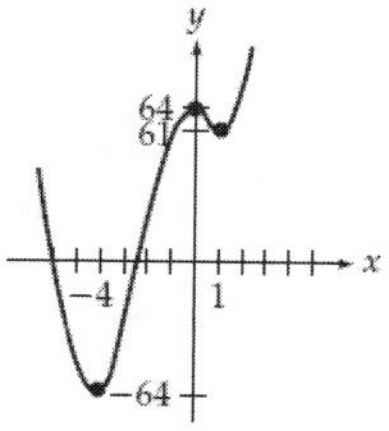

Increase: $(-4,\ 0)$ and $(1,\ \infty)$
Decrease: $(-\infty,\ -4)$ and $(0,\ 1)$

21. $f(x) = -x^4 + 4x^3 - 4x^2 + 1$

$$f'(x) = -4x^3 + 12x^2 - 8x = 0$$
$$-4x(x^2 - 3x + 2) = 0$$
$$-4x(x-2)(x-1) = 0$$
$$x = 0, 2, 1$$

The critical numbers are $x = 0, 1,$ and 2.

$f' > 0$	$f' = 0$	$f' < 0$	$f' = 0$	$f' > 0$	$f' = 0$	$f' < 0$
	$x = 0$		$x = 1$		$x = 2$	
╱	rel max (0, 1)	╲	rel min (1, 0)	╱	rel max (2, 1)	╲

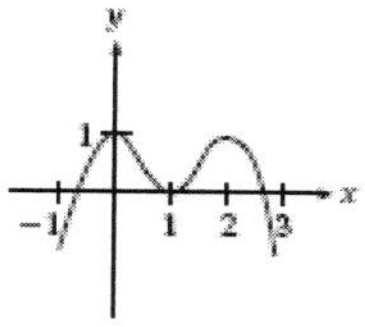

Increase: $(-\infty,\ 0)$ and $(1,\ 2)$
Decrease: $(0,\ 1)$ and $(2,\ \infty)$

23. $f(x) = 3x^4 - 8x^3 + 6x^2$

$$f'(x) = 12x^3 - 24x^2 + 12x = 0$$
$$12x(x^2 - 2x + 1) = 0$$
$$12x(x-1)^2 = 0$$
$$x = 0, 1$$

The critical numbers are $x = 0$ and 1.

$f' < 0$	$f' = 0$	$f' > 0$	$f' = 0$	$f' > 0$
	$x = 0$		$x = 1$	
╲	rel min (0, 0)	╱	neither (1, 1)	╱

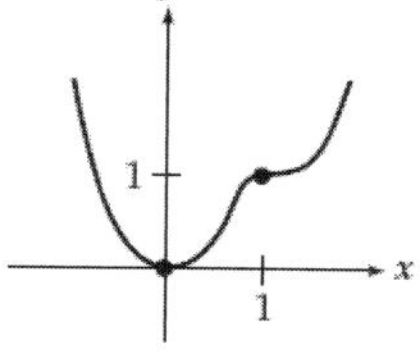

Increase: $(0,\ 1)$ and $(1,\ \infty)$
Decrease: $(-\infty,\ 0)$

25. $f(x) = (x-1)^6$

$f'(x) = 6(x-1)^5 = 0$

$x = 1$

The critical number is $x = 1$.

$f' < 0$	$f' = 0$	$f' > 0$
↘	$x = 1$	↗
	rel min (1, 0)	

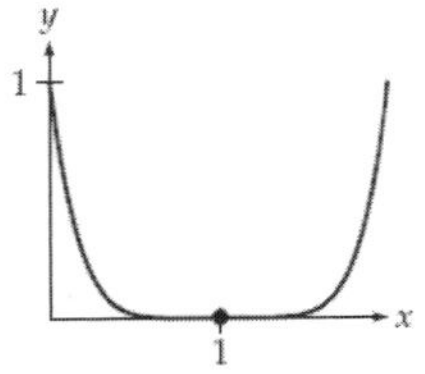

Increase: $(1, \infty)$
Decrease: $(-\infty, 1)$

27. $f(x) = (x^2 - 4)^2$

$f'(x) = 2(x^2 - 4)(2x) = 0$

$x = 0, -2, 2$

The critical numbers are $x = 0, -2$, and 2.

$f' < 0$	$f' = 0$	$f' > 0$	$f' = 0$	$f' < 0$	$f' = 0$	$f' > 0$
↘	$x = -2$	↗	$x = 0$	↘	$x = 2$	↗
	rel min (−2, 0)		rel max (0, 16)		rel min (2, 0)	

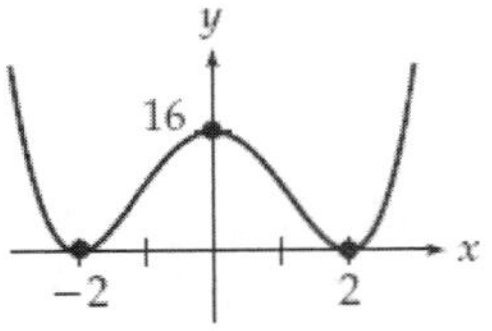

Increase: $(-2, 1)$ and $(4, \infty)$
Decrease: $(-\infty, -2)$ and $(1, 4)$

29. $f(x) = x^2(x-4)^2$

$f'(x) = 2x(x-4)^2 + x^2(x-4)(2) = 0$

$2x(x-4)[(x-4) + x] = 0$

$2x(x-4)(2x-4) = 0$

$x = 0, 4, 2$

The critical numbers are $x = 0, 2$, and 4.

$f' < 0$	$f' = 0$	$f' > 0$	$f' = 0$	$f' < 0$	$f' = 0$	$f' > 0$
↘	$x = 0$	↗	$x = 2$	↘	$x = 4$	↗
	rel min (0, 0)		rel max (2, 16)		rel min (4, 0)	

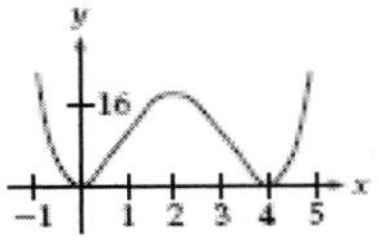

Increase: $(0, 2)$ and $(4, \infty)$
Decrease: $(-\infty, 0)$ and $(2, 4)$

31. $f(x) = x^2(x-5)^3$

$f'(x) = 2x(x-5)^3 + x^2(3)(x-5)^2 = 0$

$x(x-5)^2[2(x-5) + 3x] = 0$

$x(x-5)^2(5x - 10) = 0$

$x = 0, 5, 2$

The critical numbers are $x = 0, 2$, and 5.

$f' > 0$	$f' = 0$	$f' < 0$	$f' = 0$	$f' > 0$	$f' = 0$	$f' > 0$
↗	$x = 0$	↘	$x = 2$	↗	$x = 5$	↗
	rel max (0, 0)		rel min (2, −108)		neither (5, 0)	

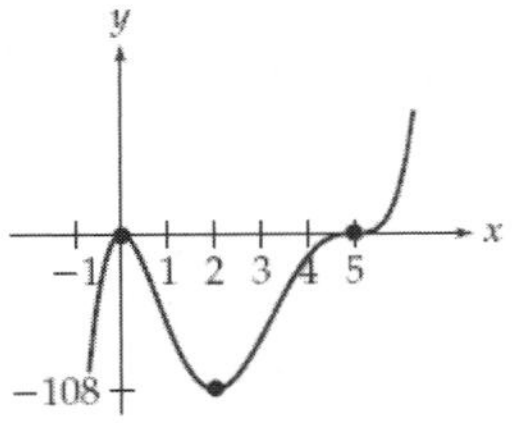

Increase: $(-\infty, 0)$ and $(2, 5)$ and $(5, \infty)$
Decrease: $(0, 2)$

33. $f(x)=\dfrac{6}{x+3}=6(x+3)^{-1}$

Vertical asymptote: $x=-3$
Horizontal asymptote: $y=0$

$$f'(x)=-6(x+3)^{-2}(1)=0$$
$$\frac{-6}{(x+3)^2}=0$$

The critical number is $x=-3$.

$f'<0$	$f'=\text{und}$	$f'<0$
↘	$x=-3$	↘

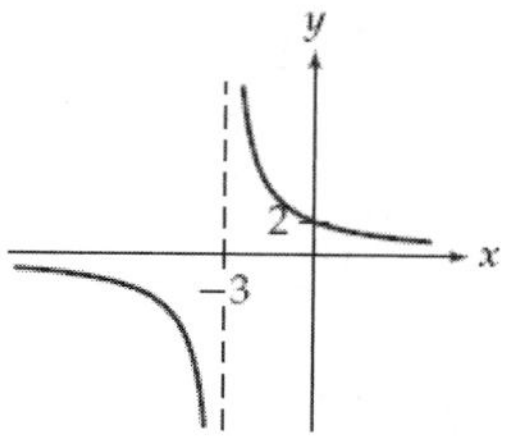

35. $f(x)=\dfrac{3x+6}{x-2}$

Vertical asymptote: $x=2$
Horizontal asymptote: $y=3$

$$f'(x)=\frac{(x-2)(3)-(3x+6)(1)}{(x-2)^2}=0$$
$$\frac{-12}{(x-2)^2}=0$$

The critical number is $x=2$.

$f'<0$	$f'=\text{und}$	$f'<0$
↘	$x=2$	↘

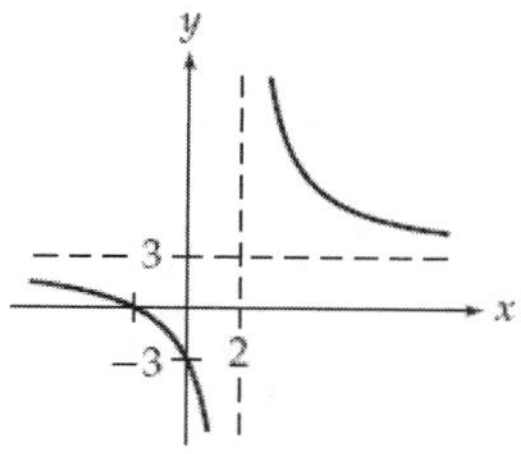

37. $f(x)=\dfrac{12x-24}{3x+6}=\dfrac{12(x-2)}{3(x+2)}=\dfrac{4(x-2)}{x+2}$

Vertical asymptote: $x=-2$
Horizontal asymptote: $y=4$

$$f'(x)=\frac{(x+2)(4)-(4x-8)(1)}{(x+2)^2}=0$$
$$\frac{16}{(x+3)^2}=0$$

The critical number is $x=-2$.

$f'>0$	$f'=\text{und}$	$f'>0$
↗	$x=-2$	↗

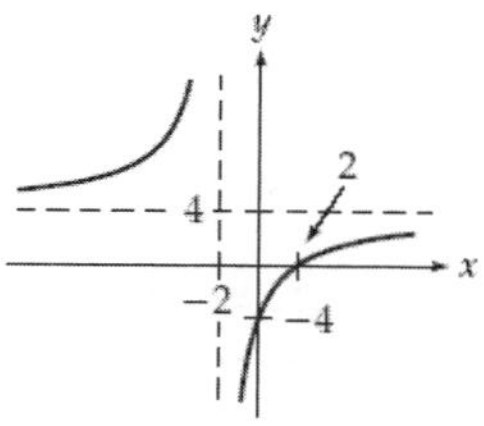

39. $f(x)=\dfrac{8}{(x-2)^2}=8(x-2)^{-2}$

Vertical asymptote: $x = 2$
Horizontal asymptote: $y = 0$

$$f'(x)=8(-2)(x-2)^{-3}(1)=0$$
$$\frac{-16}{(x-2)^3}=0$$

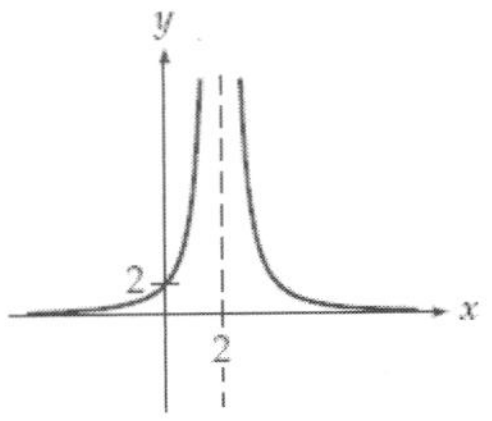

The critical number is $x = 2$.

$f'>0$	$f'=\text{und}$	$f'<0$
/	$x=2$	\

41. $f(x)=\dfrac{12}{x^2-2x-3}=\dfrac{12}{(x+1)(x-3)}$

Vertical asymptote: $x = -1, 3$
Horizontal asymptote: $y = 0$

$$f'(x)=\frac{(x^2-2x-3)(0)-12(2x-2)}{\left(x^2-2x-3\right)^2}=0$$
$$\frac{-24(x-1)}{\left(x^2-2x-3\right)^2}=0$$

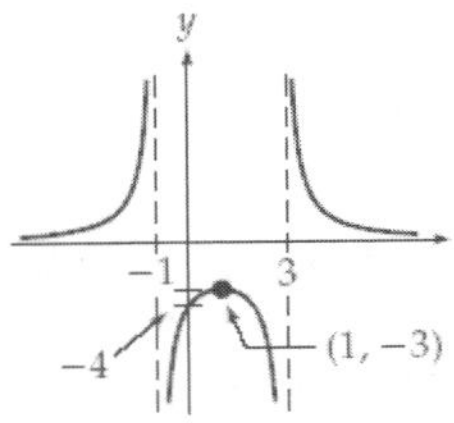

The critical numbers are $x = -1$, 1 and 3.

$f'>0$	$f'=\text{und}$	$f'<0$	$f'=0$	$f'<0$	$f'=\text{und}$	$f'<0$
/	$x=-1$	/	$x=-1$	\	$x=3$	\
			rel max (1, −3)			

43. $f(x)=\dfrac{8}{x^2+4}=8\left(x^2+4\right)^{-1}$

Vertical asymptote: none
Horizontal asymptote: $y = 0$

$$f'(x)=8(-1)(2x)\left(x^2+4\right)^{-2}=0$$
$$\frac{-16x}{\left(x^2+4\right)^2}=0$$

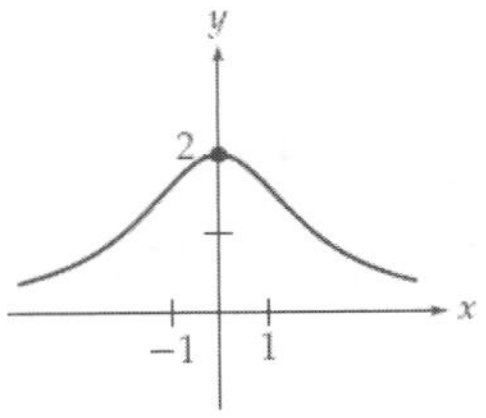

The critical number is $x = 0$.

$f'>0$	$f'=0$	$f'<0$
/	$x=0$	\
	rel max (0, 2)	

45. $f(x)=\frac{x^2}{x^2+1}$

Vertical asymptote: none

Horizontal asymptote: $y=1$

$$f'(x)=\frac{2x(x^2+1)-x^2(2x)}{(x^2+1)^2}=0$$

$$\frac{2x}{(x^2+1)^2}=0$$

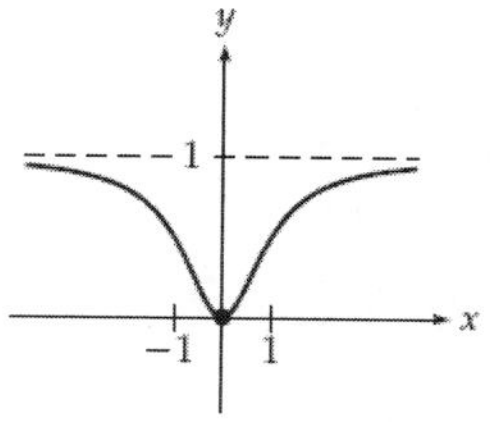

The critical number is $x=0$.

$f'<0$	$f'=0$	$f'>0$
╲	$x=0$ rel min (0, 0)	╱

47. $f(x)=\frac{2x}{x^2+1}$

Vertical asymptote: none

Horizontal asymptote: $y=0$

$$f'(x)=\frac{2(x^2+1)-2x(2x)}{(x^2+1)^2}=0$$

$$\frac{-2x^2+2}{(x^2+1)^2}=0$$

$$\frac{-2(x^2+1)}{(x^2+1)^2}=0$$

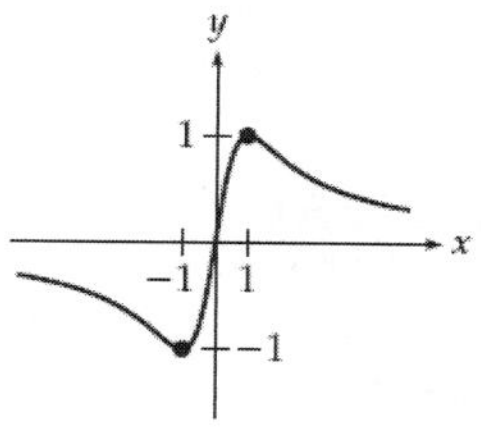

The critical numbers are $x=-1,\ 1$.

$f'<0$	$f'=0$	$f'>0$	$f'=0$	$f'<0$
╲	$x=-1$ rel min (–1, –1)	╱	$x=1$ rel max (1, 1)	╲

49. $f(x)=\frac{3}{x^2-1}=\frac{3}{(x+1)(x-1)}=3(x^2-1)^{-1}$

Vertical asymptote: $x=-1,\ 1$

Horizontal asymptote: $y=0$

$$f'(x)=-3(2x)(x^2-1)^{-2}=0$$

$$\frac{-6x}{(x^2-1)^2}=0$$

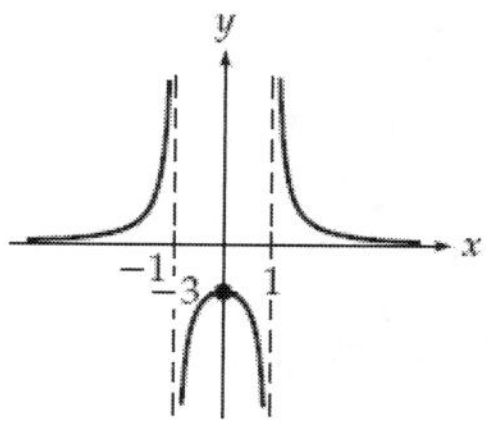

The critical numbers are $x=-1$, 0 and 1.

$f'>0$	$f'=\text{und}$	$f'>0$	$f'=0$	$f'<0$	$f'=\text{und}$	$f'<0$
╱	$x=-1$	╱	$x=0$ rel max (0, –3)	╲	$x=1$	╲

51. $f(x)=\dfrac{5x}{x^2-9}=\dfrac{5x}{(x+3)(x-3)}$

Vertical asymptote: $x=-3, 3$
Horizontal asymptote: $y=0$

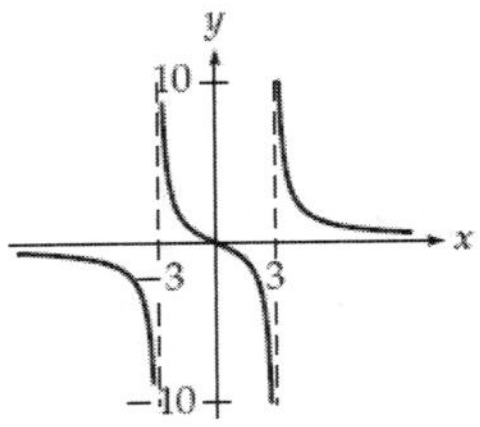

$$f'(x)=\frac{5(x^2-9)-5x(2x)}{(x^2-9)^2}=0$$
$$\frac{-5x^2-45}{(x^2-9)^2}=0$$
$$\frac{-5(x^2+9)}{(x^2-9)^2}=0$$

The critical numbers are $x=-3$, 0 and 3.

$f'<0$	$f'=$ und	$f'<0$	$f'=0$	$f'<0$	$f'=$ und	$f'<0$
↘	$x=-3$	↘	$x=0$ — neither (0, 0)	↘	$x=3$	↘

53. $f(x)=\dfrac{2x^2-4x+6}{x^2-2x-3}=\dfrac{2x^2-4x+6}{(x+1)(x-3)}$

Vertical asymptote: $x=-1, 3$
Horizontal asymptote: $y=2$

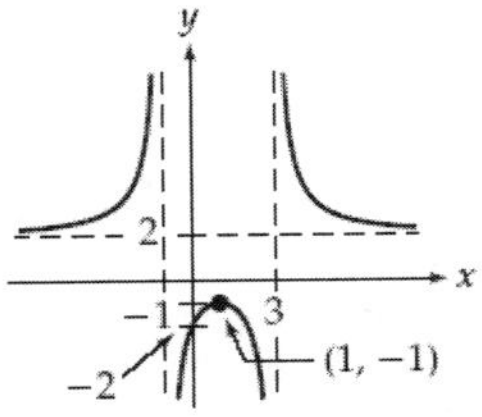

$$f'(x)=\frac{(4x-4)(x^2-2x-3)-(2x^2-4x+6)(2x-2)}{(x^2-2x-3)^2}=0$$
$$\frac{-24(x-1)}{(x^2-2x-3)^2}=0$$

The critical numbers are $x=-1$, 1 and 3.

$f'>0$	$f'=$ und	$f'>0$	$f'=0$	$f'<0$	$f'=$ und	$f'<0$
↗	$x=-1$	↗	$x=1$ — rel max (1, −1)	↘	$x=3$	↘

55. $f(x)=\dfrac{2x^2}{x^2-1}=\dfrac{2x^2}{(x+1)(x-1)}$

Vertical asymptote: $x=-1, 1$
Horizontal asymptote: $y=2$

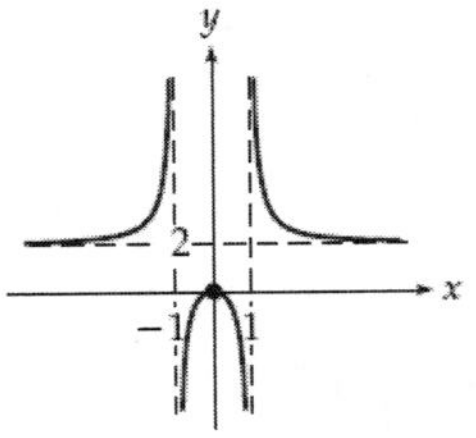

$$f'(x)=\frac{4x(x^2-1)-2x^2(2x)}{(x^2-1)^2}=0$$
$$\frac{-4x}{(x^2-4)^2}=0$$

The critical numbers are $x=-1$, 0 and 1.

$f'>0$	$f'=$ und	$f'>0$	$f'=0$	$f'<0$	$f'=$ und	$f'<0$
↗	$x=-1$	↗	$x=1$ — rel max (0, 0)	↘	$x=1$	↘

57. $f(x) = \dfrac{16}{(x+2)^3} = 16(x+2)^{-3}$

Vertical asymptote: $x = -2$
Horizontal asymptote: $y = 0$

$f'(x) = -48(x+2)^{-4} = 0$

$\dfrac{-48}{(x+2)^4} = 0$

The critical number is $x = -2$.

$f' < 0$	$f' = \text{und}$	$f' < 0$
↘	$x = -2$	↘

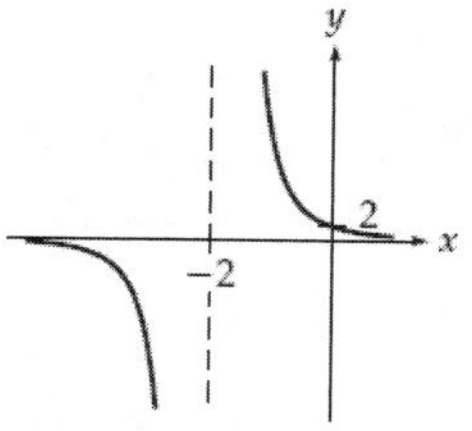

59. $f(x) = \dfrac{4}{x^2(x-3)} = 4(x^3 - 3x^2)^{-1}$

Vertical asymptote: $x = 0, 3$
Horizontal asymptote: $y = 0$

$f'(x) = -4(3x^2 - 6x)(x^3 - 3x^2)^{-2} = 0$

$\dfrac{-12(x^2 - 2x)}{x^4(x-3)^2} = 0$

$\dfrac{-12x(x-2)}{x^4(x-3)^2} = 0$

The critical numbers are $x = 0$, 2 and 3.

$f' < 0$	$f' = \text{und}$	$f' > 0$	$f' = 0$	$f' < 0$	$f' = \text{und}$	$f' < 0$
↘	$x = 0$	↗	$x = 2$	↘	$x = 3$	↘
			rel max (2, −1)			

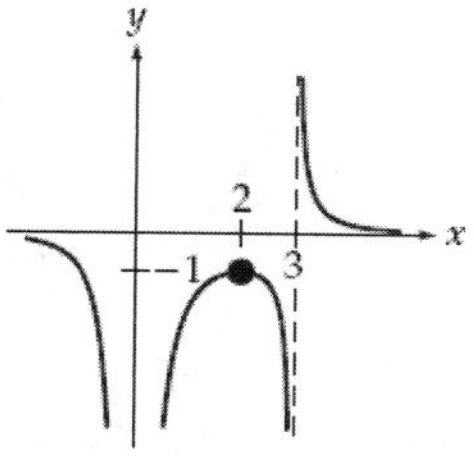

61. $f(x) = \dfrac{2x^2}{x^4 + 1}$

Vertical asymptote: none
Horizontal asymptote: $y = 0$

$f'(x) = \dfrac{4x(x^4+1) - (4x^3)(2x^2)}{(x^4+1)^2} = 0$

$\dfrac{-4x^5 + 4x}{(x^4+1)^2} = 0$

$\dfrac{-4x(x^4 - 1)}{(x^4+1)^2} = 0$

$\dfrac{-4x(x^2+1)(x^2-1)}{(x^4+1)^2} = 0$

$\dfrac{-4x(x^2+1)(x+1)(x-1)}{(x^4+1)^2} = 0$

The critical numbers are $x = -1$, 0 and 1.

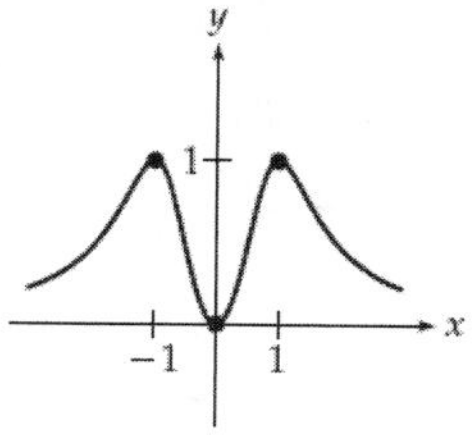

$f'>0$	$f'=0$	$f'<0$	$f'=0$	$f'>0$	$f'=0$	$f'<0$
/	$x=-1$	\	$x=0$	/	$x=1$	\
	rel max $(-1, 1)$		rel min $(0, 0)$		rel max $(1, 1)$	

63. Since the vertex occurs at a critical number, we take the derivative and set it equal to zero.

$$f(x)=ax^2+bx+c$$
$$f'(x)=2ax+b=0$$
$$2ax=-b$$
$$x=-\frac{b}{2a}$$

65. $p'(x)=3x^2-18x+24=0$

$$3(x^2-6x+8)=0$$
$$3(x-4)(x-2)=0$$
$$x=2, 4$$

The critical points are $(2,30)$ and $(4, 26)$.

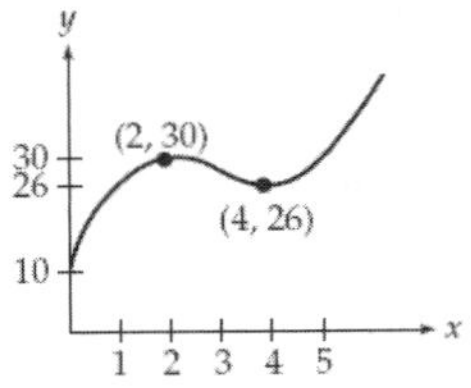

67. **i.** Since the plane is on the ground at $x=0$ $[f(0)=0]$,

$f(0)=a(0)^3=b(0)^2+c(0)+d=0$

Thus, $d=0$.

ii. Since $f'(0)=0$, we have

$f'(x)=3ax^2+2bx+c$

$f'(0)=3a(0)^2+2b(0)+c=0$

Thus, $c=0$.

iii. Since $f(100)=5$, we have $5=a(100)^3+b(100)^2$, and since $f'(100)=0$ (that is, the path is horizontal)

$$0=3a(100)^2+2b(100)$$
$$5=1,000,000a+10,000b$$
$$0=30,000a+200b$$
$$15=3,000,000a+30,000b$$
$$\underline{0=3,000,000a+20,000b}$$
$$15=10,000b$$
$$b=0.0015$$

To find a, substitute $b=0.0015$ in the equation $5=a(100)^3+b(100)^2$

$$5=a(100)^3+0.0015(100)^2$$
$$5=1,000,000a+15$$
$$-10=1,000,000a$$
$$a=-0.00001$$

Thus, the equation is $f(x)=-0.00001x^3+0.0015x^2$.

iv.

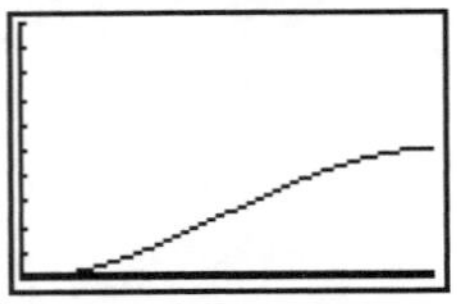

on [0,100] by [0,10]

69. **a.**

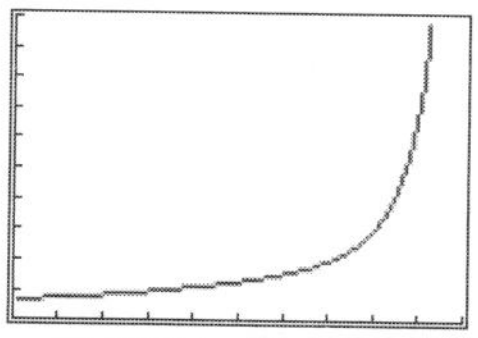

on [0, 100] by [0, 100]

b. The graph of the derivative will have small positive numbers near zero, increase slowly as x increases, and then rise steeply for numbers of x greater than 80.

c.

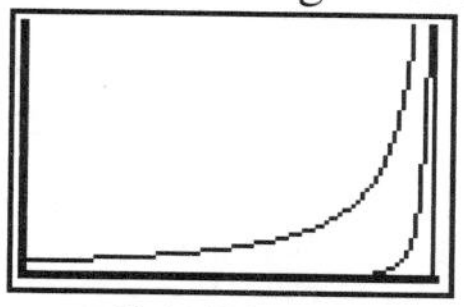

on [0,100] by [0,100]

71. **a.** $\lim\limits_{x\to\infty} AC(x) = \lim\limits_{x\to\infty} \frac{3x+50}{x} = \lim\limits_{x\to\infty}\left(3+\frac{50}{x}\right) = 3$

b.

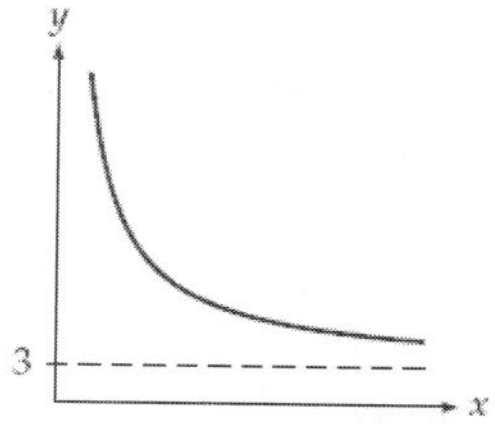

73. False: The original function must also be defined at the critical number.

75. False: The curve could have a higher relative maximum or the ends could go to infinity.

77. Decreasing on $(-\infty, 1)$ and increasing on $(1, \infty)$.

79. In a rational function, the denominator of the derivative will be the square of the denominator of the original function (from the quotient rule). Therefore, both denominators will be zero at the same x-values, making both the function and its derivative undefined at the same x-value.

81. True: The derivative must change from positive to negative for a maximum or from a negative to positive for a minimum.

83. If $p(c) = 0$ and $q(c) = 0$, there is a common factor that can be canceled out so that $\lim\limits_{x\to c} \frac{p(x)}{q(x)}$ may exist: see Example 4. Also,

$f(x) = \frac{x^2-9}{x^2-2x-3}$. $\lim\limits_{x\to 3} \frac{x^2-9}{x^2-2x-3} = \frac{0}{0}$,

however $\lim\limits_{x\to 3} \frac{(x+3)\cancel{(x-3)}}{\cancel{(x-3)}(x+1)} = \lim\limits_{x\to 3} \frac{x+3}{x+1} = \frac{6}{4} = \frac{3}{2}$.

Therefore $f(x)$ does not have a vertical asymptote.

85. **d.**

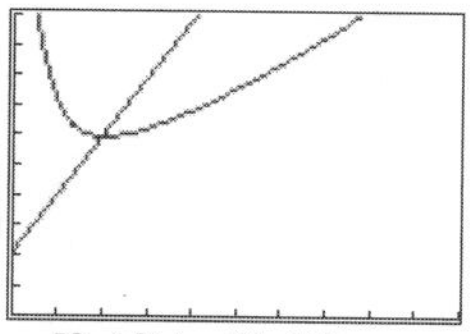

on [0, 10] by [0, 10]

EXERCISES 3.2

1. Point 2

3. Points 3 and 5

5. Points 4 and 6

7. $f(x) = x^3 + 3x^2 - 9x + 5$

$f'(x) = 3x^2 + 6x - 9 = 0$

$3(x^2 + 2x - 3) = 0$

$3(x+3)(x-1) = 0$

The critical values are $x = -3$ and 1.

$f' > 0$	$f' = 0$	$f' < 0$	$f' = 0$	$f' > 0$
	$x = -3$		$x = 1$	
/	rel max (–3, 32)	\	rel min (1, 0)	/

$f'' < 0$	$f'' = 0$	$f'' > 0$
	$x = -1$	
con dn	IP(–1, 16)	con up

$f''(x) = 6x + 6 = 6(x+1) = 0$

Inflection point at $x = -1$. Test points:

$x = 0$: $f''(0) = 6(0) + 6 = 6 > 0$, concave up

$x = -2$: $f''(-2) = 6(-2) + 6 = -6 < 0$, concave down

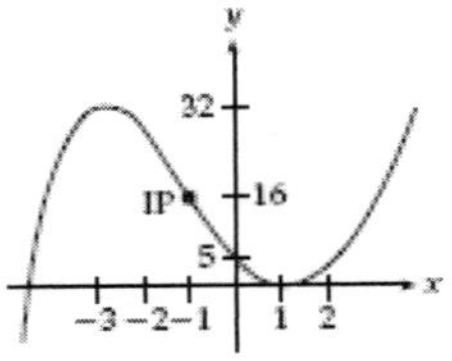

9. $f(x) = x^3 - 3x^2 + 3x + 4$

$f'(x) = 3x^2 - 6x + 3 = 0$

$3(x^2 - 2x + 1) = 0$

$3(x-1)^2 = 0$

The critical value is $x = 1$.

$f' > 0$	$f' = 0$	$f' > 0$
	$x = 1$	
/	neither (1,5)	/

$f'' < 0$	$f'' = 0$	$f'' > 0$
	$x = 1$	
con dn	IP(1, 5)	con up

$f''(x) = 6x - 6 = 6(x-1) = 0$

Inflection point at $x = 1$. Test points:

$x = 0$: $f'(0) = 6(0) - 6 = -6 < 0$, concave down

$x = 2$: $f'(2) = 6(2) - 6 = 6 > 0$, concave up

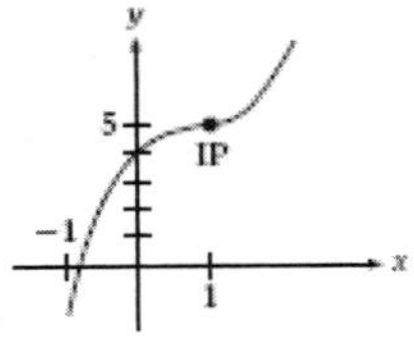

11. $f(x) = x^4 - 8x^3 + 18x^2 + 2$

$f'(x) = 4x^3 - 24x^2 + 36x = 0$

$4x(x^2 - 6x + 9) = 0$

$4x(x-3)^2 = 0$

The critical values are $x = 0$ and 3.

$f''(x) = 12x^2 - 48x + 36 = 0$

$12(x^2 - 4x + 3) = 0$

$12(x-3)(x-1) = 0$

Inflection points at $x = 1$ and 3. Test points:

$x = 0$: $f''(0) = 12(0)^2 - 48(0) + 36 = 36 > 0$, concave up

$x = 2$: $f''(2) = 12(2)^2 - 48(2) + 36 = -12 < 0$, concave down

$x = 4$: $f''(4) = 12(4)^2 - 48(4) + 36 = 36 > 0$, concave up

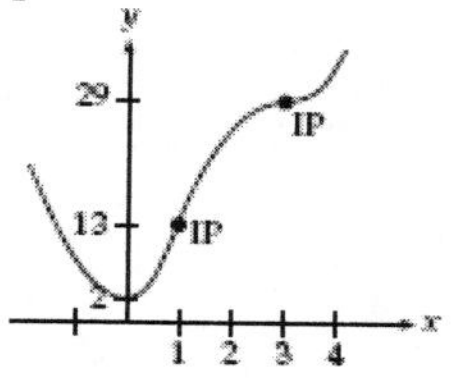

$f' < 0$	$f' = 0$	$f' > 0$	$f' = 0$	$f' > 0$
╲	$x = 0$	╱	$x = 3$	╱
	rel min (0, 2)		neither (3, 29)	

$f'' > 0$	$f'' = 0$	$f'' < 0$	$f'' = 0$	$f'' > 0$
	$x = 1$		$x = 3$	
con up	IP(1, 13)	con dn	IP(3, 29)	con up

13. $f(x) = x^4 + 4x^3 + 15$

$f'(x) = 4x^3 + 12x^2 = 0$

$4x^2(x+3) = 0$

The critical values are $x = 0$ and -3.

$f''(x) = 12x^2 + 24x = 0$

$24x(x+2) = 0$

Inflection points at $x = 0$ and -2. Test points:

$x = -3$: $f''(-3) = 12(-3)^2 + 24(-3) > 0$, concave up

$x = -1$: $f''(-1) = 12(-1)^2 + 24(-1) < 0$, concave down

$x = 1$: $f''(1) = 12(1)^2 + 24(1) > 0$, concave up

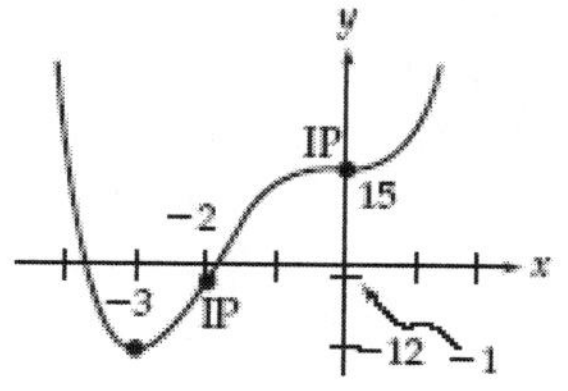

$f' < 0$	$f' = 0$	$f' > 0$	$f' = 0$	$f' > 0$
╲	$x = -3$	╱	$x = 0$	╱
	rel min (−3, −12)		neither (0, 0)	

$f'' > 0$	$f'' = 0$	$f'' < 0$	$f'' = 0$	$f'' > 0$
	$x = -2$		$x = 0$	
con up	IP(−2, −1)	con dn	IP(0, 0)	con up

15. $f(x) = 5x^4 - x^5$

$f'(x) = 20x^3 - 5x^4$

$5x^3(4 - x) = 0$

The critical values are at $x = 0$ and 4.

$f' < 0$	$f' = 0$	$f' > 0$	$f' = 0$	$f' < 0$
	$x = 0$		$x = 4$	
↘	rel min (0, 0)	↗	rel max (4, 256)	↘

$f'' > 0$	$f'' = 0$	$f'' > 0$	$f'' = 0$	$f'' < 0$
	$x = 0$		$x = 3$	
con up		con up	IP(3, 162)	con dn

$f''(x) = 60x^2 - 20x^3 = 20x^2(3 - x) = 0$

Possible inflection points at $x = 0$ and 3. Test points:

$x = -1$: $60(-1)^2 - 20(-1)^3 = 80 > 0$, concave up

$x = 1$: $60(1)^2 - 20(1)^3 = 40 > 0$, concave up

$x = 4$: $60(4)^2 - 20(4)^3 = -320 < 0$, concave down

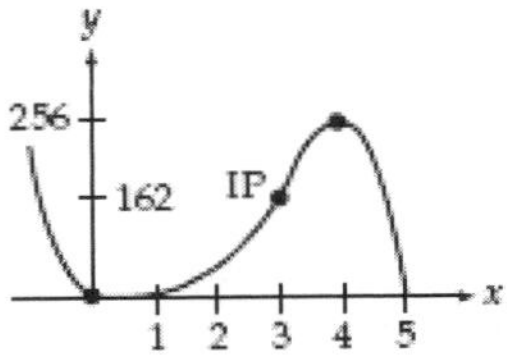

17. $f(x) = (2x + 4)^5$

$f'(x) = 5(2x + 4)^4(2) = 0$

$10(2x + 4)^4 = 0$

The critical point is at $x = -2$.

$f' > 0$	$f' = 0$	$f' > 0$
	$x = -2$	
↗	neither (–2, 0)	↗

$f'' < 0$	$f'' = 0$	$f'' > 0$
	$x = -2$	
con dn	IP(–2, 0)	con up

$f''(x) = 40(2x + 4)^3(2)$

$= 80(2x + 4)^3 = 0$

Inflection point at $x = -2$. Test points:

$x = -3$: $f''(-3) = 80[2(-3) + 4]^3 = -640 < 0$, concave down

$x = 0$: $f''(0) = 80[2(0) + 4]^3 = 5120 > 0$, concave up

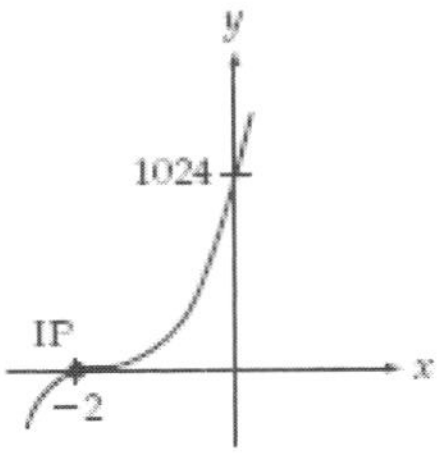

19. $f(x) = x(x-3)^2$

$$f'(x) = (x-3)^2 + x[2(x-3)] = 0$$
$$(x-3)(x-3+2x) = 0$$
$$(x-3)(3x-3) = 0$$

The critical points are at $x = 1$ and 3.

$f' > 0$	$f' = 0$	$f' < 0$	$f' = 0$	$f' > 0$
/	$x = 1$	\	$x = 3$	/
	rel max (1, 4)		rel min (3, 0)	

$f'' < 0$	$f'' = 0$	$f'' > 0$
	$x = 2$	
con dn		con up
	IP(2 2)	

$$f''(x) = 2(x-3) + 2(x-3) + 2x$$
$$= 6x - 12 = 0$$

Inflection point at $x = 2$. Test points:

$x = 0$: $f''(0) = 6(0) - 12 = -12 < 0$, concave down

$x = \frac{3}{2}$: $f''\left(\frac{3}{2}\right) = 6\left(\frac{3}{2}\right) - 12 = -3 < 0$, concave down

$x = \frac{5}{2}$: $f''\left(\frac{5}{2}\right) = 6\left(\frac{5}{2}\right) - 12 = 3 > 0$, concave up

$x = 4$: $f''(4) = 6(4) - 12 = 12 > 0$, concave up

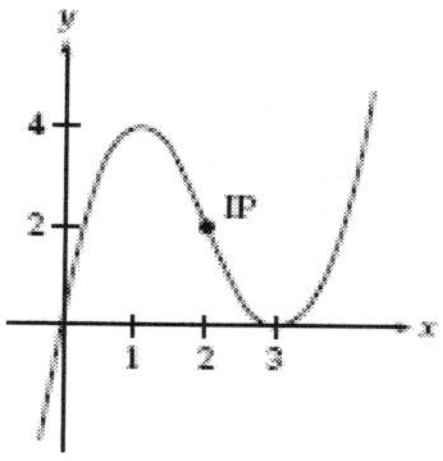

21. $f(x) = x^{3/5}$

$$f'(x) = \frac{3}{5}x^{-2/5}$$
$$= \frac{3}{5\sqrt[5]{x^2}} = 0$$

There are no critical points. f' is undefined at $x = 0$.

$f' > 0$	$f' = 0$	$f' > 0$
/	$x = 0$	/
	neither (0, 0)	

$f'' > 0$	f'' und	$f'' < 0$
	$x = 0$	
con up		con dn
	IP(0, 0)	

$$f''(x) = -\frac{6}{25}x^{-7/5} = -\frac{6}{25\sqrt[5]{x^7}} = 0$$

f'' is undefined at $x = 0$. Test points:

$x = -1$: $f''(-1) = -\frac{6}{25}(-1)^{-7/5} = \frac{6}{25} > 0$, concave up

$x = 1$: $f''(1) = -\frac{6}{25}(1)^{-7/5} = -\frac{6}{25} < 0$, concave down

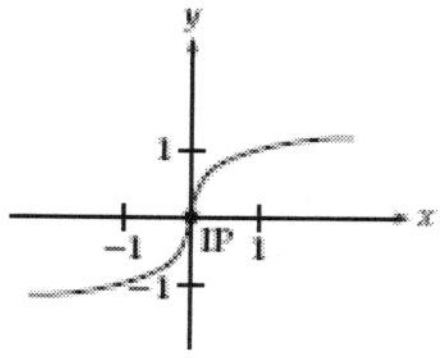

23. $f(x) = \sqrt[5]{x^4} + 2 = x^{4/5} + 2$

$f'(x) = \frac{4}{5}x^{-1/5} = 0$

f' is undefined at $x = 0$.

$f' < 0$	f' und	$f' > 0$
	$x = 0$	
↘	rel min (0, 2)	↗

$f'' < 0$	f'' und	$f'' < 0$
	$x = 0$	
con dn		con dn

$f''(x) = -\frac{4}{25}x^{-6/5} = 0$

f'' is undefined at $x = 0$. Test points:

$x = -1$: $f''(-1) = -\frac{4}{25}(-1)^{-6/5} = -\frac{4}{25} < 0$, concave down

$x = 1$: $f''(1) = -\frac{4}{25}(1)^{-6/5} = -\frac{4}{25} < 0$, concave down

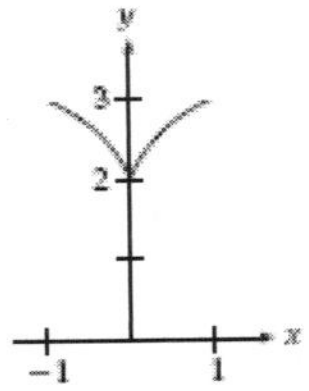

25. $f(x) = \sqrt[4]{x^3} = x^{3/4}$

$f'(x) = \frac{3}{4}x^{-1/4} = 0$

f' is undefined at $x = 0$, and f is defined only for values of $x \geq 0$.

f' und	$f' > 0$
$x = 0$	
	↗

f'' und	$f'' < 0$
$x = 0$	
	con dn

$f''(x) = -\frac{3}{16}x^{-5/4} = 0$

f'' is undefined at $x = 0$. Test point:

$x = 1$: $f''(1) = -\frac{3}{16}(1)^{-5/4} = -\frac{3}{16} < 0$, concave down

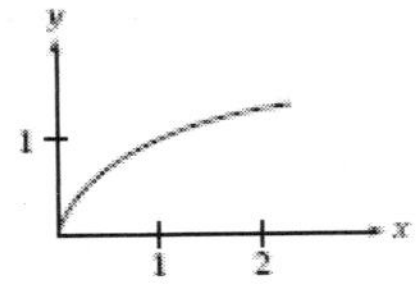

27. $f(x) = \sqrt[3]{(x-1)^2} = (x-1)^{2/3}$

$f'(x) = \frac{2}{3}(x-1)^{-1/3} = 0$

f' is undefined at $x = 1$.

$f' < 0$	f' und	$f' > 0$
	$x = 1$	
↘	rel min (1, 0)	↗

$f'' < 0$	f'' und	$f'' < 0$
	$x = 1$	
con dn		con dn

$f''(x) = -\frac{2}{9}(x-1)^{-4/3} = 0$

f'' is undefined at $x = 1$. Test points:

$x = 0$: $f''(0) = -\frac{2}{9}(0-1)^{-4/3} = -\frac{2}{9} < 0$, concave down

$x = 2$: $f''(2) = -\frac{2}{9}(2-1)^{-4/3} = -\frac{2}{9} < 0$, concave down

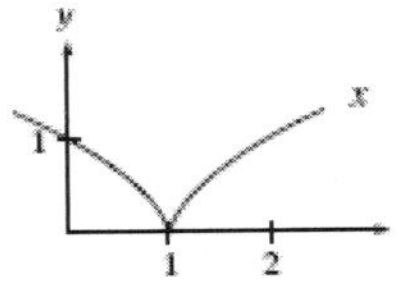

29. $f(x)=\dfrac{x}{x^2-1}=\dfrac{x}{(x+1)(x-1)}$

Vertical asymptotes: $x=-1,\ 1$.
Horizontal asymptote: $y=0$.

$$f'(x)=\frac{1(x^2-1)-x(2x)}{(x^2-1)^2}=0$$

$$\frac{-x^2-1}{(x^2-1)^2}=0$$

The critical points are $x=-1$, 0, and 1.

$$f''(x)=\frac{-2x(x^2-1)^2-(-x^2-1)2(x^2-1)2x}{(x^2-1)^4}=0$$

$$\frac{6x}{(x^2-1)^3}=0$$

f'' is undefined at $x=-1$ and 1.
Inflection point at $x=0$. Test points:

$x=-2:\ f''(-2)=\dfrac{6(-2)}{((-2)^2-1)^3}<0$, concave down

$x=-0.5:\ f''(-0.5)=\dfrac{6(-0.5)}{((-0.5)^2-1)^3}>0$, concave up

$x=0.5:\ f''(0.5)=\dfrac{6(0.5)}{((0.5)^2-1)^3}<0$, concave down

$x=2:\ f''(2)=\dfrac{6(-2)}{((-2)^2-1)^3}>0$, concave up

$f'<0$	$f'=$ und	$f'<0$	$f'=0$	$f'<0$	$f'=$ und	$f'<0$
╲	$x=-1$	╲	$x=0$	╲	$x=1$	╲
			neither (0, 0)			

$f''<0$	$f''=$ und	$f''>0$	$f''=0$	$f''<0$	$f''=$ und	$f''>0$
con dn	$x=-1$	con up	$x=0$	con dn	$x=1$	con up
			IP (0, 0)			

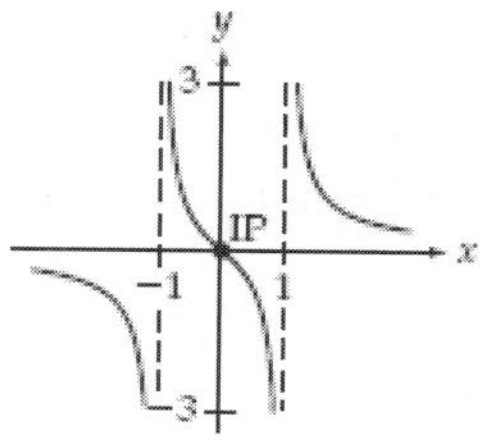

31. $f(x)=x^{1/2}$

$f'(x)=\frac{1}{2}x^{-1/2}$

f' is undefined at $x=0$.

f' und	$f'>0$
$x=0$	╱

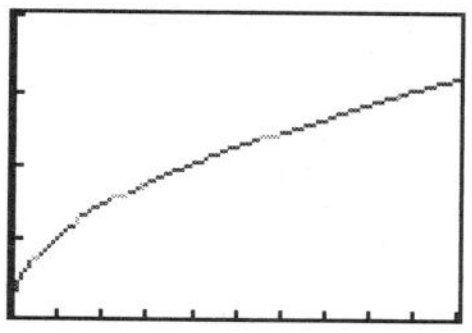
on [0, 10] by [0, 4]

33. $f(x)=x^{-1/2}$

$f'(x)=-\frac{1}{2}x^{-3/2}$

f' is undefined at $x=0$.

f' und	$f'<0$
$x=0$	╲

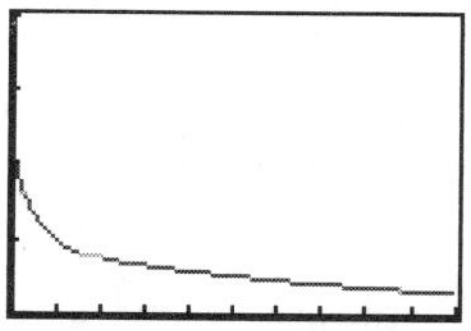
on [0, 10] by [0, 4]

35. $f(x) = 9x^{2/3} - 6x$

$f'(x) = 6x^{-1/3} - 6 = 0$

$6(x^{-1/3} - 1) = 0$

The critical point is $x = 1$, and f' is undefined at $x = 0$.

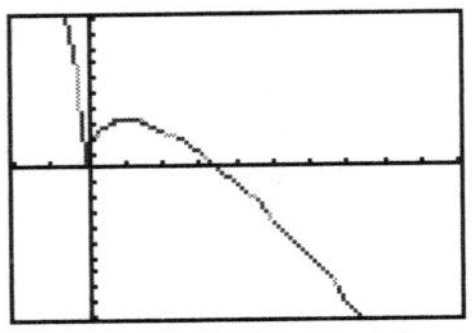

on [–2, 10] by [–10, 10]

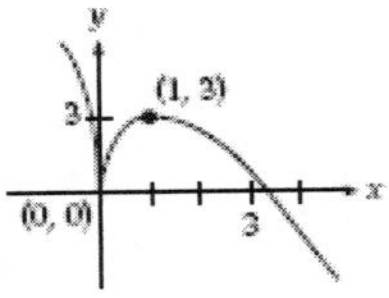

$f' < 0$	f' und	$f' > 0$	$f' = 0$	$f' < 0$
	$x = 0$		$x = 1$	
↘	rel min (0, 0)	↗	rel max (1, 3)	↘

37. $f(x) = 8x - 10x^{4/5}$

$f'(x) = 8 - 8x^{-1/5} = 0$

$8(1 - x^{-1/5}) = 0$

The critical point is $x = 1$, and f' is undefined at $x = 0$.

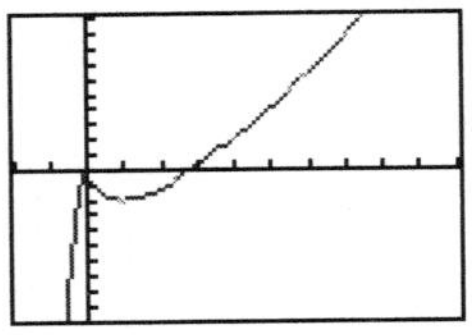

on [–2, 10] by [–10, 10]

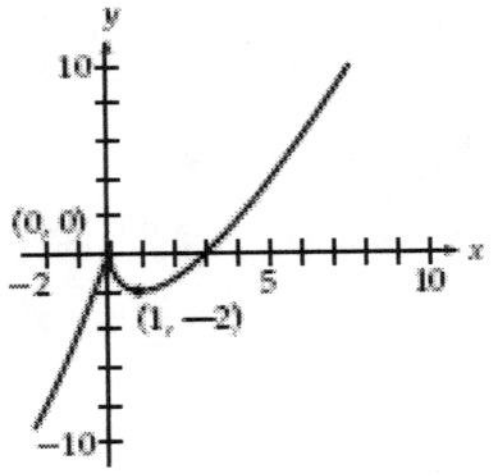

$f' > 0$	f' und	$f' < 0$	$f' = 0$	$f' > 0$
	$x = 0$		$x = 1$	
↗	rel max (0, 0)	↘	rel min (1, –2)	↗

39. $f(x) = 3x^{2/3} - x^2$

$f'(x) = 2x^{-1/3} - 2x = 0$

$2x(x^{-4/3} - 1) = 0$

The critical points are $x = \pm 1$, and f' is undefined at $x = 0$.

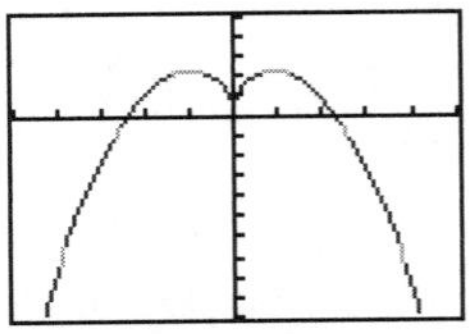

on [–5, 5] by [–10, 5]

$f' > 0$	$f' = 0$	$f' < 0$	f' und	$f' > 0$	$f' = 0$	$f' < 0$
	$x = -1$		$x = 0$		$x = 1$	
↗	rel max (–1, 2)	↘	rel min (0, 0)	↗	rel max (1, 2)	↘

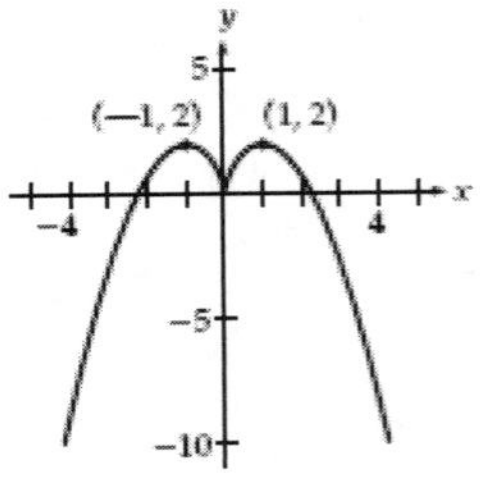

41. $f(x) = x^3 - 6x^2 + 9x - 2$

$f'(x) = 3x^2 - 12x + 9$

$3x^2 - 12x - 9 = 0$

$3(x^2 - 4x - 3) = 0$

$3(x-1)(x-3) = 0$

$x = 1,\ 3$

The critical numbers are $x = 1$ and 3.

$f''(x) = 6x - 12$

Using the second derivative test:

$f''(1) = 6(1) - 12 < 0$ relative maximum at $x = 1$

$f''(3) = 6(3) - 12 > 0$ relative minimum at $x = 3$

43. $f(x) = x^4 - 4x^3 + 4x^2 + 1$

$f'(x) = 4x^3 - 12x^2 + 8x$

$4x^3 - 12x^2 + 8x = 0$

$4x(x^2 - 3x + 2) = 0$

$4x(x-1)(x-2) = 0$

$x = 0,\ 1,\ 2$

The critical numbers are $x = 0$, 1 and 2.

$f''(x) = 12x^2 - 24x + 8$

Using the second derivative test:

$f''(0) = 12(0)^2 - 24(0) + 8 > 0$ relative minimum at $x = 0$

$f''(1) = 12(1)^2 - 24(1) + 8 < 0$ relative maximum at $x = 1$

$f''(2) = 12(2)^2 - 24(2) + 8 > 0$ relative minimum at $x = 2$

45. $f(x) = x + \frac{9}{x}$

$f'(x) = 1 - \frac{9}{x^2} = \frac{x^2 - 9}{x^2}$

$\frac{x^2 - 9}{x^2} = 0$

$\frac{(x+3)(x-3)}{x^2} = 0$

$x = -3,\ 3$; undefined at $x = 0$

The critical numbers are $x = -3$ and 3.

$f''(x) = \frac{18}{x^3}$

Using the second derivative test:

$f''(-3) = \frac{18}{(-3)^3} < 0$ relative maximum at $x = -3$

$f''(3) = \frac{18}{(3)^3} > 0$ relative minimum at $x = 3$

47. $f(x) = x^5 - 2x^3 + 3x + 4$
$f'(x) = 5x^4 - 6x^2 + 3$
$f''(x) = 20x^3 - 12x = 0$
$4x(5x^2 - 3) = 0$
Inflection points at $x = 0, -0.77$, and 0.77

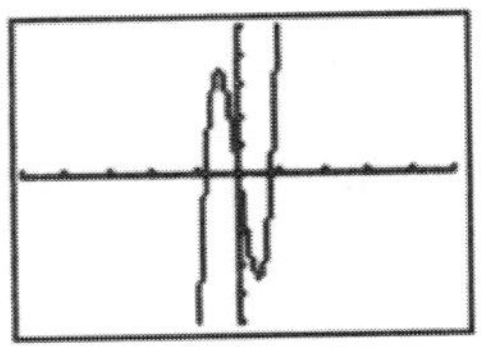
on [–5, 5] by [–5, 5]

49.

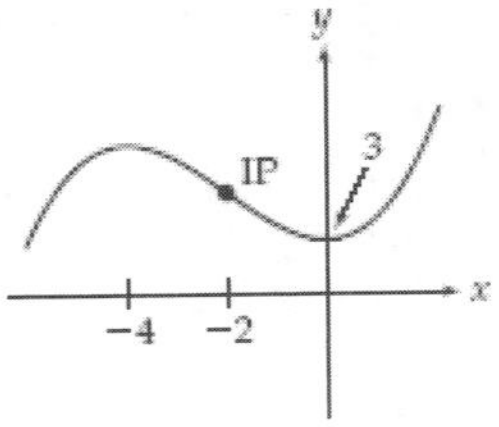

51.

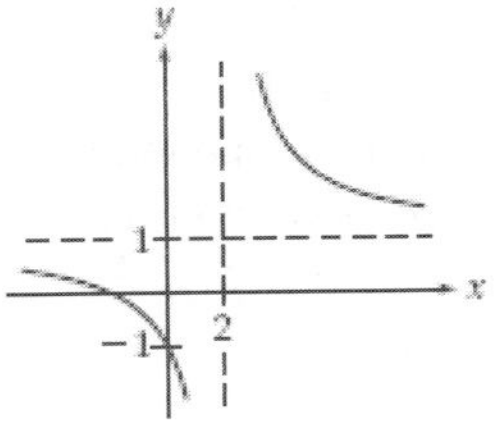

53.

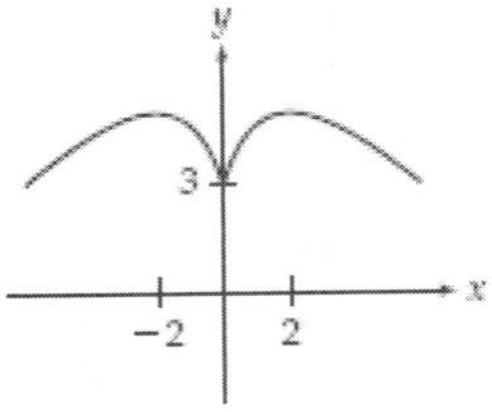

55.

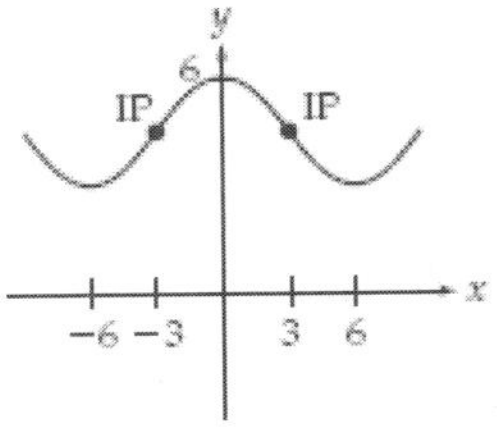

57. **a.** $f(x) = x^3 - 9x^2 + 15x + 25$
$f'(x) = 3x^2 - 18x + 15 = 0$
$3(x^2 - 6x + 5) = 0$
$3(x - 5)(x - 1) = 0$
The critical points are 1 and 5.

$f' > 0$	$f' = 0$	$f' < 0$	$f' = 0$	$f' > 0$
	$x = 1$		$x = 5$	
/	—	\	—	/
	rel max (1, 32)		rel min (5, 0)	

$f'' < 0$	$f'' = 0$	$f'' > 0$
	$x = 3$	
con dn	IP(3, 16)	con up

$f''(x) = 6x - 18 = 6(x - 3) = 0$
Inflection point at $x = 3$. Test points:
$x = 2$: $f''(2) = 6(2) - 18 = -6 < 0$, concave down
$x = 4$: $f''(4) = 6(4) - 18 = 6 > 0$, concave up

b.

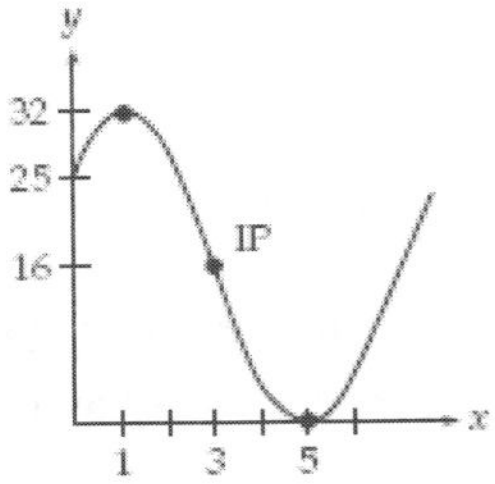

c. The decline in revenue first begins to slow at $x = 3$ years.

59. **a.** $f(x) = x^4 - 4x^3 + 112$

$f'(x) = 4x^3 - 12x^2 = 0$

$4x^2(x-3) = 0$

The critical points are 0 and 3.

$f''(x) = 12x^2 - 24x = 0$

$12x(x-2) = 0$

Possible inflection points at $x = 0$ and 2. Test points:

$x = 1$: $f''(1) = 12(1)(1-2) < 0$, concave down

$x = 3$: $f''(3) = 12(3)(3-2) > 0$, concave up

Inflection points at $x = 0$ and 2.

$f' < 0$	$f' = 0$	$f' < 0$	$f' = 0$	$f' > 0$
	$x = 0$		$x = 3$	
\	neither (0, 112)	\	rel max (3, 96)	/

$f'' > 0$	$f'' = 0$	$f'' < 0$	$f'' = 0$	$f'' > 0$
con up	$x = 0$	con dn	$x = 0$	con up
	IP(0, 112)		IP(2,85)	

b.

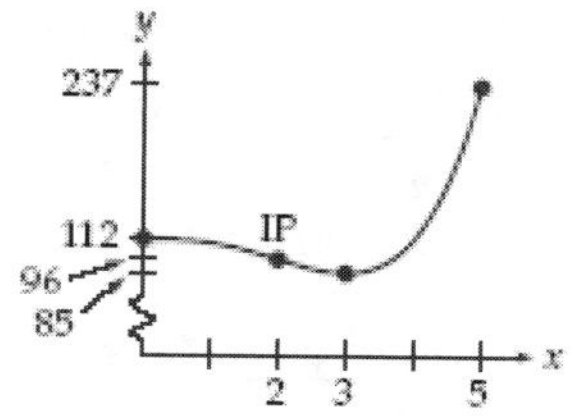

c. The decline in temperature first begins to slow at $x = 2$ hours.

61. $f(x) = x^{2/5}$

$f\ (x) = \frac{2}{5}x^{-3/5}$

f' is undefined at $x = 0$ and f is defined only for $x \geq 0$.

$f''(x) = -\frac{6}{25}x^{-8/5}$

f'' is undefined at $x = 0$. Test point:

$x = 1$: $f\,'(1) = -\frac{6}{25}(1)^{-8/5} = -\frac{6}{25} < 0$, concave down

f' und	$f' > 0$
$x = 0$ rel min (0, 0)	/

f'' und	$f'' < 0$
$x = 0$	con dn

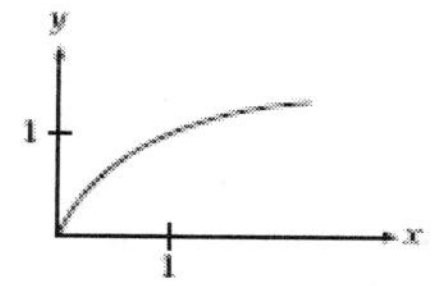

63. **a.** $S(i) = 16\sqrt{i} = 16i^{1/2}$

$S'(i) = 8i^{-1/2}$

S' is undefined at $i = 0$ and S is defined only for $i \geq 0$.

b. Concave down, so status increases more slowly at higher income levels.

$S''(i) = -4i^{-3/2}$

S'' is undefined at $i = 0$. Test point:

$i = 1$: $S\,'(1) = -4(1)^{-3/2} = -4 < 0$, concave down

S' und	$S' > 0$
$i = 0$	/

S'' und	$S'' < 0$
$i = 0$	con dn

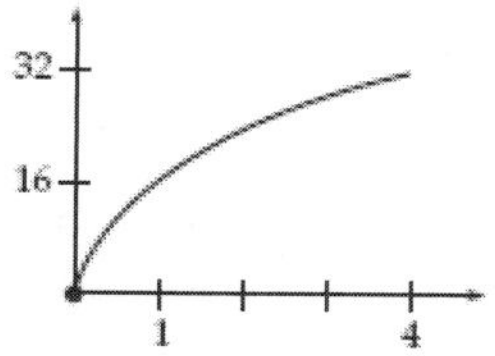

65. **a.** $I(x) = 2x^3 - 18x^2 + 48x + 220$

$I'(x) = 6x^2 - 36x + 48 = 0$

$6(x^2 - 6x + 8) = 0$

$6(x-2)(x-4) = 0$

The critical points are 2 and 4.

$I' > 0$	$I' = 0$	$I' < 0$	$I' = 0$	$I' > 0$
	$x = 2$		$x = 4$	
/	— rel max (2, 260)	\	— rel min (4, 252)	/

$I'' < 0$	$I'' = 0$	$I'' > 0$
con dn	$x = 3$	con up
	IP (3, 256)	

$I''(x) = 12x - 36 = 0$

$12(x-3) = 0$

Possible inflection point at $x = 3$. Test points:

$x = 0$: $I''(0) = 12(0) - 36 < 0$, concave down

$x = 4$: $I''(4) = 12(4) - 36 > 0$, concave up

Inflection point at $x = 3$.

b.

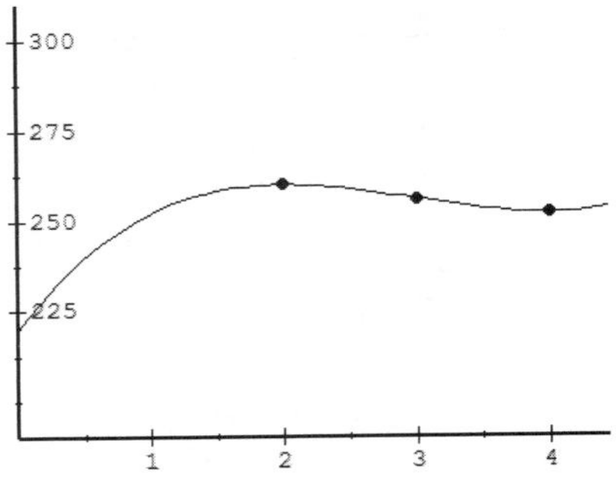

c. The inflection point in 2008 is the first sign that the decline in semiconductor sales is beginning to slow.

67. $R(x) = 10 - 3x + 24x^2 - x^3$

$R'(x) = -3 + 48x - 3x^2$

$R''(x) = 48 - 6x$

$= -6(x - 8)$

The second derivative is zero at $x = 8$. Using test points $x = 7$ and $x = 9$ on either side of $x = 8$ shows that the second derivative changes from positive to negative, so the revenue curve changes from concave up to concave down, giving a point of diminishing returns. Since x is in hundreds of thousands of dollars, the point of diminishing returns is when \$800,000 is spent on advertising.

69. $P(x) = 6x + 18x^2 - 1.5x^3$

$P'(x) = 6 + 36x - 4.5x^2$

$P''(x) = 36 - 9x$

$= -9(x - 4)$

The second derivative is zero at $x = 4$. Using test points $x = 3$ and $x = 5$ on either side of $x = 4$ shows that the second derivative changes from positive to negative, so the revenue curve changes from concave up to concave down, giving a point of diminishing returns. Since x is in millions of dollars, the point of diminishing returns is when \$4,000,000 is spent on marketing.

71. $y = -0.00001x^3 + 0.0015x^2$

$y' = -0.00003x^2 + 0.003x$

$y'' = -0.00006x + 0.003 = 0$

$x = \frac{0.003}{0.00006} = 50$

At $x = 50$,

$y = -0.00001(50)^3 + 0.0015(50)^2 = 2.5$. The inflection point is (50, 2.5). The curve is concave up before $x = 50$ and concave down after $x = 50$. Therefore, the slope is maximized at $x = 50$.

73. $f(x) = ax^2$

$f'(x) = 2ax$

$f''(x) = 2a$

For the function to be concave up, f'' must be greater than zero.

$f'' > 0$

$2a > 0$

$a > 0$

The function will be concave down when f'' is less than zero.

$f'' < 0$

$2a < 0$

$a < 0$

The function will be concave up for $a > 0$ and concave down for $a < 0$.

75. False: The sign of the second derivative must actually change (positive on one side and negative on the other) for the point to be an inflection point.

77. (1, 0) is an inflection point because the second derivative changes from negative to positive, so the concavity does change.

79. The inflection point was the sales starting to grow slower, and was when the change of sales were growing fastest.

81. True: The derivative of the slope changes from positive to negative, so the slope would have a maximum at the inflection point.

83. $f(x) = ax^2 + bx + c$
$f'(x) = 2ax + b$
$f''(x) = 2a$
If $a > 0$, then f is concave up.
If $a < 0$, then f is concave down.

85. When the curve is concave up, it lies *above* its tangent line, and where it is concave down, it lies *below* its tangent line, so at an inflection point, it must cross from above to below its tangent line.

EXERCISES 3.3

1. f is continuous and [–1, 2] is a closed interval. Find the critical values.
$f'(x) = 3x^2 - 12x + 9$
$= 3(x^2 - 4x + 3)$
The critical values are $x = 1$ and 3. Eliminate 3, which is not in the interval.
CV: $x = 1$ $\quad f(1) = (1)^3 - 6(1)^2 + 9(1) + 8$
$= 1 - 6 + 9 + 8 = 12$ maximum
EP: $x = 2$ $\quad f(2) = (2)^3 - 6(2)^2 + 9(2) + 8$
$= 8 - 24 + 18 + 8 = 10$
$x = -1$ $\quad f(-1) = (-1)^3 - 6(-1)^2 + 9(-1) + 8 = -8$ minimum

3. f is continuous and [–3, 3] is a closed interval. Find the critical values.
$f'(x) = 3x^2 - 12 = 3(x^2 - 4)$
The critical values are $x = -2$ and 2.
CV: $x = -2$ $\quad f(-2) = (-2)^3 - 12(-2) = -8 + 24 = 16$ maximum
$x = 2$ $\quad f(2) = (2)^3 - 12(2) = 8 - 24 = -16$ minimum
EP: $x = -3$ $\quad f(-3) = (-3)^3 - 12(-3) = -27 + 36 = 9$
$x = 3$ $\quad f(3) = (3)^3 - 12(3) = 27 - 36 = -9$

5. f is continuous and [–2, 1] is a closed interval. Find the critical values.
$f'(x) = 4x^3 + 12x^2 + 8x$
$= 4x(x^2 + 3x + 2) = 4x(x + 2)(x + 1)$
The critical values are $x = 0, -2,$ and -1.
CV: $x = 0$ $\quad f(0) = (0)^4 + 4(0)^3 + 4(0)^2 = 0$ minimum
$x = -2$ $\quad f(-2) = (-2)^4 + 4(-2)^3 + 4(-2)^2 = 16 - 32 + 16 = 0$ minimum
$x = -1$ $\quad f(-1) = (-1)^4 + 4(-1)^3 + 4(-1)^2 = 1 - 4 + 4 = 1$
EP: $x = 1$ $\quad f(1) = (1)^4 + 4(1)^3 + 4(1)^2 = 1 + 4 + 4 = 9$ maximum

7. f is continuous and [–4, 1] is a closed interval.
Find the critical values.

$f'(x) = 4x^3 + 12x^2$
$= 4x^2(x+3)$

The critical values are $x = 0$ and $x = -3$.

CV: $x = 0$ $f(0) = (0)^4 + 4(0)^3 + 15 = 15$
$x = -3$ $f(-3) = (-3)^4 + 4(-3)^3 + 15 = -12$ minimum
EP: $x = -4$ $f(-4) = (-4)^4 + 4(-4)^3 + 15 = 15$
$x = 1$ $f(1) = (1)^4 + 4(1)^3 + 15 = 2$ maximum

9. f is continuous and [–1,3] is a closed interval.
Find the critical values.

$f'(x) = 10x^4 - 20x^3$
$= 10x^3(x-2)$

The critical values are $x = 0$ and $x = 2$.

CV: $x = 0$ $f(0) = 2(0)^5 - 5(0)^4 = 0$
$x = 2$ $f(2) = 2(32) - 5(16) = -16$ minimum
EP: $x = -1$ $f(-1) = 2(-1) - 5(1) = -7$
$x = 3$ $f(3) = 2(243) - 5(81) = 81$ maximum

11. f is continuous and [0, 5] is a closed interval.
Find the critical values.

$f'(x) = 6x - 3x^2 = 3x(2-x)$

The critical values are $x = 0$ and $x = 2$.

CV: $x = 0$ $f(0) = 3(0)^2 - 0^3 = 0$
$x = 2$ $f(2) = 3(2)^2 - 2^3 = 12 - 8 = 4$ maximum
EP: $x = 5$ $f(5) = 3(5)^2 - 5^3 = 75 - 125 = -50$ minimum

13. f is continuous and [0, 5] is a closed interval.
Find the critical values.

$f'(x) = -1$

Since $f'(x) = -1$, there are no critical values.

EP: $x = 0$ $f(0) = 5 - 0 = 5$ maximum
$x = 5$ $f(5) = 5 - 5 = 0$ minimum

15. f is continuous and [–1, 3] is a closed interval.
Find the critical values.

$f'(x) = 6x - 3x^2 = 3x(2-x)$

The critical values are $x = 0$ and 2.

CV: $x = 0$ $f(0) = (0)^2(3-0) = 0$ minimum
$x = 2$ $f(2) = (2)^2(3-2) = 4$
EP: $x = -4$ $f(-4) = (-4)^2(3+4) = 112$ maximum
$x = 1$ $f(1) = (1)^2(3-1) = 2$

17. f is continuous and [–1, 1] is a closed interval.
Find the critical values.

$f'(x) = 2(x^2-1)(2x) = 4x(x^2-1)$

The critical values are $x = -1$, 0, and 1.

CV: $x = -1$ $f(-1) = \left[(-1)^2 - 1^2\right] = 0$ minimum
$x = 1$ $f(1) = \left[(1)^2 - 1^2\right] = 0$ minimum
$x = 0$ $f(0) = \left[(0)^2 - 1^2\right] = 1$ maximum

19. f is continuous and [–3, 3] is a closed interval.
Find the critical values.

$$f'(x) = \frac{(x^2+1)(1) - 2x(x)}{(x^2+1)^2} = \frac{1-x^2}{(x^2+1)^2}$$

The critical values are $x = -1$ and 1.

CV: $x = -1$ $f(-1) = \frac{-1}{(-1)^2+1} = -\frac{1}{2}$ minimum
$x = 1$ $f(1) = \frac{1}{(1)^2+1} = \frac{1}{2}$ maximum
EP: $x = -3$ $f(-3) = \frac{-3}{(-3)^2+1} = -\frac{3}{10}$
$x = 3$ $f(3) = \frac{3}{(3)^2+1} = \frac{3}{10}$

21. Let $x =$ the number. We wish to consider $f(x) = x - x^2$, which is continuous and [0, 3] is closed. Find the critical values.

$f'(x) = 1 - 2x$

The critical value is $x = \frac{1}{2}$.

CV: $x = \frac{1}{2}$ $f\left(\frac{1}{2}\right) = \frac{1}{2} - \frac{1}{4} = \frac{1}{4}$ maximum
EP: $x = 0$ $f(0) = 0 - 0 = 0$
$x = 3$ $f(3) = 3 - 9 = -6$ minimum

a. $\frac{1}{2}$

b. 3

23. **a.** Both at end points

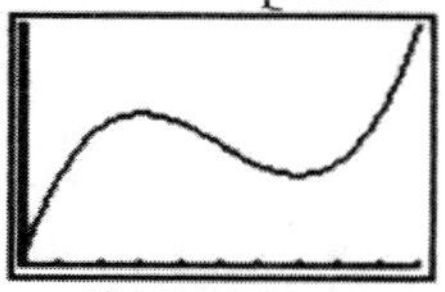

on [0,10] by [0, 130]

b. Maximum at critical value and minimum at an end point

c. Both at critical values

d. Yes; for example, [2, 10]

25. P is continuous and (0,40) is a open interval. Find the critical values.
$P(x) = 8 - 0.4x = 0.4(20 - x)$
The critical value is $x = 20$.
CV: $x = 20$: $P(20) = 8(20) - 0.2(20)^2 = 80$
The pollen count is highest on the twentieth day.

27. E is continuous and [20, 60] is a closed interval. Find the critical values.
$E'(x) = -0.02x + 0.62$
The critical value is $x = 31$.

CV: $x = 31$: $E(31) = -0.01(31)^2 + 0.62(31) + 10.4 = 20.01$

EP: $x = 20$: $E(20) = -0.01(20)^2 + 0.62(20) + 10.4 = 18.8$

$x = 60$ $E(60) = -0.01(60)^2 + 0.62(60) + 10.4 = 11.6$

Fuel economy is greatest at 31 mph.

29. f is continuous and [16, 85] is a closed interval. Find the critical values.
$$f'(x) = 0.026x - 1.35 = 0.026\left(x - 51\frac{12}{13}\right)$$
The critical value is $x = 51\frac{12}{13}$.

CV: $x = 51\frac{12}{13}$: $f\left(51\frac{12}{13}\right) = 0.013\left(51\frac{12}{13}\right)^2 - 1.35\left(51\frac{12}{13}\right) + 48 \approx 12.95$

EP: $x = 16$: $f(16) = 0.013(16)^2 - 1.35(16) + 48 = 29.728$

$x = 85$ $f(85) = 0.013(85)^2 - 1.35(85) + 48 = 27.175$

52-year olds have the least accidents.

31. f is continuous and [0, 1] is a closed interval. Find the critical values.
$$f'(x) = \frac{1}{2}(2-x)^2 + \frac{1}{2}x(2)(2-x)(-1) = 1.5x^2 - 4x + 2$$
$$= (1.5x - 1)(x - 2)$$
The critical values are $x = \frac{2}{3}$ and 2.

CV: $x = \frac{2}{3}$: $f\left(\frac{2}{3}\right) = \frac{1}{2}\left(\frac{2}{3}\right)\left(2 - \frac{2}{3}\right)^2 = \frac{16}{27}$

EP: $x = 0$: $f(0) = \frac{1}{2}(0)(2-0)^2 = 0$

$x = 1$ $f(1) = \frac{1}{2}(1)(2-1)^2 = \frac{1}{2}$

$\frac{2}{3}$ maximizes the energy extracted.

33. V is continuous and [0, 50] is a closed interval.
Find the critical values.
$V'(t) = 240t^{-1/2} - 40$
The critical values are $x = 0$ and 36.

$$\begin{array}{ll} \text{CV: } x = 0 & V(0) = 480\sqrt{0} - 40(0) = 0 \\ \quad x = 36 & V(36) = 480\sqrt{36} - 40(36) = 1440 \\ \text{EP: } x = 50 & V(50) = 480\sqrt{50} - 40(50) \approx 1394 \end{array}$$

The value is maximized after 36 years.

35. P is continuous and [0, 50] is a closed interval.
Find the critical values.
$P\,(x) = 6x - 72$
The critical value is $x = 12$.

$$\begin{array}{ll} \text{CV: } x = 12 & P(12) = 3(12)^2 - 72(12) + 576 = 144 \\ \text{EP: } x = 0 & P(0) = 3(0)^2 - 72(0) + 576 = 576 \\ \quad x = 50 & P(50) = 3(50)^2 - 72(50) + 576 \\ & \quad = 4476 \end{array}$$

Pollution is the least 12 miles away from factory A towards factory B.

37. $C(x) = 200x + 1500$
$p(x) = 600 - 5x$
$R(x) = (600 - 5x)x = 600x - 5x^2$
$P(x) = R(x) - C(x) = 600 - 5x^2 - (200x + 1500)$
$\quad = -5x^2 + 400x - 1500$
To maximize profit, we consider $P\,(x) = 0$
$P'(x) = -10x + 400 = 0$
$\quad x = 40$
To maximize profit, 40 motorbikes should be sold. The price for 40 motorbikes is
$p(40) = 600 - 5(40) = \$400$
The maximum profit is
$P(40) = R(40) - C(40)$
$\quad = 400(40) - [200(40) + 1500] = \$6500$

39. $R(x) = 240x - 3x^2$
$R'(x) = 240 - 6x = 0$
$\quad x = 40$
Rent 40 trucks per day, at \$120, with a maximum revenue of \$4800.

41. Let $x =$ the side perpendicular to the building and let $y =$ the side parallel to the building. Since only three sides will be fenced,
$2x + y = 800$
$\quad y = 800 - 2x$
We wish to maximize $A = xy = x(800 - 2x)$, so we take the derivative.
$A = x(800 - 2x)$
$\quad = 800x - 2x^2$
$A' = 800 - 4x = 0$
$x = 200$
$y = 800 - 2x = 800 - 400 = 400$
The side parallel to the building is 400 feet and the side perpendicular to the building is 200 feet.

43. $y + 4x = 1200$
$\quad xy = A$
$\quad y = 1200 - 4x$
$A = (1200 - 4x)x$
$\quad = 1200 - 4x^2$

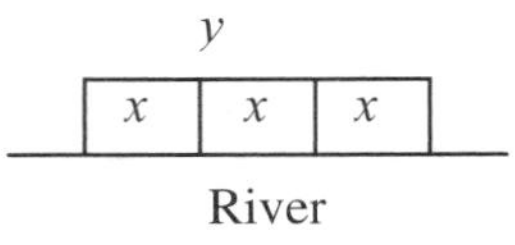

$A' = 1200 - 8x = 8(150 - x) = 0$
$x = 150$
$y = 1200 - 4(150) = 600$
Since there are three identical enclosures, the four sides perpendicular to the river are 150 yards long and the side parallel to the river is 600 yards long.

45. Suppose the sides of the square being removed have length x. Then the dimensions of the base of the box are $18-2x$ by $18-2x$.

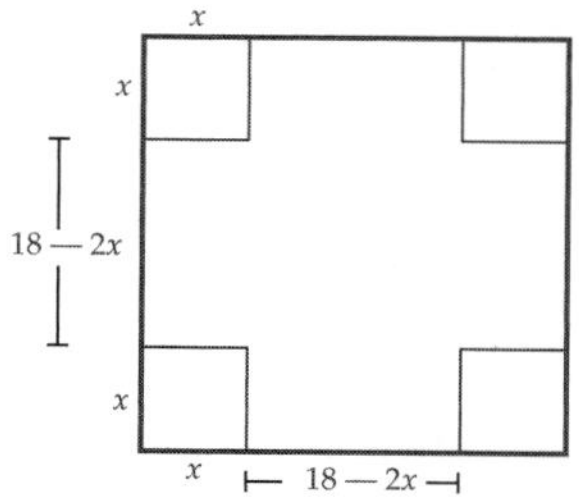

$V = l \cdot w \cdot h$
$= (18-2x)(18-2x)x$
$= (324-72x+4x^2)x$
$= 324x-72x^2+4x^3$
$V' = 324-144x+12x^2 = 0$
$= 81-36x+3x^2 = 0$
$= (3x-9)(x-9) = 0$
$x = 3, 9$

Since x cannot be 9, $x = 3$ and the base is $18-2(3) = 12$ inches by 12 inches. The maximum sized box has $V = (12)(12)(3) = 432$ cubic inches.

47. Let x and y be two numbers with a sum of 50 and a maximum product.

$x + y = 50$
$xy = \text{maximum} = M$
$x = 50 - y$
$(50-y)y = M$
$50y - y^2 = M$
$M' = 50 - 2y$
CV: $y = 25$ Thus $x = 50 - y = 25$
$M''(x) = -2$

Since $M' < 0$, the critical value is a maximum. The numbers are 25 and 25.

49. $A = (x)(2r)$ for rectangle

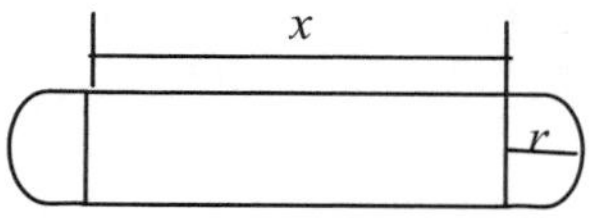

$P = 2x + 2\pi r$ $P = 440$ yards
$440 = 2x + 2\pi r$
$\frac{440-2\pi r}{2} = x$
$220 - \pi r = x$
$A = (220-\pi r)(2r)$
$= 440r - 2\pi r^2$
$A' = 440 - 4\pi r$

Setting $A' = 0$
$0 = 440 - 4\pi r$
$-440 = -4\pi r$
$\frac{110}{\pi} = r$
CV: $r = \frac{110}{\pi}$

Since $A'' = -4\pi < 0$, the area is maximized at $r = \frac{110}{\pi}$.

$440 = 2x + 2\pi\left(\frac{110}{\pi}\right)$
$440 = 2x + 220$
$x = 110$ yards

The dimensions of the largest track are $x = 110$ yards by $\frac{110}{\pi} \approx 35$ yards.

51. **a.** $N'(t) = 30t - 3t^2$

Setting $N'(t) = 0$,

$0 = 30t - 3t^2$

$3t^2 - 30t = 0$

$3t(t - 10) = 0$

$t = 0, 10$

$N''(t) = 30 - 6t$

$N''(0) = 30 > 0$, so $t = 0$ gives a min.

$N''(10) = -30 < 0$, so $t = 10$ gives a max.

The population will be the largest after 10 hours. At that time, there will be $N(10) = 1500$ thousand bacteria.

b. Setting $N''(t) = 0$,

$0 = 30 - 6t$

$6t = 30$

$t = 5$

Since $N''(t) = -6 < 0$, $t = 5$ gives a max. The population will be growing fastest after 5 hours. At that time, the population will be increasing at the rate of $N'(5) = 75$ thousand bacteria per hour. This is called an inflection point.

53. If x = the length of a side of the square that is removed from each corner, then the sides of the base are $5 - 2x$ and $7 - 2x$, and the volume V is

$V = (5 - 2x)(7 - 2x)x$

$= (35 - 24x + 4x^2)x$

$= 35x - 24x^2 + 4x^3$

To get the largest box, cut a square with a side about 0.96 inch. The volume of the box is about 15 cubic inches.

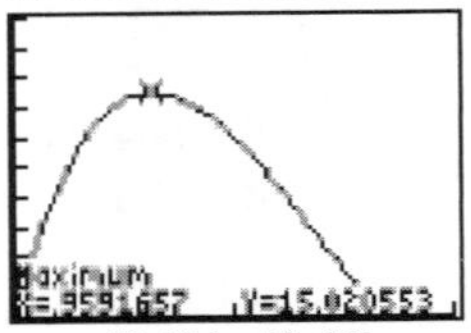

on [0, 3] by [0, 20]

55. A function can have at most one absolute maximum value (the highest value it attains), but it can have several relative maximum values (values that are high relative to their neighboring values).

57. False: The function can have absolute extreme values even on an open interval. For example: $f(x) = x^2$ has an absolute minimum at $x = 0$.

59.

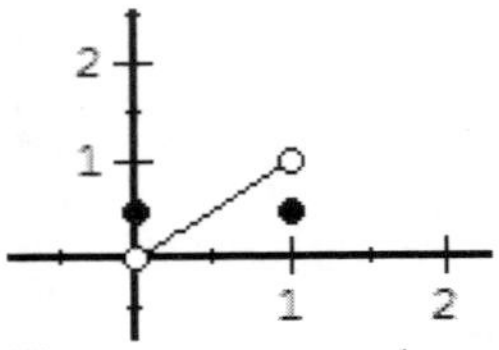

There are many other examples.

61. The slope at $x = 5$ is zero and the curve is concave up at $x = 5$, so there is a relative minimum at $x = 5$.

63. True: Multiplying by –1 will flip the entire graph causing a peak to become a valley. Also, the –1 will cause the derivative to change from positive to negative to negative to positive.

65. We wish to maximize the velocity $V(r)$. We take the derivative.

$V'(r) = c(-1)r^2 + c(3 - r)(2r)$

$= -cr^2 + 6cr - 2cr^2$

$= -3cr^2 + 6cr = 0$

$= -3cr(r - 2) = 0$

$r = 2$

The radius that maximizes the velocity is 2 cm.

EXERCISES 3.4

1. Let x = the number of \$300 price reductions. Then the price function is $p(x) = 15{,}000 - 300x$ and the quantity function is $q(x) = 12 + 2x$.

$R(x) = p(x)q(x) = (15{,}000 - 300x)(12 + 2x)$

$= 180{,}000 + 26{,}400x - 600x^2$

The cost function is the unit cost times $q(x)$ plus fixed cost.

$C(x) = 12{,}000(12 + 2x) + 1000$

$= 144{,}000 + 24{,}000x + 1000$

$= 145{,}000 + 24{,}000x$

Thus the profit function $P(x)$ is

$P(x) = R(x) - C(x)$

$= 180{,}000 + 26{,}400x - 600x^2 - (145{,}000 + 24{,}000x)$

$= 35{,}000 + 2400x - 600x^2$

$P'(x) = 2400 - 1200x = 0$

$x = 2$

$P''(x) = -1200$ shows that profit will be maximized.

To maximize profits, the price should be $p(2) = \$15{,}000 - 300(2) = \$14{,}400$, and at that price he will sell $q(x) = 12 + 2(2) = 16$ cars per day.

3. Let x = the number of \$10 price reductions.
Then $p(x) = 200 - 10x$ and $q(x) = 300 + 30x$.

$R(x) = p(x)q(x) = (200 - 10x)(300 + 30x)$

$= 60{,}000 + 3000x - 300x^2$

$R'(x) = 3000 - 600x = 0$

$x = 5$

$R''(x) = -600$ shows that revenue is maximized.

Thus, the ticket price $p(5) = 200 - 10(5) = \$150$ maximizes revenue. The quantity sold at $x = 5$ is $q(5) = 300 + 30(5) = 450$.

5. Let x = the number of \$5 price reductions.

$q(x) = 60 - 3x$

$p(x) = 80 + 5x$

Revenue $= p(x) \cdot q(x)$

$= (80 + 5x)(60 - 3x)$

$= 4800 + 60x - 15x^2$

$R'(x) = 60 - 30x$

$0 = 60 - 30x$

$-60 = -30x$

$2 = x$

$R'' = -30 < 0$

Thus revenue is maximized at $x = 2$.

$R(2) = 4800 + 60(2) - 15(2)^2$

$= 4800 + 120 - 60 = 4860$

$q(2) = 60 - 3(2) = 54$

$p(2) = 80 + 5(2) = 80 + 10 = 90$

Rent the cars for \$90 and expect to rent 54 cars (from $x = 2$ price increases).

7. Let x = the number of additional trees per acre.
The number of trees per acre is $t(x) = 20 + x$,
and the yield per tree is $y_t(x) = 90 - 3x$.
The total yield $Y(x) = t(x) \cdot y_t(x)$.

$Y(x) = (20 + x)(90 - 3x)$

$= -3x^2 + 30x + 180$

$Y'(x) = -6x + 30 = 0$

$x = 5$

$Y''(x) = -6$ shows that yield is maximized.

He should plant $t(5) = 20 + 5 = 25$ trees per acre.

9. The base is a square, so we define
x = length of side of base
y = height
The volume must equal 4 cubic feet.
$x^2 y = 4$ (1)
The box consists of a bottom (area x^2) and four sides (each of area $x \cdot y$). Minimizing the amount of materials means minimizing the surface area of the bottom and four sides.
$A = x^2 + 4xy$ (2)
Using (1), solve for y:
$y = \frac{4}{x^2}$
Now (2) becomes
$A = x^2 + 4x\left(\frac{4}{x^2}\right) = x^2 + \frac{16}{x}$
We minimize this by differentiating.
$A' = 2x - \frac{16}{x^2}$
$2x - \frac{16}{x^2} = 0$
$2x^3 = 16$
$x^3 = 8$
$x = 2$
Thus $y = \frac{4}{x^2} = \frac{4}{4} = 1$
$A'' = 2 + \frac{32}{x^3} > 0$ for $x = 2$
So the area is minimized. The base is 2 feet by 2 feet and the height is 1 foot.

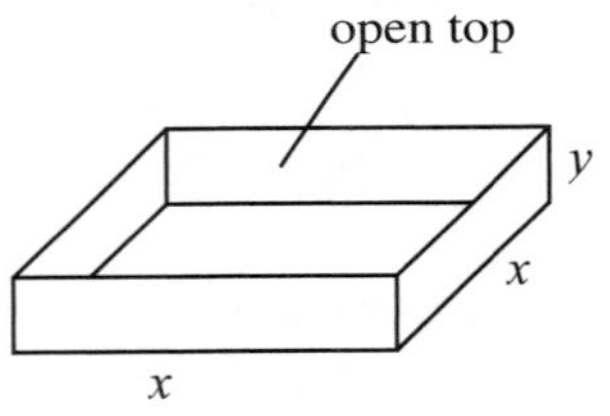

11. Let x = the side of the square base and l = the length of the box. The girth $= l + 4x = 84$. Thus, $l = 84 - 4x$. To maximize volume, take the derivative of $V(x)$.
$V(x) = x^2 l = x^2(84 - 4x) = 84x^2 - 4x^3$
$V'(x) = 168x - 12x^2 = 0$
$x = 14$
$V''(x) = 168 - 24x < 0$ for $x = 14$
Thus V is maximized. The box should have a base 14 inches by 14 inches and length $l = 84 - 4(14) = 28$ inches.

13. Let x = length (along driveway)
y = width

Since the area is to be 5000 square feet,
$A = xy = 5000$ (1)

Cost = cost along driveway + cost for other 3 sides
$C(x) = 6x + [2x + 2(2y)] = 8x + 4y$

Using (1) to solve for y, $y = \frac{5000}{x}$

Then

$$C(x) = 8x + 4\left(\frac{5000}{x}\right) = 8x + \frac{20{,}000}{x}$$

We want to minimize cost $C(x)$.

$$C'(x) = 8 - 20{,}000x^{-2}$$

$$0 = 8 - \frac{20{,}000}{x^2}$$

$$-8x^2 = -20{,}000$$

$$x^2 = 2500$$

$x = \pm 50$ or $x = 50$ since we eliminate $x = -50$

Thus the cost is minimized when $x = 50$.

$$C(50) = 8(50) + \frac{20{,}000}{50} = 800$$

The dimensions should be 50 feet along the driveway and 100 feet perpendicular to the driveway. The cost is \$800.

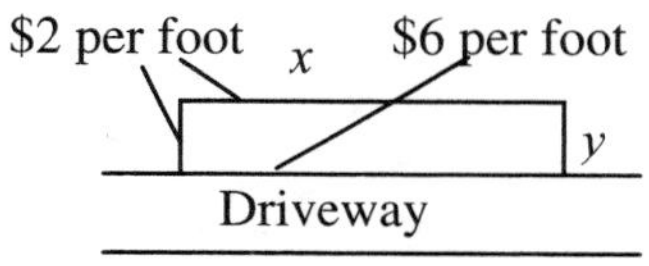

15. Revenue = tax rate times total sales $S(t)$

$$R(t) = t(8 - 15\sqrt[3]{t}) = 8t - 15t^{4/3}$$

$$R'(t) = 8 - 20t^{1/3} = 0$$

$$t^{1/3} = \frac{4}{10}$$

$$t = \frac{64}{1000} = 0.064$$

$$R''(t) = -\frac{20}{3}t^{-2/3} < 0 \quad \text{for } t = 0.064$$

showing that revenue is maximized. The tax rate is 6.4%.

17. The net value of the wine after t years is

$$V(t) = 2000 + 80t^{1/2} - 10t \quad (0 \le t \le 25)$$

$$V'(t) = 40t^{-1/2} - 10$$

$$0 = \frac{40}{\sqrt{t}} - 10$$

$$10 = \frac{40}{\sqrt{t}}$$

$$10\sqrt{t} = 40$$

$$\sqrt{t} = 4$$

CV: $t = 16$ years

$$V''(t) = -20t^{-3/2}$$

$$V''(16) = \frac{-20}{16^{3/2}} < 0$$

$$V(16) = 2160$$

EP: $V(0) = 2000$
$V(25) = 2150$

so the net value is maximized at $t = 16$ years.

19. Since the area A is fixed, then
$A = l \cdot w$

$$l = \frac{A}{w}$$

Now we take the derivative of

$$p = 2l + 2w = 2\left(\frac{A}{w}\right) + 2w$$

$$p' = -2\left(\frac{A}{w^2}\right) + 2$$

$$0 = -2\left(\frac{A}{w^2}\right) + 2$$

$$2\left(\frac{A}{w^2}\right) = 2$$

$$w^2 = A$$

$$w = \sqrt{A}$$

Since $l = \frac{A}{w}$, $l = \frac{A}{\sqrt{A}} = \sqrt{A}$. Thus, $l = w$.

21. Since we want the dimensions of the total page, let $x =$ length and $y =$ width. Then the dimensions of the (inner) print area are $x-2\left(1\frac{1}{2}\right)= x-3$ and $y-2(1)=y-2$. Thus, the (inner) print area is
$A=(x-3)(y-2)$
The page has a total area of 96 square inches.

$$xy=96$$
$$y=\frac{96}{x}$$
$$A(x)=(x-3)\left(\frac{96}{x}-2\right)$$
$$=96-2x-\frac{288}{x}+6$$
$$=102-2x-288x^{-1}$$
$$A'(x)=-2+288x^{-2}$$
$$0=-2+\frac{288}{x^2}$$
$$2x^2=288$$
$$x=12 \quad \text{(we eliminate } -12)$$
$$y=\frac{96}{x}=\frac{96}{12}=8$$
$$A''(x)=-\frac{576}{x^3}$$
$$A''(12)<0$$

so the inner area is maximized at $x=12$. The pages should be 8 inches wide and 12 inches tall.

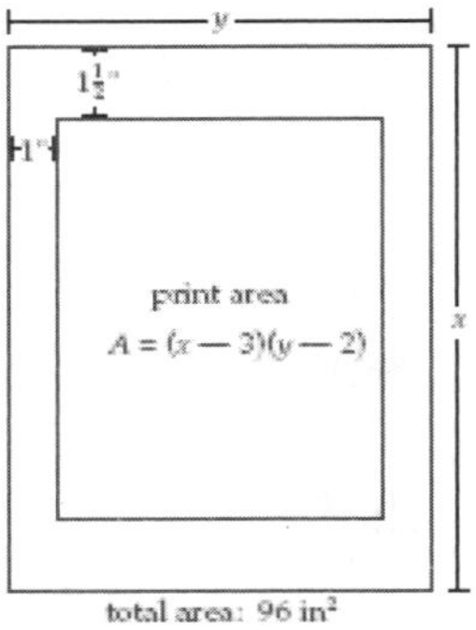

total area: 96 square inches

23. **b.** $y_3=y_1y_2=(400-10x)(20+2x)$
$$=8000+600x-20x^2$$

d. $y_5=y_3-y_4$
$$=8000+600x-20x^2-200(20+2x)$$
$$=4000+200x-20x^2$$

e. Maximum at $x=5$

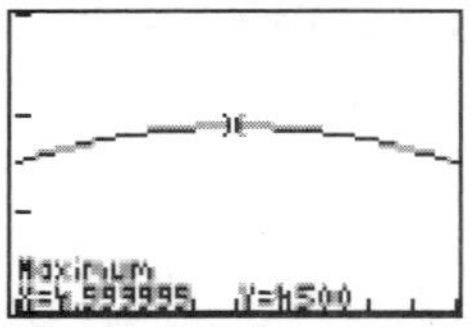

on [0, 10] by [2000, 6000]

f. $y_5=8000+600x-20x^2-150(20+2x)$
$$=5000+300x-20x^2$$
Maximum at $x=7.5$
Price = \$325 and quantity = 35 bicycles

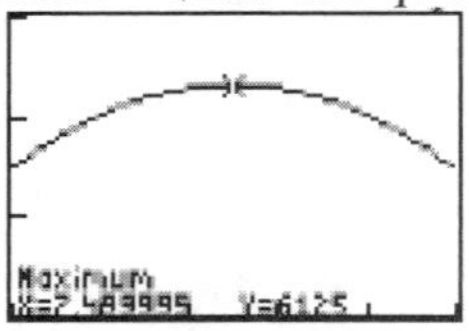

on [0, 15] by [3000, 7000]

g. Price = \$350 and quantity = 40 bicycles

25. Revenue is price times quantity.

27. \$65, because negative three price reductions means three price increases of \$5. So \$50 goes to \$65.

29. A linear cost function will have a positive slope (the unit cost), and so will be minimized at its left-hand end point. At this point, where $x = 0$, nothing is produced, which is not a realistic solution for a business situation.

31. As quantity increases, the price generally falls, which will eventually reduce profits.

33. No. Revenue and profit will generally be maximized at different points, since one has cost subtracted and the other does not. A careful look at the graphs on page 220 will show that revenue and profit are maximized at different x-values.

35.
$$R(x) = cx^2(p - x) = cpx^2 - cx^3$$
$$R'(x) = 2cpx - 3cx^2 = 0$$
$$cx(2p - 3x) = 0$$
$$x = \frac{2}{3}p$$
$$R'(x) = 2cp - 6c\left(\frac{2}{3}p\right) = 2cp - 4cp = -2cp < 0,$$
since $c, p > 0$
which shows that R is maximized.

EXERCISES 3.5

1. Let x = lot size

Storage costs = (storage per item) · (average number of items)
$$= (1)\frac{x}{2} = \frac{x}{2}$$
Cost per order $= 2x + 20$

Number of orders $= \frac{4000}{x}$

Reorder costs = (cost per order) · (number of orders)
$$= (2x + 20)\left(\frac{4000}{x}\right)$$
Total cost $C(x)$ = (storage costs) + (reorder costs)
$$= \left(\frac{x}{2}\right) + (2x + 20)\left(\frac{4000}{x}\right)$$
$$= \frac{x}{2} + 8000 + \frac{80,000}{x}$$
To minimize $C(x)$ we differentiate.
$$C'(x) = \frac{1}{2} - 80,000x^{-2}$$
$$0 = \frac{1}{2} - 80,000x^{-2}$$
$$\frac{1}{2}x^2 = 80,000$$
$$x^2 = 160,000$$
$$x = 400$$
$$C''(x) = 2(80,000)x^{-3}$$
$C''(400) > 0$ for $x = 400$

so C is minimized at $x = 400$

Lot size is 400 boxes with orders placed 10 times a year.

3. Let $x =$ order size

Storage costs $= 10\left(\frac{x}{2}\right) = 5x$

Cost per order $= 12x + 125$

Number of orders $= \frac{10{,}000}{x}$

Reorder costs $= (12x+125)\left(\frac{10{,}000}{x}\right)$

Total cost $C(x) = 5x + (12x+125)\left(\frac{10{,}000}{x}\right)$

$$= 5x + 120{,}000 + \frac{1{,}250{,}000}{x}$$

$$C'(x) = 5 - \frac{1{,}250{,}000}{x^2} = 0$$

$$5x^2 - 1{,}250{,}000 = 0$$

$$x^2 = 250{,}000$$

$$x = 500$$

$C''(x) = \frac{2{,}500{,}000}{x^3} > 0$ for $x = 500$

so C is minimized at $x = 500$

The order size is 500 bottles, and the number of orders is $\frac{10{,}000}{500} = 20$ per year.

5. Let $x =$ lot size

Storage costs $= (1000)\left(\frac{x}{2}\right) = 500x$

Cost per order $= 1000 + 9000x$

Number of orders $= \frac{800}{x}$

Reorder costs $= (1000+9000x)\left(\frac{800}{x}\right)$

Total cost $C(x) = 500x + \left(\frac{800}{x}\right)(1000+9000x)$

$$= 500x + 7{,}200{,}000 + 800{,}000x^{-1}$$

$$C(x) = 500 - 800{,}000x^{-2} = 0$$

$$0 = 500 - \frac{800{,}000}{x^2}$$

$$-500x^2 = -800{,}000$$

$$x = 40$$

$C'(x) = -2(-800{,}000)x^{-3}$

$C'(40) > 0$

so C is minimized at $x = 40$

The lot size is 40 cars per order, with 20 orders during the year.

7. Let $x =$ number manufactured at a time

Storage costs $= 2\left(\frac{x}{2}\right) = x$

Cost per run $= 3x + 500$

Number of runs $= \frac{2000}{x}$

Production costs $= (3x+500)\left(\frac{2000}{x}\right)$

Total cost $C(x) = x + (3x+500)\left(\frac{2000}{x}\right)$

$$= x + 6000 + \frac{1{,}000{,}000}{x}$$

$$C'(x) = 1 - \frac{1{,}000{,}000}{x^2} = 0$$

$$x^2 - 1{,}000{,}000 = 0$$

$$x = 1000$$

$C''(x) = \frac{2{,}000{,}000}{x^3} > 0$ for $x = 1000$

so C is minimized at $x = 1000$

The production run is 1000 games, and the number of production runs is $\frac{2000}{1000} = 2$.

9. Let x = the number of circuit boards in each run

Storage costs $= (1)\left(\frac{x}{2}\right) = \frac{x}{2}$

Cost per run $= 800 + 10x$

Number of runs $= \dfrac{1{,}000{,}000}{x}$

Production costs $= (800 + 10x)\left(\dfrac{1{,}000{,}000}{x}\right)$

Total cost $C(x) = \frac{x}{2} + (800 + 10x)\left(\dfrac{1{,}000{,}000}{x}\right)$

$= \frac{x}{2} + \dfrac{800{,}000{,}000}{x} + 10{,}000{,}000$

$C'(x) = \frac{1}{2} - 800{,}000{,}000x^{-2} = 0$

$0 = \frac{1}{2} - 8 \times 10^8 x^{-2}$

$-\frac{1}{2}x^2 = -8 \times 10^8$

$x = 40{,}000$

$C''(x) = -2(-800{,}000{,}000)(x^{-3})$

$C''(40{,}000) > 0$

so C is minimized at $x = 40{,}000$

Produce 40,000 circuit boards per run, with 25 runs for the year.

11. $f(p) = -0.01p^2 + 5p$

We set $f'(p) = 1$ to find the maximum sustainable yield.

$f'(p) = -0.02p + 5 = 1$

$-0.02p = -4$

$p = 200$ hundred

$f''(p) = -0.02 < 0$

so $f(p)$ is maximized.

The population that gives the maximum sustainable yield is 20,000. The maximum sustainable yield is

$Y(p) = f(p) - p$

$= f(200) - 200$

$= -0.01(200)^2 + 5(200) - 200$

$= -400 + 1000 - 200$

$= 400$ hundred

The maximum sustainable yield is 40,000.

13. $f(p) = -0.0004p^2 + 1.06p$

$f'(p) = -0.0008p + 1.06 = 1$

$-0.0008p = -0.06$

$p = 75$ thousand

$f''(p) = -0.0008 < 0$

so $f(p)$ is maximized.

The population that gives the maximum sustainable yield is 75,000.

$Y(p) = f(p) - p$

$Y(75) = -0.0004(75)^2 + 1.06(75) - 75$

$= -2.25 + 4.5$

$= 2.25$ thousand

The maximum sustainable yield is 2250.

15. $f(p) = -\frac{1}{8}p^2 + 7.5p$

$f'(p) = -\frac{1}{4}p + 7.5 = 1$

$p = 26$

$f''(p) = -\frac{1}{4} < 0$

so $f(p)$ is maximized.

The population that gives the maximum sustainable yield is 260,000.

$Y(p) = f(p) - p$

$Y(26) = -\frac{1}{4}(260)^2 + 7.5(260) - 260$

$= 845$ thousand

The maximum sustainable yield is 845,000 metric tons.

17. $f(p) = 50\sqrt{p} = 50p^{1/2}$

$f'(p) = 25p^{-1/2} = 1$

$25 = \sqrt{p}$

$p = 625$ thousand

$f''(p) = -\frac{25}{2}p^{-3/2} < 0$ at $x = 625$

so $f(p)$ is maximized.

The population that gives the maximum sustainable yield is 625,000.

$Y(p) = f(p) - p$

$Y(625) = 50\sqrt{625} - 625 = 1250 - 625$

$= 625$ thousand

The maximum sustainable yield is 625,000.

19. $f(p) = -0.0005p^2 + 2p$

$f'(p) = -0.001p + 2 = 1$

$p = 1000$

$f''(p) = -0.001 < 0$

so $f(p)$ is maximized.

The population that gives the maximum sustainable yield is 100,000.

$Y(p) = f(p) - p$

$Y(1000) = -0.0005(1000)^2 + 2(1000) - 1000$

$= 500$ hundred

The population size is 100,000 metric tons, with a yield of 50,000 metric tons.

21. Because of the larger number of items on hand, storage costs will increase and reorder costs will decrease.

23. Lot size would decrease because reorder costs would decrease allowing for more reorders for the same price.

25. If $f(p) < p$ for all values of p, next year's population will always be smaller than this year's population, so the animal will eventually become extinct.

27. Animal populations, such as fish in the sea, can be very difficult to estimate. Also, one year's population depends on factors other than last year's population, such as weather, predators, disease, and changing environmental conditions.

EXERCISES 3.6

1.
$$y^3 - x^2 = 4$$
$$3y^2 \frac{dy}{dx} - 2x = 0$$
$$3y^2 \frac{dy}{dx} = 2x$$
$$\frac{dy}{dx} = \frac{2x}{3y^2}$$

3.
$$x^3 = y^2 - 2$$
$$3x^2 = 2y\frac{dy}{dx}$$
$$\frac{dy}{dx} = \frac{3x^2}{2y}$$

5.
$$y^4 - x^3 = 2x$$
$$4y^3 \frac{dy}{dx} - 3x^2 = 2$$
$$\frac{dy}{dx} = \frac{3x^2+2}{4y^3}$$

7.
$$(x+1)^2 + (y+1)^2 = 18$$
$$2(x+1) + 2(y+1)\frac{dy}{dx} = 0$$
$$\frac{dy}{dx} = -\frac{x+1}{y+1}$$

9.
$$x^2 y = 8$$
$$2xy + x^2 \frac{dy}{dx} = 0$$
$$\frac{dy}{dx} = -\frac{2xy}{x^2} = -\frac{2y}{x}$$

11.
$$xy - x = 9$$
$$y + x\frac{dy}{dx} - 1 = 0$$
$$x\frac{dy}{dx} = 1 - y$$
$$\frac{dy}{dx} = \frac{1-y}{x}$$

13.
$$x(y-1)^2 = 6$$
$$(y-1)^2 + x\left[2(y-1)\frac{dy}{dx}\right] = 0$$
$$\frac{dy}{dx} = -\frac{(y-1)^2}{2x(y-1)} = -\frac{y-1}{2x}$$

15.
$$y^3 - y^2 + y - 1 = x$$
$$3y^2 \frac{dy}{dx} - 2y\frac{dy}{dx} + \frac{dy}{dx} = 1$$
$$\frac{dy}{dx} = \frac{1}{3y^2-2y+1}$$

17.
$$\frac{1}{x} + \frac{1}{y} = 2$$
$$-\frac{1}{x^2} - \frac{1}{y^2}\frac{dy}{dx} = 0$$
$$-\frac{1}{y^2}\frac{dy}{dx} = \frac{1}{x^2}$$
$$\frac{dy}{dx} = -\frac{y^2}{x^2}$$

19.
$$x^3 = (y-2)^2 + 1$$
$$3x^2 = 2(y-2)\frac{dy}{dx}$$
$$\frac{dy}{dx} = \frac{3x^2}{2(y-2)}$$

21. $y^2 - x^3 = 1$

$2y\frac{dy}{dx} - 3x^2 = 0$

$\frac{dy}{dx} = \frac{3x^2}{2y}$

At $x = 2, y = 3,$ $\frac{dy}{dx} = \frac{3(2)^2}{2(3)} = 2$.

23. $y^2 = 6x - 5$

$2y\frac{dy}{dx} = 6$

$\frac{dy}{dx} = \frac{6}{2y} = \frac{3}{y}$

At $x = 1, y = -1,$ $\frac{dy}{dx} = \frac{3}{-1} = -3$.

25. $x^2y + y^2x = 0$

$2xy + x^2\frac{dy}{dx} + 2y\frac{dy}{dx}x + y^2 = 0$

$(x^2 + 2yx)\frac{dy}{dx} = -2xy - y^2$

$\frac{dy}{dx} = -\frac{2xy+y^2}{x^2+2yx}$

At $x = -2, y = 2,$

$\frac{dy}{dx} = -\frac{2(-2)(2)+(2)^2}{(-2)^2+2(2)(-2)} = -\frac{-4}{-4} = -1$

27. $x^2 + y^2 = xy + 7$

$2x + 2y\frac{dy}{dx} = y + x\frac{dy}{dx}$

$(2y - x)\frac{dy}{dx} = y - 2x$

$\frac{dy}{dx} = \frac{y-2x}{2y-x}$

At $x = 3, y = 2,$

$\frac{dy}{dx} = \frac{2-2(3)}{2(2)-3} = \frac{-4}{1} = -4$

29. $p^2 + p + 2x = 100$

$2p\frac{dp}{dx} + \frac{dp}{dx} + 2 = 0$

$\frac{dp}{dx} = -\frac{2}{2p+1}$

31. $12p^2 + 4p + 1 = x$

$24p\frac{dp}{dx} + 4\frac{dp}{dx} = 1$

$\frac{dp}{dx} = \frac{1}{24p+4}$

33. $xp^3 = 36$

$p^3 + x\left(3p^2\frac{dp}{dx}\right) = 0$

$\frac{dp}{dx} = -\frac{p}{3x}$

35. $(p+5)(x+2) = 120$

$\frac{dp}{dx}(x+2) + p + 5 = 0$

$\frac{dp}{dx} = -\frac{p+5}{x+2}$

37. $y^2 = x^3 - 4x + 1$

Use implicit differentiation.

$2y\frac{dy}{dx} = 3x^2 - 4$

$\frac{dy}{dx} = \frac{3x^2 - 4}{2y}$

At (–2, 1), the slope is

$\frac{dy}{dx} = \frac{3(-2)^2 - 4}{2(1)} = 4$

The tangent line has slope 4 at (–2, 1).

$y - 1 = 4(x + 2)$

$y = 4x + 9$

39. $\left(x^2+y^2-y\right)^2=x^2+y^2$

Use implicit differentiation.

$$2\left(x^2+y^2-y\right)\left(2x+2y\frac{dy}{dx}-\frac{dy}{dx}\right)=2x+2y\frac{dy}{dx}$$

At (–1, 0), the slope is

$$2\left((-1)^2+0^2-0\right)\left(2(-1)+2(0)\frac{dy}{dx}-\frac{dy}{dx}\right)=2(-1)+2(0)\frac{dy}{dx}$$
$$2(1)\left(-2-\frac{dy}{dx}\right)=-2$$
$$\frac{dy}{dx}=-1$$

The tangent line has slope –1 at (–1, 0).

$$y-0=-1(x+1)$$
$$y=-x-1$$

41. $x=\sqrt{2000-p^2}$

$$1=\frac{1}{2}\left(2000-p^2\right)^{-1/2}(-2p)\frac{dp}{dx}$$
$$\frac{dp}{dx}=-\frac{\sqrt{2000-p^2}}{p}$$

When $p = 40$,

$$\frac{dp}{dx}=-\frac{\sqrt{2000-(40)^2}}{40}=-\frac{\sqrt{400}}{40}=-\frac{1}{2}$$

The rate of change of price with respect to quantity is $-\frac{1}{2}$. Prices decrease by 50¢ for each increase of 1 in quantity.

43. $2x^2=15{,}000-p^2$

$$4x\frac{dx}{dp}=-2p$$
$$\frac{dx}{dp}=-\frac{p}{2x}$$

When $p = 100$, $x = 50$

$$\frac{dx}{dp}=-\frac{100}{2(50)}=-1$$

Sales will decrease by 1 for each \$1 increase in price.

45. $60h^5+2{,}000{,}000=p^3$

$$300h^4\frac{dh}{dp}=3p^2$$
$$\frac{dh}{dp}=\frac{p^2}{100h^4}$$

When $p = 200$, $h = 10$

$$\frac{dh}{dp}=\frac{(200)^2}{100(10)^4}=\frac{1}{25}=0.04$$

A person will work for an extra $\frac{1}{25}$ hour (2.4 minutes) for an extra \$1 in pay (at the given pay rate.)

47. **a.** $s^2=r^3-55$

$$2s\frac{ds}{dr}=3r^2$$
$$\frac{ds}{dr}=\frac{3r^2}{2s}$$

At $r=4$, $s=3$, $\frac{ds}{dr}=\frac{3(4)^2}{2(3)}=\frac{48}{6}=8$

47. **b.** $s^2 = r^3 - 55$

$2s = 3r^2 \frac{dr}{ds}$

$\frac{dr}{ds} = \frac{2s}{3r^2}$

At $r = 4$, $s = 3$, $\frac{dr}{ds} = \frac{2(3)}{3(4)^2} = \frac{1}{8}$

c. $ds / dr = 8$ means that the rate of change of sales with respect to research expenditures is 8, so that increasing research by \$1 million will increase sales by about \$8 million. (at these levels of r and s).
$dr / ds = 1/8$ means that the rate of change of research expenditures with respect to sales is 1/8, so that increasing sales by \$1 million will increase research by about \$1/8 million. (at these levels of r and s).

49. $144 = 12L^{2/3}K^{1/3}$

Use implicit differentiation and solving for $\frac{dK}{dL}$.

$$0 = 12\left(\frac{2}{3}L^{-1/3}K^{1/3} + L^{2/3}\frac{1}{3}K^{-2/3}\frac{dK}{dL}\right)$$

$$0 = \frac{2}{3}L^{-1/3}K^{1/3} + L^{2/3}\frac{1}{3}K^{-2/3}\frac{dK}{dL}$$

$$\frac{dK}{dL} = -\frac{2L^{-1/3}K^{1/3}}{L^{2/3}K^{-2/3}}$$

$$\frac{dK}{dL} = -\frac{2K}{L}$$

For $L = 8$ and $K = 27$

$$\frac{dK}{dL} = -\frac{2(27)}{8} = -6.75$$

$$MRST = \left|\frac{dK}{dL}\right| = |-6.75| = 6.75$$

At a production level of 144 units using 8 units of labor and 27 units of capital, each additional unit of labor reduces the capital requirement by 6.75 units.

51. $x^3 + y^2 = 1$

$3x^2 \frac{dx}{dt} + 2y\frac{dy}{dt} = 0$

53. $x^2 y = 80$

$2xy\frac{dx}{dt} + x^2\frac{dy}{dt} = 0$

55. $3x^2 - 7xy = 12$

$6x\frac{dx}{dt} - 7y\frac{dx}{dt} - 7x\frac{dy}{dt} = 0$

57. $x^2 + xy = y^2$

$2x\frac{dx}{dt} + x\frac{dy}{dt} + y\frac{dx}{dt} = 2y\frac{dy}{dt}$

59. The volume of a sphere is $V = \frac{4}{3}\pi r^3$.

Both the volume and the radius decrease with time, so both V and r are functions of t.

$\frac{dV}{dt} = V'$, $\frac{dr}{dt} = r'$; thus, $V' = 4\pi r^2 r'$

The radius is decreasing at the rate of 2 inches per hour ($i.e,\ r' = -2$) when $r = 3$.

$$V' = 4\pi(3)^2(-2)$$
$$= -72\pi$$

Therefore, at the moment the radius is 3 inches, the volume is decreasing by $72\pi \approx 226$ in^3 per hour.

61. Since both the volume and the radius are decreasing with time, we differentiate with respect to t.

$V = \frac{4}{3}\pi r^3$

$\frac{dV}{dt} = 4\pi r^2 \frac{dr}{dt}$

Since $\frac{dr}{dt} = \frac{1}{2}$ and $r = 4$, we have

$\frac{dV}{dt} = 4\pi(4)^2\left(\frac{1}{2}\right) = 32\pi \approx 101\text{ cm}^3$ per week

63. $R = 1000x - x^2$

$R' = 1000x' - 2xx'$

$= x'(1000 - 2x)$

The quantity sold, x, is increasing at 80 per day ($i.e.,\ x' = 80$) when $x = 400$.

$R' = 80[1000 - 2(400)]$

$= 16,000$

Therefore the company's revenue is increasing at \$16,000 per day.

65. Since population and the number of cases increase with time, differentiate with respect to t.

$W = 0.003p^{4/3}$

$\frac{dW}{dt} = 0.004p^{1/3}\frac{dp}{dt}$

Since $\frac{dp}{dt} = 1000$ and $p = 1,000,000$,

$\frac{dW}{dt} = 0.004(1,000,000)^{1/3}(1000)$

$= 400$ cases per year

67. $V = c(R^2 - r^2);\ \frac{dR}{dt} = -0.01$ mm per year

$V = cR^2 - cr^2 \quad c = 500$

$\frac{dV}{dt} = 2cR\frac{dR}{dt} - 0$

When $R = 0.05$ mm,

$\frac{dV}{dt} = 2(500)(0.05)(-0.01)$

$= -0.5$

The rate at which blood flow in an artery is being reduced is slowing by $\frac{1}{2}$ mm/sec per year.

69. Since the distances are increasing with time, differentiate with respect to time. We use the Pythagorean Theorem to obtain an equation for x.

$x^2 + (0.25)^2 = z^2$

$2x\frac{dx}{dt} = 2z\frac{dz}{dt}$

$\frac{dx}{dt} = \frac{2z}{2x}\frac{dz}{dt} = \frac{z}{x}\frac{dz}{dt}$

Since

$\frac{dz}{dt} = 57,\ z = 0.5,$ and $x = \sqrt{(0.5)^2 - (0.25)^2}$,

then $\frac{dx}{dt} \approx 65.8$ mph.

Yes, the car is speeding.

71. $x^2 = 4500 - 5p^2$

$2x\frac{dx}{dt} = -10p\frac{dp}{dt}$

$\frac{dx}{dt} = \frac{-5p}{x}\frac{dp}{dt}$

When $\frac{dp}{dt} = -1$ and $p = 20$, $x = 250$,

$\frac{dx}{dt} = \frac{-5(20)(-1)}{50} = 2$

The sales are increasing by 2 per week.

73. $y = \sqrt[3]{x^2 - 1}$

$y^3 = x^2 - 1$

$y^3 - x^2 + 1 = 0$

75. $\frac{dy}{dx}$ would be the distance driven per gallon of gas, or miles per gallon. $\frac{dx}{dy}$ would be the amount of gas used per mile of driving, or gallons per mile (which, you would hope, is a small fraction).

77. The slope is not -1 at the point (10, 10) because $x = 10$ is not part of the domain.

79. Let p = population and t = time in years.

$\frac{dp}{dt} = -1500$

81. Let R = revenue and P = profit, and t = time in years.

$\frac{dR}{dt} = 2\frac{dP}{dt}$

EXERCISES 3.7

1. $$\begin{aligned}dy &= (x^2-4x+5)'dx\\ &= (2x-4)dx\\ &= (2\cdot 3-4)\cdot 0.25\\ &= 0.5\end{aligned}$$

3. $$\begin{aligned}dy &= \left(\frac{x+1}{x-1}\right)'dx\\ &= \frac{(x-1)(1)-(1)(x+1)}{(x-1)^2}dx\\ &= \frac{-2}{(x-1)^2}dx\\ &= \frac{-2}{(2-1)^2}\cdot(-0.15)\\ &= 0.3\end{aligned}$$

5. $$\begin{aligned}dy &= \left(\frac{x+3}{x^2+1}\right)'dx\\ &= \frac{(x^2+1)(1)-(2x)(x+3)}{(x^2+1)^2}dx\\ &= \frac{-x^2-6x+1}{(x^2+1)^2}dx\\ &= \frac{-(2)^2-6(2)+1}{((2)^2+1)^2}\cdot 0.25\\ &= -0.15\end{aligned}$$

7. $$\begin{aligned}dy &= (x^3+x^2+3)'dx\\ &= (3x^2+2x)dx\\ &= (3\cdot 1^2+2(1))\cdot(-0.05)\\ &= -0.25\\ \Delta y &= f(x+\Delta x)-f(x)\\ &= \left((1-0.05)^3+(1-0.05)^2+3\right)-(1^3+1^2+3)\\ &\approx -0.24\end{aligned}$$

9. $$\begin{aligned}dy &= \left(\frac{x+5}{x+1}\right)'dx\\ &= \frac{(x+1)(1)-(1)(x+5)}{(x+1)^2}dx\\ &= \frac{-4}{(x+1)^2}dx\\ &= \frac{-4}{(3+1)^2}\cdot 0.4\\ &= -0.1\\ \Delta y &= f(x+\Delta x)-f(x)\\ &= \left(\frac{3+0.4+5}{3+0.4+1}\right)-\left(\frac{3+5}{3+1}\right) = -\frac{1}{11}\\ &\approx -0.09\end{aligned}$$

11. $$\begin{aligned}dy &= \left(\frac{3x+5}{x^2+1}\right)'dx\\ &= \frac{(x^2+1)(3)-(2x)(3x+5)}{(x^2+1)^2}dx\\ &= \frac{-3x^2-10x+3}{(x^2+1)^2}dx\\ &= \frac{-3(1)^2-10(1)+3}{(1^2+1)^2}\cdot 0.2\\ &= -0.5\\ \Delta y &= f(x+\Delta x)-f(x)\\ &= \left(\frac{3(1+0.2)+5}{(1+0.2)^2+1}\right)-\left(\frac{3(1)+5}{1^2+1}\right)\\ &\approx -0.48\end{aligned}$$

13. Find $f(x+\Delta x)$.

$$\begin{aligned}f(x+\Delta x) &= (x+\Delta x)^2\\ &= x^2+(x^2)'dx\\ &= x^2+2xdx\end{aligned}$$

Evaluate at $x = 10$ and $dx = 2$.

$$\begin{aligned}(x+\Delta x)^2 &= x^2+2xdx\\ &\approx (10)^2+2(10)(2)\\ &\approx 140\end{aligned}$$

15. Find $f(x+\Delta x)$.

$$\begin{aligned}f(x+\Delta x) &= \frac{1}{x+\Delta x}\\ &= \frac{1}{x}+\left(\frac{1}{x}\right)'dx\\ &= \frac{1}{x}-\frac{1}{x^2}dx\end{aligned}$$

Evaluate at $x = 5$ and $dx = 1$.

$$\begin{aligned}(x+\Delta x)^2 &= \frac{1}{x}-\frac{1}{x^2}dx\\ &\approx \frac{1}{5}-\frac{1}{5^2}\cdot 1\\ &\approx 0.16\end{aligned}$$

17. Approximate $\sqrt{50}$ using $\sqrt{49}$.
Then $x = 49$ and $\Delta x = dx = 1$.
From Exercise 14,

$$f(x+\Delta x) = \sqrt{x} + \frac{1}{2\sqrt{x}}dx$$

Evaluate at $x = 49$ and $dx = 1$.

$$\sqrt{49+1} \approx \sqrt{49} + \frac{1}{2\sqrt{49}}\cdot 1$$
$$\approx 7 + \frac{1}{14} \approx 7.0714$$

19. Approximate $\sqrt[3]{9}$ using $\sqrt[3]{8}$.
Then $x = 8$ and $\Delta x = dx = 1$.
Find $f(x+\Delta x)$.

$$\begin{aligned} f(x+\Delta x) &= \sqrt[3]{x+\Delta x} \\ &= \sqrt[3]{x} + \left(\sqrt[3]{x}\right)' dx \\ &= \sqrt[3]{x} + \frac{1}{3\sqrt[3]{x^2}}dx \end{aligned}$$

Evaluate at $x = 8$ and $dx = 1$.

$$\sqrt[3]{8+1} \approx \sqrt[3]{8} + \frac{1}{3\sqrt[3]{8^2}}\cdot 1$$
$$\approx 2 + \frac{1}{12} \approx 2.0833$$

21.
$$\begin{aligned} dI &= (45.4u^{-1.54} - 1)'du \\ &= -1.54(45.4u^{-2.54})du \\ &= -1.54(45.4\cdot 3^{-2.54})\cdot 0.5 \\ &\approx -2.146 \end{aligned}$$

At “low” unemployment, increasing unemployment by just 0.5% lowers the inflation by about 2.1%.

23.
$$\begin{aligned} dV &= (1000(1+0.01r)^5)'dr \\ &= 1000(5(1+0.01r)^4(0.01))dr \\ &= 50(1+0.01r)^4 dr \\ &= 50(1+0.01(6))^4\cdot 0.75 \\ &\approx \$47.34 \end{aligned}$$

$$\begin{aligned} \Delta V &= V(r+\Delta r) - V(r) \\ &= 1000(1+0.01(6+0.75))^5 \\ &\quad - 1000(1+0.01(6))^5 \\ &\approx \$48.02 \end{aligned}$$

The difference, $\Delta V - dV = \$48.02 - \$47.34$ is just 68 cents.

25.
$$\begin{aligned} dR &= (5x^2 - 35x + 104)'dx \\ &= (10x - 35)dx \\ &= (10(2) - 35)\cdot 0.4 \\ &= -6 \end{aligned}$$

$$\begin{aligned} \Delta R &= R(x+\Delta x) - R(x) \\ &= (5(2+0.4)^2 - 35(2+0.4) + 104) \\ &\quad - (5(2)^2 - 35(2) + 104) \\ &= -5.2 \end{aligned}$$

The difference between dR and ΔR is fairly close.

27.
$$\begin{aligned} dV &= (2000 + 80\sqrt{t} - 10t)'dt \\ &= \left(\frac{40}{\sqrt{t}} - 10\right)dt \\ &= \left(\frac{40}{\sqrt{9}} - 10\right)\cdot 1 \\ &\approx \$3.33 \end{aligned}$$

$$\begin{aligned} V(t+\Delta t) &\approx V(t) + V'(t)dt \\ &= (2000 + 80\sqrt{t} - 10t) + \left(\frac{40}{\sqrt{t}} - 10\right)dt \\ &= (2000 + 80\sqrt{9} - 10(9)) + \left(\frac{40}{\sqrt{9}} - 10\right)\cdot 1 \\ &\approx \$2153.33 \end{aligned}$$

29. First find the error for the volume function.

$$\begin{aligned} dV &= (x^3)'dx \\ &= 3x^2 dx \\ &= 3(10)^2\cdot \pm 0.01 \\ &= \pm 3 \text{ in}^3 \end{aligned}$$

Find the relative error.

$$V(10) = 10^3 = 1000 \text{ in}^3$$
$$\frac{dV}{V} = \frac{\pm 3}{1000} = \pm 0.003 \text{ or } \pm 0.3\%$$

31. First find the error for the function.

$$\begin{aligned} dS &= (9.4L^{0.37})'dL \\ &= 3.478L^{-0.63}dL \\ &= 3.478(150)^{-0.63}\cdot \pm 10 \\ &\approx \pm 1.48 \text{ mph} \end{aligned}$$

Find the relative error.

$$S(150) = 9.4(150)^{0.37} \approx 60 \text{ mph}$$
$$\frac{dS}{S} = \frac{\pm 1.48}{60} \approx \pm 0.025 \text{ or } \pm 2.5\%$$

33. $C(13) = \dfrac{125(13)+375}{13+7} = 100$

$$MC(x) = \left(\frac{125x+375}{x+7}\right)' = \frac{(x+7)(125)-(1)(125x+375)}{(x+7)^2} = \frac{500}{(x+7)^2}$$

$$MC(13) = \frac{500}{(13+7)^2} = 1.25$$

$$\Delta x = \frac{60}{100} = 0.6$$

$$C(13.60) \approx C(13) + MC(13)\cdot 0.6 = 100 + 1.25\cdot 0.6 = 100.75$$

The cost of producing 1360 items is \$100.75. At this level of production, the cost increases about \$1.25 per additional hundred items.

35. $R(x) = p(x)\cdot x = \left(10-\sqrt{3x}\right)x = 10x - \sqrt{3}x^{3/2}$

$$R(12) = 10(12) - \sqrt{3}(12)^{3/2} = 48$$

$$MR(x) = \left[10x - \sqrt{3}x^{3/2}\right]' = 10 - \frac{3\sqrt{3}}{2}x^{1/2} = 10 - \frac{3\sqrt{3x}}{2}$$

$$MR(12) = 10 - \frac{3\sqrt{3(12)}}{2} = 1$$

$$\Delta x = \frac{20}{100} = 0.2$$

$$R(12.2) \approx R(12) + MR(12)\cdot 0.2 = 48 + 1\cdot 0.2 = 48.20$$

The revenue is \$4820 when 1220 items are sold. At this level of sales, the revenue increases about \$100 per additional hundred items.

37. $R(x) = p(x)\cdot x = (-0.5x+15)x = -0.5x^2 + 15x$

$$P(x) = R(x) - C(x) = -0.5x^2 + 15x - \frac{80x+160}{x+12}$$

$$P(8) = -0.5(8)^2 + 15(8) - \frac{80(8)+160}{8+12} = 48$$

$$MP(x) = \left[-0.5x^2 + 15x - \frac{80x+160}{x+12}\right]' = -x + 15 - \frac{(x+12)(80)-(1)(80x+160)}{(x+12)^2} = -x + 15 - \frac{800}{(x+12)^2}$$

$$MP(8) = -8 + 15 - \frac{800}{(8+12)^2} = 5$$

$$\Delta x = \frac{30}{100} = 0.3$$

$$P(8.3) \approx P(8) + MP(8)\cdot 0.3 = 48 + 5\cdot 0.3 = 49.50$$

The profit is \$4950 when 830 items are sold.
At this level of production and sales, the profit increases about \$500 per additional hundred items.

39. $\Delta y = dy$ when $y = f(x)$ is a linear function.

41. For $y = ax^b$,

$$\Delta y \approx dy = a\cdot bx^{b-1}dx$$

When $\Delta x = \dfrac{1}{100}x = dx$, then

$$\Delta y \approx \left(a\cdot bx^{b-1}\right)\cdot\left(\frac{x}{100}\right) = a\left(x^{b-1}\cdot x\right)\cdot\frac{b}{100} = (ax^b)\cdot\frac{b}{100} = y\cdot\frac{b}{100}$$

which is a $b\%$ change in y.

43. For $N(t) = 0.00437t^{3.2}$,

$$\Delta N \approx dN = 3.2(0.00437t^{2.2})dt$$

When $\Delta t = \frac{10}{100}t = dt$, then

$$\begin{aligned}\Delta N &\approx 3.2(0.00437t^{2.2})\cdot\left(\frac{t}{10}\right)\\ &= 0.00437\left(t^{2.2}\cdot t\right)\cdot\frac{3.2}{10}\\ &= 0.00437(t^{3.2})\cdot\frac{32}{100}\\ &= N\cdot\frac{32}{100}\end{aligned}$$

which is a 32% increase.

45. The du's cancel in

$$\frac{dy}{dx} = \frac{dy}{du}\cdot\frac{du}{dx}$$

only if $du = g'(x)dx$ is not zero.

REVIEW EXERCISES AND CHAPTER TEST FOR CHAPTER 3

1.

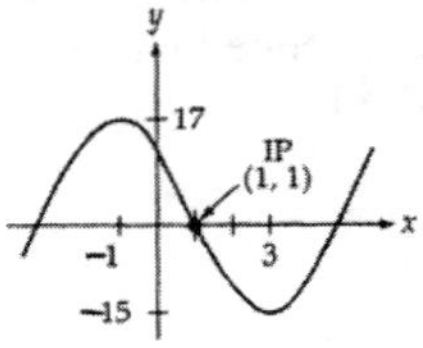

2.

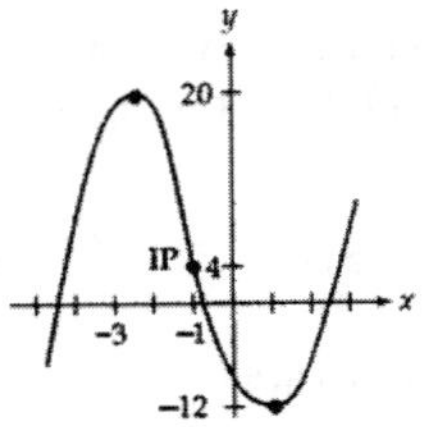

3.

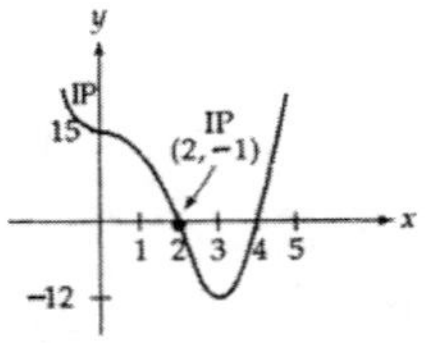

4.

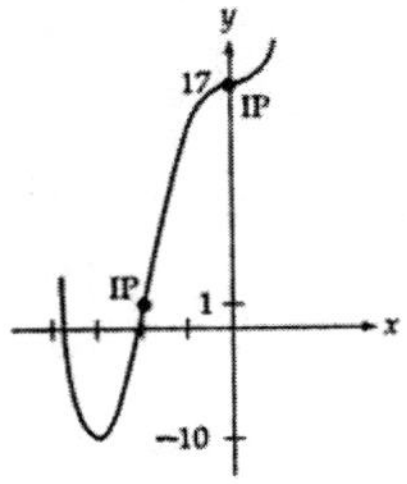

5.

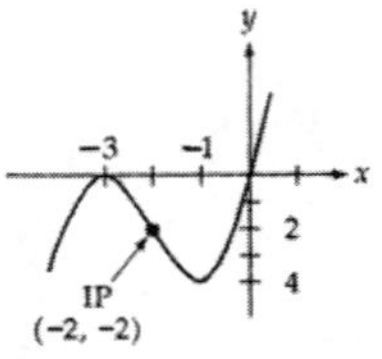

6.

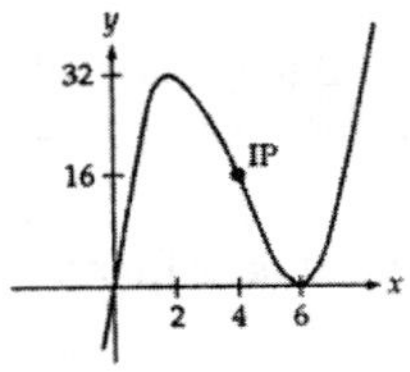

7.

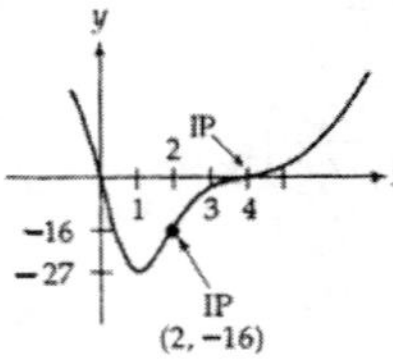

8.

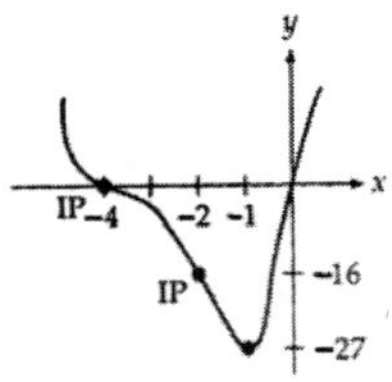

9.

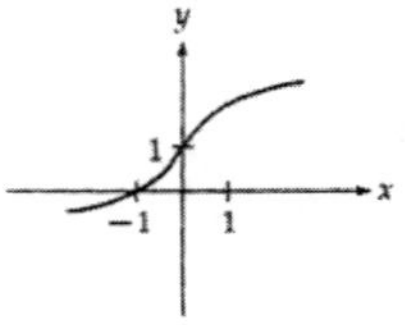

10.

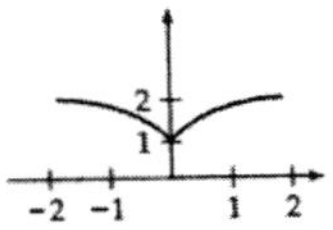

11.

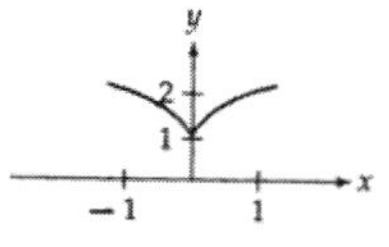

12.

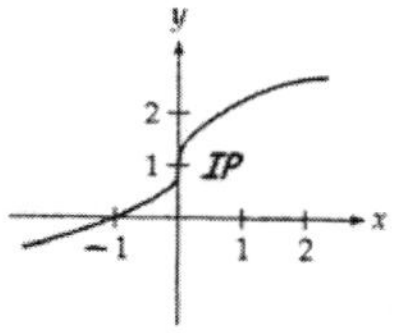

13.

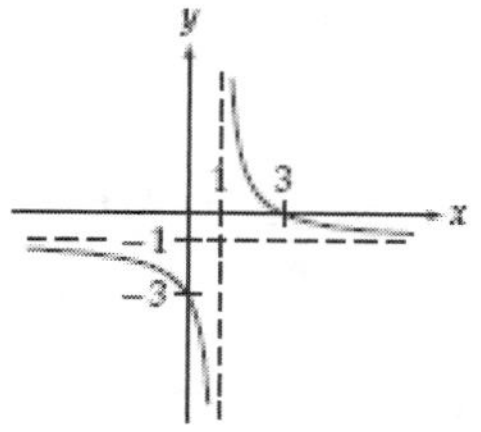

14.

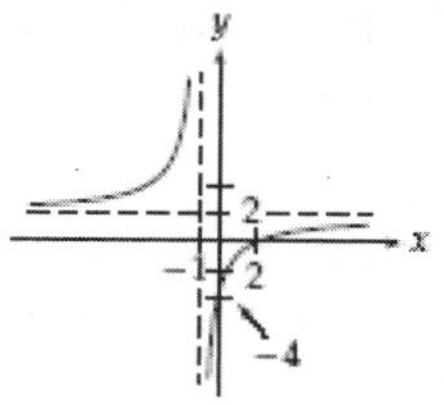

15.

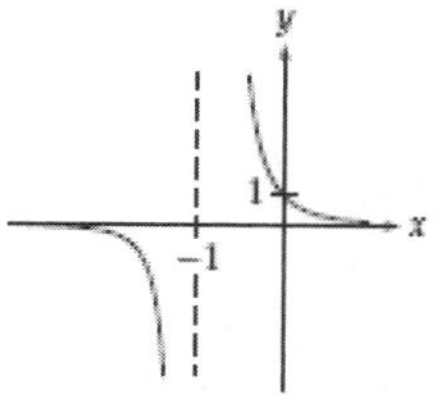

16.

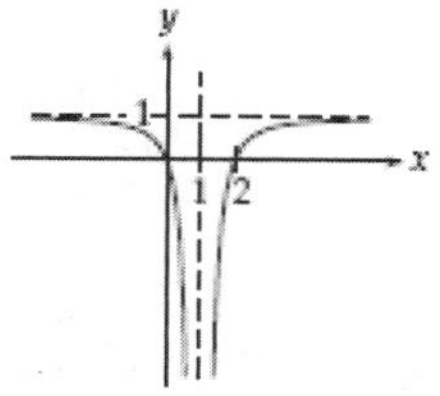

17.

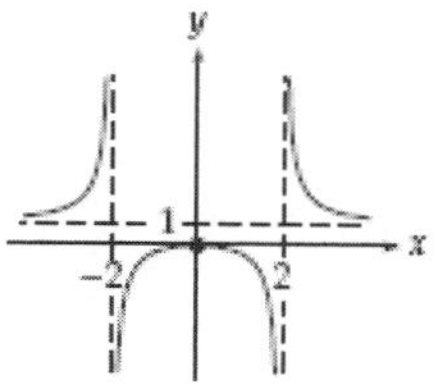

18.

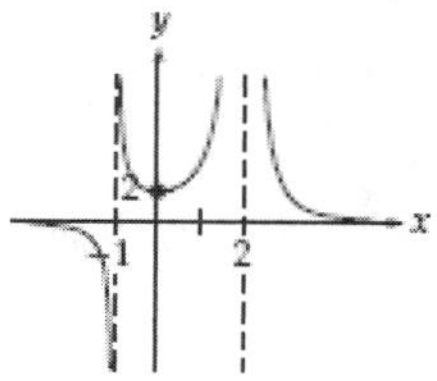

19.

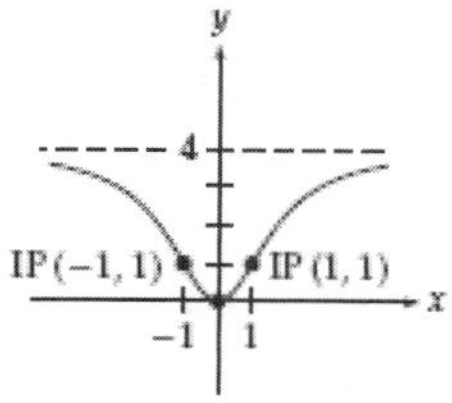

20.

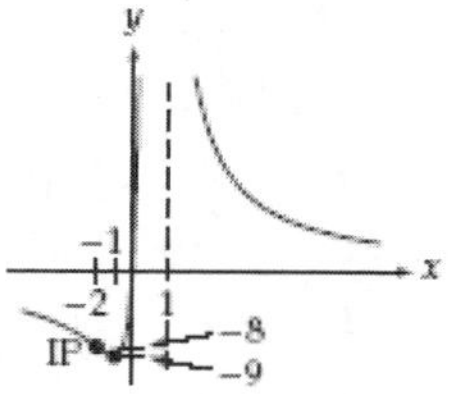

21. $f(x) = 2x^3 - 6x$

$f'(x) = 6x^2 - 6$

$x = \pm 1$

Eliminate $x = -1$ because it is not in the interval. Since f is continuous and [0, 5] is a closed interval, we consider the CV and EP.

CV: $x = 1$ $\quad f(1) = 2(1)^3 - 6(1)$

$= -4$ minimum

EP: $x = 0$ $\quad f(0) = 2(0)^3 - 6(0) = 0$

$x = 5$ $\quad f(5) = 2(5)^3 - 6(5)$

$= 220$ maximum

22. $f(x) = 2x^3 - 24x$

$f'(x) = 6x^2 - 24$

$x = \pm 2$

Eliminate $x = -2$ because it is not in the interval. Since f is continuous and [0, 5] is a closed interval, we consider the CV and EP.

CV: $x = 2$ $\quad f(2) = 2(2)^3 - 24(2)$

$= -32$ minimum

EP: $x = 0$ $\quad f(0) = 2(0)^3 - 24(0) = 0$

$x = 5$ $\quad f(5) = 2(5)^3 - 24(5)$

$= 130$ maximum

23. $f(x)=x^4-4x^3-8x^2+64$

$f'(x)=4x^3-12x^2-16x=0$

$4x(x^2-3x-4)=0$

$x=0,4,-1$

Since f is continuous and [–1, 5] is a closed interval, we consider the CV and EP.

CV: $x=0$ $f(0)=(0)^4-4(0)^3-8(0)^2+64$ $=64$ maximum

$x=4$ $f(4)=(4)^4-4(4)^3-8(4)^2+64$ $=-64$ minimum

$x=-1$ $f(-1)=(-1)^4-4(-1)^3-8(-1)^2+64$ $=61$

EP: $x=5$ $f(5)=(5)^4-4(5)^3-8(5)^2+64$ $=-11$

24. $f(x)=x^4-4x^3+4x^2+1$

$f'(x)=4x^3-12x^2+8x=0$

$4x(x^2-3x+2)=0$

$x=0,1,2$

Since f is continuous and [0, 10] is a closed interval, we consider the CV and EP.

CV: $x=0$ $f(0)=1$ minimum

$x=1$ $f(1)=(1)^4-4(1)^3+4(1)^2+1=2$

$x=2$ $f(2)=(2)^4-4(2)^3+4(2)^2+1$ $=1$ minimum

EP: $x=10$ $f(10)=(10)^4-4(10)^3+4(10)^2+1$ $=6401$ maximum

25. $h(x)=(x-1)^{2/3}$

$h'(x)=\frac{2}{3}(x-1)^{-1/3}$

The critical value is at $x=1$.

Since h is continuous and [0, 9] is a closed interval, we consider the CV and EP.

CV: $x=1$ $h(1)=(1-1)^{2/3}=0$ minimum

EP: $x=0$ $h(0)=(0-1)^{2/3}=1$

$x=9$ $h(9)=(9-1)^{2/3}=4$ maximum

26. $f(x)=\sqrt{100-x^2}=(100-x^2)^{1/2}$

$f'(x)=\frac{1}{2}(100-x^2)^{-1/2}(2x)=0$

The critical values are at $x=0,\pm 10$.

Since f is continuous and [–10, 10] is a closed interval, we consider the CV and EP.

CV: $x=0$ $f(0)=10$ maximum

$x=-10$ $f(-10)=0$ minimum

$x=10$ $f(10)=0$ minimum

27. $g(w)=(w^2-4)^2$

$g'(w)=2(w^2-4)(2w)$

$=4w(w^2-4)=0$

The critical values are at $w=0,\pm 2$.

Since g is continuous and [–3, 3] is a closed interval, we consider the CV and EP.

CV: $w=0$ $g(0)=(0^2-4)^2=16$

$w=2$ $g(2)=(2^2-4)^2=0$ minimum

$w=-2$ $g(-2)=[(-2)^2-4]^2=0$ minimum

EP: $w=-3$ $g(-3)=[(-3)^2-4]^2=25$ maximum

$w=3$ $g(3)=(3^2-4)^2=25$ maximum

28. $g(x)=x(8-x)=8x-x^2$

$g'(x)=8-2x=0$

The critical value is at $x=4$.

Since g is continuous and [0, 8] is a closed interval, we consider the CV and EP.

CV: $x=4$ $g(4)=16$ maximum

EP: $x=0$ $g(0)=0$ minimum

$x=8$ $g(8)=0$ minimum

29. $f(x)=\frac{x}{x^2+1}$

$f'(x)=\frac{x^2+1-2x(x)}{(x^2+1)^2}=\frac{1-x^2}{(x^2+1)^2}=0$

The critical values are at $x=\pm 1$.

Since f is continuous and [–3, 3] is a closed interval, we consider the CV and EP.

CV: $x=1$ $f(1)=\frac{1}{2}$ maximum

$x=-1$ $f(-1)=-\frac{1}{2}$ minimum

EP: $x=-3$ $f(-3)=-\frac{3}{10}$

$x=3$ $f(3)=\frac{3}{10}$

30. $f(x)=\frac{x}{x^2+4}$

$f'(x)=\frac{x^2+4-2x(x)}{(x^2+4)^2}=\frac{4-x^2}{(x^2+4)^2}=0$

The critical values are at $x=\pm 2$.

Since f is continuous and [–4, 4] is a closed interval, we consider the CV and EP.

CV: $x=2$ $f(2)=\frac{1}{4}$ maximum

$x=-2$ $f(-2)=-\frac{1}{4}$ minimum

EP: $x=-4$ $f(-4)=-\frac{1}{5}$

$x=4$ $f(4)=\frac{1}{5}$

31. $E(v) = \frac{v^2}{v-c}$

$E'(v) = \frac{(v-c)(2v)-v^2}{(v-c)^2} = \frac{v^2-2vc}{(v-c)^2} = 0$

$v(v-2c) = 0$

$v = 0, 2c$

$E''(v) = \frac{(v-c)^2(2v-2c)-2(v-c)(v^2-2vc)}{(v-c)^2}$

$E''(2c) = 2c > 0$

So $E(v)$ is minimized.

The tugboat's velocity is $2c$.

32. Since weight is constant and P varies with v, we differentiate P with respect to v.

$P = \frac{aw^2}{v} + bv^3$

$P = -\frac{aw^2}{v^2} + 3bv^2 = 0$

$-aw^2 + 3bv^4 = 0$

$v = \sqrt[4]{\frac{aw^2}{3b}}$

$P' = \frac{2aw^2 + 6bv^4}{v^3} > 0$

since $a, w, v > 0$

so P is minimized.

33. Let $x =$ the length of the fence perpendicular to the wall

Let $y =$ the length of the fence parallel to the wall.

We know that $4x + y = 240$, and we wish to maximize the area.

$A = xy = x(240 - 4x) = 240x - 4x^2$

$A' = 240 - 8x = 0$

$x = 30$

$y = 120$

$A''(x) = -8$

so A is maximized.

The maximum area is 3600 square feet.

34. Let $x =$ length and $y =$ width. Since only 240 feet of fence are used,

$4x + 2y = 240 \qquad (1)$

Area $= xy$

From (1),

$x = \frac{240-2y}{4}$

$= \frac{120-y}{2}$

$A = \left(\frac{120-y}{2}\right)(y)$

$= \frac{120y-y^2}{2}$

We want to maximize the area.

$A' = 60 - y$

$0 = 60 - y$

$y = 60$

The critical value is $y = 60$.

$A'' = -1 < 0$

so area is maximized.

$240 = 4x + 2(60)$

$240 = 4x + 120$

$120 = 4x$

$30 = x$

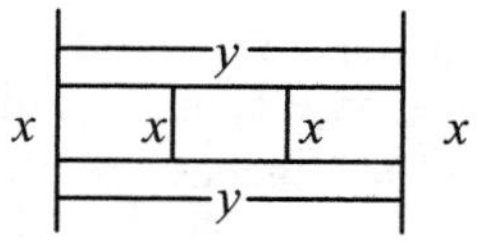

$y = 60$ feet and $x = 30$ feet or each adjacent pen is 20 feet by 30 feet. The total area is 1800 square feet.

35. Let $x =$ the length of the longer side and let $y =$ the length of the shorter side.

Since $xy = 2025$, $x = \dfrac{2025}{y}$.

To minimize the cost, the shorter side should be gilded.

Thus

$$C(x) = 3(2x + y) + 51y$$
$$= 3\left[2\left(\frac{2025}{y}\right) + y\right] + 51y$$
$$= \frac{12{,}150}{y} + 3y + 51y$$
$$= \frac{12{,}150}{y} + 54y$$
$$C'(x) = -\frac{12{,}150}{y^2} + 54 = 0$$
$$y = 15$$
$$C''(x) = \frac{12{,}150(2)}{y^3} > 0 \quad \text{for } y = 15$$

so C is minimized.

The dimensions are 15 cubits by $\frac{2025}{15} = 135$ cubits.

36. Let $x =$ the length of a side of the base and $h =$ the height.

Since $V = 500$,

$$V = x^2 h = 500$$
$$h = \frac{500}{x^2}$$

We wish to minimize the surface area.

$$S = 4(xh) + x^2$$
$$= 4x\left(\frac{500}{x^2}\right) + x^2$$
$$= \frac{2000}{x} + x^2$$
$$S' = -\frac{2000}{x^2} + 2x = 0$$
$$-2000 + 2x^3 = 0$$
$$x^3 = 1000$$
$$x = 10$$
$$S'' = \frac{4000}{x^3} > 0 \quad \text{for } x = 10$$

so S is minimized.

The base is 10 inches by 10 inches and the height is $\frac{500}{10^2} = 5$ inches.

37.
$$V = 16\pi = \pi r^2 h$$
$$h = \frac{16}{r^2}$$
$$A = 2\pi r^2 + 2\pi r h$$
$$= 2\pi r^2 + 2\pi r\left(\frac{16}{r^2}\right) = 2\pi r^2 + \frac{32\pi}{r}$$
$$A' = 4\pi r - \frac{32\pi}{r^2} = 0$$
$$r = 2$$
$$A'' = 4\pi + \frac{64\pi}{r^3} > 0 \quad \text{for } r = 2$$

so A is minimized.

The base radius is 2 inches and the height is $\frac{16}{2^2} = 4$ inches.

38. Since the volume is 8π,

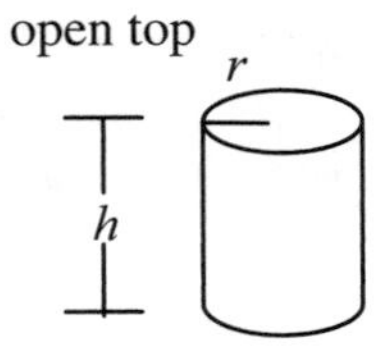

$$V = \pi r^2 h = 8\pi$$
$$h = \frac{8\pi}{\pi r^2} = \frac{8}{r^2}$$
$$A = \pi r^2 + 2\pi r h$$

Then,

$$A = \pi r^2 + (2\pi r)\left(\frac{8}{r^2}\right)$$
$$= \pi r^2 + \frac{16\pi}{r}$$
$$A' = 2\pi r - 16\pi r^{-2}$$
$$0 = 2\pi r - \frac{16\pi}{r^2}$$
$$0 = 2\pi r^3 - 16\pi$$
$$2\pi r^3 = 16\pi$$
$$r^3 = 8$$
$$r = 2$$
$$A'' = 2\pi + 32\pi r^{-3} > 0 \quad \text{for } r > 0$$

so area is maximized.

$$h = \frac{8}{r^2} = \frac{8}{4} = 2$$

The radius of the base is 2 inches and the height is 2 inches.

39. Let $x =$ the number of \$400 decreases. The price function is $p(x) = 2000 - 400x$, and the quantity function is $q(x) = 12 + 3x$.

$R(x) = p(x)q(x) = (2000 - 400x)(12 + 3x)$

$= 24,000 + 1200x - 1200x^2$

The cost function is unit cost times $q(x)$.

$C(x) = 1200q(x)$

$= 1200(12 + 3x) = 14,400 + 3600x$

$P(x) = R(x) - C(x)$

$= 24,000 + 1200x - 1200x^2 - (14,400 + 3600x)$

$= 9600 - 2400x - 1200x^2$

$P(x) = -2400 - 2400x = 0$

$x = -1$

$P'(x) = -2400$

so profit is maximized.

The price is $p(-1) = 2000 - 400(-1) = \$2400$, and the quantity is $q(-1) = 12 + 3(-1) = 9$ computers per week.

40. Let $x =$ the number of additional weeks. The yield is $Y(x) = 100 + 10x$, and the price is $p(x) = 40 - 2x$.

$R(x) = Y(x)p(x)$

$= (100 + 10x)(40 - 2x)$

$= 4000 + 200x - 20x^2$

$R'(x) = 200 - 40x = 0$

$x = 5$ weeks

$R''(x) = -40 < 0$

so R is maximized.

41. Total cost = cost of cable underwater + cost of cable underground

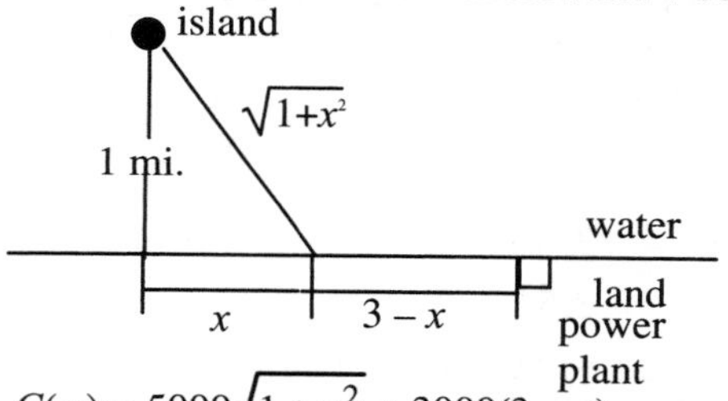

$C(x) = 5000\sqrt{1 + x^2} + 3000(3 - x)$

$= 5000(1 + x^2)^{1/2} + 3000(3 - x)$

We want to minimize cost.

$C''(x) = 2500(1 + x^2)^{-1/2}(2x) - 3000$

$= \dfrac{5000x}{\sqrt{1 + x^2}} - 3000 = 0$

$5000x - 3000\sqrt{1 + x^2} = 0$

$\sqrt{1 + x^2} = \dfrac{5}{3}x$

$1 + x^2 = \dfrac{25}{9}x^2$

$x^2 - \dfrac{25}{9}x^2 + 1 = 0$

$-\dfrac{16}{9}x^2 = -1$

$x^2 = \dfrac{9}{16}$

$x = \dfrac{3}{4}$

$$C''(x) = \frac{\left(\sqrt{1 + x^2}\right)(5000) - 5000x\left(\frac{1}{2}\right)(1 + x^2)^{-1/2}(2x)}{1 + x^2}$$

At $x = \frac{3}{4}$, $C''(x) > 0$ so cost is minimized. Thus, the distance downshore from the island where the cable should meet the land is $\frac{3}{4}$ mile.

42. **a.** $0.624 = 62.4\%$

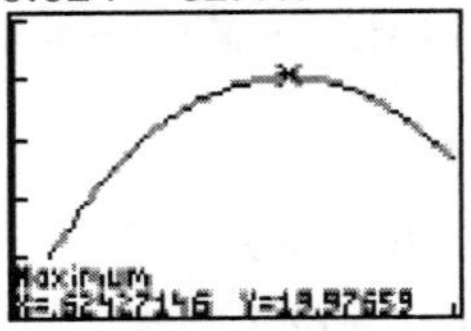

on [0, 1] by [0, 25]

b. \$1.26

43. Since $V = 21.66$,

$V = \pi r^2 h = 21.66$

$h = \dfrac{21.66}{\pi r^2}$

Since the top and bottom are twice as thick as the sides, the area is

$$A = 2(2\pi r^2) + 2\pi rh$$
$$= 4\pi r^2 + 2\pi r\left(\frac{21.66}{\pi r^2}\right)$$
$$= 4\pi r^2 + \frac{43.32}{r}$$
$$A' = 8\pi r - \frac{43.32}{r^2} = 0$$
$$8\pi r^3 - 43.32 = 0$$
$$r \approx 1.2$$
$$A'' = 8\pi + \frac{2(43.32)}{r^3} > 0 \quad \text{at } r = 1.2$$

so A is minimized. The radius is about 1.2 inches, and the height is $\frac{21.66}{\pi(1.2)^2} \approx 4.8$ inches.

44. Let height $= x$. Then length $= 5.5 - x$ and width $= 8.5 - 2x$.

$$V = lwh = x(5.5 - x)(8.5 - 2x)$$
$$= 46.75x - 19.5x^2 + 2x^3$$
$$V' = 46.75 - 39x + 6x^2$$
$$x \approx 1.59$$
$$V'' = -39 + 12x < 0 \quad \text{at } x = 1.59$$

so V is maximized.

The height $x \approx 1.59$ inches and $V \approx 33.07$ cubic inches.

45. Let height $= x$. Then length $= 4 - x$ and width $= 6 - 2x$.

$$V = lwh = x(4 - x)(6 - 2x) = x(24 - 14x + 2x^2)$$
$$= 24x - 14x^2 + 2x^3$$
$$V' = 24 - 28x + 6x^2 = 0$$
$$x \approx 1.13, 3.54$$
$$V'' = -28 + 12x < 0 \quad \text{at } x \approx 1.13$$

and $V''(3.54) > 0$

so V is maximized at $x \approx 1.13$.

Evaluating $x \approx 1.13$, we find that 1.13 inches gives the largest volume, about 12.13 cubic inches.

46. Let $x =$ the number in the print run.

Storage costs $= 4\left(\frac{x}{2}\right) = 2x$

Cost per run $= 200x + 800$

Number of runs $= \frac{900}{x}$

Production costs $= (200x + 800)\left(\frac{900}{x}\right)$

Total cost $C(x) = 2x + (200x + 800)\left(\frac{900}{x}\right)$

$$= 2x + 180{,}000 + \frac{720{,}000}{x}$$

$$C'(x) = 2 - \frac{720{,}000}{x^2} = 0$$

$$x^2 = 360{,}000$$

$$x = 600 \text{ rolls}$$

$$C''(x) = \frac{1{,}440{,}000}{x^3} > 0 \quad \text{for } x = 600$$

so C is minimized.
There should be 600 rolls per run, and the number of press runs is $\frac{900}{600} = 1.5$.

47. Let $x =$ order size

Storage costs $= 200\left(\frac{x}{2}\right) = 100x$

Cost per order $= 300x + 500$

Number of orders $= \frac{500}{x}$

Reorder costs $= (300x + 500)\left(\frac{500}{x}\right)$

Total cost $C(x) = 100x + (300x + 500)\left(\frac{500}{x}\right)$

$$= 100x + 150{,}000 + \frac{250{,}000}{x}$$

$$C'(x) = 100 - \frac{250{,}000}{x^2} = 0$$

$$100x^2 - 250{,}000 = 0$$

$$x^2 = 2500$$

$$x = 50$$

$$C''(x) = \frac{500{,}000}{x^3} > 0 \quad \text{for } x = 50$$

so C is minimized.
There should be 50 motorbikes per order, and $\frac{500}{50} = 10$ orders per year.

48. $f(p) = -0.02p^2 + 7p$

$f'(p) = -0.04p + 7 = 1$

$p = 150$ thousand

$f''(p) = -0.04 < 0$ so $f(p)$ is maximized.

$Y(p) = f(p) - p$

$Y(150) = f(150) - 150$

$-0.02(150)^2 + 7(150) - 150$

$= 450$ thousand

The population that gives the maximum sustainable yield is 150,000, and the maximum sustainable yield is 450,000.

49. $f(p) = 60\sqrt{p} = 60p^{1/2}$

$f'(p) = 30p^{-1/2} = \frac{30}{\sqrt{p}} = 1$

$\sqrt{p} = 30$

$p = 900$ thousand

$f''(p) = -15p^{-3/2} < 0$ for $p = 900$ so $f(p)$ is maximized.

$Y(p) = f(p) - p$

$Y(900) = f(900) - 900$

$= 60\sqrt{900} - 900$

$= 1800 - 900 = 900 = 900$ thousand

The population that gives the maximum sustainable yield is 900,000, and the maximum sustainable yield is 900,000.

50. $6x^2 + 8xy + y^2 = 100$

$$12x + 8y + 8x\frac{dy}{dx} + 2y\frac{dy}{dx} = 0$$

$$(8x + 2y)\frac{dy}{dx} = -(12x + 8y)$$

$$\frac{dy}{dx} = -\frac{6x+4y}{4x+y}$$

51. $8xy^2 - 8y = 1$

$$8y^2 + 8x\left(2y\frac{dy}{dx}\right) - 8\frac{dy}{dx} = 0$$

$$16xy\frac{dy}{dx} - 8\frac{dy}{dx} = -8y^2$$

$$\frac{dy}{dx} = -\frac{y^2}{2xy-1}$$

52. $2xy^2 - 3x^2y = 0$

$$2y^2 + 2x\left(2y\frac{dy}{dx}\right) - 6xy - 3x^2\frac{dy}{dx} = 0$$

$$4xy\frac{dy}{dx} - 3x^2\frac{dy}{dx} = 6xy - 2y^2$$

$$\frac{dy}{dx} = \frac{6xy-2y^2}{4xy-3x^2}$$

53. $\sqrt{x} - \sqrt{y} = 10$

$$x^{1/2} - y^{1/2} = 10$$

$$\frac{1}{2}x^{-1/2} - \frac{1}{2}y^{-1/2}\frac{dy}{dx} = 0$$

$$\frac{dy}{dx} = \frac{x^{-1/2}}{y^{-1/2}} = \frac{y^{1/2}}{x^{1/2}}$$

54. $x + y = xy$

$$1 + \frac{dy}{dx} = y + x\frac{dy}{dx}$$

$$\frac{dy}{dx} = \frac{1-y}{x-1}$$

At $x = 2$, $y = 2$,

$$\frac{dy}{dx} = -\frac{1}{1} = -1$$

55. $y^3 - y^2 - y = x$

$$3y^2\frac{dy}{dx} - 2y\frac{dy}{dx} - \frac{dy}{dx} = 1$$

$$(3y^2 - 2y - 1)\frac{dy}{dx} = 1$$

$$\frac{dy}{dx} = \frac{1}{3y^2-2y-1}$$

At $x = 2$, $y = 2$,

$$\frac{dy}{dx} = \frac{1}{3(2)^2-2(2)-1} = \frac{1}{7}$$

56. $xy^2 = 81$

$$y^2 + x\left(2y\frac{dy}{dx}\right) = 0$$

$$\frac{dy}{dx} = \frac{-y^2}{2xy} = -\frac{y}{2x}$$

At $x = 9$, $y = 3$,

$$\frac{dy}{dx} = -\frac{3}{18} = -\frac{1}{6}$$

57. $x^2y^2 - xy = 2$

$$2xy^2 + x^2\left(2y\frac{dy}{dx}\right) - y - x\frac{dy}{dx} = 0$$

$$2xy^2 - y + (2x^2y - x)\frac{dy}{dx} = 0$$

$$\frac{dy}{dx} = \frac{y-2xy^2}{2x^2y-x}$$

At $x = -1$, $y = 1$,

$$\frac{dy}{dx} = \frac{1-2(-1)(1)^2}{2(-1)^2(1)-(-1)} = \frac{1+2}{2+1} = 1$$

58. $y^2 = x^3 - 4x + 4$

Use implicit differentiation.

$$2y\frac{dy}{dx} = 3x^2 - 4$$

$$\frac{dy}{dx} = \frac{3x^2-4}{2y}$$

At (2, 2), the slope is

$$\frac{dy}{dx} = \frac{3(2)^2-4}{2(2)} = 2$$

The tangent line has slope 2 at (2, 2).

$$y - 2 = 2(x-2)$$

$$y = 2x - 2$$

59. $2x^2+3y^2=2xy+15$

Use implicit differentiation.

$$4x+6y\frac{dy}{dx}=2\left(y+x\frac{dy}{dx}\right)$$

At (–2, 1), the slope is

$$4(-2)+6(1)\frac{dy}{dx}=2\left(1+(-2)\frac{dy}{dx}\right)$$
$$-8+6\frac{dy}{dx}=2-4\frac{dy}{dx}$$
$$\frac{dy}{dx}=1$$

The tangent line has slope 1 at (–2, 1).

$$y-1=1(x+2)$$
$$y=x+3$$

60. Since the edge and the volume are decreasing with time, we differentiate with respect to t.

$V=x\cdot x\cdot x=x^3$

$$\frac{dV}{dt}=3x^2\frac{dx}{dt}$$

Since $\frac{dx}{dt}=2$ at $x=10$,

$\frac{dV}{dt}=3(10)^2(2)=600$ cubic inches per hour.

61. Since sales and profit are increasing with time, we differentiate with respect to t.

$P(x)=2x^2-20x$

$$\frac{dP}{dt}=4x\frac{dx}{dt}-20\frac{dx}{dt}$$

Since $\frac{dx}{dt}=30$ and $x=40$,

$$\frac{dP}{dt}=4(40)(30)-20(30)$$

$=4800-600=\$4200$ per day

62. Since revenue and sales are increasing with time, we differentiate with respect to t.

$R=x^2+500x$

$$\frac{dR}{dt}=2x\frac{dx}{dt}+500\frac{dx}{dt}$$

Since $\frac{dx}{dt}=50$ and $x=200$,

$\frac{dR}{dt}=2(200)(50)+500(50)=\$45{,}000$ per month.

63. **a.** $\frac{dr}{dt}=-0.1$ cm per minute

$$V=\frac{4}{3}\pi r^3$$

Differentiating with respect to t we get

$$\frac{dV}{dt}=4\pi r^2\frac{dr}{dt}$$

when $r=0.5$ cm,

$$\frac{dV}{dt}=4\pi(0.5)^2(-0.1)$$

$=-0.1\pi\approx-0.31$ cm^3 per minute

b. when $r=0.2$ cm,

$$\frac{dV}{dt}=4\pi(0.2)^2(-0.1)$$

$=-0.016\pi\approx-0.05$ cm^3 per minute

64.
$$dy=(x-\sqrt{x})'dx$$
$$=\left(1-\frac{1}{2\sqrt{x}}\right)dx$$
$$=\left(1-\frac{1}{2\sqrt{4}}\right)\cdot0.02$$
$$=0.015$$

65.
$$dy=\left(x^2+\frac{4}{x}\right)'dx$$
$$=\left(2x-\frac{4}{x^2}\right)dx$$
$$=\left(2(2)-\frac{4}{2^2}\right)\cdot0.01$$
$$=0.03$$
$$\Delta y=f(x+\Delta x)-f(x)$$
$$=\left((2+0.01)^2+\frac{4}{2+0.01}\right)-\left(2^2+\frac{4}{2}\right)$$
$$\approx0.0301$$

66. Approximate $\sqrt{35}$ using $\sqrt{36}$.
Then $x = 36$ and $\Delta x = dx = -1$.
Find $f(x+\Delta x)$.

$$f(x+\Delta x) = \sqrt{x+\Delta x}$$
$$= \sqrt{x} + \left(\sqrt{x}\right)' dx$$
$$= \sqrt{x} + \frac{1}{2\sqrt{x}} dx$$

Evaluate at $x = 36$ and $dx = -1$.

$$\sqrt{36-1} \approx \sqrt{36} + \frac{1}{2\sqrt{36}} \cdot (-1)$$
$$\approx 6 - \frac{1}{12} \approx 5.917$$

67. First find the error for the surface area function.

$$dS = (6x^2)' dx$$
$$= 12x dx$$
$$= 12(10) \cdot \pm 0.01$$
$$= \pm 1.2 \text{ in}^2$$

Find the relative error.

$$S(10) = 600 \text{ in}^2$$
$$\frac{dS}{S} = \frac{\pm 1.2}{600} = \pm 0.002 \text{ or } \pm 0.2\%$$

68. $C(12) = 135\sqrt[3]{2(12)+3} = 405$

$$MC(x) = \left(135\sqrt[3]{2x+3}\right)' = \frac{135}{3}(2x+3)^{-2/3}(2) = \frac{90}{\sqrt[3]{(2x+3)^2}}$$

$$MC(12) = \frac{90}{\sqrt[3]{(2(12)+3)^2}} = 10$$

$$\Delta x = \frac{-400}{1000} = -0.4$$

$$C(11.6) \approx C(12) + MC(12) \cdot (-0.4)$$
$$= 405 + 10(-0.4)$$
$$= \$401$$

At this level of production, the cost increases about \$10 per additional thousand items.

CUMULATIVE REVIEW—CHAPTERS 1–3

1. First, find the slope.

$$m = \frac{-2-3}{6-(-4)} = -\frac{5}{10} = -\frac{1}{2}$$

Now use the point-slope form with (–4, 3).

$$y - 3 = -\frac{1}{2}[x - (-4)]$$
$$y - 3 = -\frac{1}{2}x - 2$$
$$y = -\frac{1}{2}x + 1$$

2. $\left(\frac{4}{25}\right)^{-1/2} = \left(\frac{25}{4}\right)^{1/2} = \frac{5}{2}$

3.

x	$(1+3x)^{1/x}$
1×10^{-1}	13.7858
1×10^{-2}	19.2186
1×10^{-4}	20.0765
1×10^{-8}	20.0855
1×10^{-12}	20.0855

$$\lim_{x \to 0} (1+3x)^{1/x} = 20.086$$

4. **a.**

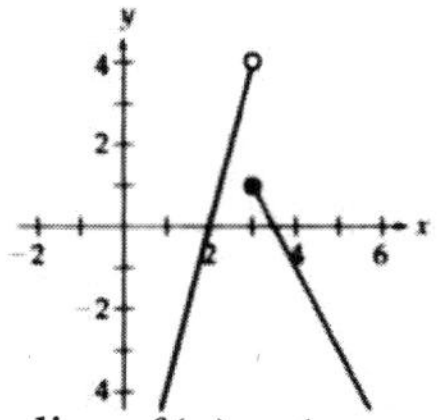

b. $\lim_{x\to 3^-} f(x) = 4$

c. $\lim_{x\to 3^+} f(x) = 1$

d. $\lim_{x\to 3} f(x)$ does not exist

e. $f(x)$ is discontinuous at $x = 3$.

5. $$\lim_{h\to 0}\frac{f(x+h)-f(x)}{h} = \lim_{h\to 0}\frac{2(x+h)^2-5(x+h)+7-(2x^2-5x+7)}{h}$$
$$= \lim_{h\to 0}\frac{2x^2+4xh+2h^2-5x-5h+7-2x^2+5x-7}{h}$$
$$= \lim_{h\to 0} 4x+2h-5 = 4x-5$$

6. $f(x) = 8\sqrt{x^3} - \frac{3}{x^2} + 5 = 8x^{3/2} - 3x^{-2} + 5$

$f'(x) = 12x^{1/2} + 6x^{-3}$

7. $f(x) = (x^5-2)(x^4+2)$

$f'(x) = (5x^4)(x^4+2) + (x^5-2)(4x^3)$
$= 5x^8 + 10x^4 + 4x^8 - 8x^3$
$= 9x^8 + 10x^4 - 8x^3$

8. $f(x) = \frac{2x-5}{3x-2}$

$f'(x) = \frac{(3x-2)(2)-3(2x-5)}{(3x-2)^2} = \frac{6x-4-6x+15}{(3x-2)^2}$
$= \frac{11}{(3x-2)^2}$

9. $y = \sqrt{x} - \frac{16}{x^2} = x^{1/2} - 16x^{-2}$

$y(4) = \sqrt{4} - \frac{16}{4^2} = 1$

The point is (4, 1).

$y'(x) = \frac{1}{2}x^{-1/2} - 16(-2)x^{-3} = \frac{1}{2\sqrt{x}} + \frac{32}{x^3}$

$y'(4) = \frac{1}{2\sqrt{4}} + \frac{32}{4^3} = \frac{3}{4}$

The tangent line has slope $\frac{3}{4}$ at (4, 1).

$y - 1 = \frac{3}{4}(x-4)$
$y = \frac{3}{4}x - 2$

10. $P(x) = 3600x^{2/3} + 250{,}000$

$P'(x) = 2400x^{-1/3}$ $\quad P'(8) = \frac{2400}{2} = 1200$

$P''(x) = -800x^{-4/3}$ $\quad P''(8) = -\frac{800}{2(8)} = -50$

$P'(8) = 1200$, so in 8 years the population will be increasing by 1200 people per year.
$P''(8) = -50$, so in 8 years the rate of growth will be slowing by 50 people per year per year.

11. $\frac{d}{dx}\left(\sqrt{2x^2-5}\right) = \frac{d}{dx}(2x^2-5)^{1/2}$
$= \frac{1}{2}(2x^2-5)^{-1/2}(4x)$
$= \frac{2x}{\sqrt{2x^2-5}}$

12. $\frac{d}{dx}[(3x+1)^4(4x+1)^3] = 4(3x+1)^3(3)(4x+1)^3 + (3x+1)^4(3)(4x+1)^2(4)$

$= 12(3x+1)^3(4x+1)^3 + 12(3x+1)^4(4x+1)^2$

$= 12(3x+1)^3(4x+1)^2[(4x+1)+(3x+1)]$

$= 12(3x+1)^3(4x+1)^2(7x+2)$

13. $\frac{d}{dx} = \left(\frac{x-2}{x+2}\right)^3$

$= 3\left(\frac{x-2}{x+2}\right)^2\left[\frac{x+2-(x-2)}{(x+2)^2}\right] = \frac{12(x-2)^2}{(x+2)^4}$

14. $y = \frac{4(x+3)}{\sqrt{x^2+3}}$

$y(-1) = \frac{4(-1+3)}{\sqrt{(-1)^2+3}} = 4$

The point is (–1, 4).

$y'(x) = \frac{\sqrt{x^2+3}(4) - 4(x+3)\frac{1}{2}(x^2+3)^{-1/2}(2x)}{x^2+3}$

$= \frac{4\sqrt{x^2+3} - 4x(x+3)(x^2+3)^{-1/2}}{x^2+3}$

$y'(-1) = \frac{4\sqrt{(-1)^2+3} - 4(-1)(-1+3)((-1)^2+3)^{-1/2}}{(-1)^2+3}$

$= 3$

The tangent line has slope 3 at (–1, 4).

$y - 4 = 3(x+1)$

$y = 3x + 7$

15. $f(x) = x^3 - 12x^2 - 60x + 400$

$f'(x) = 3x^2 - 24x - 60 = 0$

$x^2 - 8x - 20 = 0$

$(x-10)(x+2) = 0$

The critical values are $x = 10$ and $x = -2$.

$f' > 0$ | $f' = 0$ | $f' < 0$ | $f' = 0$ | $f' > 0$

$x = -2$ | $x = 10$

↗ → ↘ → ↗

$f''(x) = 6x - 24$

$f'' < 0$ | $f'' > 0$

$x = 4$

con dn | con up

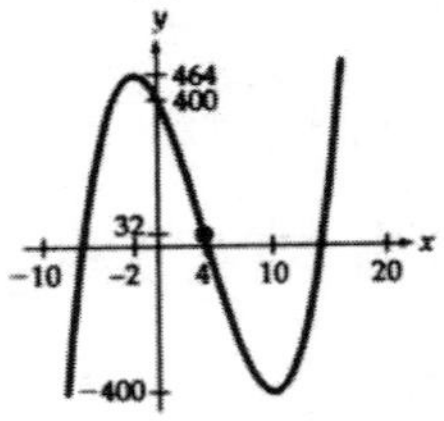

16. $f(x) = \sqrt[3]{x^2} - 1 = x^{2/3} - 1$

$f'(x) = \frac{2}{3}x^{-1/3} = 0$

The critical value is $x = 0$.

$f' < 0$ | $f' > 0$

$x = 0$

↘ → ↗

$f''(x) = -\frac{2}{9}x^{-4/3}$

The critical value is $x = 0$.

$f'' < 0$ | $f'' < 0$

$x = 0$

con dn | con dn

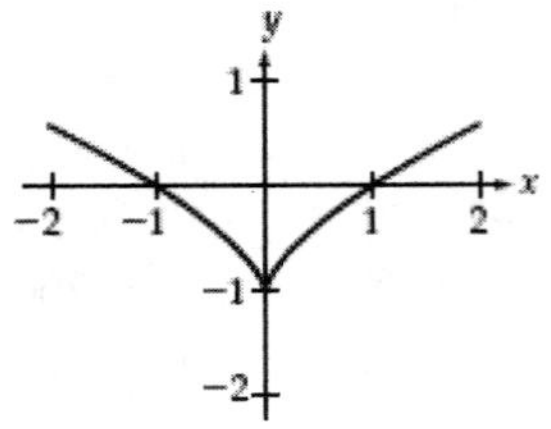

17. Let x = the length of a pen and y = the width of a pen. Then

$$3x + 4y = 600$$
$$x = 200 - \frac{4}{3}y$$
$$A = 2xy$$
$$= 2y\left(200 - \frac{4}{3}y\right)$$
$$= 400y - \frac{8}{3}y^2$$
$$A' = 400 - \frac{16}{3}y = 0$$
$$\frac{16}{3}y = 400$$
$$y = 75 \text{ feet}$$
$$x = 200 - \frac{4}{3}(75) = 100 \text{ feet}$$

Total area = $2xy = 2(100)(75) = 15{,}000$ square feet

18. Let x = the number of \$10 reductions.

$$p(x) = 200 - 10x$$
$$q(x) = 12 + 2x$$
$$R(x) = p(x)q(x) = (200 - 10x)(12 + 2x)$$
$$= 2400 - 120x + 400x - 20x^2$$
$$= 2400 + 280x - 20x^2$$
$$C(x) = \text{unit cost} \cdot q(x) = 80(12 + 2x) = 960 + 160x$$

$$P(x) = R(x) - C(x)$$
$$= 1440 + 120x - 20x^2$$
$$P'(x) = 120 - 40x = 0$$
$$x = 3$$

$P''(x) = -40$, so P is maximized.

Price = $p(3) = 200 - 30 = \$170$.

Quantity = $q(3) = 12 + 6 = 18$ answering machines per day.

19.

$$x^3 + 9xy^2 + 3y = 43$$
$$3x^2 + 9y^2 + 9x\left(2y\frac{dy}{dx}\right) + 3\frac{dy}{dx} = 0$$
$$(18xy + 3)\frac{dy}{dx} = -3x^2 - 9y^2$$
$$\frac{dy}{dx} = \frac{-x^2 - 3y^2}{6xy + 1}$$

At $x = 1, y = 2$.

$$\frac{dy}{dx} = \frac{-1^2 - 3(2)^2}{6(1)(2) + 1} = -\frac{13}{13} = -1$$

20.

$$V = \frac{4}{3}\pi r^3$$
$$\frac{dV}{dr} = 4\pi r^2$$
$$dr = \frac{dV}{4\pi r^2}$$

Since $dV = 32$ and $r = 2$, we have

$$dr = \frac{32}{4\pi(2)^2} = \frac{32}{16\pi} = \frac{2}{\pi}$$

The radius is increasing at a rate of $\frac{2}{\pi}$ feet per minute.

21.

$$dy = \left(\sqrt{x^3 + 6x - 11}\right)' dx$$
$$= \left(\frac{1}{2}(x^3 + 6x - 11)^{-1/2}(3x^2 + 6)\right)dx$$
$$= \frac{3x^2 + 6}{2\sqrt{x^3 + 6x - 11}}dx$$
$$= \frac{3(2)^2 + 6}{2\sqrt{2^3 + 6(2) - 11}} \cdot (-0.04)$$
$$= -0.12$$
$$\Delta y = f(x + \Delta x) - f(x)$$
$$= \sqrt{(2 - 0.04)^3 + 6(2 - 0.04) - 11}$$
$$- \sqrt{2^3 + 6(2) - 11}$$
$$\approx -0.1208$$

22. First find the error for the function.

$$dD = (16t^2)'dt$$
$$= 32t\,dt$$
$$= 32(2.5) \cdot \pm 0.1$$
$$\approx \pm 8 \text{ feet}$$

Find the relative error.

$$D(2.5) = 16(2.5)^2 = 100 \text{ feet}$$
$$\frac{dD}{D} = \frac{\pm 8}{100} = \pm 0.08 \text{ or } \pm 8\%$$

Chapter 4: Exponential and Logarithmic Functions

EXERCISES 4.1

1. **a.** $e^2 \approx 7.389$

b. $e^{-2} \approx 0.135$

c. $e^{1/2} \approx 1.649$

3. **a.** $e^5 e^{-2} = e^{5+(-2)} = e^3$

b. $\dfrac{e^5}{e^3} = e^{5-3} = e^2$

c. $\dfrac{e^5 e^{-1}}{e^{-2} e} = \dfrac{e^5 e^{-1}}{e^{-2} e^1} = \dfrac{e^{5+(-1)}}{e^{-2+1}} = \dfrac{e^4}{e^{-1}} = e^{4-(-1)} = e^5$

5.

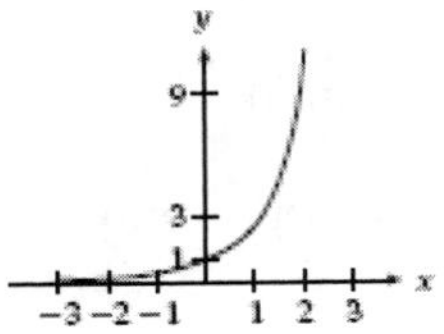

7.

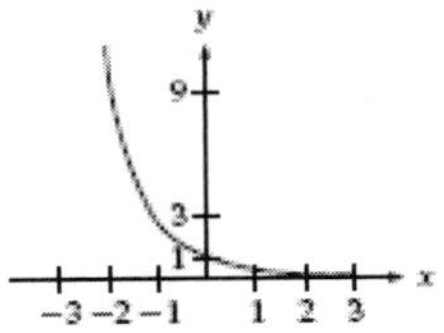

9. 5.697

11. **a.**

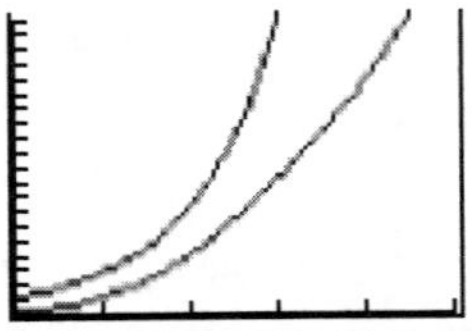

on [0, 5] by [0, 20]

e^x is higher than x^2.

b.

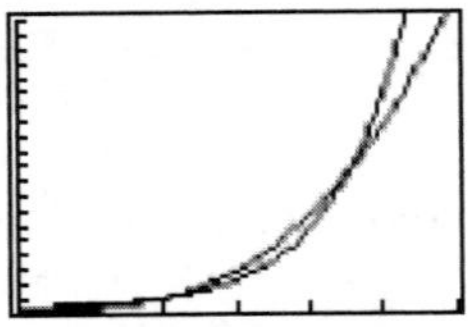

on [0, 6] by [0, 200]

e^x is higher than x^3 for large values of x.

c.

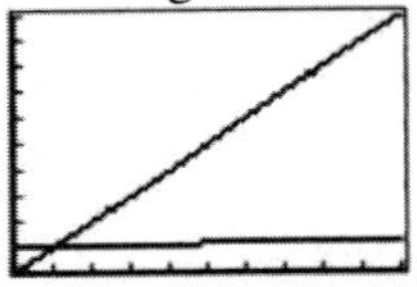

on [0, 10] by [0, 10,000]

e^x is higher than x^4 for large values of x.

d.

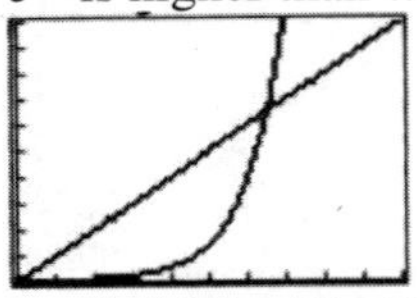

on [0, 15] by [0, 1,000,000]

e^x is higher than x^5 for large values of x.

e. e^x will exceed any power of x for large enough values of x.

13. **a.** For $m = 1$ (annual compounding),

$P\left(1+\dfrac{r}{m}\right)^{mt}$ simplifies to $P(1+r)^t$.

When $P = 1000$, $r = 0.1$, and $t = 8$, the value is

$1000(1+0.1)^8 = 1000(1.1)^8 = 2143.59$

The value is \$2143.59.

b. For quarterly compounding, $m = 4$, $P = 1000$, $r = 0.1$, and $t = 8$. Thus,

$$1000(1+\tfrac{0.1}{4})^{4\cdot 8} = 1000(1+0.025)^{4\cdot 8}$$
$$= 1000(1.025)^{32} = 2203.76$$

The value is \$2203.76.

c. For continuous compounding, $P = 1000$, $r = 0.10$, and $t = 8$. Thus,

$Pe^{rt} = 1000e^{0.1\cdot 8} - 1000e^{0.8} = 2225.54$

The value is \$2225.54.

15. **a.** $P = 100$, $r = 0.02$, and $t = 3$, compounded weekly, which is 52 times per year. This gives a value of

$$\begin{aligned} 100(1+0.02)^{52\cdot 3} &= 100(1.02)^{52\cdot 3} \\ &= 100(1.02)^{156} \\ &= \$2196 \end{aligned}$$

b. The "vig" is equal to the amount owed after three years minus the amount loaned. This is $\$2196 - \$100.00 = \$2096$.

17. The stated rate of 9.25% (compounded daily) is the nominal rate of interest. To determine the effective rate of interest, use the compound interest formula, $P(1+r)^m$, with $r = \frac{9.25\%}{\text{number of days}}$ and m = number of days in a year. Since some banks use 365 days and some use 360 in a year, we will try both ways.
If $m = 365$ days, then

$$r = \frac{9.25\%}{365} = \frac{0.0925}{365} \approx 0.0002534$$

Then $P(1+r)^m = P(1.0002534)^{365} \approx 1.0969P$.
Subtracting 1 gives 0.0969, which expressed as a percent gives the effective rate of interest as 9.69%.

If $m = 360$ days, then $r = \frac{0.0925}{360} \approx 0.0002569$.

Then $P(1+r)^m = P(1.0002569)^{360} \approx 1.0969P$ and the effective rate is also 9.69%.
Thus, the error in the advertisement is 9.825%. The annual yield should be 9.69% (based on the nominal rate of 9.25%).

19. If the amount of money P invested at 8% compounded quarterly yields \$1,000,000 in 60 years, then $\frac{r}{m} = \frac{0.08}{4} = 0.02$ and $mt = 60 \cdot 4 = 240$.

$$\begin{aligned} 1{,}000{,}000 &= P(1+0.02)^{240} \\ P &= \frac{1{,}000{,}000}{(1+0.02)^{240}} \approx \$8629 \end{aligned}$$

21. For compounding annually, $r = 0.06$, $m = 1$, $t = 45$.

$$\text{Present value} = \frac{P}{(1+r)^t} = \frac{1000}{(1+0.06)^{45}} \approx \$72.65$$

23. For 6% compounded annually, $r = 0.06$ $m = 1$, and $t = 12$.

$$\text{Present value} = \frac{P}{(1+r)^t} = \frac{1000}{(1+0.06)^{12}} \approx \$496.97$$

25. To compare two interest rates that are compounded differently, convert them both to annual yields.
10% compounded quarterly:

$$P(1+r)^m = P(1.025)^4 \approx P(1.1038)$$

Subtracting 1, $1.1038 - 1 = 0.1038$
The effective rate of interest is 10.38%.
9.8% compounded continuously:

$$Pe^{rt} = Pe^{0.098} \approx P(1.1030)$$

Subtracting 1: $1.1030 - 1 = 0.1030$
The effective rate of interest is 10.30%.
Thus, 10% compounded quarterly is better than 9.8% compounded continuously.

27. Since the depreciation is 35% per year, $r = -0.35$.

a. $P(1+r)^t = 19{,}700(1-0.35)^4 \approx \$3517$

b. $P(1+r)^t = 19{,}700(1-0.35)^{0.5} \approx \$15{,}883$

29. Since 2030 is 20 years after 2010, $x = 20$.
$6.9e^{0.00743(20)} \approx 8$ billion

31. **a.** $1-(0.9997)^{100(25)} \approx 0.5277$

b. $1-(0.9997)^{100(40)} \approx 0.6989$

33. The proportion of light that penetrates to a depth of x feet is given by $e^{-0.44x}$.

a. If the depth is 3 feet, $e^{-0.44x} = e^{-0.44(3)} = e^{-1.32} \approx 0.267$ or 26.7%

b. If the depth is 10 feet, $e^{-0.44x} = e^{-0.44(10)} = e^{-4.4} \approx 0.0123$ or 1.23%

35. **a.** $f(24) = 2e^{-0.018(24)} \approx 1.3$ mg

b. $f(48) = 2e^{-0.018(48)} \approx 0.8$ mg

37. $S(10) = 100 + 800e^{-0.2(10)} \approx 208$

39. **a.** After 15 minutes, $t = \frac{15}{60} = \frac{1}{4} = 0.25$ and

$T(0.25) = 70 + 130e^{-1.8(0.25)} \approx 153$ degrees

b. After 30 minutes, $t = \frac{30}{60} = 0.5$ and

$T(0.5) = 70 + 130e^{-1.8(0.5)} \approx 123$ degrees

41. If $S = 400$, $x = 10$, and $r = 0.01$, then the Reed-Frost model is $I = 400(1 - e^{-(0.01)(10)}) \approx 38$. The model estimates there will be about 38 newly infected people.

43. We have $P = \$8000$, $mt = 4$, and the value after 4 years is \$10,291.73.

$$10{,}291.73 = 8000(1+r)^4$$

$$\sqrt[4]{\frac{10{,}291.73}{8000}} = 1 + r$$

$$r = \sqrt[4]{\frac{10{,}291.73}{8000}} - 1 \approx 0.065 = 6.5\%$$

45. The air pressure is $(1-0.004)^{a/100}$ percent of the original pressure, where a is the change in altitude.

$$(1-0.004)^{\left(\frac{7347-30}{100}\right)} \approx .75 = 75\%$$

So the air pressure decreased by about 100 – 75 = 25%.

47. **a.**

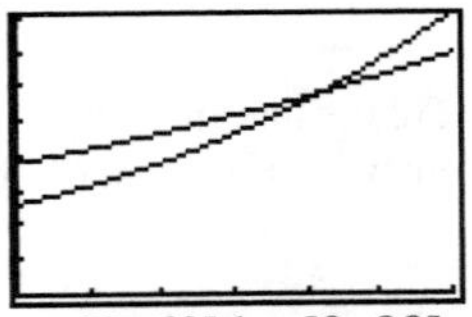

on [0, 60] by [0, 80]

b. During the year 2054 $(x \approx 42)$

49. Find the annual interest rate. For $m = 1$, $t = 365$ (1% per day compounding 365 days), $r = 0.01$,

$$\left(1 + \frac{0.01}{1}\right)^{1 \cdot 365} = (1+r)^1$$

$$r = (1.01)^{365} - 1 \approx 36.78$$

The interest rate is about 3678% per year.

51. $\$40{,}000e^{0.195(2)} \approx \$59{,}079$

53. $\lim_{x\to\infty} e^x = \infty$ and $\lim_{x\to-\infty} e^x = 0$. The x-axis ($y = 0$) is a horizontal asymptote.

55. For very large values of x, e^x will be largest because it will have the largest exponent.

57. That its growth is proportional to its size.

59. With 5% monthly, you must wait until the end of the month to receive interest, while for 5% compounded continuously you begin receiving interest right away, so the interest begins earning interest without any delay.

61. Depreciation by a fixed percentage gives the bigger decrease in the first year, and the straight line gives the bigger decrease in the last year.

EXERCISES 4.2

1. **a.** $\log_5 25 = \log_5 5^2 = 2$

b. $\log_3 81 = \log_3 3^4 = 4$

c. $\log_3 \frac{1}{3} = \log_3 3^{-1} = -1$

d. $\log_3 \frac{1}{9} = \log_3 3^{-2} = -2$

e. $\log_4 2 = \log_4 4^{1/2} = \frac{1}{2}$

f. $\log_4 \frac{1}{2} = \log_4 4^{-1/2} = -\frac{1}{2}$

3. **a.** $\ln(e^{10}) = 10$

b. $\ln\sqrt{e} = \ln e^{1/2} = \frac{1}{2}$

c. $\ln\sqrt[3]{e^4} = \ln e^{4/3} = \frac{4}{3}$

d. $\ln 1 = 0$

e. $\ln(\ln(e^e)) = \ln e = 1$ because $\ln(e^e) = e$

f. $\ln\left(\frac{1}{e^3}\right) = \ln e^{-3} = -3$

5. $f(x) = \ln(9x) - \ln 9 = \ln 9 + \ln x - \ln 9 = \ln x$

7. $f(x) = \ln(x^3) - \ln x = 3\ln x - \ln x = 2\ln x$

9. $f(x) = \ln\left(\frac{x}{4}\right) + \ln 4 = \ln x - \ln 4 + \ln 4 = \ln x$

11. $f(x) = \ln(e^{5x}) - 2x - \ln 1 = 5x - 2x - 0 = 3x$

13. $f(x) = 8x - e^{\ln x} = 8x - x = 7x$

15. The domain of $\ln(x^2 - 1)$ is the values of x such that $x^2 - 1 > 0$

$$x^2 > 1$$
$$x > 1 \text{ or } x < -1$$

The domain is $\{x \mid x > 1 \text{ or } x < -1\}$.

The range is $\mathbb{R}$.

17. **a.** We use the formula $P(1+r)^n$ with monthly interest rate $r = \frac{1}{12} \cdot 24\% = 2\% = 0.02$.

Since double P dollars is $2P$ dollars, we solve

$$P(1+0.02)^n = 2P$$
$$1.02^n = 2$$
$$\ln(1.02^n) = \ln 2$$
$$n \ln 1.02 = \ln 2$$
$$n = \frac{\ln 2}{\ln 1.02} \approx \frac{0.6931}{0.0198} \approx 35$$

Since n is in months, we divide by 12 to convert to years.

A sum at 24% compounded monthly doubles in about 2.9 years.

b. To find how many years it will take for the investment to increase by 50%:

$$P(1+r)^n = 1.5P$$
$$(1+r)^n = 1.5$$
$$\ln(1+r)^n = \ln 1.5$$
$$n \ln(1+r) = \ln 1.5$$

Now, $r = \frac{1}{12} \cdot 24\% = 2\% = 0.02$

thus, $n = \frac{\ln 1.5}{\ln(1+r)} = \frac{\ln 1.5}{\ln 1.02} = \frac{0.4055}{0.0198} \approx 20.5$

Since n is in months, we divide by 12 to convert to years.

$\frac{20.5}{12} \approx 1.7$ years

A sum at 24% compounded monthly increases by 50% in about 1.7 years.

19. **a.** We use Pe^{rn} with $r = 0.07$. Since triple P dollars is $3P$ dollars, we solve

$$Pe^{0.07n} = 3P$$
$$e^{0.07n} = 3$$
$$0.07n = \ln 3$$
$$n = \frac{\ln 3}{0.07} \approx 15.7 \text{ years}$$

b. If P increases by 25%, the total amount is $1.25P$. We solve

$$Pe^{0.07n} = 1.25P$$
$$e^{0.07n} = 1.25$$
$$0.07n = \ln 1.25$$
$$n = \frac{\ln 1.25}{0.07} \approx 3.2 \text{ years}$$

21. If the depreciation is 30% = 0.3 per year, then $r = -0.3$. We use the interest formula $P(1+r)^n$ and we solve

$$P(1-0.3)^n = 0.5P$$
$$0.7^n = 0.5$$
$$\ln 0.7^n = \ln 0.5$$
$$n = \frac{\ln 0.5}{\ln 0.7} \approx 1.9 \text{ years}$$

23. We use the interest formula $P(1+r)^n$ with $r = 2.4\% = 0.024$. Since the increase is 50%, we must find the number of years to reach $1.5P$. We solve

$$P(1+0.024)^n = 1.5P$$
$$1.024^n = 1.5$$
$$n = \frac{\ln 1.5}{\ln 1.024} \approx 17.1 \text{ years}$$

25. We want to find the value of t that produces $p(t) = 0.9$.

$$0.9 = 1 - e^{-0.03t}$$
$$e^{-0.03t} = 0.1$$
$$-0.03t = \ln 0.1$$
$$t = \frac{\ln 0.1}{-0.03} \approx 77 \text{ days}$$

27. $p(5) = 0.9 - 0.2 \ln 5 \approx 0.58$ or 58%

29.
$$s(t) = 100(1 - e^{-0.4t})$$
$$100(1 - e^{-0.4t}) = 80$$
$$1 - e^{-0.4t} = 0.8$$
$$-e^{-0.4t} = -0.2$$
$$e^{-0.4t} = 0.2$$
$$-0.4t = \ln 0.2$$
$$t = \frac{\ln 0.2}{-0.4} \approx 4 \text{ weeks}$$

31. To find the number of years after which 2.3% = 0.023 of the original carbon 14 is left, we solve

$$e^{-0.00012t} = 0.023$$
$$-0.00012t = \ln 0.023$$
$$t = \frac{\ln 0.023}{-0.00012} \approx 31{,}400 \text{ years}$$

33. The proportion of potassium 40 remaining after t million years is $e^{-0.00054t}$. If the skeleton contained 99.76% of its original potassium 40, then

$$e^{-0.00054t} = 0.9976$$
$$\ln e^{-0.00054t} = \ln 0.9976$$
$$-0.00054t = \ln 0.9976$$
$$t = \frac{\ln 0.9976}{-0.00054} \approx \frac{-0.002403}{-0.00054} \approx 4.45$$

Therefore, the estimate of the age of the skeleton is approximately 4.45 million years.

35. To find when the radioactivity decreases to 0.001, we solve $e^{-0.087t} = 0.001$

$$-0.087t = \ln 0.001$$
$$t = \frac{\ln 0.001}{-0.087} \approx 79 \text{ days}$$

37.
$$1.16^x = 2$$
$$x = \frac{\ln 2}{\ln 1.16} \approx 4.7 \text{ years}$$

39.

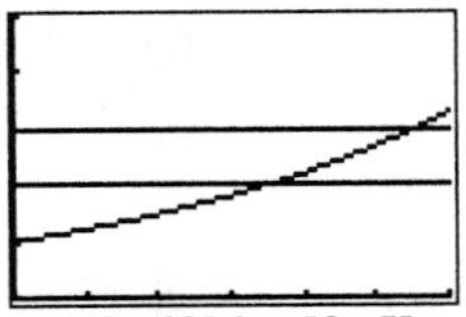

on [0, 60] by [0, 5]

a. 35 years

b. 55.5 years

41. Let t = number of years. Since the rate is 6% compounded quarterly, then $r = \frac{0.06}{4} = 0.015$ and the amount of money is $(1+0.015)^{4t} = 1.015^{4t}$.

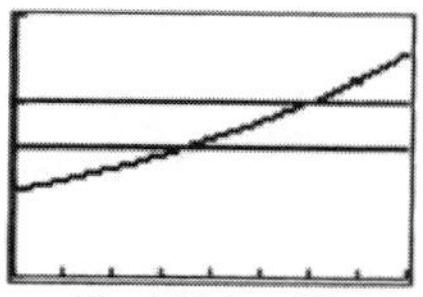

on [0, 16] by [0, 3]

a. 11.6 years

b. 6.8 years

43.

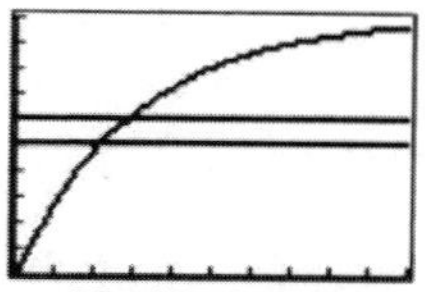

on [0, 40] by [0, 1]

a. About 9 days

b. About 11 days

45.

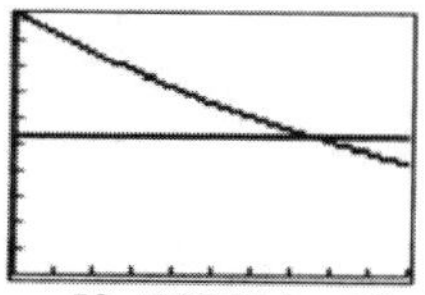

on [0, 7000] by [0, 1]

About 5300 years

47. If the amount of radioactive waste is growing by 11.3% annually since 2010, then the amount of waste x years after 2010 is $800,000(1.113)^x$. We must find the value of x such that $(1.113)^x = 2$.

$$1.113^x = 2$$
$$\ln 1.113^x = \ln 2$$
$$x \ln 1.113 = \ln 2$$
$$x = \frac{\ln 2}{\ln 1.113} \approx 6.5$$

The amount will double in about 6.5 years.

49. We must find the value of x such that $(1.0925)^x = 2$.

$$1.0925^x = 2$$
$$\ln 1.0925^x = \ln 2$$
$$x \ln 1.0925 = \ln 2$$
$$x = \frac{\ln 2}{\ln 1.0925} \approx 7.83 \text{ years}$$

0.83 of a year is approximately 10 months, so the doubling time for the five-year bond is 7 years and 10 months.

51. $e^0 = 1$

$\ln 1 = 0$

53. $\ln \frac{x}{y} = \ln x - \ln y$

55. $\log_{10} y$ or $\log y$

57. If $\ln(-2) = x$, then $e^x = -2$, but e^x is always positive. So $\ln(-2)$ is undefined.

59. Continuous compounding would give the shortest doubling time. Annual compounding would give the longest doubling time.

61. **a.**

$$e^{-kt} = \frac{1}{2}$$
$$-kt = \ln \frac{1}{2}$$
$$-kt = \ln 1 - \ln 2$$
$$t = \frac{\ln 2}{k}$$

The half-life is $\frac{\ln 2}{k}$.

b. If $k = 0.018$, half-life $= \frac{\ln 2}{0.018} \approx 39$ hours.

63. $$f(x) = H_0\left(1 - \frac{1}{2N}\right)^x$$
Substitute 500 for N.
$$f(x) = H_0\left(1 - \frac{1}{2(500)}\right)^x = H_0(0.999)^x$$
Reducing the frequency by 6%,
$$\begin{aligned} H_0 - 0.06H_0 &= H_0(0.999)^x \\ 0.94H_0 &= H_0(0.999)^x \\ 0.94 &= (0.999)^x \\ \ln 0.94 &= \ln(0.999)^x \\ \ln 0.94 &= x\ln(0.999) \\ x &= \frac{\ln 0.94}{\ln 0.999} \approx 62 \text{ generations} \end{aligned}$$

65. **a.** For $r = 6\%$, the doubling time is approximately $\frac{72}{6} = 12$ years.

b. $$\begin{aligned} P(1+0.06)^n &= 2P \\ 1.06^n &= 2 \\ \ln 1.06^n &= \ln 2 \\ n &= \frac{\ln 2}{\ln 1.06} \approx 11.9 \text{ years} \end{aligned}$$

67. If the interest rate r is compounded annually, the formula is
$$\begin{aligned} P(1+r)^n &= kP \\ (1+r)^n &= k \\ \ln(1+r)^n &= \ln k \\ n &= \frac{\ln k}{\ln(1+r)} \end{aligned}$$

EXERCISES 4.3

1. $$\begin{aligned} \frac{d}{dx}(x^2 \ln x) &= 2x\ln x + \frac{x^2}{x} \\ &= 2x\ln x + x \end{aligned}$$

3. $$\frac{d}{dx}\ln x^2 = \frac{2x}{x^2} = \frac{2}{x}$$

5. $$\frac{d}{dx}\ln\sqrt{x} = \frac{d}{dx}\ln x^{1/2} = \frac{\frac{1}{2}x^{-1/2}}{x^{1/2}} = \frac{1}{2}x^{-1}$$

7. $$\frac{d}{dx}\ln(x^2+1)^3 = \frac{3(x^2+1)^2(2x)}{(x^2+1)^3} = \frac{6x}{x^2+1}$$

9. $$\frac{d}{dx}\ln(-x) = \frac{-1}{-x} = \frac{1}{x}$$

11. $$\frac{d}{dx}\left(\frac{e^x}{x^2}\right) = \frac{x^2e^x - 2xe^x}{(x^2)^2} = \frac{xe^x - 2e^x}{x^3}$$

13. $$\frac{d}{dx}e^{x^3+2x} = e^{x^3+2x}(3x^2+2)$$

15. $$\frac{d}{dx}e^{x^3/3} = e^{x^3/3}\left(\frac{3x^2}{3}\right) = x^2e^{x^3/3}$$

17. $$\frac{d}{dx}(x - e^{-x}) = 1 - e^{-x}(-1) = 1 + e^{-x}$$

19. $$\frac{d}{dx}\ln e^{2x} = \frac{d}{dx}(2x) = 2 \quad \text{because } \ln e^{2x} = 2x$$

21. $$\frac{d}{dx}e^{1+e^x} = e^{1+e^x}\frac{d}{dx}(1+e^x) = e^{1+e^x}e^x$$

23. $$\frac{d}{dx}x^e = ex^{e-1}$$

25. $$\frac{d}{dx}e^3 = 0 \quad \text{because } e^3 \text{ is a constant}$$

27. $$\frac{d}{dx}[\ln(x^4+1) - 4e^{x/2} - x] = \frac{4x^3}{x^4+1} - 4e^{x/2}\left(\frac{1}{2}\right) - 1 = \frac{4x^3}{x^4+1} - 2e^{x/2} - 1$$

29. $$\begin{aligned} \frac{d}{dx}\left(x^2\ln x - \frac{1}{2}x^2 + e^{x^2} + 5\right) &= 2x\ln x + x^2\left(\frac{1}{x}\right) - \frac{1}{2}(2x) + e^{x^2}(2x) + 0 \\ &= 2x\ln x + x - x + 2xe^{x^2} = 2x(\ln x + e^{x^2}) \end{aligned}$$

31. $\frac{d}{dx}(e^x \ln x^2) = e^x \ln x^2 + e^x\left(\frac{2}{x}\right) = e^x\left(\ln x^2 + \frac{2}{x}\right) = 2e^x\left(\ln x + \frac{1}{x}\right)$

33. $\frac{d}{dx}\left(\ln\frac{1}{x^2}\right) = \frac{d}{dx}(\ln x^{-2}) = -2\frac{d}{dx}(\ln x) = \frac{-2}{x}$

35. $\frac{d}{dt}(e^{2t}+1)^3 = 3(e^{2t}+1)^2(2e^{2t}) = 6e^{2t}(e^{2t}+1)^2$

37. $\frac{d}{dt}(t^2+2\ln t)^{1/2} = \frac{1}{2}(t^2+2\ln t)^{-1/2}\left(2t+\frac{2}{t}\right) = \frac{t+1/t}{\sqrt{t^2+2\ln t}}$

39. $\frac{d}{dz}\left(\frac{e^z}{z^2-1}\right) = \frac{(z^2-1)e^z - e^z(2z)}{(z^2-1)^2} = \frac{e^z(z^2-2z-1)}{(z^2-1)^2}$

41. $\frac{d}{dz}\left(\frac{10}{1+e^{-2z}}\right) = \frac{d}{dz}10(1+e^{-2z})^{-1} = -10(1+e^{-2z})^{-2}(-2e^{-2z}) = \frac{20e^{-2z}}{(1+e^{-2z})^2}$

43. $\frac{d}{dx}\left(\frac{e^x+e^{-x}}{e^x-e^{-x}}\right) = \frac{(e^x-e^{-x})(e^x-e^{-x})-(e^x+e^{-x})(e^x+e^{-x})}{(e^x-e^{-x})^2} = \frac{e^{2x}-2+e^{-2x}-(e^{2x}+2+e^{-2x})}{(e^x-e^{-x})^2} = \frac{-4}{(e^x-e^{-x})^2}$

45. **a.** $f(x) = \frac{\ln x}{x^5}$

$f'(x) = \frac{x^5\left(\frac{1}{x}\right)-5x^4\ln x}{(x^5)^2} = \frac{x^4-5x^4\ln x}{x^{10}} = \frac{1-5\ln x}{x^6}$

b. $f'(1) = \frac{1-\ln 1}{(1)^6} = \frac{1-0}{1} = 1$

47. **a.** $f(x) = \ln(x^4+48)$

$f'(x) = \frac{4x^3}{x^4+48}$

b. $f'(2) = \frac{4(2)^3}{(2)^4+48} = \frac{32}{16+48} = \frac{32}{64} = \frac{1}{2}$

49. **a.** $f(x) = \ln(e^x-3x)$

$f'(x) = \frac{e^x-3}{e^x-3x}$

b. $f'(0) = \frac{e^0-3}{e^0-3(0)} = \frac{1-3}{1-0} = -2$

51. $f(x) = x\ln(x-1)$

$f(2) = 2\ln(2-1) = 0$

The point is (2, 0).

$f'(x) = \ln(x-1) + \frac{x}{x-1}$

$f'(2) = \ln(2-1) + \frac{2}{2-1} = 2$

The tangent line has slope 2 at (2, 0).

$y-0 = 2(x-2)$

$y = 2x-4$

53. $f(x) = e^{x^2-1}$

$f(1) = e^{1^2-1} = 1$

The point is (1, 1).

$f'(x) = 2xe^{x^2-1}$

$f'(1) = 2(1)e^{1^2-1} = 2$

The tangent line has slope 2 at (1, 1).

$y-1 = 2(x-1)$

$y = 2x-1$

55. **a.** $f(x) = 5x\ln x$

$f'(x) = 5\ln x + 5x\left(\frac{1}{x}\right) = 5\ln x + 5$

b. $f'(2) = 5\ln 2 + 5 \approx 8.466$

57. **a.** $f(x)=\dfrac{e^x}{x}$

$f'(x)=\dfrac{xe^x-(1)e^x}{x^2}=\dfrac{xe^x-e^x}{x^2}$

b. $f'(3)=\dfrac{3e^3-e^3}{(3)^2}=\dfrac{2e^3}{9}$

$f'(3)\approx 4.463$

59. $\dfrac{d}{dx}e^{-x^5/5}=e^{-x^5/5}\left[-5\left(\dfrac{x^4}{5}\right)\right]=-x^4e^{-x^5/5}$

$\dfrac{d^2}{dx^2}e^{-x^5/5}=\dfrac{d}{dx}(-x^4e^{-x^5/5})$

$=-4x^3e^{-x^5/5}-x^4e^{-x^5/5}\left[5\left(-\dfrac{x^4}{5}\right)\right]$

$=-4x^3e^{-x^5/5}+x^8e^{-x^5/5}$

61. $\dfrac{d}{dx}(e^{kx})=e^{kx}(k)=ke^{kx}$

$\dfrac{d^2}{dx^2}(e^{kx})=\dfrac{d}{dx}(ke^{kx})=ke^{kx}(k)=k^2e^{kx}$

$\dfrac{d^3}{dx^3}(e^{kx})=\dfrac{d}{dx}(k^2e^{kx})=k^2e^{kx}(k)=k^3e^{kx}$

$\dfrac{d^n}{dx^n}(e^{kx})=k^ne^{kx}$

63. $dy=(e^{-2x})'dx$

$=\left(-2e^{-2x}\right)dx$

$=\left(-2e^{-2(3)}\right)(0.5)\qquad x=3,\ dx=0.5$

$=-e^{-6}\approx -0.00248$

65. $dy=(xe^x)'dx$

$=\left(e^x+xe^x\right)dx$

$=\left(e^1+(1)e^1\right)(0.1)\qquad x=1,\ dx=0.1$

$=0.2e\approx 0.544$

67. $dy=(x^2\ln x)'dx$

$=\left(2x\ln x+x^2\cdot\dfrac{1}{x}\right)dx$

$=(2x\ln x+x)\,dx$

$=(2e\ln e+e)(0.01)\qquad x=e,\ dx=0.01$

$=0.03e\approx 0.0815$

69.

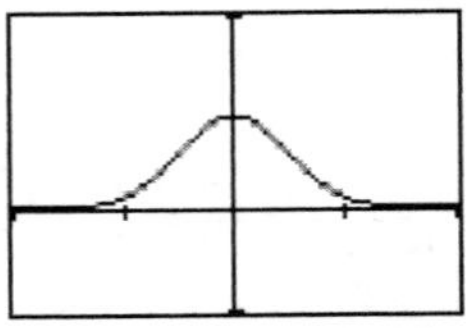

on [–2, 2] by [–1, 2]
There is a relative maximum at (0, 1); no relative minima. There are inflection points at about (0.5, 0.61) and (–0.5, 0.61).

71.

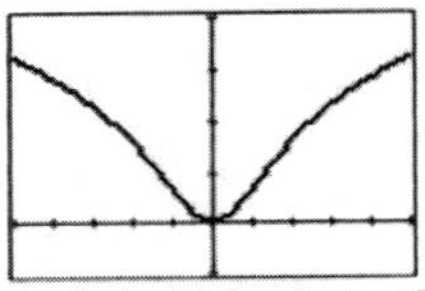

on [–5, 5] by [–1, 4]
There is a relative minimum at (0, 0); no relative maxima. There are inflection points at about (1, 0.69) and (–1, 0.69).

73.

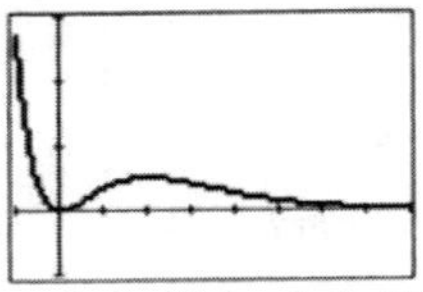

on [–1, 8] by [–1, 3]
There is a relative maximum at about (2, 0.54); relative minimum at (0, 0). There are inflection points at about (0.59, 0.19) and (3.41, 0.38).

75.

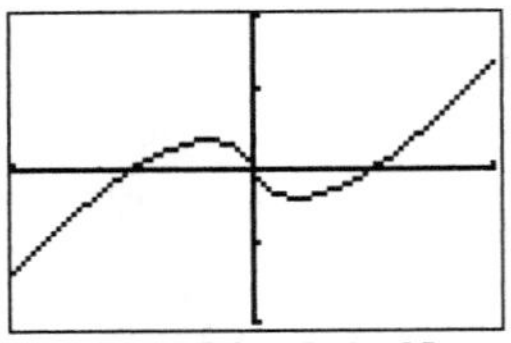

on [–2, 2] by [–2, 2]
There is a relative maximum at about (–0.37, 0.37); relative minimum at about (0.37, –0.37). There are no inflection points. (The function is not defined at $x=0$.)

77. $y^2-ye^x=12$

$\frac{d}{dx}\left(y^2-ye^x\right)=\frac{d}{dx}(12)$

$2yy'-(y'e^x+ye^x)=0$

$y'(2y-e^x)=ye^x$

$y'=\dfrac{ye^x}{2y-e^x}$

79.
$f(x) = 40{,}000e^{0.195x}$
$f'(x) = 40{,}000(0.195)e^{0.195x}$
$f'(x) = 7800e^{0.195x}$
$f'(1) = 7800e^{0.195(1)}$
$f'(1) = 9479.425695$
Annual salary increases by about \$9480 per extra year of calculus.

81. **a.** To find the rate of growth after 0 years, evaluate $V'(0)$.
$V(t) = 1000e^{0.05t}$
$V'(t) = 1000e^{0.05t}(0.05) = 50e^{0.05t}$
$V'(0) = 50e^{0.05(0)} = 50e^0 = 50$
The rate of growth after 0 years is \$50 per year.

b. To find the rate of growth after 10 years, evaluate $V'(10)$.
$V'(t) = 50e^{0.05t}$
$V'(10) = 50e^{0.05(10)} = 50e^{0.5} \approx 82.44$
The rate of growth after 10 years is \$82.44 per year.

83.
$P(t) = 317e^{0.01t}$
$P'(t) = 317e^{0.01t}(0.01) = 3.17e^{0.01t}$
In 2023, $t = 2023 - 2013 = 10$.
$P'(10) = 3.17e^{0.01(10)} \approx 3.5$
In 2023, the United States population will be growing by about 3.5 million people per year.

85.
$A(t) = 1.2e^{-0.05t}$
$A'(t) = 1.2e^{-0.05t}(-0.05) = -0.06e^{-0.05t}$

a. $A'(0) = -0.06e^{-0.05(0)} = -0.06e^0 = -0.06$
The amount remaining after 0 hours is decreasing by 0.06 mg per hour.

b. $A'(2) = -0.06e^{-0.05(2)} = -0.06e^{-0.1} \approx -0.054$
The amount remaining after 2 hours is decreasing by 0.054 mg per hour.

87.
$S(x) = 1000 - 900e^{-0.1x}$ ← weekly sales (in thousands)
$S'(x) = -900e^{-0.1x}(-0.1)$ ← rate of change of sales per week
$= 90e^{-0.1x}$

a. $S'(1) = 90e^{-0.1}$ ← rate of change of sales per week after 1 week
$\approx 90(0.9048)$
≈ 81.4 thousand sales per week

b. $S'(10) = 90e^{-1} \approx 90(0.3679) \approx 33$ thousand sales per week after 10 weeks

89. To find the maximum consumer expenditure, solve $E'(p) = 0$, where $E(p) = pD(p)$.
$E(p) = pD(p) = 5000pe^{-0.01p}$
$E'(p) = 5000e^{-0.01p} + 5000pe^{-0.01p}(-0.01)$
$= 5000e^{-0.01p} - 50pe^{-0.01p}$
Setting $E'(p) = 0$, we get
$5000e^{-0.01p} - 50pe^{-0.01p} = 0$
$50e^{-0.01p}(100 - p) = 0$
$p = \$100$
We use the second derivative test to show that $p = 100$ is a maximum.
$E''(p) = 5000e^{-0.01p}(-0.01) - 50e^{-0.01p} - 50pe^{-0.01p}(-0.01)$
$= -50e^{-0.01p} - 50e^{-0.01p} + 0.5pe^{-0.01p}$
$= -100e^{-0.01p} + 0.5pe^{-0.01p}$
$E''(100) = -100e^{-1} + 50e^{-1} = -50e^{-1} < 0$
so $E(p)$ is maximized.

91. $p(x) = 400e^{-0.20x}$ $\leftarrow$ price function (in dollars)

a. $R(x) = xp(x)$ $\leftarrow$ revenue function

$= 400xe^{-0.20x}$

b. To maximize $R(x)$, differentiate.

$R'(x) = 400e^{-0.20x} + 400xe^{-0.20x}(-0.20) = 400e^{-0.20x}(1-0.2x)$

$R'(x) = 0$ when $x = \dfrac{1}{0.2} = 5$, which is the only critical value.

We calculate $R''(x)$ for the second derivative test.

$$\begin{aligned} R''(x) &= 400e^{-0.20x}(-0.20)(1-0.2x) + 400e^{-0.20x}(-0.20) \\ &= 400e^{-0.20x}(-0.20)(1-0.2x+1) \\ &= -80e^{-0.20x}(2-0.2x) \\ &= -160e^{-0.20x}(1-0.1x) \end{aligned}$$

At the critical value of $x = 5$,

$R''(5) = -80e^{-1} < 0$,

so $R(x)$ is maximized at $x = 5$..

At $x = 5$, $p(x) = 400e^{-0.20x}$

$p(5) = 400e^{-1} \approx 147.15$

The revenue is maximized at quantity $x = 5000$ and price $p = \$147.15$.

93. a.

$$\begin{aligned} T(x) &= 70 - 30e^{-3.5t} \\ T(0.25) &= 70 - 30e^{-3.5(0.25)} \approx 57.5 \\ T'(x) &= -30e^{-3.5t}(-3.5) = 105e^{-3.5t} \\ T'(0.25) &= 105e^{-3.5(0.25)} \approx 43.8 \end{aligned}$$

After 15 minutes, the temperature of the beer is 57.5 degrees and is increasing at the rate of 43.8 degrees per hour.

b.

$$\begin{aligned} T(x) &= 70 - 30e^{-3.5t} \\ T(1) &= 70 - 30e^{-3.5(1)} \approx 69.1 \\ T'(x) &= -30e^{-3.5t}(-3.5) = 105e^{-3.5t} \\ T'(1) &= 105e^{-3.5(1)} \approx 3.2 \end{aligned}$$

After 1 hour, the temperature of the beer is 69.1 degrees and is increasing at the rate of 3.2 degrees per hour.

95.

$$\begin{aligned} D(x) &= 0.96 + 0.011e^x + \ln(x) \\ D'(x) &= 0.011e^x + \frac{1}{x} \\ D'(5) &= 0.011e^5 + \frac{1}{5} \approx 1.83 \end{aligned}$$

In the 2000s, the percentage of the U.S. population will diabetes was increasing by about 1.83% per year.

97 A speed of 10 meters per second after 2.31 seconds.

99 a.

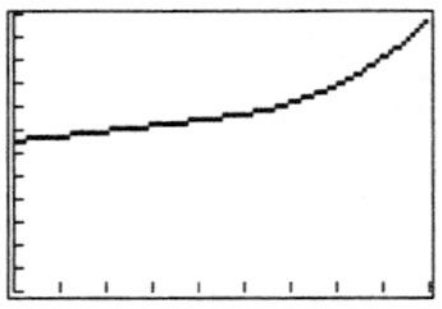

on [35, 80] by [0, 240]

b. $f(35) \approx 132.4$; $f'(35) \approx 0.834$

At age 35, the fastest man's time is 2 hours 12.4 minutes and increasing at about 0.834 minutes per year.

c. $f(80) \approx 234.9$; $f'(80) \approx 7.76$

At age 80, the fastest man's time is 3 hours 54.9 minutes and increasing at about 7.76 minutes per year.

101. b. $\frac{d}{dx}e^x = e^x$

103. c. $\frac{d}{dx}e^5 = \frac{d}{dx}5 \cdot e^5$

$\frac{d}{dx}e^5 = 0 \cdot e^5$

$\frac{d}{dx}e^5 = 0$

105. b. $\frac{d}{dx}\ln x = \frac{1}{x}$

107. c. $\frac{d}{dx}\ln 5 = \frac{\frac{d}{dx}5}{5}$

$\frac{d}{dx}\ln 5 = \frac{0}{5}$

$\frac{d}{dx}\ln 5 = 0$

109. $e^{\ln x} = x$ and $\frac{d}{dx}x = 1$

111. $N(t) = Ke^{-ae^{-bt}}$ for $t \geq 0, k > 0, b > 0.$

$$\frac{dN}{dt} = Ke^{-ae^{-bt}} \frac{d}{dt}\left(-ae^{-bt}\right)$$
$$= Ke^{-ae^{-bt}}\left(-ae^{-bt}(-b)\right)$$
$$= bN(ae^{-bt})$$
$$= bN(\ln e^{ae^{-bt}})$$
$$= bN\ln\left(\frac{K}{Ke^{-ae^{-bt}}}\right)$$
$$= bN\ln\left(\frac{K}{N}\right)$$

If $N > K$, $\ln\left(\frac{K}{N}\right) < 0$ and the population will be decreasing. If $N < K$, $\ln\left(\frac{K}{N}\right) > 0$ and the population will be increasing.

113. To maximize $R(r)$, solve $R'(r) = 0$.

$R'(r) = \frac{a}{r} - b = 0$

$a - br = 0$

$r = \frac{a}{b}$

Using the second derivative test,

$R''(r) = -\frac{a}{r^2} < 0$

so $R(r)$ is maximized.

115 **a.** $y = \log_a x$ for $a > 0$ and $x > 0$

$x = a^y$

$\ln x = \ln a^y$

$\ln x = y\ln a$

$\frac{\ln x}{\ln a} = y$

$\frac{\ln x}{\ln a} = \log_a x$

b. $y = \frac{\ln x}{\ln a}$

$y\ln a = \ln x$

$\ln a^y = \ln x$

$a^y = x$

$\log_a x = y$

$\log_a = \frac{\ln x}{\ln a}$

117. $2.23^x = \left(e^{\ln 2.23}\right)^x = e^{(\ln 2.23)x} \approx e^{0.802x}$

119. **a.** $\frac{d}{dx}10^x = (\ln 10)10^x$

b. $\frac{d}{dx}3^{x^2+1} = (\ln 3)3^{x^2+1}(2x) = 2(\ln 3)x\,3^{x^2+1}$

c. $\frac{d}{dx}2^{3x} = (\ln 2)2^{3x}(3) = (3\ln 2)2^{3x}$

d. $\frac{d}{dx}5^{3x^2} = (\ln 5)5^{3x^2}(6x) = 6(\ln 5)x\,5^{3x^2}$

e. $\frac{d}{dx}2^{4-x} = (\ln 2)2^{4-x}(-1) = (-\ln 2)\,2^{4-x}$

121. **a.** $\frac{d}{dx}\log_2 x = \frac{1}{(\ln 2)x}$

b. $\frac{d}{dx}\log_{10}(x^2 - 1) = \frac{2x}{(\ln 10)(x^2 - 1)}$

c. $\frac{d}{dx}\log_3(x^4 - 2x) = \frac{4x^3 - 2}{(\ln 3)(x^4 - 2x)}$

EXERCISES 4.4

1. **a.** $\ln f(t) = \ln t^2 = 2\ln t$

$\frac{d}{dt}\ln f(t) = \frac{d}{dt}2\ln t = \frac{2}{t}$

b. For $t = 1$, the relative rate of change is $\frac{2}{1} = 2$.

For $t = 10$, the relative rate of change is $\frac{2}{10} = 0.2$.

3. **a.** $\ln f(t) = \ln 100e^{0.2t}$

$= \ln 100 + \ln e^{0.2t}$

$= \ln 100 + 0.2t$

$\frac{d}{dt}\ln f(t) = 0.2$

b. For $t = 5$, the relative rate of change is 0.2.

5. **a.** $\ln f(t) = \ln e^{t^2} = t^2$

$\frac{d}{dt}\ln f(t) = 2t$

b. For $t = 10$, the relative rate of change is $2(10) = 20$.

7. **a.** $\ln f(t) = \ln e^{-t^2} = -t^2$

$\frac{d}{dt}\ln f(t) = -2t$

b. For $t = 10$, the relative rate of change is $-2(10) = -20$.

9. **a.** $\ln f(t) = \ln 25\sqrt{t-1}$

$= \ln 25(t-1)^{1/2}$

$= \ln 25 + \frac{1}{2}\ln(t-1)$

$\frac{d}{dt}\ln f(t) = 0 + \frac{1}{2(t-1)}$

b. For $t = 6$, the relative rate of change is $\frac{1}{2(6-1)} = \frac{1}{10}$.

11. $A(x) = 4.5e^{0.507x}$

$\ln A(x) = \ln(4.5e^{0.507x}) = \ln 4.5 + \ln e^{0.507x}$

$= \ln 4.5 + 0.507x$

$\frac{d}{dx}\ln A(x) = 0.507$

Data traffic is growing by about 51% per year.

13. $\ln N(t) = \ln(0.5 + 1.1e^{0.01t})$

$\frac{d}{dt}\ln N(t) = \frac{1.1e^{0.01t}(0.01)}{0.5+1.1e^{0.01t}}$

For $t = 10$, the relative rate of change is

$\frac{1.1e^{0.01(10)}(0.01)}{0.5+1.1e^{0.01(10)}} \approx 0.0071$ or 0.71%.

15. **a.** $\ln P(t) = \ln(4 + 1.3e^{0.04t})$

$\frac{d}{dt}\ln P(t) = \frac{1.3e^{0.04t}(0.04)}{4+1.3e^{0.04t}} = \frac{0.052e^{0.04t}}{4+1.3e^{0.04t}}$

For $t = 8$, the relative rate of change is

$\frac{0.052e^{0.04(8)}}{4+1.3e^{0.04(8)}} \approx 0.012$ or 1.2%.

b. If the relative rate of change is 1.5% = 0.0015, then

$\frac{0.052e^{0.04t}}{4+1.3e^{0.04t}} = 0.015$

$0.052e^{0.04t} = 0.06 + 0.0195e^{0.04t}$

$0.0325e^{0.04t} = 0.06$

$e^{0.04t} = \frac{0.06}{0.0325}$

$0.04t = \ln\left(\frac{0.06}{0.0325}\right)$

$t = \frac{1}{0.04}\left[\ln\left(\frac{0.06}{0.0325}\right)\right] \approx 15.3$

The relative rate of change will reach 1.5% in about 15.3 years.

17. $D(p) = 200 - 5p,\ p = 10$

a. Elasticity of demand is

$$E(p) = \frac{-pD'(p)}{D(p)} = \frac{-p(-5)}{200-5p} = \frac{5p}{200-5p} = \frac{p}{40-p}$$

b. Evaluating at $p = 10$,

$$E(10) = \frac{10}{40-10} = \frac{10}{30} = \frac{1}{3}$$

The elasticity is less than 1, and so the demand is inelastic at $p = 10$..

19. **a.** $E(p) = \frac{-p(-2p)}{300-p^2} = \frac{2p^2}{300-p^2}$

b. $E(10) = \frac{2(10)^2}{300-(10)^2} = \frac{200}{200} = 1$

Since $E(10) = 1$, the demand is unitary elastic.

21. **a.** $E(p) = \frac{-p(-\frac{300}{p^2})}{\frac{300}{p}} = \frac{\frac{300}{p}}{\frac{300}{p}} = 1$

b. Since $E(4) = 1$, the demand is unitary elastic.

23. **a.** $E(p) = \frac{-p[\frac{1}{2}(175-3p)^{-1/2}](-3)}{(175-3p)^{1/2}} = \frac{3p}{2(175-3p)}$

b. $E(50) = \frac{3(50)}{2[175-3(50)]} = \frac{150}{50} = 3$

Since $E(50) > 1$, the demand is elastic.

25. **a.** $E(p) = \frac{-p\left(-\frac{200}{p^3}\right)}{\frac{100}{p^2}} = \frac{\frac{200}{p^2}}{\frac{100}{p^2}} = 2$

b. Since $E(40) > 1$, the demand is elastic.

27. **a.** $E(p) = \frac{-p(4000e^{-0.01p})(-0.01)}{4000e^{-0.01p}} = 0.01p$

b. $E(200) = 0.01(200) = 2$

Since $E(200) > 1$, the demand is elastic.

29. The demand function is $D(p) = 2(15 - 0.001p)^2$

To determine whether the dealer needs to raise or lower the price to increase revenue, we need to determine the elasticity of demand.

$$E(p) = \frac{-pD'(p)}{D(p)} = \frac{-p \cdot 4(15-0.001p)(-0.001)}{2(15-0.001p)^2} = \frac{0.002p}{15-0.001p}$$

When the cars sell at a price of \$12,000,

$$E(12{,}000) = \frac{0.002(12{,}000)}{15-0.001(12{,}000)} = \frac{24}{15-12} = \frac{24}{3} = 8$$

Since $E = 8 > 1$, to increase revenues the dealer should lower prices.

31. To determine the elasticity of demand, we consider

$$E(p) = \frac{-p\{\frac{1}{2}[150{,}000(1.75-p)^{-1/2}(-1)]\}}{150{,}000(1.75-p)^{1/2}} = \frac{p}{2(1.75-p)}$$

When the fare is 75 cents,

$$E(1.25) = \frac{1.25}{2(1.75-1.25)} = \frac{1.25}{1} = 1.25$$

Since demand is elastic, raising the fare will not succeed.

33. $D(p) = \frac{120}{10+p}$

$$E(p) = \frac{-pD'(p)}{D(p)} = \frac{-p \cdot \frac{-120}{(10+p)^2}}{\frac{120}{10+p}} = \frac{p}{10+p}$$

$$E(6) = \frac{6}{10+6} = \frac{6}{16} = \frac{3}{8} = 0.375$$

Since $E = \frac{3}{8} < 1$, increasing prices should increase revenues. Yes, the commission should grant the request.

35. To determine the elasticity of demand, we consider

$$E(p) = \frac{-p(3.5e^{-0.06p})(-0.06)}{3.5e^{-0.06p}} = 0.06p$$

When the price is \$120 per barrel,

$E(120) = 0.06(120) = 7.2$

Since demand is elastic, it should lower prices.

37.
$$E(p) = \frac{-p(7.881)(-0.112p^{-1.112})}{7.881p^{-0.112}} = \frac{0.112p}{p} = 0.112$$

39. **a.** $E = 0.35$

b. Since demand is inelastic, raising prices will raise revenue.

c. \$20,401

41.
$$E = \frac{\text{percent change in demand}}{\text{percent change in price}} = \frac{\frac{52-47.5}{52}}{\frac{2.80-2.40}{2.40}} \approx \frac{0.08654}{0.1667} = \frac{8.65\%}{16.7\%}$$
$$\approx 0.52 \text{ or } 52\%$$

43.
$$f(x) = e^{1.2x}$$
$$\frac{d}{dx}\ln f(x) = \frac{d}{dx}\ln e^{1.2x} = \frac{d}{dx}1.2x = 1.2$$

45.
$$f(x) = x^n$$
$$\frac{d}{dx}\ln f(x) = \frac{d}{dx}\ln x^n = \frac{\frac{d}{dx}x^n}{x^n} = \frac{nx^{n-1}}{x^n} = \frac{n}{x}$$

47.
$$E(p) = \frac{-p\left(\frac{d}{dx}e^{-2p}\right)}{e^{-2p}}$$
$$E(p) = \frac{-p(-2)(e^{-2p})}{e^{-2p}}$$
$$E(p) = 2p$$

49. Demand will increase strongly.

51. Inelastic for cigarettes (they are habit-forming) and elastic for jewelry.

53.
$$E(p) = \frac{-p(ae^{-cp})(-c)}{ae^{-cp}} = cp$$

55. If $S(p) = ap^n$, then
$$E_s(p) = \frac{pS'(p)}{S(p)} = \frac{p(anp^{n-1})}{ap^n} = \frac{np^n}{p^n} = n$$

REVIEW EXERCISES AND CHAPTER TEST FOR CHAPTER 4

1. **a.** For quarterly compounding, $r = \frac{0.08}{4} = 0.02$ and $n = 8 \cdot 4 = 32$.
$$P(1+r)^n = 10{,}000(1+0.02)^{32} = 10{,}000(1.02)^{32} \approx \$18{,}845.41$$

b. For compounding continuously, $r = 0.08$ and $n = 8 \cdot 4 = 32$.
$$Pe^{rn} = 10{,}000e^{(0.08)(8)} \approx \$18{,}964.81$$

2. For 6% compounded quarterly, $r = \frac{0.06}{4} = 0.015$ and $n = 4 \cdot 1 = 4$.
$$P(1+r)^n = 1(1+0.015)^4 \approx 1.0614$$
For 5.98% compounded continuously, $r = 0.0598$ and $n = 1$.
$$Pe^{rn} = e^{0.0598} \approx 1.0616$$
5.98% compounded continuously has a higher yield.

3. **a.** If the depreciation is 20% per year, $r = -20\% = -0.2$ and $n = t$. The formula for the value is
$$V = 800{,}000(1-0.2)^t = 800{,}000(0.8)^t$$

b. After 4 years, its value is
$$V = 800{,}000(0.8)^4 = \$327{,}680$$

4. For Drug A, $C(t) = 2e^{-0.2t}$. After 4 hours,
$$C(4) = 2e^{-0.8} \approx 0.899$$
For Drug B, $C(t) = 3e^{-0.25t}$. After 4 hours,
$$C(4) = 3e^{-1} \approx 1.103$$
Drug B has a higher concentration.

5. Graph both functions and find the intersection.

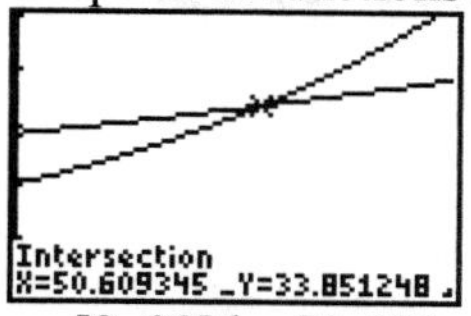

on [0, 90] by [0, 50]
Since $x \approx 50.6$, São Paulo will overtake Tokyo in about 50 years, or during the year 2060.

6. The value $t = 2017 - 1987 = 30$ corresponds to 2017 and
$C(30) = 4^{30/3} = 4^{10} = 1{,}048{,}576$ megabits, which is enough to hold more than 200,000 books on one chip.

7. $r = \frac{1}{2}10\% = 5\% = 0.05$

a.
$$P(1+0.05)^n = 2P$$
$$1.05^n = 2$$
$$\ln 1.05^n = \ln 2$$
$$n \ln 1.05 = \ln 2$$
$$n = \frac{\ln 2}{\ln 1.05} \approx \frac{0.6931}{0.0488} \approx 14.2$$
Since n is in half years, we divide by 2 to convert to years. About 7.1 years.

b.
$$P(1.05)^n = 1.5P$$
$$1.05^n = 1.5$$
$$\ln 1.05^n = \ln 1.5$$
$$n \ln 1.05 = \ln 1.5$$
$$n = \frac{\ln 1.5}{\ln 1.05} \approx \frac{0.4055}{0.0488} \approx 8.31 \text{ half years}$$
$$\frac{8.31}{2} \approx 4.2 \text{ years}$$

8. **a.** For compounding continuously, $r = 7\% = 0.07$.
$$Pe^{0.07n} = 2P$$
$$e^{0.07n} = 2$$
$$0.07n = \ln 2$$
$$n = \frac{\ln 2}{0.07} \approx 9.9 \text{ years}$$

b. To increase by 50%,
$$Pe^{0.07n} = 1.5P$$
$$e^{0.07n} = 1.5$$
$$n = \frac{\ln 1.5}{0.07} \approx 5.8 \text{ years}$$

9. The proportion of potassium 40 remaining after t million years is $e^{-0.00054t}$.
Since $97.3\% = 0.973$ remains,
$$e^{-0.00054t} = 0.973$$
$$-0.00054t = \ln 0.973$$
$$t = \frac{\ln 0.973}{-0.00054} \approx 50.7 \text{ million years old}$$

10. Since $99.9\% = 0.999$ remains,
$$e^{-0.00054t} = 0.999$$
$$t = \frac{\ln 0.999}{-0.00054} \approx 1.85 \text{ million years old}$$

11. We want to solve $N(t) = 1{,}000{,}000(1 - e^{-0.3t})$ with $N(t) = 500{,}000$.
$$1{,}000{,}000(1 - e^{-0.3t}) = 500{,}000$$
$$1 - e^{-0.3t} = 0.5$$
$$-e^{-0.3t} = -0.5$$
$$e^{-0.3t} = 0.5$$
$$-0.3t = \ln 0.5$$
$$t = \frac{\ln 0.5}{-0.3} \approx \frac{-0.6931}{-0.3} \approx 2.3 \text{ hours}$$

12. Since the rate of increase is $3\% = 0.03$ per year, $r = 0.03$. To find when demand will increase by $50\% = 0.5$,
$$1.5P = P(1+0.03)^n$$
$$1.5 = (1.03)^n$$
$$n = \frac{\ln 1.5}{\ln 1.03} \approx 13.7 \text{ years}$$

13. **a.** If the interest rate is 6.5% = 0.065 compounded quarterly, $r = \frac{0.065}{4}$.

$$1000\left(1+\tfrac{0.065}{4}\right)^n = 1.5(1000)$$
$$\left(1+\tfrac{0.065}{4}\right)^n = 1.5$$
$$n = \frac{\ln 1.5}{\ln\left(1+\frac{0.065}{4}\right)}$$
$$\approx 25.2 \text{ quarters}$$
$$\approx 6.3 \text{ years}$$

b. If the interest rate is 6.5% = 0.065 compounded continuously, $r = 0.065$.

$$1000e^{0.065n} = 1.5(1000)$$
$$e^{0.065n} = 1.5$$
$$0.065n = \ln 1.5$$
$$n = \tfrac{\ln 1.5}{0.065} \approx 6.24 \text{ years}$$

14. **a.** To reach 30% = 0.3 of the people, $p(t) = 0.3$.

$$0.3 = 1 - e^{-0.032t}$$
$$e^{-0.032t} = 0.7$$
$$-0.032t = \ln 0.7$$
$$t = \tfrac{\ln 0.7}{-0.032} \approx 11 \text{ days}$$

b. To reach 40% = 0.4 of the people, $p(t) = 0.4$.

$$0.4 = 1 - e^{-0.032t}$$
$$e^{-0.032t} = 0.6$$
$$t = \tfrac{\ln 0.6}{-0.032} \approx 16 \text{ days}$$

15. $\frac{d}{dx}\ln 2x = \frac{2}{2x} = \frac{1}{x}$

16. $\frac{d}{dx}\ln(x^2-1)^2 = \frac{2(x^2-1)(2x)}{(x^2-1)^2} = \frac{4x}{x^2-1}$

17. $\frac{d}{dx}\ln(1-x) = -\frac{1}{1-x}$

18. $\frac{d}{dx}\ln\sqrt{x^2+1} = \frac{\frac{1}{2}(x^2+1)^{-1/2}(2x)}{(x^2+1)^{1/2}} = \frac{x}{x^2+1}$

19. $\frac{d}{dx}\ln\sqrt[3]{x} = \frac{d}{dx}\ln x^{1/3} = \frac{\frac{1}{3}x^{-2/3}}{x^{1/3}} = \frac{1}{3x}$

20. $\frac{d}{dx}\ln e^x = \frac{d}{dx}(x) = 1$

21. $\frac{d}{dx}\ln x^2 = \frac{d}{dx}(2\ln x) = \frac{2}{x}$

22. $\frac{d}{dx}(x\ln x - x) = \ln x + x\left(\frac{1}{x}\right) - 1 = \ln x$

23. $\frac{d}{dx}e^{-x^2} = e^{-x^2}(-2x) = -2xe^{-x^2}$

24. $\frac{d}{dx}e^{1-x} = e^{1-x}(-1) = -e^{1-x}$

25. $\frac{d}{dx}\ln e^{x^2} = \frac{d}{dx}(x^2) = 2x$

26. $\frac{d}{dx}(e^{x^2\ln x - x^2/2}) = e^{x^2\ln x - x^2/2}\left[2x\ln x + x^2\left(\frac{1}{x}\right) - x\right] = 2x\ln x\, e^{x^2\ln x - x^2/2}$

27. $\frac{d}{dx}(5x^2 + 2x\ln x + 1) = 10x + 2\ln x + 2x\left(\frac{1}{x}\right) = 10x + 2\ln x + 2$

28. $\frac{d}{dx}(2x^3 + 3x\ln x - 1) = 6x^2 + 3\ln x + 3x\left(\frac{1}{x}\right) = 6x^2 + 3\ln x + 3$

29. $\frac{d}{dx}(2x^3 - 3xe^{2x}) = 6x^2 - 3e^{2x} - 3xe^{2x}(2)$
$= 6x^2 - 3e^{2x} - 6xe^{2x}$

30. $\frac{d}{dx}(4x - 2x^2e^{2x}) = 4 - 4xe^{2x} - 2x^2e^{2x}(2)$
$= 4 - 4xe^{2x} - 4x^2e^{2x}$

31. $f(x) = [x + \ln(x+2)]^2$
$f(-1) = [-1 + \ln(-1+2)]^2 = 1$
The point is (–1, 1).
$f'(x) = 2[x + \ln(x+2)]\left(1 + \frac{1}{x+2}\right)$
$f'(-1) = 2[-1 + \ln(-1+2)]\left(1 + \frac{1}{-1+2}\right) = -4$
The tangent line has slope –4 at (–1, 1).
$y - 1 = -4(x+1)$
$y = -4x - 3$

32. $f(x) = \frac{x^2 - 3}{e^{x-2}}$
$f(2) = \frac{2^2 - 3}{e^{2-2}} = 1$
The point is (2, 1).
$f'(x) = \frac{e^{x-2}(2x) - e^{x-2}(x^2 - 3)}{(e^{x-2})^2}$
$f'(2) = \frac{e^{2-2}(2\cdot 2) - e^{2-2}(2^2 - 3)}{(e^{2-2})^2} = 3$
The tangent line has slope 3 at (2, 1).
$y - 1 = 3(x - 2)$
$y = 3x - 5$

33.

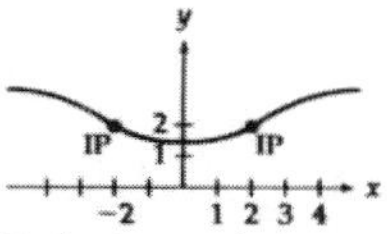

Relative minimum at (0, 1.4)

34.

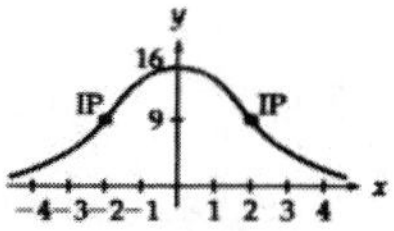

Relative maximum at (0, 16).

35. **a.** $S(x) = 2000 - 1500e^{-0.1x}$
$S'(x) = -1500e^{-0.1x}(-0.1) = 150e^{-0.1x}$
$S'(1) = 150e^{-0.1(1)} \approx 136$
Sales are increasing by 136,000 after 1 week.

b. $S'(10) = 150e^{-0.1(10)} = 150e^{-1} \approx 55$
Sales are increasing by 55,000 after 10 weeks.

36. **a.** $A(t) = 1.5e^{-0.08t}$
$A'(t) = 1.5e^{-0.08t}(-0.08) = -0.12e^{-0.08t}$
$A'(0) = -0.12e^{-0.08(0)} = -0.12$
The amount of the drug in the bloodstream immediately after the injection is decreasing by 0.12 mg per hour.

b. $A'(5) = -0.12e^{-0.08(5)} \approx -0.08$
The amount of the drug in the bloodstream after 5 hours is decreasing by 0.08 mg per hour.

37. $P(t) = 100 - 200\ln(t+1)$
$P'(t) = \frac{-200}{t+1}$ ← rate of change
$P'(5) = \frac{-200}{5+1} = \frac{-200}{6}$ ← rate of change after 5 seconds
$= \frac{-100}{3} = -33\frac{1}{3}$

The rate of change after 5 seconds is decreasing by $33\frac{1}{3}\%$ per second.

38. **a.** $T(t) = 70 - 35e^{-0.1t}$
$T'(x) = -35e^{-0.1t}(-0.1) = 3.5e^{-0.1t}$
$T'(0) = 3.5e^0 = 3.5$ degrees per hour

b. $T'(5) = 3.5e^{-0.5} \approx 2.1$ degrees per hour

39. **a.** $N(t) = 30{,}000(1 - e^{-0.3t})$

$N'(t) = -30{,}000e^{-0.3t}(-0.3) = 9000e^{-0.3t}$

$N'(1) = 9000e^{-0.3(1)} \approx 6667$

After 1 hour, the rate of change in the number of informed people is increasing by 6667 per hour.

b. $N'(8) = 9000e^{-0.3(8)} \approx 816$

After 8 hours, the rate of change in the number of informed people is increasing by 816 per hour.

40. $V(t) = 50t^2e^{-0.08t}$

$V'(t) = 100te^{-0.08t} + 50t^2e^{-0.08t}(-0.08)$

$= 100te^{-0.08t} - 4t^2e^{-0.08t}$

We set $V'(t) = 0$ to maximize $V(t)$.

$0 = 100te^{-0.08t} - 4t^2e^{-0.08t}$

$0 = 4te^{-0.08t}(25 - t)$

A critical value is $t = 25$. Now we use the second derivative test.

$V''(t) = 100e^{-0.08t} + 100te^{-0.08t}(-0.08) - 8te^{-0.08t} - 4t^2e^{-0.08t}(-0.08)$

$= 100e^{-0.08t} - 8te^{-0.08t} - 8te^{-0.08t} + 0.32t^2e^{-0.08t}$

$= 100e^{-0.08t} - 16te^{-0.08t} + 0.32t^2e^{-0.08t}$

$V''(25) = 100e^{-2} - 400e^{-2} + 200e^{-2} = -100e^{-2} < 0$

so V is maximized.

The present value is maximized in 25 years.

41. $p(x) = 200e^{-0.25x}$

a. $R(x) = xp(x) = 200xe^{-0.25x}$

b. $R'(x) = 200e^{-0.25x} + 200xe^{-0.25x}(-0.25)$

$= 200e^{-0.25x}(1 - 0.25x)$

$R'(x) = 0$ when $x = \dfrac{1}{0.25} = 4$, which is the only critical value.

$R''(x) = 200e^{-0.25x}(-0.25)(1 - 0.25x) + 200e^{-0.25x}(-0.25)$

$= 200e^{-0.25x}(-0.25)(1 - 0.25x + 1)$

$= -50e^{-0.25x}(2 - 0.25x)$

$R''(4) = -50e^{-1} < 0$,

so $R(x)$ is maximized at $x = 4$.

Thus, quantity $x = 4000$ and price $p(4) = 200e^{-1} \approx \$73.58$ maximize revenue.

42. **a.** $R(x) = p \cdot x = (5 - \ln x)x = 5x - x\ln x$

b. $R'(x) = 5 - \ln x - x\left(\dfrac{1}{x}\right) = 4 - \ln x$

$R'(x) = 0$ when $x = e^4$, which is the only critical value.

$R''(x) = -\dfrac{1}{x} < 0 \quad \text{when } x = e^4$

so R is maximized.

$p(e^4) = 5 - \ln(e^4) = 1$

Thus, the quantity $x = e^4$ and the price \$1 maximize revenue.

43. To find the maximum consumer expenditure, solve $E'(p) = 0$, where $E(p) = pD(p)$.

$$E(p) = p(25{,}000e^{-0.02p})$$
$$E'(p) = 25{,}000e^{-0.02p} + 25{,}000e^{-0.02p}(-0.02)p$$
$$= 25{,}000e^{-0.02p} - 500pe^{-0.02p}$$

$E'(p) = 0$ when $p = 50$, which is the only critical value.

We use the second derivative test to show that $p = 50$ is a maximum.

$$E''(p) = 25{,}000e^{-0.02p}(-0.02) - 500e^{-0.02p} - 500pe^{-0.02p}(-0.02)$$
$$= -500e^{-0.02p} - 500e^{-0.02p} + 10pe^{-0.02p}$$
$$= -1000e^{-0.02p} + 10pe^{-0.02p}$$
$$E''(50) = -1000e^{-0.02(50)} + 10(50)e^{-0.02(50)}$$
$$= -1000e^{-1} + 500e^{-1} < 0$$

so E is maximized.

The price \$50 maximizes consumer expenditure.

44.

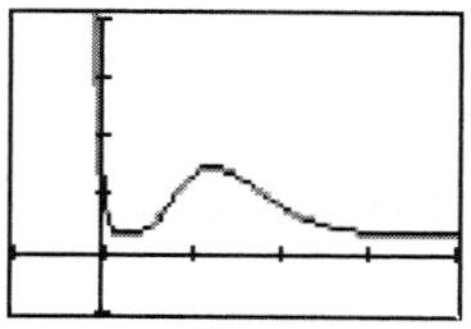

on [–5, 15] by [–5, 15]

The function has a relative maximum at about (4, 4.69) and a relative minimum at (0, 0). There are inflection points at about (2, 2.17) and (6, 3.21).

45.

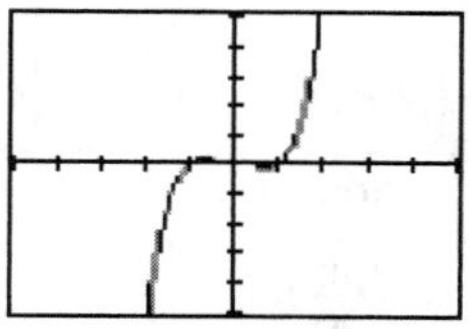

on [–5, 5] by [–5, 5]

The function has a relative maximum at about (–0.72, 0.12) and a relative minimum at about (0.72, –0.12). There are inflection points at about (–0.43, 0.07) and (0.43, –0.07).

46. To find the maximum consumer expenditure, solve $E'(p) = 0$, where $E(p) = pD(p)$.

$$E(p) = p(200e^{-0.0013p}) = 200pe^{-0.0013p}$$
$$E'(p) = 200e^{-0.0013p} + 200pe^{-0.0013p}(-0.0013)$$
$$= 200e^{-0.0013p} - 0.26pe^{-0.0013p}$$

$E'(p) = 0$ when the price is about 769, which is the only critical value.

$$E''(p) = -0.26e^{-0.0013p} - 0.26e^{-0.0013p} - 0.26pe^{-0.0013p}(-0.0013)$$
$$= -0.52e^{-0.0013p} + 0.000338pe^{-0.0013p}$$
$$E''(769) = -0.52e^{-0.0013(769)} + 0.000338(769)e^{-0.0013(769)}$$
$$= -0.26e^{-0.0013(769)} < 0$$

so E is maximized.

The price \$769 maximizes consumer expenditure.

47. $R(x) = xp(x) = x(20e^{-0.0077x}) = 20xe^{-0.0077x}$

$$R'(x) = 20e^{-0.0077x} + 20xe^{-0.0077x}(-0.0077)$$
$$= 20e^{-0.0077x} - 0.154xe^{-0.0077x}$$

$R'(x) = 0$ when the quantity is about 130.

$$R''(x) = 20e^{-0.0077x}(-0.0077) - 0.154e^{-0.0077x} + 0.0012xe^{-0.0077x}$$
$$R''(130) = -0.154e^{-0.0077(130)} - 0.154e^{-0.0077(130)} + 0.0012(130)e^{-0.0077(130)}$$
$$= -0.154e^{-0.0077(130)} < 0$$

so R is maximized.

The quantity 130 maximizes revenue.

48. $G(t) = 5 + 2e^{0.01t}$

$$\frac{d}{dt}\ln G(t) = \frac{2e^{0.01t}(0.01)}{5+2e^{0.01t}} = \frac{0.02e^{0.01t}}{5+2e^{0.01t}}$$

If $t = 20$, then the relative rate of change is $\frac{0.02e^{0.2}}{5+2e^{0.2}} \approx 0.0033 = 0.33\%$

49. For $t = 10$, $\frac{G'(t)}{G(t)} = \frac{0.02e^{0.01(10)}}{5+2e^{0.01(10)}} = \frac{0.02e^{0.1}}{5+2e^{0.1}} \approx 0.0031 = 0.31\%$

50. $E(p) = \frac{-pD'(p)}{D(p)} = \frac{-p(-4p)}{63-2p^2} = \frac{4p^2}{63-2p^2}$

$E(3) = \frac{4\cdot 9}{63-18} = \frac{36}{45} = \frac{4}{5}$

Since $E(p) < 1$, demand is inelastic. Raising prices will increase revenue.

51. $E(p) = \frac{-pD'(p)}{D(p)} = \frac{-p\left[200(600-p)^{-1/2}\right](-1)}{400(600-p)^{1/2}} = \frac{200p}{400(600-p)} = \frac{p}{2(600-p)}$

$E(350) = \frac{350}{2(600-350)} = \frac{350}{500} = 0.7$

Since $E(p) < 1$, demand is inelastic. Raising prices will increase revenue.

52. $E(p) = \frac{-pD'(p)}{D(p)} = \frac{-p(-0.528p^{-1.44})}{1.2p^{-0.44}} = \frac{0.528}{1.2} = 0.44$

Demand is inelastic.

53. Relative rate of change of $P(x) = \frac{d}{dx}\ln P(x)$.

$$\frac{d}{dx}\ln P(x) = \frac{0.04+0.006x^2}{3.25+0.04x+0.002x^3}$$

At $x = 9$, $\frac{0.04+0.006(9)^2}{3.25+0.04(9)+0.002(9)^3} \approx 0.104 = 10.4\%$

54. **a.**

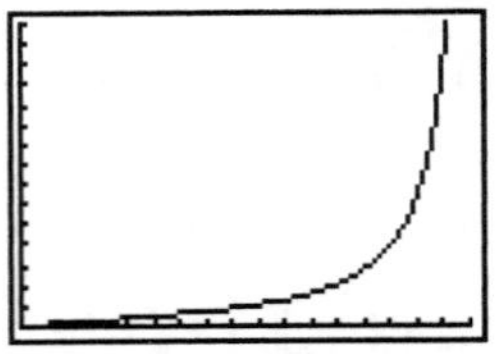

on [0, 17] by [0, 15]

$$E(p) = \frac{-pD'(p)}{D(p)}$$
$$= \frac{-p(-20+2p-0.09p^2)}{200-20p+p^2-0.03p^3}$$
$$= \frac{0.09p^3-2p^2+20p}{200-20p+p^2-0.03p^3}$$

The demand is elastic at \$10,000 because $E(10) \approx 1.29$.

b. Since demand is elastic, the dealer should lower the price.

c. $E(8.7) \approx 1$; thus at \$8700 elasticity equals 1.

Chapter 5: Integration and Its Applications

EXERCISES 5.1

1. $\int x^4 dx = \frac{x^5}{5} + C$ (with $n = 4$, $n + 1 = 5$)

3. $\int x^{2/3} dx = \frac{x^{5/3}}{\frac{5}{3}} = \frac{3}{5}x^{5/3} + C$ (with $n = 2/3$, $n + 1 = 5/3$)

5. $\int \sqrt{u}\, du = \int u^{1/2} du = \frac{u^{3/2}}{\frac{3}{2}} + C = \frac{2}{3}u^{3/2} + C$ (with $n = 1/2$, $n + 1 = 3/2$)

7. $\int \frac{dw}{w^4} = \int w^{-4} dw = \frac{w^{-3}}{-3} + C = -\frac{1}{3}w^{-3} + C$ (with $n = -4$, $n + 1 = -3$)

9. $\int \frac{dz}{\sqrt{z}} = \int z^{-1/2} dz = \frac{z^{1/2}}{\frac{1}{2}} = 2\sqrt{z} + C$ (with $n = -1/2$, $n + 1 = 1/2$)

9. $\int 6x^5 dx = 6\int x^5 dx = 6 \cdot \frac{1}{6}x^6 + C = x^6 + C$

13. $\int (8x - 5)dx = \frac{8x^2}{2} - 5x + C = 4x^2 - 5x + C$

15.
$$\begin{aligned}\int (8x^3 - 3x^2 + 2)dx &= 8\int x^3 dx - 3\int x^2 dx + 2\int 1 dx \\ &= 8 \cdot \frac{1}{4}x^4 - 3 \cdot \frac{1}{3}x^3 + 2x + C \\ &= 2x^4 - x^3 + 2x + C\end{aligned}$$

17.
$$\begin{aligned}\int \left(6\sqrt{x} + \frac{1}{\sqrt[3]{x}}\right) dx &= 6\int x^{1/2} dx + \int x^{-1/3} dx \\ &= 6 \cdot \frac{1}{\frac{3}{2}}x^{3/2} + \frac{x^{2/3}}{\frac{2}{3}} + C \\ &= 4x^{3/2} + \frac{3}{2}x^{2/3} + C\end{aligned}$$

19.
$$\begin{aligned}\int \left(16\sqrt[3]{x^5} - \frac{16}{\sqrt[3]{x^5}}\right) dx &= 16\int x^{5/3} dx - 16\int x^{-5/3} dx \\ &= 16 \cdot \frac{3}{8}x^{8/3} - 16 \cdot \frac{1}{-\frac{2}{3}}x^{-2/3} + C \\ &= 6x^{8/3} + 24x^{-2/3} + C\end{aligned}$$

21.
$$\begin{aligned}\int \left(10\sqrt[3]{t^2} + \frac{1}{\sqrt[3]{t^2}}\right) dt &= 10\int t^{2/3} dt + \int t^{-2/3} dt \\ &= 10 \cdot \frac{3}{5}t^{5/3} + \frac{1}{\frac{1}{3}}t^{1/3} + C \\ &= 6t^{5/3} + 3t^{1/3} + C\end{aligned}$$

23. $\int(4z^{-3}+z^{-1/2})\,dz = \frac{4z^{-2}}{-2}+2z^{1/2}+C$
$= \frac{-2}{z^2}+2\sqrt{z}+C$

25. $\int(x-1)^2\,dx = \int(x^2-2x+1)\,dx$
$= \int x^2dx - 2\int x\,dx + \int 1\,dx$
$= \frac{x^3}{3}-x^2+x+C$

27. $\int(12x^3-12x^2)\,dx = \frac{12x^4}{4}-\frac{12x^3}{3}+C$
$= 3x^4-4x^3+C$

29. $\int(1+10w)\sqrt{w}\,dw = \int(w^{1/2}+10w^{3/2})\,dw$
$= \int w^{1/2}dw + 10\int w^{3/2}dw$
$= \frac{w^{3/2}}{\frac{3}{2}}+10\frac{w^{5/2}}{\frac{5}{2}}+C = \frac{2}{3}w^{3/2}+4w^{5/2}+C$

31. $\int\frac{6x^3-6x^2+x}{x}\,dx = \int\frac{x(6x^2-6x+1)}{x}\,dx$
$= \int(6x^2-6x+1)dx$
$= 6\int x^2dx - 6\int xdx + \int 1dx$
$= 2x^3-3x^2+x+C$

33. $\int(x-2)(x+4)dx = \int(x^2+2x-8)\,dx$
$= \int x^2dx + 2\int x\,dx - 8\int 1\,dx$
$= \frac{x^3}{3}+x^2-8x+C$

35. $\int(r-1)(r+1)dr = \int(r^2-1)\,dr = \int r^2dr - \int 1\,dr$
$= \frac{r^3}{3}-r+C$

37. $\int\frac{x^2-1}{x+1}\,dx = \int\frac{(x+1)(x-1)}{x+1}\,dx = \int(x-1)dx = \int x\,dx - \int 1dx = \frac{x^2}{2}-x+C$

39. $\int(t+1)^3\,dt = \int(t^3+3t^2+3t+1)\,dt = \int t^3dt + 3\int t^2dt + 3\int t\,dt + \int 1\,dt$
$= \frac{t^4}{4}+t^3+\frac{3}{2}t^2+t+C$

41. **b.**

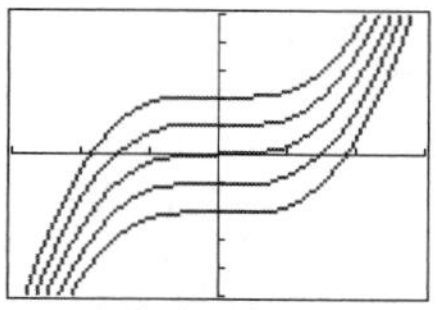

on [–3, 3] by [–5, 5]

Each increase in C of 1 shifts the curve up 1 unit.

c. For $x = 2$, the slope of each curve is 4. Thus, the derivative of each curve is x^2.

43.
$$C(x) = \int MC(x)dx = \int\left(20x^{3/2} - 15x^{2/3} + 1\right)dx$$
$$= 20\int x^{3/2}dx - 15\int x^{2/3}dx + \int 1\,dx$$
$$= 20\cdot\frac{1}{\frac{5}{2}}x^{5/2} - 15\cdot\frac{1}{\frac{5}{3}}x^{5/3} + x + K = 8x^{5/2} - 9x^{5/3} + x + K$$

Since fixed costs are \$4000,
$$4000 = C(0) = 8(0)^{5/2} - 9(0)^{5/3} + 0 + K$$
$$4000 = K$$

The cost function is $C(x) = 8x^{5/2} - 9x^{5/3} + x + 4000$.

45.
$$R(x) = \int MR(x)dx = \int\left(120 - 12x^{1/3}\right)dx = \int 120dx - 12\int x^{1/3}dx$$
$$= 120x - 12\cdot\frac{1}{\frac{4}{3}}x^{4/3} + C = 120x - 9x^{4/3} + C$$

Since $R(0) = 0$,
$$120(0) - 9(0)^{4/3} + C = 0$$
$$C = 0$$

The revenue function is $R(x) = 120x - 9x^{4/3}$

47. $v(t) = -0.24t^2 + 18t$ feet per second after t seconds (for $0 \le t < 40$).

a.
$$D(t) = \int v(t)dt = \int\left(-0.24t^2 + 18t\right)dt$$
$$= \int\left(-0.24t^2\right)dt + \int 18t\,dt = -0.24\int t^2dt + 18\int t\,dt$$
$$= -0.24\frac{t^3}{3} - 18\frac{t^2}{2} + C = -0.08t^3 + 9t^2 + C$$

Evaluating $D(t)$ at $t = 0$ gives $D(0) = C$. Since the distance is 0 at time $t = 0$, $C = 0$. Thus,
$$D(t) = -0.08t^3 + 9t^2$$

b. The distance the car will travel in the first 10 seconds is given by $D(t)$.
$$D(10) = -0.08(10)^3 + 9(10)^2$$
$$= -80 + 900 = 820 \text{ ft}$$

49. **a.** The formula for the number of words memorized in t minutes is found by integrating the rate.

$$N(t) = \int \frac{3}{\sqrt{t}}\,dt = 3\int t^{-1/2}\,dt$$
$$= 3\cdot\frac{1}{\frac{1}{2}}t^{1/2} + C = 6t^{1/2} + C$$

Since $N(0) = 0$, $6(0)^{1/2} + C = 0$, and thus $C = 0$. The formula is $N(t) = 6t^{1/2}$.

b. After 25 minutes, the total number of words memorized is

$N(25) = 6(25)^{1/2} = 30$ words

51. **a.** Total amount of pollution =

$$P(t) = \int 40\sqrt{t^3}\,dt$$
$$P(t) = \int 40t^{3/2}dt = 40\left(\frac{2}{5}\right)t^{5/2} + C$$
$$= 16t^{5/2} + C$$

Evaluate $P(0)$ to find C:

since $P(0) = 0$. Thus, $P(t) = 16t^{5/2}$

b. $P(4) = 16(4)^{5/2} = 512$ tons

Thus, 512 tons of pollution will enter the lake in the first four years of the plant's operation.

c. No, since $512 > 400$.

53. To find the young-adult population formula

$$P(x) = \int(-x^2 + 50x - 300)\,dx$$
$$= -\frac{x^3}{3} + \frac{50x^2}{2} - 300x + C$$
$$= -\frac{x^3}{3} + 25x^2 - 300x + C$$

Since $P(0) = 31{,}000$, then,

$$-\frac{(0)^3}{3} + 25(0)^2 - 300(0) + C = 31{,}000$$

and thus $C = 31{,}000$. The formula is

$$P(x) = -\frac{x^3}{3} + 25x^2 - 300x + 31{,}000$$

In the year 2040, $x = 30$

$$P(30) = -\frac{(30)^3}{3} + 25(30)^2 - 300(30) + 31{,}000$$
$$= 35{,}500 \text{ thousand}$$

55. To find the income formula

$$I(x) = \int(3.2x + 17.4)\,dx$$
$$= \frac{3.2x^2}{2} + 17.4x + C$$
$$= 1.6x^2 + 17.4x + C$$

Since $I(0) = 74$, then,

$$1.6(0)^2 + 17.4(0) + C = 74$$

and thus $C = 74$. The formula is

$$I(x) = 1.6x^2 + 17.4x + 74$$

In the year 2020, $x = 7$

$$I(7) = 1.6(7)^2 + 17.4(7) + 74$$
$$= \$274.2 \text{ billion}$$

57. $\frac{d}{dx}\ln(-x) = \frac{1}{-x}(-1) = \frac{1}{x}$

59. C (a constant)

61. $f(x) = g(x) + C$

63. **a.** $\int x^2 dx = \frac{x^3}{3} + C$

b.
$$\int x\,dx \cdot \int x\,dx = \left(\frac{x^2}{2} + C_1\right)\left(\frac{x^2}{2} + C_2\right)$$
$$= \frac{x^4}{4} + Cx^2 + C$$

c. No, so the integral of the product is not the product of the integral.

65. **a.** $\int 1\,dx = x + C$

b.
$$\frac{\int x\,dx}{\int x\,dx} = \frac{\frac{x^2}{2} + C}{\frac{x^2}{2} + C}$$

c. No, so the integral of a quotient is not the quotient of the integrals.

67. Cost

69. **a.** differentiate

b. integrate

c. differentiate

d. integrate

EXERCISES 5.2

1. $\int e^{3x}\,dx = \frac{1}{3}e^{3x} + C$ ($a = 3$; $1/a$ indicated)

3. $\int e^{\frac{1}{4}x}\,dx = \frac{1}{\frac{1}{4}}e^{\frac{1}{4}x} + C = 4e^{\frac{1}{4}x} + C$ ($a=1/4$; $1/a$ indicated)

5. $\int e^{0.05x}\,dx = \frac{1}{0.05}e^{0.05x} + C$ ($a=0.05$; $1/a$ indicated)

$=20e^{0.05x}+C$

7. $\int e^{-2y}\,dy = \frac{1}{-2}e^{-2y} + C = -\frac{1}{2}e^{-2y} + C$ ($a=-2$; $1/a$ indicated)

9. $\int e^{-0.5x}\,dx = \frac{1}{-0.5}e^{-0.5x} + C$ ($a=-0.5$; $1/a$ indicated)

$=-2e^{-0.5x}+C$

11. $\int 6e^{\frac{2}{3}x}\,dx = 6\int e^{\frac{2}{3}x}\,dx = 6\left(\frac{1}{\frac{2}{3}}e^{\frac{2}{3}x}\right)+C$ ($a=2/3$; $1/a$ indicated)

$= 9e^{\frac{2}{3}x}+C$

13. $\int(-5x^{-1})\,dx = -5\int x^{-1}\,dx = -5\ln|x| + C$

15. $\int \frac{3\,dx}{x} = \int \frac{3}{x}\,dx = 3\int \frac{1}{x}\,dx = 3\ln|x| + C$

17. $\int \frac{3}{2v}\,dv = \int \frac{3}{2}\frac{1}{v}\,dv = \frac{3}{2}\int \frac{1}{v}\,dv = \frac{3}{2}\ln|v| + C$

19. $\int\left(4e^{2x} - 6x\right)dx = \frac{4e^{2x}}{2} - \frac{6x^2}{2} + C = 2e^{2x} - 3x^2 + C$

21. $\int\left(e^{3x} - \frac{3}{x}\right)dx = \int e^{3x}\,dx - \int \frac{3}{x}\,dx$

$= \int e^{3x}\,dx - 3\int \frac{1}{x}\,dx = \frac{1}{3}e^{3x} - 3\ln|x| + C$

23. $\int\left(3e^{0.5t} - 2t^{-1}\right)dt = \int 3e^{0.5t}dt - \int 2t^{-1}dt = 3\int e^{0.5t}dt - 2\int t^{-1}dt = \frac{3}{0.5}e^{0.5t} - 2\ln|t| + C$

$= 6e^{0.5t} - 2\ln|t| + C$

25. $\int\left(x^2 + x + 1 + x^{-1} + x^{-2}\right)dx = \int x^2dx + \int x\,dx + \int 1\,dx + \int x^{-1}dx + \int x^{-2}dx$

$= \frac{x^3}{3} + \frac{x^2}{2} + x + \ln|x| + \frac{x^{-1}}{-1} + C = \frac{x^3}{3} + \frac{x^2}{2} + x + \ln|x| - x^{-1} + C$

27. $\int\left(5e^{0.02t} - 2e^{0.01t}\right)dt = \int 5e^{0.02t}\,dt - \int 2e^{0.01t}\,dt$

$= 5\int e^{0.02t}\,dt - 2\int e^{0.01t}\,dt$

$= \frac{5}{0.02}e^{0.02t} - \frac{2}{0.01}e^{0.01t} + C$

$= 250e^{0.02t} - 200e^{0.01t} + C$

29. $\frac{1}{2}\int(e^w - w)\,dw = \frac{1}{2}\left(e^w - \frac{w^2}{2}\right) + C = \frac{1}{2}e^w - \frac{1}{4}w^4 + C$

31. $\int\left(z + \frac{1}{z}\right)dz = \frac{z^2}{2} + \ln|z| + C$

33. $\int\left(e^x + \frac{1}{x}\right)dx = e^x + \ln|x| + C$

35
$$\begin{aligned}\int\frac{(x+1)^2}{x}dx &= \int\frac{x^2+2x+1}{x}dx\\ &= \int\left(\frac{x^2}{x}+\frac{2x}{x}+\frac{1}{x}\right)dx\\ &= \int x dx + \int 2\, dx + \int\frac{1}{x}dx\\ &= \frac{x^2}{2} + 2x + \ln|x| + C\end{aligned}$$

37
$$\begin{aligned}\int\frac{(t-1)(t+3)}{t^2}dt &= \int\frac{t^2+2t-3}{t^2}dt = \int\frac{t^2}{t^2}dt + \int\frac{2t}{t^2}dt - 3\int t^{-2}dt\\ &= t + 2\ln|t| - 3\left(\frac{1}{-1}\right)^{-1} + C = t + 2\ln|t| + \frac{3}{t} + C\\ &= t + 2\ln|t| + 3t^{-1} + C\end{aligned}$$

39
$$\begin{aligned}\int\frac{(x-2)^3}{x}dx &= \int\frac{x^3-6x^2+12x-8}{x}dx = \int\left(x^2 - 6x + 12 - \frac{8}{x}\right)dx\\ &= \int x^2 dx - 6\int x dx + 12\int 1\, dx - 8\int\frac{1}{x}dx\\ &= \frac{x^3}{3} - \frac{6x^2}{2} + 12x - 8\ln|x| + C\\ &= \frac{x^3}{3} - 3x^2 + 12x - 8\ln|x| + C\end{aligned}$$

41 **a.** To find a formula for the number of cases of flu, integrate $r(t)$.

$$\begin{aligned}\int r(t)\,dt &= \int 18e^{0.05t}\,dt = 18\int e^{0.05t}\,dt\\ &= 18\left(\frac{1}{0.05}e^{0.05t}\right) + C\\ &= 360e^{0.05t} + C\end{aligned}$$

Since there are 5 cases on day $t = 0$,

$$\begin{aligned}360e^{0.05(0)} + C &= 5\\ 360 + C &= 5\\ C &= -355\end{aligned}$$

The formula for the number of cases of flu is $F(t) = 360e^{0.05t} - 355$.

b. To find for the number of cases in the first 20 days, evaluate

$$\begin{aligned}F(20) &= 360e^{0.05(20)} - 355\\ &= 360e^1 - 355 \approx 624 \text{ cases}\end{aligned}$$

43 **a.** To find the total number of discs sold, integrate the rate.

$\int \frac{50}{t}\,dt = 50\int \frac{1}{t}\,dt = 50\ln t + C$

(We can drop the absolute value symbols because $t \geq 1$.) Since 0 were sold at $t = 1$,

$50\ln 1 + C = 0$

$50(0) + C = 0$

$C = 0$

The formula is $50\ln t$.

b. To find the number of discs sold by $t = 30$, evaluate

$50\ln 30 \approx 170$ records

Thus, not all the inventory will be sold in 30 days.

45. **a.** To find the total amount of tin consumed, integrate the rate.

$C(t) = \int 342e^{0.02t}\,dt = \frac{342}{0.02}e^{0.02t} + C$

$= 17{,}100e^{0.02t} + C$

The total consumed in the first zero years is zero, so $C(0) = 0$.

$17{,}100e^{0.02(0)} + C = 0$

$17{,}100 + C = 0$

$C = -17{,}100$

The formula is $C(t) = 17{,}100e^{0.02t} - 17{,}100$.

b. To find the value of t that exhausts the 4900 thousand metric tons, solve

$17{,}100e^{0.02t} - 17{,}100 = 4900$

$17{,}100e^{0.02t} = 22{,}000$

$e^{0.02t} = \frac{22{,}000}{17{,}100}$

$0.02t = \ln\left(\frac{22{,}000}{17{,}100}\right)$

$t = \frac{1}{0.02}\ln\left(\frac{22{,}000}{17{,}100}\right)$

$t \approx 12.6$ years

Thus, the world supply will be exhausted in 2027 (13 years from 2014).

47. To find a formula for the total maintenance cost during the first x years, we integrate the rate of cost, $r(x) = 1800e^{0.05x}$ (dollars per year).

a. $M(x) = \int r(x)\,dx = \int 1800e^{0.05x}\,dx = 1800\int e^{0.05x}\,dx = 1800\left(\frac{1}{0.05}\right)e^{0.05x} + C$

$= 36{,}000e^{0.05x} + C$

Since the total maintenance cost should be zero at $x = 0$, we can find C when $M(0) = 0$.

$36{,}000e^{0.05(0)} + C = 0$

$C = -36{,}000$

Thus, $M(x) = 36{,}000e^{0.05x} - 36{,}000$.

b. Evaluate $M(5)$ to find the total maintenance cost during the first 5 years.

$M(5) = 36{,}000e^{0.05(5)} - 36{,}000 \approx \$10{,}225$

49. **a.** To find the temperature, integrate the rate.

$\int -12e^{-0.2t}\,dt = -12\int e^{-0.2t}\,dt$

$= -12\left(\frac{1}{-0.2}e^{-0.2t}\right) + C$

$= 60e^{-0.2t} + C$

Since the water was 70 degrees at $t = 0$,

$60e^{-0.2(0)} + C = 70$

$60 + C = 70$

$C = 10$

The formula is $60e^{-0.2t} + 10$.

b. To find when the temperature is 32 degrees, solve

$60e^{-0.2t} + 10 = 32$

$e^{-0.2t} = \frac{22}{60}$

$-0.2t = \ln\left(\frac{22}{60}\right)$

$t = -\frac{1}{0.2}\ln\left(\frac{22}{60}\right) \approx 5$ hours

51. **a.** To find the total savings that the equipment will generate, integrate the rate.

$$\int 800e^{-0.2t}dt = 800\int e^{-0.2t}dt$$
$$= \frac{800}{-0.2}e^{-0.2t} + C$$
$$= -4000e^{-0.2t} + C$$

Since no savings is generated at $t = 0$,

$$-4000e^{-0.2(0)} + C = 0$$
$$-4000(1) + C = 0$$
$$C = 4000$$

The formula is $-4000e^{-0.2t} + 4000$

b. To find when the savings is 2000, solve

$$-4000e^{-0.2t} + 4000 = 2000$$
$$e^{-0.2t} = 0.5$$
$$t = \frac{\ln 0.5}{-0.2} \approx 3.5 \text{ years}$$

53. **a.** $V(0) = 5000$

$$V'(t) = 400e^{0.05t}$$
$$V(t) = \frac{400}{0.05}e^{0.05t} + C$$
$$= 8000e^{0.05t} + C$$

Using the initial condition,

$$5000 = 8000e^{0.05(0)} + C$$
$$5000 = 8000 + C$$
$$C = -3000$$
$$V(t) = 8000e^{0.05t} - 3000$$

b. $V(10) \approx \$10{,}190$

55. **a.** To find the total amount of lead consumed, integrate the rate.

$$C(t) = \int 5e^{0.05t}dt = 5\int e^{0.05t}dt$$
$$= \frac{5}{0.05}e^{0.05t} + C = 100e^{0.05t} + C$$

Since there was 0 consumption at $t = 0$,

$$100e^{0.05(0)} + C = 0$$
$$C = -100$$

The formula is

$$C(t) = 100e^{0.05t} - 100$$

b. To find when the resources will be exhausted, solve

$$100e^{0.05t} - 100 = 89$$

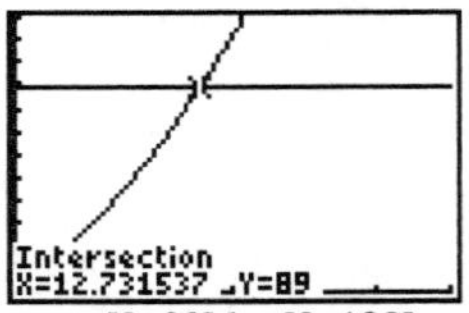

on [0, 30] by [0, 120]

The known resources of lead will be exhausted in about 2027 (approximately 13 years after 2014).

57. **a.** $A(t) = \int 2.15e^{0.086t}dt = \frac{2.15}{0.086}e^{0.086t} + C$

$$= 25e^{0.086t} + C$$

Since there was 0 increase at $t = 0$,

$$25e^{0.086(0)} + C = 0$$
$$C = -25$$

The formula is $25e^{0.086t} - 25$.

b. To find the increase between 2014 and 2020,

$$A(6) = 25e^{0.086(6)} - 25 = 25e^{0.516} - 25$$
$$A(0) = 25e^{0.086(0)} - 25 = 25 - 25 = 0$$

Then,

$$A(6) - A(0) = 25e^{0.516} - 25 - 0$$
$$\approx \$16.9 \text{ million}$$

59. b

61. a

63. c

65.

$$\int e^{ax+b}dx = \int e^{ax}e^{b}dx = e^{b}\int e^{ax}dx$$
$$= e^{b}\left(\frac{e^{ax}}{a} + C\right) = \frac{1}{a}e^{ax+b} + C$$

67. No, the box applies only to $\int \frac{1}{x}\,dx$ and not to $\int \frac{1}{x^5}\,dx$.

EXERCISES 5.3

1.

Left Endpoint	Value at Left Endpoint	Width	Area
x	$y = 2x$		$y \times width$
1	2	0.25	0.5
1.25	2.5	0.25	0.625
1.5	3	0.25	0.75
1.75	3.5	0.25	0.875

The total area is 2.75 square units.

3.

Left Endpoint	Value at Left Endpoint	Width	Area
x	$y = \sqrt{x}$		$y \times width$
1	1	1	1
2	1.4142	1	1.4142
3	1.7321	1	1.7321

The total area is approximately 4.15 square units.

5.

Left Endpoint	Value at Left Endpoint	Width	Area
x	$y = \frac{1}{x}$		$y \times width$
1	1	0.25	0.25
1.25	0.8	0.25	0.2
1.5	0.6667	0.25	0.1667
1.75	0.5714	0.25	0.1429

The total area is approximately 0.76 square units.

7. **i.** For five rectangles, divide the distance from $a = 1$ to $b = 2$ into five equal parts, so that each width is $\Delta x = \frac{2-1}{5} = 0.2$

Mark the points 1, 1.2, 1.4, 1.6, 1.8, and 2 using $\Delta x = 0.2$, and draw the rectangles.

$$\text{Area of 5 rectangles } = 2(1)(0.2)+2(1.2)(0.2)+2(1.4)(0.2)+2(1.6)(0.2)+2(1.8)(0.2)$$
$$= 0.4+0.48+0.56+0.64+0.72 = 2.8 \text{ square units}$$

ii. $\int_1^2 2x\,dx = 2\int_1^2 x\,dx = 2\left(\frac{1}{2}x^2\right)\Big|_1^2 = x^2\Big|_1^2 = (2)^2-(1)^2 = 4-1 = 3$ square units

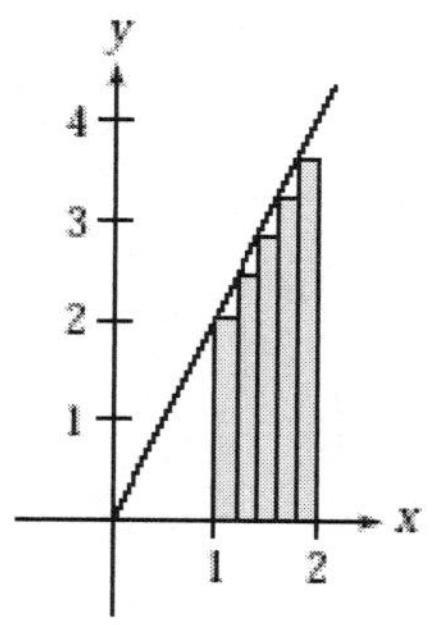

9. **i.** For six rectangles, divide the distance $a = 1$ to $b = 4$ into six equal parts, so that each width is

$$\Delta x = \frac{4-1}{6} = \frac{3}{6} = \frac{1}{2}$$

Mark the points 1, 1.5, 2, 2.5, 3, 3.5, and 4, and draw the rectangles.

$$\text{Area of 6 rectangles } = (\sqrt{1})\left(\frac{1}{2}\right) + (\sqrt{1.5})\left(\frac{1}{2}\right) + (\sqrt{2})\left(\frac{1}{2}\right) + (\sqrt{2.5})\left(\frac{1}{2}\right) + (\sqrt{3})\left(\frac{1}{2}\right) + (\sqrt{3.5})\left(\frac{1}{2}\right)$$
$$\approx 4.411 \text{ square units}$$

ii.
$$\int_1^4 \sqrt{x}\, dx = \frac{2}{3}x^{3/2}\Big|_1^4 = \frac{2}{3}x\sqrt{x}\Big|_1^4$$
$$= \frac{2}{3}(4)(2) - \frac{2}{3}(1)(1) = \frac{14}{3} \approx 4.667 \text{ square units}$$

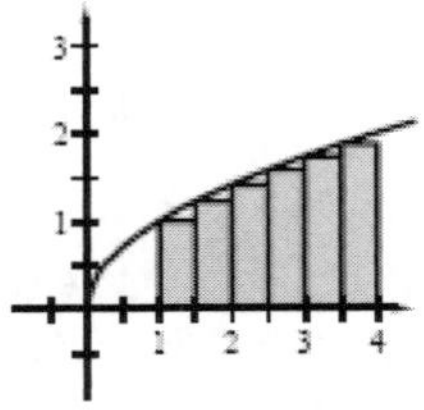

11. **i.** For 10 rectangles, divide the distance from $a = 1$ to $b = 2$ into 10 equal parts, so that each width is

$$\Delta x = \frac{2-1}{10} = 0.1$$

Mark the points 1.0, 1.1, 1.2, 1.3, 1.4, 1.5, 1.6, 1.7, 1.8, 1.9, and 2.0, and draw in the rectangles.

$$\text{Area of 10 rectangles } = \left(\frac{1}{1.0}\right)(0.1) + \left(\frac{1}{1.1}\right)(0.1) + \left(\frac{1}{1.2}\right)(0.1) + \left(\frac{1}{1.3}\right)(0.1) + \left(\frac{1}{1.4}\right)(0.1)$$
$$+ \left(\frac{1}{1.5}\right)(0.1) + \left(\frac{1}{1.6}\right)(0.1) + \left(\frac{1}{1.7}\right)(0.1) + \left(\frac{1}{1.8}\right)(0.1) + \left(\frac{1}{1.9}\right)(0.1)$$
$$\approx (1 + 0.909 + 0.833 + 0.769 + 0.714 + 0.667 + 0.625 + 0.588 + 0.556 + 0.526)(0.1)$$
$$= 0.719$$

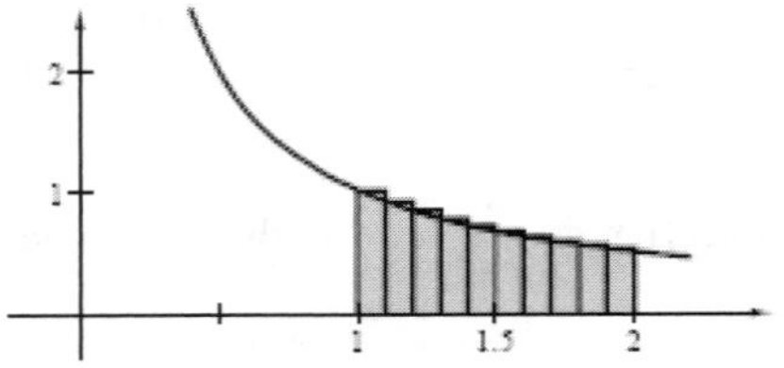

ii.
$$\int_1^2 \frac{1}{x}\, dx = \ln x \Big|_1^2 = \ln 2 - \ln 1 = \ln 2 - 0 \approx 0.693$$

13. Using left rectangles.

i.

n	Left	Midpoint	Right
10	2.9	3	3.1
100	2.99	3	3.01
1000	2.999	3	3.001

ii.
$$\int_1^2 2x\, dx = 2\int_1^2 x\, dx$$
$$= 2\left(\frac{x^2}{2}\right)\Big|_1^2 = x^2\Big|_1^2$$
$$= 4 - 1 = 3 \text{ square units}$$

15. Using left rectangles.

i.

n	Left	Midpoint	Right
10	4.515	4.667	4.815
100	4.652	4.667	4.682
1000	4.665	4.667	4.668

ii.
$$\int_1^4 \sqrt{x}\, dx = \int_1^4 x^{1/2} dx$$
$$= \frac{2}{3}x^{3/2}\Big|_1^4$$
$$= \frac{2}{3}(4)^{3/2} - \frac{2}{3}(1)^{3/2}$$
$$= \frac{16}{3} - \frac{2}{3} = \frac{14}{3}$$
$$\approx 4.667 \text{ square units}$$

17. Using left rectangles.

i.

n	Left	Midpoint	Right
10	0.719	3	3.1
100	0.696	3	3.01
1000	0.693	3	3.001

ii.
$$\int_1^2 \frac{1}{x}\,dx = \ln x\,\Big|_1^2$$
$$= \ln 2 - \ln 1$$
$$= \ln 2$$
$$\approx 0.693 \text{ square units}$$

19.
$$\int_0^3 x^2 dx = \frac{x^3}{3}\,\Big|_0^3$$
$$= \frac{(3)^3}{3} - \frac{(0)^3}{3}$$
$$= 9 \text{ square units}$$

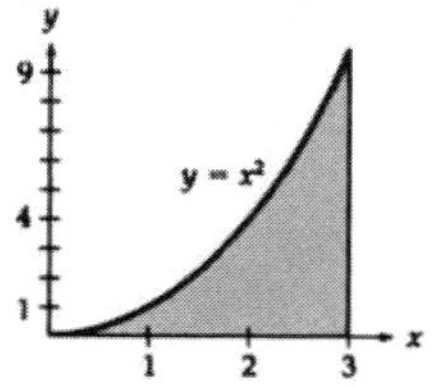

21.
$$\int_0^4 (4-x)\,dx = \left(4x - \frac{x^2}{2}\right)\Bigg|_0^4$$
$$= 4(4) - \frac{(4)^2}{2} - \left[4(0) - \frac{(0)^2}{2}\right]$$
$$= 16 - 8 - (0-0)$$
$$= 8 \text{ square units}$$

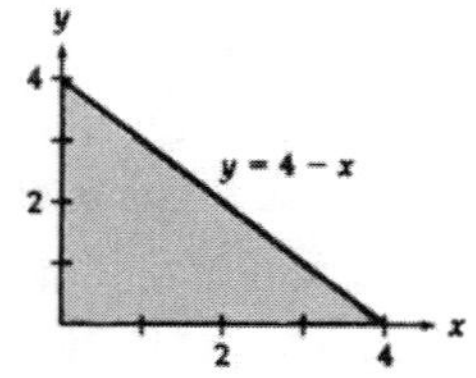

23.
$$\int_1^2 \frac{1}{x}\,dx = (\ln x)\,\Big|_1^2 = \ln 2 - \ln 1$$
$$= \ln 2 \text{ square units}$$

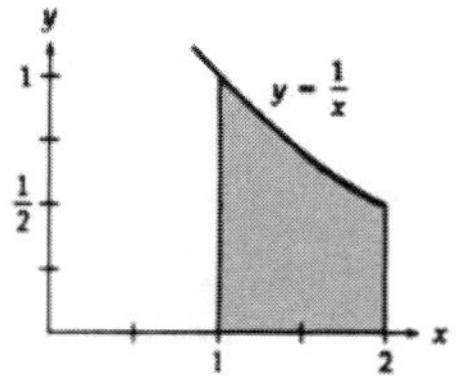

25.
$$\int_1^3 8x^3\,dx = 8\int_1^3 x^3 dx = 8\left(\frac{x^4}{4}\right)\Bigg|_1^3 = 2x^4\Big|_1^3$$
$$= 2(3)^4 - 2(1)^4 = 162 - 2$$
$$= 160 \text{ square units}$$

27.
$$\int_1^2 \left(6x^2+4x-1\right)dx = \left[6\left(\frac{x^3}{3}\right)+4\left(\frac{x^2}{2}\right)-x\right]\Bigg|_1^2 = \left(2x^3+2x^2-x\right)\Big|_1^2$$
$$= 2(2)^3 + 2(2)^2 - 2 - \left[2(1)^3 + 2(1)^2 - 1\right]$$
$$= 16 + 8 - 2 - 2 - 2 + 1 = 19 \text{ square units}$$

29.
$$\int_4^9 \frac{1}{\sqrt{x}}\,dx = \int_4^9 x^{-1/2}\,dx = \frac{x^{1/2}}{\frac{1}{2}}\,\Bigg|_4^9 = 2x^{1/2}\,\Big|_4^9 = 2(9)^{1/2} - 2(4)^{1/2}$$
$$= 2(3) - 2(2) = 2 \text{ square units}$$

31.
$$\int_0^8 \left(8 - 4\sqrt[3]{x}\right)dx = \int_0^8 \left(8 - 4x^{1/3}\right)dx = \left[8x - 4\left(\frac{x^{4/3}}{\frac{4}{3}}\right)\right]\Bigg|_0^8$$
$$= \left(8x - 3x^{4/3}\right)\Big|_0^8 = 8(8) - 3(8)^{4/3} - \left[8(0) - 3(0)^{4/3}\right]$$
$$= 64 - 48 - (0-0) = 16 \text{ square units}$$

33. $\int_1^5 \frac{1}{x}\,dx = \ln x\,\Big|_1^5 = \ln 5 - \ln 1 = \ln 5 - 0 = \ln 5$ square units

35. $\int_1^2 \left(x^{-1} + x^2\right)\,dx = \left(\ln x + \frac{x^3}{3}\right)\Big|_1^2$

$= \ln 2 + \frac{2^3}{3} - \left(\ln 1 + \frac{1^3}{3}\right) = \ln 2 - \frac{7}{3}$ square units

37. $\int_0^{\ln 3} 2e^x\,dx = 2\int_0^{\ln 3} e^x dx = 2e^x\,\Big|_0^{\ln 3} = 2e^{\ln 3} - 2e^0 = 2(3) - 2 = 4$ square units

39. $\int_0^2 e^{\frac{1}{2}x}\,dx = \frac{1}{\frac{1}{2}}e^{\frac{1}{2}x}\,\Big|_0^2 = 2e^{\frac{1}{2}x}\,\Big|_0^2 = 2e^{\frac{1}{2}(2)} - 2e^{\frac{1}{2}(0)} = 2e - 2$ square unit

41. **a.** $\int_1^2 \left(12 - 3x^2\right)\,dx = \left[12x - 3\left(\frac{x^3}{3}\right)\right]\Big|_1^2 = \left(12x - x^3\right)\Big|_1^2$

$= 12(2) - (2)^3 - \left[12(1) - (1)^3\right]$

$= 24 - 8 - 12 + 1 = 5$ square units

b. Using FnInt, the definite integral is 5 square units.

43. **a.** $\int_1^4 \frac{1}{x^3}\,dx = \int_1^4 x^{-3}\,dx = \frac{x^{-2}}{-2}\,\Big|_1^4 = -\frac{1}{2x^2}\,\Big|_1^4 = -\frac{1}{2(4)^2} - \left[-\frac{1}{2(1)^2}\right]$

$= -\frac{1}{32} + \frac{1}{2} = \frac{15}{32}$ square unit

b. Using FnInt, the definite integral is $\frac{15}{32}$.

45. **a.** $\int_1^2 \left(2x + 1 + x^{-1}\right)\,dx = \left[2\left(\frac{x^2}{2}\right) + x + \ln x\right]\Big|_1^2 = \left(x^2 + x + \ln x\right)\Big|_1^2$

$= (2)^2 + 2 + \ln 2 - \left[(1)^2 + 1 + \ln 1\right]$

$= 4 + 2 + \ln 2 - 1 - 1 - 0 = 4 + \ln 2$ square units ≈ 4.6931 square units

b. Using FnInt, the definite integral is approximately 4.6931.

47. $\int_0^1 \left(x^{99} + x^9 + 1\right)\,dx = \left(\frac{x^{100}}{100} + \frac{x^{10}}{10} + x\right)\Big|_0^1 = \left[\frac{(1)^{100}}{100} + \frac{(1)^{10}}{10} + 1\right] - \left[\frac{(0)^{100}}{100} + \frac{(0)^{10}}{10} + 0\right]$

$= \frac{1}{100} + \frac{1}{10} + 1 - 0 - 0 - 0 = \frac{111}{100}$

49. $\int_1^2 \left(6t^2 - 2t^{-2}\right)\,dt = \left[6\left(\frac{t^3}{3}\right) - 2\left(\frac{t^{-1}}{-1}\right)\right]\Big|_1^2 = \left(2t^3 + \frac{2}{t}\right)\Big|_1^2$

$= 2(2)^3 + \frac{2}{2} - \left[2(1)^3 + \frac{2}{1}\right] = 16 + 1 - 2 - 2 = 13$

51. $\int_1^4 \frac{1}{y^2}\,dy = \int_1^4 y^{-2}\,dy = \left(\frac{y^{-1}}{-1}\right)\Big|_1^4 = -\frac{1}{y}\,\Big|_1^4 = -\frac{1}{4} + 1 = \frac{3}{4}$

53. $\int_1^e \frac{dx}{x} = \int_1^e \frac{1}{x}\,dx = \ln x \,\Big|_1^e = 1 - 0 = 1$

55. $\int_1^3 \left(9x^2 + x^{-1}\right) dx = \left[9\left(\frac{x^3}{3}\right) + \ln x\right]\Big|_1^3 = \left(3x^3 + \ln x\right)\Big|_1^3$
$= 3(3)^3 + \ln 3 - \left(3(1)^3 + \ln 1\right) = 78 + \ln 3$

57. $\int_{-2}^{-1} 3x^{-1}\,dx = 3\int_{-2}^{-1} x^{-1}\,dx = 3\ln|x| \,\Big|_{-2}^{-1} = 3\ln|-1| - 3\ln|-2| = -3\ln 2$

59. $\int_0^1 12e^{3x}\,dx = 12\int_0^1 e^{3x}\,dx = 12\left(\frac{1}{3}e^{3x}\right)\Big|_0^1 = 4e^{3x}\,\Big|_0^1 = 4e^3 - 4e^0 = 4e^3 - 4$

61. $\int_{-1}^1 5e^{-x}\,dx = 5\int_{-1}^1 e^{-x}\,dx = 5\left(\frac{1}{-1}e^{-x}\right)\Big|_{-1}^1 = -5e^{-x}\,\Big|_{-1}^1 = -5e^{-1} - \left(-5e^{-(-1)}\right) = 5e - 5e^{-1}$

63. $\int_{\ln 2}^{\ln 3} e^x\,dx = e^x\,\Big|_{\ln 2}^{\ln 3} = e^{\ln 3} - e^{\ln 2} = 3 - 2 = 1$

65. $\int_1^2 \frac{(x+1)^2}{x} = \int_1^2 \left(\frac{x^2+2x+1}{x}\right) dx = \int_1^2 \left(x + 2 + \frac{1}{x}\right) dx = \left(\frac{x^2}{2} + 2x + \ln x\right)\Big|_1^2$
$= \frac{(2)^2}{2} + 2(2) + \ln 2 - \left[\frac{(1)^2}{2} + 2(1) + \ln 1\right] = 2 + 4 + \ln 2 - \frac{1}{2} - 2 - 0 = \frac{7}{2} + \ln 2$

67. Using the FnInt function, the definite integral is about 1.107.

69. Using the FnInt function, the definite integral is about 2.925.

71. Using the FnInt function, the definite integral is about 92.744.

73. **a.** $\int_0^3 x^2\,dx = \frac{x^3}{3}\,\Big|_0^3 = \frac{(3)^3}{3} - \frac{(0)^3}{3} = 9$

b. $\int_0^3 x^2\,dx = \left(\frac{1}{3}x^3 + C\right)\Big|_0^3$
$= \frac{(3)^3}{3} + C - \left[\frac{(0)^3}{3} + C\right]$
$= 9 + C - 0 - C = 9$

c. C always cancels because it is always added and subtracted in the evaluation.

75. We calculate the area under the curve. The area of the quarter circle with radius 1 is $\frac{\pi(1)^2}{4} = \frac{\pi}{4}$ square units. The area of the rectangle is $3 \cdot 1 = 3$ square units. The area of the triangle is $\frac{1}{2}(1)(1) = \frac{1}{2}$ square unit. Thus,

$$\int_0^4 f(x)\,dx = \frac{\pi}{4} + 3 + \frac{1}{2} = \frac{7}{2} + \frac{\pi}{4}$$

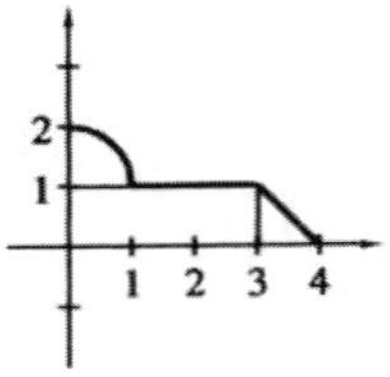

77. To find the total consumption of electricity, integrate the rate.

$$\int_1^5 \left(-3t^2 + 18t + 10\right) dt = \left[-3\left(\tfrac{t^3}{3}\right) + 18\left(\tfrac{t^2}{2}\right) + 10t\right]\Big|_1^5$$
$$= \left(-t^3 + 9t^2 + 10t\right)\Big|_1^5 = -(5)^3 + 9(5)^2 + 10(5) - \left[-(1)^3 + 9(1)^2 + 10(1)\right]$$
$$= -125 + 225 + 50 + 1 - 9 - 10 = 132 \text{ units}$$

79. To find the total number of checks processed, integrate $r(t)$.

$$\int_0^3 \left(-t^2 + 90t + 5\right) dt = \left[-\left(\tfrac{t^3}{3}\right) + 90\left(\tfrac{t^2}{2}\right) + 5t\right]\Big|_0^3 = \left(-\tfrac{t^3}{3} + 45t^2 + 5t\right)\Big|_0^3$$
$$= -\tfrac{(3)^3}{3} + 45(3)^2 + 5(3) - \left[-\tfrac{(0)^3}{3} + 45(0)^2 + 5(0)\right]$$
$$= -9 + 405 + 15 = 411 \text{ checks}$$

81. To find the total cost of the first 100 units, integrate $MC(x)$ of 0 to 100.

$$\int_0^{100} MC(x)\,dx = \int_0^{100} 6e^{-0.02x}\,dx = 6\int_0^{100} e^{-0.02x}\,dx$$
$$= 6\left(\tfrac{1}{-0.02}e^{-0.02x}\right)\Big|_0^{100} = -300e^{-0.02x}\Big|_0^{100}$$
$$= -300e^{-0.02(100)} - \left(-300e^{-0.02(0)}\right) = -300e^{-2} + 300 \approx \$259.40$$

83. To find the total increase in M1, integrate the rate from $x = 0$ to $x = 5$ (2014 is 5 decades since 1964).

$$\int_0^5 94e^{0.56x}dx = 94\int_0^5 e^{0.56x}dx = \frac{94}{0.56}e^{0.56x}\Big|_0^5 = \frac{94}{0.56}(e^{0.56(5)} - e^{0.56(0)}) \approx 2592$$

The total increase in M1 from 1964 to 2014 will be about \$2592 billion.

85. To find the total price change, integrate the rate from $t = 0$ (the year 2014) to $t = 10$.

$$\int_0^{10} 15e^{0.05t}dt = 15\int_0^{10} e^{0.05t}dt = \frac{15}{0.05}e^{0.05t}\Big|_0^{10} = 300(e^{0.05(10)} - e^{0.05(0)}) \approx 195$$

The total price increase will be about \$1.95.

87.

$$\int_0^{10} 342e^{0.02t}dt = 342\int_0^{10} e^{0.02t}dt = \frac{342}{0.02}e^{0.02t}\Big|_0^{10} = 17{,}100(e^{0.02(10)} - e^{0.02(0)}) \approx 3786$$

The total consumption of tin from 2014 to 2024 will be approximately 3786 thousand metric tons.

89. To find the number of words, integrate the rate from $t = 0$ to $t = 10$.

$$\int_0^{10} 6e^{-t/5}\,dt = 6\left(\frac{1}{-\frac{1}{5}}e^{-t/5}\right)\Bigg|_0^{10} = -30e^{-t/5}\Big|_0^{10}$$
$$= -30e^{-10/5} - \left(-30e^{-0/5}\right) = -30e^{-2} + 30 \approx 26 \text{ words}$$

91. **a.** Using the FnInt function,

$$\int_0^1 x\,dx = 0.5 = \tfrac{1}{2} \text{ square unit}$$
$$\int_0^1 x^2\,dx = 0.333 = \tfrac{1}{3} \text{ square unit}$$
$$\int_0^1 x^3\,dx = 0.25 = \tfrac{1}{4} \text{ square unit}$$
$$\int_0^1 x^4\,dx = 0.2 = \tfrac{1}{5} \text{ square unit}$$

b. $\int_0^1 x^n\,dx = \frac{1}{n+1}$ square unit

c. $$\int_0^1 x^n\,dx = \frac{x^{n+1}}{n+1}\Bigg|_0^1 = \frac{(1)^{n+1}}{n+1} - \frac{(0)^{n+1}}{n+1} = \frac{1}{n+1}$$

93. $$\int_0^{30} 5e^{-0.04t}dt = 5\int_0^{30} e^{-0.04t}dt$$
$$= 5\left(\frac{1}{-0.04}\right)e^{-0.04t}\Bigg|_0^{30}$$
$$= -125e^{-0.04(30)} - (-125e^{-0.04(0)})$$
$$= -125e^{-1.2} + 125$$
$$\approx 87 \text{ mg}$$

95. To find the total sales in the first half year, integrate from $x = 0$ to $x = 26$.

$$\int_0^{26} 8xe^{-0.1x}\,dx \approx 586 \text{ cars}$$

97. To find the total number of arrivals in the first two weeks, integrate $100e^{0.4\sqrt{x}}$ from $t = 0$ to $t = 14$.

$$\int_0^{14} 100e^{0.4\sqrt{x}}\,dx \approx 4023 \text{ arrivals}$$

99. The Riemann Sum will underestimate the actual area.

101. The Riemann Sum will underestimate the actual area.

103. The second friend is right: indefinite integrals are antiderivatives plus constants, and definite integrals are limits of Riemann sums. The Fundamental Theorem of Integral Calculus shows how to evaluate one using the other, but the two ideas are not the same.

105. 0 (since limits are 1 to 1)

107. The function is not defined at $x = 0$, so the Fundamental Theorem does not apply.

109. $$N = \int_A^B ax^{-b}\,dx = \frac{a}{-b+1}x^{-b+1}\Bigg|_A^B = \frac{a}{-b+1}B^{-b+1} - \frac{a}{-b+1}A^{-b+1} = \frac{a}{1-b}\left(B^{1-b} - A^{1-b}\right)$$

111. For interest rate $i = 0.06$ and $r(t) = 240,000$, the capital value for $T = 10$ years is

$$(\text{Capital value}) = \int_0^{10} 240,000e^{-0.06t}dt = \frac{240,000}{-0.06}e^{-0.06t}\Bigg|_0^{10}$$
$$= -4,000,000e^{-0.06t}\Big|_0^{10} = -4,000,000e^{-0.06(10)} - \left(-4,000,000e^{-0.06(0)}\right)$$
$$= -4,000,000e^{-0.6} + 4,000,000 \approx \$1,804,753$$

EXERCISES 5.4

1. $$(\text{Average Value}) = \tfrac{1}{3-0}\int_0^3 x^2\,dx = \tfrac{1}{3}\int_0^3 x^2\,dx = \tfrac{1}{3}\left(\tfrac{x^3}{3}\right)\Big|_0^3$$
$$= \tfrac{1}{3}\left[\frac{(3)^3}{3}\right] - \tfrac{1}{3}\left[\frac{(0)^3}{3}\right] = \tfrac{27}{9} = 3$$

3. $$(\text{Average Value}) = \tfrac{1}{4-0}\int_0^4 3\sqrt{x}\,dx = \tfrac{3}{4}\int_0^4 x^{1/2}\,dx = \tfrac{3}{4}\left(\tfrac{2}{3}x^{3/2}\right)\Big|_0^4$$
$$= \tfrac{1}{2}x^{3/2}\Big|_0^4 = \tfrac{1}{2}(4)^{3/2} - \tfrac{1}{2}(0)^{3/2} = \tfrac{1}{2}(8) = 4$$

5. $$(\text{Average Value}) = \tfrac{1}{5-1}\int_1^5 \tfrac{1}{x^2}\,dx = \tfrac{1}{4}\int_1^5 x^{-2}\,dx = \tfrac{1}{4}\left(\tfrac{x^{-1}}{-1}\right)\Big|_1^5 = -\tfrac{1}{4x}\Big|_1^5$$
$$= -\tfrac{1}{4(5)} - \left[-\tfrac{1}{4(1)}\right] = -\tfrac{1}{20} + \tfrac{1}{4} = -\tfrac{1}{20} + \tfrac{5}{20} = \tfrac{4}{20} = \tfrac{1}{5}$$

7. $$(\text{Average Value}) = \tfrac{1}{4-0}\int_0^4 (2x+1)\,dx = \tfrac{1}{4}\left(2\tfrac{x^2}{2} + x\right)\Big|_0^4 = \tfrac{1}{4}\left(x^2 + x\right)\Big|_0^4$$
$$= \tfrac{1}{4}\left[(4)^2 + 4\right] - \tfrac{1}{4}\left[(0)^2 + 0\right] = \tfrac{1}{4}(20) = 5$$

9. $$(\text{Average Value}) = \tfrac{1}{2-(-2)}\int_{-2}^2 \left(36 - x^2\right)dx = \tfrac{1}{4}\left(36x - \tfrac{x^3}{3}\right)\Big|_{-2}^2$$
$$= \tfrac{1}{4}\left[36(2) - \frac{(2)^3}{3}\right] - \tfrac{1}{4}\left[36(-2) - \frac{(-2)^3}{3}\right] = \tfrac{1}{4}\left(72 - \tfrac{8}{3}\right) - \tfrac{1}{4}\left(-72 + \tfrac{8}{3}\right)$$
$$= 18 - \tfrac{2}{3} + 18 - \tfrac{2}{3} = \tfrac{104}{3}$$

11. $$(\text{Average Value}) = \tfrac{1}{2-(-1)}\int_{-1}^2 \left(3z^2 - 2z\right)dz = \tfrac{1}{3}\left(z^3 - z^2\right)\Big|_{-1}^2$$
$$= \tfrac{1}{3}\left[\left(2^3 - 2^2\right) - \left((-1)^3 - (-1)^2\right)\right] = \tfrac{1}{3}(4+2) = 2$$

13. $$(\text{Average Value}) = \tfrac{1}{50-10}\int_{10}^{50} 3\,dx = \tfrac{1}{40}(3x)\Big|_{10}^{50}$$
$$= \tfrac{1}{40}(3\cdot 50 - 3\cdot 10) = \tfrac{1}{40}(150 - 30) = 3$$

15. $$(\text{Average Value}) = \tfrac{1}{2-0}\int_0^2 e^{\frac{1}{2}x}\,dx = \tfrac{1}{2}(2e^{\frac{1}{2}x})\Big|_0^2$$
$$= e^{2/2} - e^{\frac{1}{2}(0)} = e - 1$$

17. $$(\text{Average Value}) = \tfrac{1}{10-0}\int_0^{10} e^{0.01t}\,dt = \tfrac{1}{10}\frac{e^{0.01t}}{0.01}\Big|_0^{10}$$
$$= \tfrac{1}{10}\left(\frac{e^{0.01(10)}}{0.01} - \frac{e^0}{0.01}\right) \approx 1.05$$

19. $$(\text{Average Value}) = \tfrac{1}{2-1}\int_1^2 \tfrac{1}{x}\,dx = \ln x\big|_0^2 = \ln 2 - \ln 1 = \ln 2$$

21. $(\text{Average Value}) = \frac{1}{1-0}\int_0^1 x^n\,dx = \frac{x^{n+1}}{n+1}\Big|_0^1$

$= \frac{(1)^{n+1}}{n+1} - \frac{(0)^{n+1}}{n+1} = \frac{1}{n+1}$

23. $(\text{Average Value}) = \frac{1}{2-0}\int_0^2 (ax+b)\,dx = \frac{1}{2}\left(\frac{ax^2}{2}+bx\right)\Big|_0^2$

$= \frac{1}{2}\left[\frac{a(2)^2}{2}+b(2)\right] - \frac{1}{2}\left[\frac{a(0)^2}{2}+b(0)\right]$

$= \frac{1}{2}(2a+2b) = a+b$

25. $(\text{Average Value}) = \frac{1}{1-(-1)}\int_{-1}^1 e^{-x^4}\,dx$

$\approx \frac{1}{2}(1.690)$

$= 0.845$

27. To find the average sales during the first 3 days, calculate the average value of $S(x)$ for $x = 0$ to $x = 3$.

$\frac{1}{3-0}\int_0^3 S(x)\,dx = \frac{1}{3}\int_0^3 \left(200x+6x^2\right)dx = \frac{1}{3}\left(200\cdot\frac{x^2}{2}+6\cdot\frac{x^3}{3}\right)\Big|_0^3$

$= \frac{1}{3}(100x^2+2x^3)\Big|_0^3 = \frac{1}{3}\left[100(3)^2+2(3)^3\right] - \frac{1}{3}\left[100(0)^2+2(0)^3\right]$

$= \frac{1}{3}(900+54) = 318$

29. To find the average temperature, calculate the average value of $T(t)$ for $t = 0$ to $t = 10$.

$\frac{1}{10-0}\int_0^{10} T(t)\,dt = \frac{1}{10}\int_0^{10}\left(-0.3t^2+4t+60\right)dt = \frac{1}{10}\left(-0.1t^3+2t^2+60t\right)\Big|_0^{10}$

$= \frac{1}{10}\left[-0.1(10)^3+2(10)^2+60(10)\right] - \frac{1}{10}\left[-0.1(0)^3+2(0)^2+60(0)\right]$

$= \frac{1}{10}(-100+200+600) = 70$

31. To find the average amount of pollution, calculate the average value of $P(x)$ from $x = 1$ to $x = 10$.

$\frac{1}{10-1}\int_1^{10} P(x)\,dx = \frac{1}{9}\int_1^{10}\frac{100}{x}\,dx = \frac{100}{9}\ln x\Big|_1^{10} = \frac{100}{9}\ln 10 - \frac{100}{9}\ln 1 = \frac{100}{9}\ln 10 \approx 25.6 \text{ tons}$

33. To find the average value, calculate the average value of $V(t)$ from $t = 0$ to $t = 40$.

$\frac{1}{40-0}\int_0^{40} V(t)\,dt = \frac{1}{40}\int_0^{50} 1000e^{0.05t}\,dt = \frac{1000}{40(0.05)}e^{0.05t}\Big|_0^{40}$

$= 500e^{0.05t}\Big|_0^{40} = 500e^2 - 500e^0 = 500e^2 - 500 \approx \$3194.53$

35. To find the average revenue, calculate the average value of $14.2x^2+120x+474$, from $x = 0$ to $x = 10$.

$\frac{1}{10-0}\int_0^{10}(14.2x^2+120x+474)dx = \frac{1}{10}\left[\frac{14.2x^3}{3}+\frac{120x^2}{2}+474x\right]_0^{10}$

$= \frac{1}{10}\left[\frac{14.2}{3}(10)^3+60(10)^2+474(10)-0\right]$

$\approx \$1547 \text{ million} = \$1.547 \text{ billion}$

37. We integrate the "upper" curve minus the "lower" curve.

$$\int_0^3 [(x^2+1)-(2x-1)]\,dx = \int_0^3 (x^2-2x+2)\,dx = \left(\frac{x^3}{3}-x^2+2x\right)\Big|_0^3$$
$$= \frac{(3)^3}{3}-(3)^2+2(3)-\left[\frac{(0)^3}{3}-(0)^2+2(0)\right] = 6 \text{ square units}$$

39. $$\int_0^2 (e^{2x}-e^x)\,dx = \left(\frac{1}{2}e^{2x}-e^x\right)\Big|_0^2 = \frac{1}{2}e^4-e^2-\left(\frac{1}{2}e^0-e^0\right) = \frac{1}{2}e^4-e^2+\frac{1}{2} \text{ square units}$$

41. **a.**

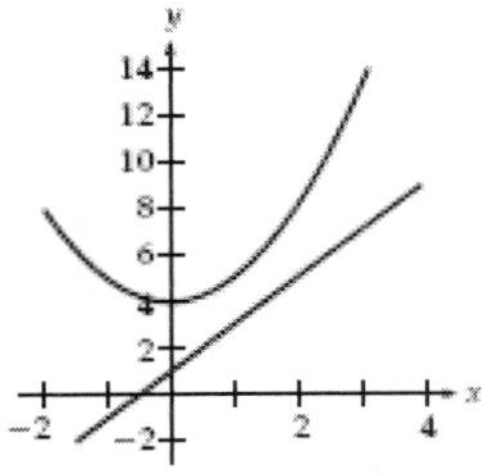

b. $$\int_0^3 [(x^2+4)-(2x+1)]\,dx = \int_0^3 (x^2-2x+3)\,dx = \left(\frac{x^3}{3}-x^2+3x\right)\Big|_0^3$$
$$= \frac{(3)^3}{3}-(3)^2+3(3)-\left[\frac{(0)^3}{3}-(0)^2+3(0)\right] = 9 \text{ square units}$$

43. **a.**

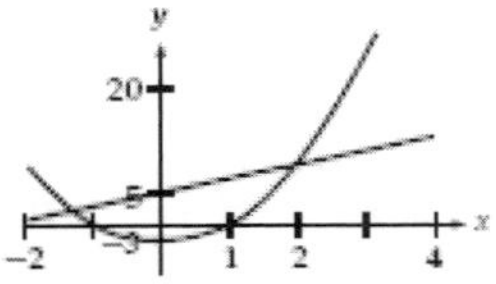

b. $$\int_0^2 [2x+5-(3x^2-3)]\,dx + \int_2^3 [3x^2-3-(2x+5)]\,dx$$
$$= \int_0^2 (-3x^2+2x+8)\,dx + \int_2^3 (3x^2-2x-8)\,dx$$
$$=(-x^3+x^2+8x)\Big|_0^2 + (x^3-x^2-8x)\Big|_2^3$$
$$= -(2)^3+(2)^2+8(2)-[(0)^3-(0)^2-8(0)]+[(3)^3-(3)^2-8(3)]-[(2)^3-(2)^2-8(2)]$$
$$= -8+4+16+27-9-24-8+4+16 = 18 \text{ square units}$$

45. Set the functions equal to each other and solve to find the points of intersection.

$$x^2-1 = 2-2x^2$$
$$3x^2-3=0$$
$$3(x-1)(x+1)=0$$
$$x=-1,\ 1$$

Substitute a value between –1 and 1 to determine the upper and lower curves. Use $x = 0$.

$$(0)^2-1=-1$$
$$2-2(0)^2=2$$

Thus, $y=2-2x^2$ is the upper curve and $y=x^2-1$ is the lower curve.

$$\text{Area} = \int_{-1}^1 [2-2x^2-(x^2-1)]\,dx = \int_{-1}^1 (3-3x^2)\,dx = (3x-x^3)\Big|_{-1}^1$$
$$= 3(1)-(1)^3-[3(-1)-(-1)^3] = 3-1+3-1 = 4 \text{ square units}$$

47. Set the functions equal to each other and solve.

$$6x^2 - 10x - 8 = 3x^2 + 8x - 23$$
$$3x^2 - 18x + 15 = 0$$
$$3(x^2 - 6x + 5) = 0$$
$$3(x-5)(x-1) = 0$$
$$x = 5,\ 1$$

Substitute a value in the interval between 1 and 5 to determine the upper and lower curves. Use $x = 2$.

$$6(2)^2 - 10(2) - 8 = -4$$
$$3(2)^2 + 8(2) - 23 = 5$$

Thus, $y = 3x^2 + 8x - 23$ is the upper curve and $y = 6x^2 - 10x - 8$ is the lower curve.

$$\text{Area} = \int_1^5 [(3x^2 + 8x - 23) - (6x^2 - 10x - 8)]\,dx$$
$$= \int_1^5 (-3x^2 + 18x - 15)\,dx = (-x^3 + 9x^2 - 15x)\Big|_1^5$$
$$= (-125 + 225 - 75) - (-1 + 9 - 15) = 25 + 7 = 32 \text{ square units}$$

49. Set the functions equal to each other and solve.

$$3x^2 - 12 = 0$$
$$3x^2 = 12$$
$$x^2 = 4$$
$$x = -2,\ 2$$

Substitute the value $x = 0$ to determine the upper and lower curves.

$$3(0)^2 - 12 = -12$$

Thus, $y = 0$ is the upper curve and $y = 3x^2 - 12$ is the lower curve.

$$\text{Area} = -\int_0^4 (3x^2 - 12)\,dx = -(x^3 - 6x^2)\Big|_0^4$$
$$= -\left[(4)^3 - 6(4)^2\right] + \left[(0)^3 - 6(0)^2\right]$$
$$= 32 \text{ square units}$$

51. Set the functions equal to each other and solve.

$$3x^2 = 12$$
$$x^2 = 4$$
$$x = -2,\ 2$$

Substitute the value $x = 0$ to determine the upper and lower curves.

$$3(0)^2 = 0$$

Thus, $y = 12$ is the upper curve and $y = 3x^2$ is the lower curve.

$$\text{Area} = \int_{-2}^{2} (12 - 3x^2)\,dx = (12x - x^3)\Big|_{-2}^{2}$$
$$= \left[12(2) - (2)^3\right] + \left[12(-2) - (-2)^3\right]$$
$$= 32 \text{ square units}$$

53. Set the functions equal to each other and solve.

$$x^2 = x^3$$
$$x^3 - x^2 = 0$$
$$x^2(x-1) = 0$$
$$x = 0, 1$$

Substitute $x = \frac{1}{2}$ to determine the upper and lower curves.

$$\left(\tfrac{1}{2}\right)^2 = \tfrac{1}{4}$$
$$\left(\tfrac{1}{2}\right)^3 = \tfrac{1}{8}$$

Thus, $y = x^2$ is the upper curve and $y = x^3$ is the lower curve.

$$\text{Area} = \int_0^1 (x^2 - x^3)\,dx = \left(\frac{x^3}{3} - \frac{x^4}{4}\right)\Bigg|_0^1 = \frac{1}{3} - \frac{1}{4} - \left(\frac{0^3}{3} - \frac{0^4}{4}\right) = \frac{1}{12} \text{ square unit}$$

55. Set the functions equal to each other and solve.

$$4x^3 + 3 = 4x + 3$$
$$x^3 - x = 0$$
$$x^2(x-1) = 0$$
$$x = 0,\ 1$$

Substitute $x = \frac{1}{2}$ to determine the upper and lower curves.

$$4\left(\tfrac{1}{2}\right)^3 + 3 = 3\tfrac{1}{2}$$
$$4\left(\tfrac{1}{2}\right) + 3 = 5$$

Thus, $y = 4x + 3$ is the upper curve and $y = 4x^3 + 3$ is the lower curve.

$$\text{Area} = \int_{-1}^{0} (4x^3 + 3 - 4x - 3)\,dx + \int_0^1 (4x + 3 - 4x^3 - 3)\,dx$$
$$= \left(x^4 - 2x^2\right)\Big|_{-1}^{0} + \left(2x^2 - x^4\right)\Big|_0^1$$
$$= \left[0^4 - 2(0)^2 - \left((-1)^4 - 2(-1)^2\right)\right] + \left[2(1)^2 - 1^4 - \left(2(0)^2 - (0)^4\right)\right]$$
$$= 1 + 1$$
$$= 2 \text{ square units}$$

57. Set the functions equal to each other and solve.

$$7x^3 - 36x = 3x^3 + 64x$$
$$4x^3 - 100x = 0$$
$$4x(x-5)(x+5) = 0$$
$$x = 0, \ -5, 5$$

Substitute $x = -1$ to determine the upper and lower curves on [–5, 0].

$$7(-1)^3 - 36(-1) = -7 + 36 = 29$$
$$3(-1)^3 + 64(-1) = -3 - 64 = -67$$

Thus, $y = 7x^3 - 36x$ is the upper curve and $y = 3x^3 + 64x$ is the lower curve on [–5, 0].
Substitute $x = 1$ to determine the upper and lower curves on [0, 5].

$$7(1)^3 - 36(1) = 7 - 36 = -29$$
$$3(1)^3 + 64(1) = 3 + 64 = 67$$

Thus, $y = 3x^3 + 64x$ is the upper curve and $y = 7x^3 - 36x$ is the lower curve on [0, 5]

$$\text{Area} = \int_{-5}^{0} [7x^3 - 36x - (3x^3 + 64x)]\, dx + \int_0^5 [3x^3 + 64x - (7x^3 - 36x)]\, dx$$
$$= \int_{-5}^{0} (4x^3 - 100x)\, dx + \int_0^5 (-4x^3 + 100x)\, dx$$
$$= (x^4 - 50x^2)\Big|_{-5}^{0} + (-x^4 + 50x^2)\Big|_0^5$$
$$= -[(-5)^4 - 50(-5)^2] + [-(5)^4 + 50(5)^2]$$
$$= -625 + 1250 - 625 + 1250 = 1250 \text{ square units}$$

59. The points of intersection are –2.95 and 1.51, and $y = x + 3$ is the upper curve and $y = e^x$ is the lower curve.

$$\text{Area} = \int_{-2.95}^{1.51} (x + 3 - e^x)\, dx \approx 5.694 \text{ square unit}$$

61. To find the total increase in population, calculate the area between the curves $22e^{0.02t}$ and $17e^{0.02t}$ from $t = 0$ to $t = 50$.

$$\text{Area} = \int_0^{50} (22e^{0.02t} - 17e^{0.02t})\, dt = \int_0^{50} 5e^{0.02t}\, dt$$
$$= 250e^{0.02t}\Big|_0^{50} = 250e - 250 \approx 430$$

The population increase is approximately 430 million.

63. **a.** To find the year x when $C(x) = S(x)$, set the functions equal to each other and solve.

$$100x = 1200 - 20x$$
$$120x = 12000$$
$$x = 10 \text{ years after installation}$$

b. To find the accumulated net savings, integrate $S(x) - C(x)$ from $x = 0$ to $x = 10$.

$$\int_0^{10} [S(x) - C(x)]dx = \int_0^{10} (1200 - 20x - 100x)dx = \int_0^{10} (1200 - 120x)dx$$
$$= (1200x - 60x^2)\Big|_0^{10} = 12{,}000 - 60(10)^2$$
$$= \$6000$$

65. Imports: $I(t) = 30e^{0.2t}$; Exports: $E(t) = 25e^{0.1t}$ (billion dollars)

$$(\text{Accumulated trade deficit}) = \int_{t=0}^{t=10} [I(t) - E(t)]\, dt$$
$$= \int_0^{10} (30e^{0.2t} - 25e^{0.1t})\, dt$$
$$= (150e^{0.2t} - 250e^{0.1t})\Big|_0^{10}$$
$$= (150e^2 - 250e^1) - (150 - 250)$$
$$\approx (1108.4 - 679.6) - (150 - 250)$$
$$\approx 429 + 100 \approx 529$$

Therefore, the country's accumulated trade deficit is about \$529 billion.

67. $$(\text{Total profit}) = \int_{100}^{200} [MR(x) - MC(x)]\, dx$$
$$= \int_{100}^{200} \left(\frac{700}{x} - \frac{500}{x}\right) dx$$
$$= (700 \ln x - 500 \ln x)\Big|_{100}^{200} = 200 \ln x\Big|_{100}^{200}$$
$$= 200(\ln 200 - \ln 100) \approx \$139 \text{ thousand}$$

69. $$(\text{Lives saved}) = \int_0^{20} \left(43e^{-0.1x} - 31e^{-0.1x}\right) dx = \int_0^{20} 12e^{-0.1x}\, dx$$
$$= \frac{12}{-0.1} e^{-0.1x}\Big|_0^{20} \approx 104{,}000 \text{ lives}$$

71. $\frac{d}{dx}(e^{x^2+5x}) = e^{x^2+5x}(2x+5)$

73. $\frac{d}{dx}[\ln(x^2 + 5x)] = \frac{1}{x^2+5x}(2x+5) = \frac{2x+5}{x^2+5x}$

75. The curve rises more steeply in the beginning and so spend "more of the time" higher up, and so its average should be higher than the "middle" y-value.

77. $\frac{y_1 + y_2}{2}$ (since the average height of a line is at its midpoint)

79. The average value will be greater than 7.

81. $-c$

83. The additional sales during the month resulting from the advertising.

EXERCISES 5.5

1. (Consumers' surplus) = $\int_0^A [d(x) - B]\, dx$, where A is the demand level and $B = d(A)$. Since the demand level is $x = 100$, $B = d(100) = 4000 - 12(100) = 2800$.

$$(\text{Consumers' surplus}) = \int_0^{100} (4000 - 12x - 2800)\, dx = \int_0^{100} (1200 - 12x)\, dx$$
$$= (1200x - 6x^2)\Big|_0^{100} = 120{,}000 - 60{,}000$$
$$= \$60{,}000$$

3. $B = d(200) = 300 - \frac{1}{2}(200) = 200$

$$(\text{Consumers' surplus}) = \int_0^{200} (300 - \tfrac{1}{2}x - 200)\, dx = \int_0^{200} \left(100 - \tfrac{1}{2}x\right) dx$$
$$= (100x - \tfrac{x^2}{4})\Big|_0^{200} = 20{,}000 - \tfrac{40{,}000}{4} = \$10{,}000$$

5. $B = d(50) = 350 - 0.09(50)^2 = 350 - 225 = 125$

$$\text{(Consumers' surplus)} = \int_0^{50} (350 - 0.09x^2 - 125)\, dx = \int_0^{50} (225 - 0.09x^2)\, dx$$
$$= (225x - 0.03x^3)\Big|_0^{50} = 11{,}250 - 3750 = \$7500$$

7. $B = d(100) = 200e^{-0.01(100)} = 200e^{-1}$

$$\text{(Consumers' surplus)} = \int_0^{100} (200e^{-0.01x} - 200e^{-1})\, dx = (-20{,}000e^{-0.01x} - 200e^{-1}x)\Big|_0^{100}$$
$$= -20{,}000e^{-1} - 20{,}000e^{-1} + 20{,}000e^0 \approx \$5285$$

9. The market price $B = s(x)$ for $x = 100$. Thus, $B = s(100) = 0.02(100) = 2$.

$$\text{(Producers' surplus)} = \int_0^{100} [B - s(x)]\, dx = \int_0^{100} (2 - 0.02x)\, dx$$
$$= (2x - 0.01x^2)\Big|_0^{100} = 200 - 0.01(100)^2 = \$100$$

11. $B = s(200) = 0.03(200)^2 = 1200$

$$\text{(Producers' surplus)} = \int_0^{200} (1200 - 0.03x^2)\, dx = (1200x - 0.01x^3)\Big|_0^{200}$$
$$= 240{,}000 - 80{,}000 = \$160{,}000$$

13. **a.** To find the market demand, solve $d(x) = s(x)$.

$$300 - 0.4x = 0.2x$$
$$0.6x = 300$$
$$x = 500$$

b. At $x = 500$, the market price is $d(500) = 300 - 0.4(500) = 100$.

$$\text{(Consumers' surplus)} = \int_0^{500} (300 - 0.4x - 100)\, dx = \int_0^{500} (200 - 0.4x)\, dx$$
$$= (200x - 0.2x^2)\Big|_0^{500} = 100{,}000 - 0.2(500)^2 = \$50{,}000$$

c. Since $d(500) = s(500) = 100$,

$$\text{(Producers' surplus)} = \int_0^{500} (100 - 0.2x)\, dx = (100x - 0.1x^2)\Big|_0^{500}$$
$$= 50{,}000 - 25{,}000 = \$25{,}000$$

15. **a.** To find the market demand, solve $d(x) = s(x)$.

$$300 - 0.03x^2 = 0.09x^2$$
$$0.12x^2 - 300 = 0$$
$$0.12(x^2 - 2500) = 0$$
$$0.12(x + 50)(x - 50) = 0$$

Since $x \neq -50$, the market demand occurs at $x = 50$.

b. At $x = 50$, the market price is $d(50) = s(50) = 0.09(50)^2 = 225$.

$$\text{(Consumers' surplus)} = \int_0^{50} [d(x) - \text{market price}]\, dx$$
$$= \int_0^{50} (300 - 0.03x^2 - 225)\, dx$$
$$= \int_0^{50} (75 - 0.03x^2)\, dx = (75x - 0.01x^3)\Big|_0^{50}$$
$$= 3750 - 1250 = \$2500$$

c. $(\text{Producers' surplus}) = \int_0^{50} [\text{market price} - s(x)]\, dx$

$$= \int_0^{50} (225 - 0.09x^2)\, dx = (225x - 0.03x^3)\Big|_0^{50}$$

$$= 11{,}250 - 3750 = \$7500$$

17. a. To find the market demand, solve $d(x) = s(x)$.

$300e^{-0.01x} = 100 - 100e^{-0.02x}$

Using the intersection function, the functions intersect at $x \approx 119.48$. The market demand is $x = 119.48$.

b. At $x = 119.48$, the market price is $d(119.48) = 300e^{-0.01(119.48)} \approx 90.83$

$$(\text{Consumers' surplus}) = \int_0^{119.48} (300e^{-0.01x} - 90.83)\, dx$$

$$= \left(-\frac{300}{0.01}e^{-0.01x} - 90.83x\right)\Big|_0^{119.48} = -30{,}000e^{-0.01(119.48)} - 90.83(119.48) + 30{,}000$$

$$\approx -30{,}000e^{-1.1948} - 10{,}852.4 + 30{,}000 \approx \$10{,}065$$

c. Since $d(119.48) = s(119.48) = 90.83$,

$$(\text{Producers' surplus}) = \int_0^{119.48} [90.83 - (100 - 100e^{-0.02x})]\, dx$$

$$= (-9.17x - 5000e^{-0.02x})\Big|_0^{119.48} \approx -1095.6 - 458.3 + 5000 \approx \$3446$$

19. $(\text{Gini index}) = 2\int_0^1 [x - L(x)]\, dx$

$$= 2\int_0^1 [x - x^{3.2}]\, dx$$

$$= 2\left(\frac{x^2}{2} - \frac{x^{4.2}}{4.2}\right)\Big|_0^1 = 2\left(\frac{1}{2} - \frac{1}{4.2}\right)$$

$$\approx 0.52$$

21. $(\text{Gini index}) = 2\int_0^1 [x - L(x)]\, dx$

$$= 2\int_0^1 [x - x^{2.1}]\, dx$$

$$= 2\left(\frac{x^2}{2} - \frac{x^{3.1}}{3.1}\right)\Big|_0^1 = 2\left(\frac{1}{2} - \frac{1}{3.1}\right)$$

$$\approx 0.35$$

23. $(\text{Gini index}) = 2\int_0^1 [x - (0.4x + 0.6x^2)]\, dx$

$$= 2\int_0^1 (0.6x - 0.6x^2)\, dx$$

$$= 2(0.3x^2 - 0.2x^3)\Big|_0^1 = 2(0.3 - 0.2) = 0.2$$

25. $L(x) = x^n$ (for $n > 1$)

$(\text{Gini index}) = 2\int_0^1 [x - L(x)]dx$

$$= 2\int_0^1 (x - x^n)\, dx = 2\left(\frac{x^2}{2} - \frac{x^{n+1}}{n+1}\right)\Big|_0^1$$

$$= 2\left[\frac{1}{2} - \frac{1^{n+1}}{n+1} - 0\right] = 2\left(\frac{1}{2} - \frac{1}{n+1}\right)$$

$$= 1 - \frac{2}{n+1} = \frac{n-1}{n+1}$$

27. $(\text{Gini index}) = 2\int_0^1 \left(x - \frac{e^{x^2} - 1}{e - 1}\right) dx \approx 0.46$

(using FnInt)

29. $(\text{Gini index}) = 2\int_0^1 \left(x - \frac{x + x^2 + x^3}{3}\right) dx \approx 0.28$

(using FnInt)

31.

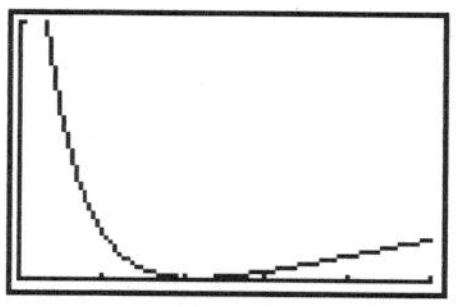

on [0, 5] by [0, 1]

The minimum occurs at $x \approx 2.13$, and the Lorenz function for this income distribution is $L(x) = x^{2.13}$.

$$(\text{Gini index}) = 2\int_0^1 (x - x^{2.13})\, dx \approx 0.36$$

(using FnInt)

33. $\frac{d}{dx}(x^5 - 3x^3 + x - 1)^4$
$= 4(x^5 - 3x^3 + x - 1)^3(5x^4 - 9x^2 + 1)$

35. $\frac{d}{dx}\ln(x^4 + 1) = \frac{1}{x^4+1}(4x^3) = \frac{4x^3}{x^4+1}$

37. $\frac{d}{dx}(e^{x^3}) = e^{x^3}(3x^2) = 3x^2 e^{x^3}$

39. When demand is 0, the price is 1000, or equivalently, at a price of 1000 the demand is 0, so 1000 is such a high price that *no one* will buy the product.

41. When demand is 0, the price is 0, or equivalently, at a price of 0 the demand is 0, so whatever they are, you can't even *give* them away.

43. Upward, since at a higher price producers should be willing to make more of them.

45. No. Such a curve would mean, for example, that the lowest paid 10% of the population is making *more* than 10% of the income, but then they can't be the lowest paid 10%.

47. Good thing: everyone has the same income, so on one is poorer than anyone else. Bad thing: everyone is paid the same, no matter whether they are lazy or hard working, so ambition is not rewarded. (Other answers are possible.)

EXERCISES 5.6

1. Use the formula 5 with $u = x^2 + 1$ and $n = 9$. For $u = x^2 + 1$, the differential is $du = 2x\, dx$.

$$\int (x^2+1)^9\, 2x\, dx = \int u^9\, du = \frac{u^{10}}{10} + C$$

Substituting $u = x^2 + 1$,

$$\int (x^2+1)^9\, 2x\, dx = \frac{(x^2+1)^{10}}{10} + C$$

3.
$$\begin{aligned}\int (x^2+1)^9\, x\, dx &= \frac{1}{2}\int (x^2+1)^9\, 2x\, dx \\ &= \frac{1}{2}\int u^9\, du \\ &= \frac{1}{2}\left(\frac{u^{10}}{10}\right) + C \\ &= \frac{u^{10}}{20} + C = \frac{(x^2+1)^{10}}{20} + C\end{aligned}$$

Let $u = x^2 + 1$
$du = 2x\, dx$

5. $\int e^{x^5} x^4\, dx = \frac{1}{5}\int e^{x^5}\, 5x^4\, dx$ Let $u = x^5$, $du = 5x^4\, dx$

$= \frac{1}{5}\int e^u\, du = \frac{1}{5}e^u + C$

$= \frac{1}{5}e^{x^5} + C$

7. $\int \frac{x^5\, dx}{x^6+1} = \frac{1}{6}\int \frac{1}{u}\, du$ Let $u = x^6 + 1$, $du = 6x^5\, dx$

$= \frac{1}{6}\ln|u| + C$

$= \frac{1}{6}\ln\left|x^6 + 1\right| + C$

9. $\int \sqrt{x^3+1}\; x\, dx$

If $u = x^3 + 1$, then $du = 3x^2\, dx$. To use the substitution formula $\int u^n\, du = \frac{1}{n+1}u^{n+1} + C$, du must contain a variable raised to a power of 2, not 1.

11. $\int e^{x^4} x^5\, dx$

If $u = x^4$, $du = 4x^3$. To use the formula $\int e^u\, du = e^u + C$, du must contain a variable raised to a power of 3, not 5.

13. $\int (x^4 - 16)^5 x^3\, dx = \frac{1}{4}\int u^5\, du = \frac{1}{4}\cdot\frac{u^6}{6} + C$ Let $u = x^4 - 16$, $du = 4x^3\, dx$

$= \frac{1}{24}u^6 + C = \frac{1}{24}(x^4 - 16)^6 + C$

15. $\int e^{-x^2} x\, dx = -\frac{1}{2}\int e^u\, du$ Let $u = -x^2$, $du = -2x\, dx$

$= -\frac{1}{2}e^u + C$

$= -\frac{1}{2}e^{-x^2} + C$

17. $\int e^{3x}\, dx = \frac{1}{3}\int e^u\, du$ Let $u = 3x$, $du = 3\, dx$

$= \frac{1}{3}e^u + C$

$= \frac{1}{3}e^{3x} + C$

19. $\int e^{x^2} x^2\, dx$

The integral cannot be found by the substitution formulas. If $u = x^2$, then du must contain a variable raised to a power of 1, not 2, to use the formula $\int e^u\, du = e^u + C$.

21. $\int \frac{dx}{1+5x} = \frac{1}{5}\int \frac{du}{u} = \frac{1}{5}\ln|u| + C$ Let $u = 1 + 5x$, $du = 5\, dx$

$= \frac{1}{5}\ln|1 + 5x| + C$

23. $\int (x^2+1)^9\, 5x\, dx = \frac{5}{2}\int u^9 du$ — Let $u = x^2+1$, $du = 2x\, dx$

$= \frac{5}{2}\left(\frac{u^{10}}{10}\right) + C$

$= \frac{(x^2+1)^{10}}{4} + C$

25. $\int \sqrt[4]{z^4+16}\, z^3\, dz = \frac{1}{4}\int u^{1/4}\, du$ — Let $u = z^4+16$, $du = 4z^3\, dz$

$= \left(\frac{1}{4}\right)\left(\frac{4}{5}\right)u^{5/4} + C$

$= \frac{1}{5}(z^4+16)^{5/4} + C$

27. The integral cannot be found by the substitution formulas. If $u = x^4+16$, then du must contain a variable raised to a power of 3, not 2, to use the formula $\int u^n\, du = \frac{u^{n+1}}{n+1} + C$.

29. $\int (2y^2+4y)^5 (y+1)\, dy = \int u^5 \frac{du}{4}$ — Let $u = 2y^2+4y$, $du = (4y+4)\, dy = 4(y+1)\, dy$

$= \frac{1}{4}\int u^5\, du$

$= \frac{1}{4}\left(\frac{u^6}{6}\right) + C$

$= \frac{1}{24}(2y^2+4y)^6 + C$

31. $\int e^{x^2+2x+5}(x+1)\, dx = \frac{1}{2}\int e^u\, du$ — Let $u = x^2+2x+5$, $du = (2x+2)\, dx = 2(x+1)\, dx$

$= \frac{1}{2}e^u + C$

$= \frac{1}{2}e^{x^2+2x+5} + C$

33. $\int \frac{x^3+x^2}{3x^4+4x^3}\, dx = \frac{1}{12}\int \frac{1}{u}\, du$ — Let $u = 3x^4+4x^3$, $du = (12x^3+12x^2)\, dx = 12(x^3+x^2)\, dx$

$= \frac{1}{12}\ln|u| + C$

$= \frac{1}{12}\ln\left|3x^4+4x^3\right| + C$

35. $\int \frac{x^3+x^2}{(3x^4+4x^3)^2}\, dx = \frac{1}{12}\int \frac{1}{u^2}\, du$ — Let $u = 3x^4+4x^3$, $du = (12x^3+12x^2)\, dx = 12(x^3+x^2)\, dx$

$= \frac{1}{12}\left(-\frac{1}{u}\right) + C$

$= -\frac{1}{12}\left(\frac{1}{3x^4+4x^3}\right) + C$

37. $\int \frac{x}{1-x^2}\, dx = -\frac{1}{2}\int \frac{du}{u} = -\frac{1}{2}\ln|u| + C$ — Let $u = 1-x^2$, $du = -2x\, dx$

$= -\frac{1}{2}\ln\left|1-x^2\right| + C$

39. $\int (2x-3)^7\, dx = \frac{1}{2}\int u^7\, du$ — Let $u = 2x-3$, $du = 2\, dx$

$= \frac{1}{2}\left(\frac{u^8}{8}\right) + C$

$= \frac{1}{16}(2x-3)^8 + C$

41. $\int \frac{e^{2x}}{e^{2x}+1}\,dx = \frac{1}{2}\int \frac{du}{u} = \frac{1}{2}\ln|u| + C$
$= \frac{1}{2}\ln\left|e^{2x}+1\right| + C$

Let $u = e^{2x}+1$
$du = 2e^{2x}\,dx$

43. $\int \frac{\ln x}{x}\,dx = \int u\,du = \frac{u^2}{2} + C$
$= \frac{1}{2}(\ln x)^2 + C$

Let $u = \ln x$
$du = \frac{1}{x}\,dx$

45. $\int \frac{e^{\sqrt{x}}}{\sqrt{x}}\,dx = 2\int e^u\,du = 2e^u + C$
$= 2e^{\sqrt{x}} + C$

Let $u = x^{1/2}$
$du = \frac{1}{2}x^{-1/2}\,dx$

47. $\int (x+1)x^2\,dx = \int (x^3+x^2)\,dx = \frac{x^4}{4} + \frac{x^3}{3} + C$

49. $\int (x+1)^2x^3\,dx = \int (x^2+2x+1)x^3\,dx$
$= \int (x^5+2x^4+x^3)\,dx$
$= \frac{x^6}{6} + \frac{2x^5}{5} + \frac{x^4}{4} + C$

51. **a.** $\int_0^3 e^{x^2}x\,dx = \frac{1}{2}\int_0^9 e^u\,du$
$= \frac{1}{2}e^u\Big|_0^9 = \frac{1}{2}e^9 - \frac{1}{2}e^0 = \frac{1}{2}e^9 - \frac{1}{2}$
≈ 4051.04

Let $u = x^2$
$du = 2x\,dx$
For $x = 0,\ u = 0$
$x = 3,\ u = 9$

b. Using FnInt, the integral is about 4051.04.

53. **a.** $\int_0^1 \frac{x}{x^2+1}\,dx = \frac{1}{2}\int_1^2 \frac{du}{u} = \frac{1}{2}\ln u\Big|_1^2$
$= \frac{1}{2}(\ln 2 - \ln 1) = \frac{1}{2}\ln 2$
≈ 0.347

Let $u = x^2+1$
$du = 2x\,dx$
For $x = 0,\ u = 1$
$x = 1,\ u = 2$

b. Using FnInt, the integral is about 0.347.

55. **a.** $\int_0^4 \sqrt{x^2+9}\,x\,dx = \frac{1}{2}\int_9^{25}\sqrt{u}\,du$
$= \frac{1}{2}\left(\frac{2}{3}u^{3/2}\right)\Big|_9^{25}$
$= \frac{1}{3}(125) - \frac{1}{3}(27) \approx 32.667$

Let $u = x^2+9$
$du = 2x\,dx$
For $x = 0,\ u = 9$
$x = 4,\ u = 25$

b. Using FnInt, the integral is about 32.667.

57. **a.** $\int_2^3 \frac{dx}{1-x} = -\int_{-1}^{-2}\frac{du}{u} = -\ln|u|\Big|_{-1}^{-2}$
$= -\ln 2 \approx -0.693$

Let $u = 1-x$
$du = -dx$
For $x = 2,\ u = -1$
$x = 3,\ u = -2$

b. Using FnInt, the integral is about –0.693.

59. **a.** $\int_1^8 \frac{e^{\sqrt[3]{x}}}{\sqrt[3]{x^2}}\,dx = 3\int_1^2 e^u\,du = 3e^u\Big|_1^2$
$= 3e^2 - 3e \approx 14.0123$

Let $u = \sqrt[3]{x}$
$du = \frac{1}{3}x^{-2/3}\,dx$
For $x = 1,\ u = 1$
$x = 8,\ u = 2$

b. Using FnInt, the integral is about 14.0123.

61. **a.** $\frac{d}{dx}\left(\frac{1}{n+1}u^{n+1}+C\right)=\frac{n+1}{n+1}u^{(n+1)-1}\frac{du}{dx}+0$

$=u^n\frac{du}{dx}=u^n u'$

b. $du = u'\,dx$ Thus, $\int u^n\,du=\int u^n u'\,dx=\frac{1}{n+1}u^{n+1}+C$

63. **a.** $\frac{d}{dx}(\ln u+C)=\frac{1}{u}\frac{du}{dx}=\frac{1}{u}u'$

b. $\frac{du}{u}=\frac{1}{u}u'\,dx$. Thus, $\int\frac{du}{u}=\ln u+C$.

65. $MC(x)=\frac{1}{2x+1}$

Let $u=2x+1$
$du=2\,dx$

$C(x)=\int MC(x)\,dx=\int\left(\frac{1}{2x+1}\right)dx$

$C(u)=\frac{1}{2}\int\frac{du}{u}=\frac{1}{2}\ln|u|+K$

$C(x)=\frac{1}{2}|2x+1|+K$

To find K, let $x = 0$. $C(0) = 50$, so

$50=\frac{1}{2}\ln|2x+1|+K$ for $x=0$

$50=\frac{1}{2}\ln|0+1|+K$

$50=K$

Thus, $C(x)=\frac{1}{2}\ln|2x+1|+50$

67. To find the average population, calculate the average value of $P(x)$ between $x = 0$ and $x = 8$.

$(\text{Average population})=\frac{1}{8-0}\int_0^8 P(x)\,dx=\frac{1}{8}\int_0^8 x(x^2+36)^{-1/2}\,dx$

$=\frac{1}{16}\int_{36}^{100}u^{-1/2}\,du=\frac{1}{8}u^{1/2}\Big|_{36}^{100}$

$=\frac{1}{8}(10-6)=\frac{4}{8}=\frac{1}{2}$ million

Let $u=x^2+36$
$du=2x\,dx$
For $x=0,\ u=36$
$x=8,\ u=100$

69. $S(x)=\frac{1}{x+1}$

Average sales $=\frac{1}{4-1}\int_1^4\left(\frac{1}{x+1}\right)dx$

$\frac{1}{3}\int_2^5\frac{du}{u}=\frac{1}{3}\ln u\Big|_2^5$

$=\frac{1}{3}\ln 5-\frac{1}{3}\ln 2=\frac{1}{3}(\ln 5-\ln 2)\approx 0.305$ million

Let $u=x+1$
$du=1\,dx$
For $x=1,\ u=2$
$x=4,\ u=5$

71. To find the total change during the first 3 days, integrate the rate from $t = 0$ to $t = 3$.

$\int_0^3\sqrt{25-t^2}\,dt$

$=-\frac{1}{2}\int_{25}^{16}u^{1/2}\,du=-\frac{1}{3}u^{3/2}\Big|_{25}^{16}$

$=-\frac{1}{3}(64-125)\approx\frac{61}{3}$ units

Let $u=25-t^2$
$du=-2t\,dt$
For $t=0,\ u=25$
$t=3,\ u=16$

73. To find the number sold, integrate the rate from 0 to 8.

$$\int_0^8 100e^{-x/4}\,dx$$

$$-4\int_0^{-2} 100e^u\,du$$

$$= -400e^u \Big|_0^{-2} = -400e^{-2} - (-400)$$

$$\approx -54.13 + 400 \approx 346$$

Let $u = -\frac{1}{4}x$
$du = -\frac{1}{4}\,dx$
For $x = 0,\ u = 0$
$x = 8,\ u = -2$

75. To find the amount of pollution in the first 3 years, integrate $r(t)$ from $t = 0$ to $t = 3$.

$$\int_0^3 r(t)\,dt = \int_0^3 \frac{t}{t^2+1}\,dx$$

$$= \frac{1}{2}\int_1^{10} \frac{1}{u}\,du = \frac{1}{2}\ln u \Big|_1^{10}$$

$$= \frac{1}{2}\ln 10 \approx 1.15 \text{ tons}$$

Let $u = t^2 + 1$
$du = 2t\,dt$
For $t = 0,\ u = 1$
$t = 3,\ u = 10$

77. To find the amount of weight gain from age 11 to 19, integrate from $x = 11$ to $x = 19$.

$$\int_{11}^{19} 14(x-10)^{-1/2}\,dx = 14(2)(x-10)^{1/2}\Big|_{11}^{19} = 28(19-10)^{1/2} - 28(11-10)^{1/2}$$

$$= 56 \text{ lbs}$$

79. Yes – one is easy (or as hard) as the other.

81. It simply changes the "name" of the variable, changing, for example, $\int e^{x^2}\,dx$ into $\int e^{u^2}\,du$, which is just as difficult.

83. It ignores the fact that dx and $du = 2x\,dx$ are different, so one cannot be substituted for the other.

85. $$\int (u+6)u^4\,du = \int (u^5 + 6u^4)\,du = \frac{u^6}{6} + \frac{6u^5}{5} + C$$

$$= \frac{1}{6}(x-5)^6 + \frac{6}{5}(x-5)^5 + C$$

Let $u = x - 5$
$du = dx$

87. $$\int (u+2)u^6\,du = \int (u^7 + 2u^6)\,du = \frac{u^8}{8} + \frac{2u^7}{7} + C$$

$$= \frac{1}{8}(x-2)^8 + \frac{2}{7}(x-2)^7 + C$$

Let $u = x - 2$
$du = dx$

89. $$\int \frac{u+1}{u}\,du = \int \left(1 + \frac{1}{u}\right)du = u + \ln|u| + C$$

$$= (x-2) + \ln|x-2| + C$$

$$= x + \ln|x-2| + C_1$$

Let $u = x - 2$
$du = dx$

91. $$\int (u+4)u^{1/3}\,du = \int (u^{4/3} + 4u^{1/3})\,du = \frac{3u^{7/3}}{7} + 4\left(\frac{3u^{4/3}}{4}\right) + C$$

$$= \frac{3}{7}(x-4)^{7/3} + 3(x-4)^{4/3} + C$$

Let $u = x - 4$
$du = dx$

93. $$\int \frac{u-2}{u^{1/2}}\,du = \int (u^{1/2} - 2u^{-1/2})\,du = \frac{2}{3}u^{3/2} - 2\left(\frac{2}{1}\right)u^{1/2} + C$$

$$= \frac{2}{3}(x+2)^{3/2} - 4(x+2)^{1/2} + C$$

Let $u = x + 2$
$du = dx$

REVIEW EXERCISES AND CHAPTER TEST FOR CHAPTER 5

1. $$\int(24x^2-8x+1)\,dx = \int 24x^2\,dx - \int 8x\,dx + \int dx$$
$$= 8x^3-4x^2+x+C$$

2. $$\int(12x^3+6x-3)\,dx = \int 12x^3\,dx + \int 6x\,dx - \int 3\,dx$$
$$= 3x^4+3x^2-3x+C$$

3. $$\int(6\sqrt{x}-5)\,dx = \int 6\sqrt{x}\,dx - \int 5\,dx = 6\int x^{1/2}dx - 5\int dx$$
$$= (6)\left(\tfrac{2}{3}\right)x^{3/2}-5x+C = 4x^{3/2}-5x+C$$

4. $$\int(8\sqrt[3]{x}-2)\,dx = \int 8x^{1/3}\,dx - \int 2\,dx$$
$$= 6x^{4/3}-2x+C$$

5. $$\int(10\sqrt[3]{x^2}-4x)\,dx = 6x^{5/3}-2x^2+C$$

6. $$\int(5\sqrt{x^3}-6x)\,dx = 2x^{5/2}-3x^2+C$$

7. $$\int(x+4)(x-4)\,dx = \int(x^2-16)\,dx$$
$$= \frac{x^3}{3}-16x+C$$

8. $$\int\frac{3x^3+2x^2+4x}{x}\,dx = \int(3x^2+2x+4)\,dx$$
$$= x^3+x^2+4x+C$$

9. To find the cost function, integrate $MC(x)$.

$$C(x) = \int MC(x)\,dx = \int(x^{-1/2}+4)\,dx = 2x^{1/2}+4x+K$$

To find K, let $x = 0$ because $C(0) = 20{,}000$ (fixed costs).

$$20{,}000 = 2(0)^{1/2}+4(0)+K$$
$$K = 20{,}000$$

Thus, $C(x) = 2x^{1/2}+4x+20{,}000$

10. **a.** The population t years from now is the integral of the rate.

$$\int 300\sqrt{t}\,dt$$
$$= 300\left(\tfrac{2}{3}t^{3/2}\right)+C$$
$$= 200t^{3/2}+C$$

To find C, let $t = 0$, because $r(0) = 40{,}000$.

$$40{,}000 = 200(0)^{3/2}+C$$
$$C = 40{,}000$$

The formula for the population of the town is $200t^{3/2}+40{,}000$.

b. The population 16 years from now will be

$$200(16)^{3/2}+40{,}000 = 200(64)+40{,}000$$
$$= 52{,}800 \text{ people}$$

11. $$\int e^{\frac{1}{2}x}\,dx = \frac{1}{\frac{1}{2}}e^{\frac{1}{2}x}+C = 2e^{\frac{1}{2}x}+C$$

12. $$\int e^{-2x}\,dx = -\tfrac{1}{2}e^{-2x}+C$$

13. $\int 4x^{-1}\,dx = 4\int \frac{1}{x}\,dx = 4\ln|x| + C$

14. $\int \frac{2}{x}\,dx = 2\ln|x| + C$

15. $\int\left(6e^{3x} - \frac{6}{x}\right)dx = \int 6e^{3x}dx - \int \frac{6}{x}\,dx = 6\int e^{3x}dx - 6\int \frac{1}{x}dx$
$= 6\left(\frac{1}{3}\right)e^{3x} - 6\ln|x| + C = 2e^{3x} - 6\ln|x| + C$

16. $\int(x - x^{-1})dx = \int\left(x - \frac{1}{x}\right)dx = \frac{x^2}{2} - \ln|x| + C$

17. $\int(9x^2 + 2x^{-1} + 6e^{3x})\,dx = \int 9x^2\,dx + \int \frac{2}{x}\,dx + \int 6e^{3x}dx$
$= 9\left(\frac{x^3}{3}\right) + 2\ln|x| + 6\left(\frac{1}{3}e^{3x}\right) + C$
$= 3x^3 + 2\ln|x| + 2e^{3x} + C$

18. $\int\left(\frac{1}{x^2} + \frac{1}{x} + e^{-x}\right)dx = \frac{x^{-1}}{-1} - \ln|x| + \frac{1}{-1}e^{-x} + C$
$= -x^{-1} + \ln|x| - e^{-x} + C$

19. **a.** To find to formula the total amount of antimony consumed, integrate the rate.

$\int 146e^{0.02t}\,dt = \frac{146}{0.02}e^{0.02t} + C = 7300e^{0.02t} + C$

Since consumption since 2014 is 0 at $t = 0$ (the year 2014),

$0 = 7300e^{0.02(0)} + C$
$0 = 7300 + C$
$C = -7300$

The formula is $7300e^{0.02t} - 7300$.

b. To find the year when the known resources will be exhausted, solve $7300e^{0.02t} - 7300 = 1800$.

$7300e^{0.02t} - 7300 = 1800$
$7300e^{0.02t} = 9100$
$e^{0.02t} = \frac{9100}{7300}$
$t = \frac{1}{0.02}\ln\left(\frac{9100}{7300}\right) \approx 11.0$ years

Current supplies will be exhausted about 11 years from 2014, that is, by 2025.

20. **a.** To find the formula for total savings, integrate the rate.

$\int 200e^{0.1t}\,dt = 2000e^{0.1t} + C$

Since savings is 0 at $t = 0$,

$0 = 2000e^{0.1(0)} + C$
$0 = 2000 + C$
$C = -2000$

The formula is $2000e^{0.1t} - 2000$.

b. $1500 = 2000e^{0.1t} - 2000$
$e^{0.1t} = 1.75$
$t = 10\ln 1.75 \approx 5.6$ years

21. **a.** To find the formula for the total amount of zinc consumed, integrate the rate.

$$\int 15e^{0.06t}\,dt = \frac{15}{0.06}e^{0.06t} + C = 250e^{0.06t} + C$$

Since consumption is 0 at $t = 0$ (the year 2014),

$$0 = 250e^{0.06(0)} + C$$
$$0 = 250 + C$$
$$C = -250$$

The formula is $250e^{0.06t} - 250$.

b. To find the year when the known resources will be exhausted, solve

$$250e^{0.06t} - 250 = 250$$
$$e^{0.06t} = \frac{500}{250}$$
$$t = \frac{1}{0.06}\ln(2) \approx 11.6 \text{ years}$$

Current supplies will be exhausted about 12 years from 2014, that is, by 2026.

22. **a.** To find total growth in profit, integrate the rate.

$$\int 200x^{-1}\,dx = 200\ln x + C$$

Since growth in profit since month 1 is 0 at $t = 1$ (month 1)

$$200\ln 1 + C = 0$$
$$C = 0$$

The formula is $200\ln x$.

b. Growth in profit will be 600 when $200\ln x = 600$.

$$200\ln x = 600$$
$$\ln x = 3$$
$$x = e^3 \approx 20 \text{ months}$$

23. $$\int_1^9 \left(x - \frac{1}{\sqrt{x}}\right) dx = \int_1^9 (x - x^{-1/2})\,dx = \left(\frac{x^2}{2} - 2x^{1/2}\right)\Big|_1^9$$
$$= \frac{(9)^2}{2} - 2(9)^{1/2} - \left[\frac{(1)^2}{2} - 2(1)^{1/2}\right] = \frac{81}{2} - 6 - \frac{1}{2} + 2 = 36$$

24. $$\int_2^5 (3x^2 - 4x + 5)\,dx = \left[3\left(\frac{x^3}{3}\right) - 4\left(\frac{x^2}{2}\right) + 5x\right]\Big|_2^5 = (x^3 - 2x^2 + 5x)\Big|_2^5$$
$$= 125 - 50 + 25 - (8 - 8 + 10) = 90$$

25. $$\int_1^{e^4} \frac{dx}{x} = \ln x\Big|_1^{e^4} = \ln e^4 - \ln 1 = 4 - 0 = 4$$

26. $$\int_1^5 \frac{dx}{x} = \ln x\Big|_1^5 = \ln 5$$

27. $$\int_0^2 e^{-x}\,dx = \frac{1}{-1}e^{-x}\Big|_0^2 = -e^{-x}\Big|_0^2 = -e^{-2} - (-e^0)$$
$$= 1 - e^{-2}$$

28. $$\int_0^2 e^{\frac{1}{2}x}\,dx = 2e^{\frac{1}{2}x}\Big|_0^2 = 2e - 2$$

29. $$\int_0^{100} (e^{0.05x} - e^{0.01x})\,dx = \left(\frac{1}{0.05}e^{0.05x} - \frac{1}{0.01}e^{0.01x}\right)\Big|_0^{100} = (20e^{0.05x} - 100e^{0.01x})\Big|_0^{100}$$
$$= 20e^5 - 100e^1 - (20e^0 - 100e^0) = 20e^5 - 100e + 80$$

30. $$\int_0^{10}(e^{0.04x}-e^{0.02x})\,dx=\left(\frac{1}{0.04}e^{0.04x}-\frac{1}{0.02}e^{0.02x}\right)\Big|_0^{10}$$
$$=(25e^{0.04x}-50e^{0.02x})\Big|_0^{10}$$
$$=25e^{0.4}-50e^{0.2}-(25e^0-50e^0)=25e^{0.4}-50e^{0.2}+25$$

31. a. $$\text{Area}=\int_1^2(6x^2-1)\,dx=\left[6\left(\tfrac{x^3}{3}\right)-x\right]\Big|_1^2$$
$$=(2x^3-x)\Big|_1^2=16-2-(2-1)$$
$$=13\text{ square units}$$

b. Using FnInt, Area = 13 square units.

32. a. $$\text{Area}=\int_{-3}^3(9-x^2)\,dx$$
$$=\left(9x-\tfrac{x^3}{3}\right)\Big|_{-3}^3$$
$$=27-9-(-27+9)=36\text{ square units}$$

b. Using FnInt, Area = 36 square units.

33. a. $$\text{Area}=\int_0^3 12e^{2x}\,dx=\tfrac{12}{2}e^{2x}\Big|_0^3=6e^{2x}\Big|_0^3$$
$$=6e^6-6\approx 2414.6\text{ square units}$$

b. Using FnInt, Area $\approx$ 2414.6 square units.

34. a. $$\text{Area}=\int_0^4 e^{x/2}\,dx=2e^{x/2}\Big|_0^4=2e^2-2$$
$$\approx 12.8\text{ square units}$$

b. Using FnInt, Area $\approx$ 12.8 square units.

35. a. $$\text{Area}=\int_1^{100}\tfrac{1}{x}\,dx=\ln x\Big|_1^{100}=\ln 100$$
$$\approx 4.6\text{ square units}$$

b. Using FnInt, Area $\approx$ 4.6 square units.

36. a. $$\text{Area}=\int_1^{1000}x^{-1}\,dx=\ln x\Big|_1^{1000}=\ln 1000$$
$$\approx 6.9\text{ square units}$$

b. Using FnInt, Area $\approx$ 6.9 square units.

37.

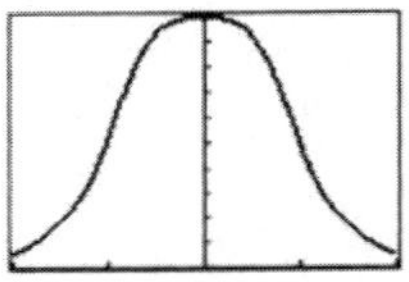

on [–2, 2] by [0, 10]

Area $\approx$ 21.4 square units

38.

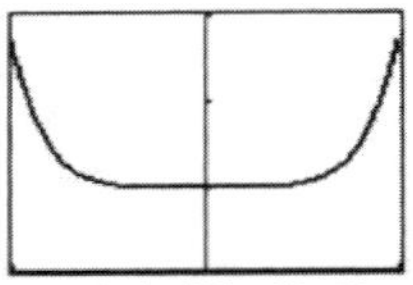

on [–1, 1] by [0, 3]

Area $\approx$ 2.54 square units

39. To find the total weight gain, integrate the rate from $t = 11$ to $t = 19$.

$$\int_{11}^{19}16(x-10)^{-1/2}dx$$
$$=16(2)(x-10)^{1/2}\Big|_{11}^{19}$$
$$=32(19-10)^{1/2}-32(11-10)^{1/2}$$
$$=64\text{ lbs}$$

40. To find the number of words memorized from time $t = 1$ to time $t = 8$, integrate the rate from $t = 0$ to $t = 8$.

$$\int_1^8\frac{2}{\sqrt[3]{t}}\,dt=\int_1^8 2t^{-1/3}\,dt=3t^{2/3}\Big|_1^8$$
$$=3(8)^{2/3}-3(1)^{2/3}=9\text{ words}$$

41. To find the rise in temperature over the next 100 years, integrate the rate from $t = 0$ to $t = 100$.

$$\int_0^{100}0.014t^{0.4}dt=\left(\tfrac{0.014}{0.4}t^{1.4}\right)\Big|_0^{100}=0.01t^{1.4}\Big|_0^{100}$$
$$=0.01(100^{1.4})-0.01(0^{1.4})\approx 6.3\text{ degrees}$$

42. To find the cost of the first 400 units, integrate $MC(x)$ from $x = 0$ to $x = 400$.

$$\int_0^{400}MC(x)\,dx=\int_0^{400}(x^{-1/2}+4)\,dx=(2x^{1/2}+4x)\Big|_0^{400}$$
$$=2(400)^{1/2}+4(400)=\$1640$$

43. To find the cost of the first 100 units, integrate $MC(x)$ from $x = 0$ to $x = 100$.

$$\int_0^{100} MC(x)\,dx = \int_0^{100} 22e^{-\sqrt{x}/5}\,dx$$
$$\approx \$653.39 \quad \text{(using FnInt)}$$

44. To find the number of pages proofread in 8 hours, integrate the rate from $x = 0$ to $x = 8$.

$$\int_0^{8} 15xe^{-0.25x}\,dx \approx 143 \text{ pages} \quad \text{(using FnInt)}$$

45. **a.** For 10 rectangles, divide the distance from 0 to 2 into 10 equal parts, so that each width is $\Delta x = \frac{2-0}{10} = 0.2$

Mark the points 0, 0.2, 0.4, 0.6, 0.8, 1.0, 1.2, 1.4, 1.6, 1.8, and 2.0 and draw rectangles.

Area $\approx (0)^2(0.2)+(0.2)^2(0.2)+(0.4)^2(0.2)+(0.6)^2(0.2)+(0.8)^2(0.2)+(1.0)^2(0.2)+(1.2)^2(0.2)$
$+(1.4)^2(0.2)+(1.6)^2(0.2)+(1.8)^2(0.2) = 2.28$ square units

b. $\int_0^2 x^2 dx = \left.\frac{x^3}{3}\right|_0^2 = \frac{(2)^3}{3} = \frac{8}{3}$ square units

46. **a.** Divide the distance from 0 to 4 into 10 equal parts, so that each width is $\Delta x = \frac{4-0}{10} = 0.4$

Mark the points 0, 0.4, 0.8, 1.2, 1.6, 2.0, 2.4, 2.8, 3.2, 3.6, and 4.0.

Area $\approx \sqrt{0}(0.4)+\sqrt{0.4}(0.4)+\sqrt{0.8}(0.4)+\sqrt{1.2}(0.4)+\sqrt{1.6}(0.4)+\sqrt{2.0}(0.4)+\sqrt{2.4}(0.4)$
$+\sqrt{2.8}(0.4)+\sqrt{3.2}(0.4)+\sqrt{3.6}(0.4) = 4.884$ square units

b. $\int_0^4 \sqrt{x}\,dx = \left.\frac{2}{3}x^{3/2}\right|_0^4 = \frac{2}{3}(4)^{3/2}$
$= \frac{2}{3}(8) = \frac{16}{3}$ square units

47. Using left rectangles:

a.

n	Left	Midpoint	Right
10	5.899	7.206	8.801
100	7.110	7.253	7.400
1000	7.239	7.254	7.268

b. $\int_{-2}^{2} e^x dx = \left.e^x\right|_{-2}^{2} = e^2 - e^{-2} \approx 7.254$ square units

48. Using left rectangles:

a.

n	Left	Midpoint	Right
10	1.506	1.383	1.281
100	1.398	1.386	1.375
1000	1.387	1.386	1.385

b. $\int_1^4 \frac{1}{x}\,dx = \left.\ln x\right|_1^4 = \ln 4 \approx 1.386$ square units

49. To find the points of intersection, equate the two functions and solve.

$$x^2+3x=3x+1$$
$$x^2-1=0$$
$$x=-1,\ 1$$

Substitute a value between −1 and 1 to find the upper and lower curves.

$$(0)^2+3(0)=0$$
$$3(0)+1=1$$

Thus, $y=3x+1$ is the upper curve and $y=x^2+3x$ is the lower curve.

$$\int_{-1}^{1}[3x+1-(x^2+3x)]\,dx=\int_{-1}^{1}(-x^2+1)\,dx=\left(-\tfrac{x^3}{3}+x\right)\Big|_{-1}^{1}$$
$$=-\tfrac{1}{3}+1-\left[-\tfrac{(-1)}{3}-1\right]=\tfrac{4}{3}\text{ square units}$$

50. To find the points of intersection, equate the two functions and solve.

$$12x-3x^2=6x-24$$
$$3x^2-6x-24=0$$
$$3(x^2-2x-8)=0$$
$$3(x-4)(x+2)=0$$
$$x=4,\ -2$$

To find the upper and lower curves, substitute $x = 0$.

$$12(0)-3(0)=0$$
$$6(0)-24=-24$$

Thus, $y=12x-3x^2$ is the upper curve and $y=6x-24$ is the lower curve.

$$\int_{-2}^{4}[12x-3x^2-(6x-24)]\,dx=\int_{-2}^{4}(-3x^2+6x+24)\,dx=\left(-x^3+3x^2+24x\right)\Big|_{-2}^{4}$$
$$=-(4)^3+3(4)^2+24(4)-[-(-2)^3+3(-2)^2+24(-2)]$$
$$=-64+48+96-(8+12-48)=108\text{ square units}$$

51. To find the points of intersection, equate the two functions and solve.

$$x^2=x$$
$$x(x-1)=0$$
$$x=0,\ 1$$

To find the upper and lower curves, substitute $x=\frac{1}{2}$.

$$\left(\tfrac{1}{2}\right)^2=\tfrac{1}{4}$$
$$\tfrac{1}{2}=\tfrac{1}{2}$$

Thus, $y=x$ is the upper curve and $y=x^2$ is the lower curve.

$$\int_{0}^{1}(x-x^2)\,dx=\left(\tfrac{x^2}{2}-\tfrac{x^3}{3}\right)\Big|_{0}^{1}=\tfrac{1}{2}-\tfrac{1}{3}$$
$$=\tfrac{1}{6}\text{ square units}$$

52. To find the points of intersection, equate the two functions and solve.

$$x^4=x$$
$$x(x^3-1)=0$$
$$x=0,\ 1$$

To find the upper and lower curves, substitute $x=\frac{1}{2}$.

$$\left(\tfrac{1}{2}\right)^4=\tfrac{1}{16}$$
$$\tfrac{1}{2}=\tfrac{1}{2}$$

Thus, $y=x$ is the upper curve and $y=x^4$ is the lower curve.

$$\int_{0}^{1}(x-x^4)\,dx=\left(\tfrac{x^2}{2}-\tfrac{x^5}{5}\right)\Big|_{0}^{1}=\tfrac{1}{2}-\tfrac{1}{5}$$
$$=\tfrac{3}{10}\text{ square units}$$

53. To find the points of intersection, equate the two functions and solve.

$$\begin{aligned} 4x^3 &= 12x^2 - 8x \\ 4x^3 - 12x^2 + 8x &= 0 \\ 4x(x-1)(x-2) &= 0 \\ x &= 0,\ 1,\ 2 \end{aligned}$$

To find the upper and lower curves in [0, 1], substitute $x = \frac{1}{2}$.

$$4\left(\tfrac{1}{2}\right)^3 = \tfrac{1}{2}$$
$$12\left(\tfrac{1}{2}\right)^2 - 8\left(\tfrac{1}{2}\right) = -1$$

Thus, $y = 4x^3$ is the upper curve and $y = 12x^2 - 8x$ is the lower curve in [0, 1].

To find the upper and lower curves in [1, 2], substitute $x = \frac{3}{2}$.

$$4\left(\tfrac{3}{2}\right)^3 = 13.5$$
$$12\left(\tfrac{3}{2}\right)^2 - 8\left(\tfrac{3}{2}\right) = 15$$

Thus, $y = 12x^2 - 8x$ is the upper curve and $y = 4x^3$ is the lower curve in [1, 2].

$$\begin{aligned} \text{Area} &= \int_0^1 [4x^3 - (12x^2 - 8x)]\,dx + \int_1^2 (12x^2 - 8x - 4x^3)\,dx \\ &= (x^4 - 4x^3 + 4x^2)\Big|_0^1 + (4x^3 - 4x^2 - x^4)\Big|_1^2 \\ &= 2 \text{ square units} \end{aligned}$$

54. To find the points of intersection, equate the two functions and solve.

$$\begin{aligned} x^3 + x^2 &= x^2 + x \\ x^3 - x &= 0 \\ x(x^2 - 1) &= 0 \\ x &= 0, -1, 1 \end{aligned}$$

To find the upper and lower curves in [−1, 0], substitute $x = -\frac{1}{2}$.

$$\left(-\tfrac{1}{2}\right)^3 + \left(-\tfrac{1}{2}\right)^2 = -\tfrac{1}{8} + \tfrac{1}{4} = \tfrac{1}{8}$$
$$\left(-\tfrac{1}{2}\right)^2 + \left(-\tfrac{1}{2}\right) = \tfrac{1}{4} - \tfrac{1}{2} = -\tfrac{1}{4}$$

Thus, $y = x^3 + x^2$ is the upper curve and $y = x^2 + x$ is the lower curve in [−1, 0].

For [0, 1], substitute $x = \dfrac{1}{2}$.

$$\left(\tfrac{1}{2}\right)^3 + \left(\tfrac{1}{2}\right)^2 = \tfrac{1}{8} + \tfrac{1}{4} = \tfrac{3}{8}$$
$$\left(\tfrac{1}{2}\right)^2 + \left(\tfrac{1}{2}\right) = \tfrac{1}{4} + \tfrac{1}{2} = \tfrac{3}{4}$$

Thus, $y = x^2 + x$ is the upper curve and $y = x^3 + x^2$ is the lower curve in [0, 1].

$$\begin{aligned} \text{Area} &= \int_{-1}^0 [x^3 + x^2 - (x^2 + x)]\,dx + \int_0^1 [x^2 + x - (x^3 + x^2)]\,dx \\ &= \int_{-1}^0 (x^3 - x)\,dx + \int_0^1 (x - x^3)\,dx \\ &= \left(\frac{x^4}{4} - \frac{x^2}{2}\right)\Bigg|_{-1}^0 + \left(\frac{x^2}{2} - \frac{x^4}{4}\right)\Bigg|_0^1 \\ &= \frac{(0)^4}{4} - \frac{(0)^2}{2} - \left(\frac{(-1)^4}{4} - \frac{(-1)^2}{2}\right) + \left(\frac{1^2}{2} - \frac{1^4}{4}\right) - \left(\frac{(0)^2}{2} - \frac{(0)^4}{4}\right) \\ &= -\left(\frac{1}{4} - \frac{1}{2}\right) + \left(\frac{1}{2} - \frac{1}{4}\right) = \frac{1}{4} + \frac{1}{4} = \frac{1}{2} \text{ square unit} \end{aligned}$$

55. Equating the two functions, we get $e^x = x + 5$. Using the intersection function, the points of intersection are –4.993 and 1.937. Using FnInt,

$$\text{Area} = \int_{-4.993}^{1.937} (x + 5 - e^x)\,dx \approx 17.13 \text{ square units}$$

56. Equating the two functions, we get $\ln x = \frac{x^2}{10}$.

Using the intersection function, the points of intersection are 1.138 and 3.566. Using FnInt,

$$\text{Area} = \int_{1.138}^{3.566} (\ln x - \frac{x^2}{10})\,dx \approx 0.496 \text{ square unit}$$

57. $$(\text{Average value}) = \frac{1}{4-1}\int_1^4 \frac{1}{x^2}\,dx$$
$$= \frac{1}{3}\ln x\Big|_1^4 = \frac{1}{3}(\ln 4 - \ln 1) = \frac{1}{3}\ln 4$$

58. $$(\text{Average value}) = \tfrac{1}{4-1}\int_1^4 6\sqrt{x}\,dx = \tfrac{1}{3}(4x^{3/2})\Big|_1^4 = \tfrac{1}{3}(32-4) = \tfrac{28}{3}$$

59. $$(\text{Average value}) = \tfrac{1}{5-0}\int_0^5 \sqrt{x^3+1}\,dx \approx \tfrac{1}{5}(23.596) \approx 4.72 \quad \text{(using FnInt)}$$

60. $$\text{Average value} = \tfrac{1}{2-(-2)}\int_{-2}^{2} \ln(e^{x^2}+10)\,dx \approx 2.77 \quad \text{(using FnInt)}$$

61. To find the average population between 2010 and 2110, the find average value of $P(t)$ between $t = 0$ and $t = 100$.

$$\tfrac{1}{100-0}\int_0^{100} P(t)\,dt = \tfrac{1}{100}\int_0^{100} 310e^{0.0073t}\,dt$$
$$= \frac{310}{100(0.0073)}e^{0.0073t}\Big|_0^{100}$$
$$= \frac{310}{100(0.0073)}\left(e^{0.73} - 1\right)$$
$$\approx 457 \text{ million}$$

62. To find the average value during the first 20 years, find the average value of $V(t)$ between $t = 0$ and $t = 20$.

$$\tfrac{1}{20-0}\int_0^{20} V(t)\,dt = \tfrac{1}{20}\int_0^{20} 3000e^{0.06t}\,dt$$
$$= \frac{3000}{20(0.06)}e^{0.06t}\Big|_0^{20}$$
$$= 2500e^{1.2} - 2500 \approx \$5800$$

63. To find the average value, use FnInt to find the average value of the function from $x = 0$ to $x = 10$.

$$\tfrac{1}{10-0}\int_0^{10} 4.3e^{0.01x^2}\,dx \approx 6.29 \text{ hundred thousand} = \$629{,}000$$

64. To find the average price, use FnInt to find the average value of the function between $x = 0$ and $x = 52$.

$$\tfrac{1}{52-0}\int_0^{52} 28e^{0.01x^{1.2}}\,dx \approx \$49.95$$

65. $$\text{Area} = \int_0^9 (3\sqrt{x} - \tfrac{1}{9}x^2)\,dx = \left(2x^{3/2} - \frac{x^3}{27}\right)\Big|_0^9$$
$$= 2(27) - 27 = 27 \text{ square meters}$$

66. $$\begin{aligned}(\text{Accumulated surplus}) &= \int_0^{10} (40e^{0.2t} - 20e^{0.1t})\, dt \\ &= 200e^{0.2t} - 200e^{0.1t} \Big|_0^{10} \\ &= 200e^2 - 200e^1 - (200 - 200) \approx \$934 \text{ billion}\end{aligned}$$

67. First, we must find the market price B at $x = 200$.

$$\begin{aligned}B = d(200) &= 8000 - 24(200) \\ &= 8000 - 4800 = 3200\end{aligned}$$

$$\begin{aligned}(\text{Consumers' surplus}) &= \int_0^{200} (8000 - 24x - 3200)\, dx \\ &= \int_0^{200} (4800 - 24x)\, dx \\ &= 4800x - 12x^2 \Big|_0^{200} \\ &= 4800(200) - 12(200)^2 \\ &\approx \$480{,}000\end{aligned}$$

68. Find the market price at $x = 200$.

$$B = d(200) = 1800 - 0.03(200)^2 = 600$$

$$\begin{aligned}(\text{Consumers' surplus}) &= \int_0^{200} (1800 - 0.03x^2 - 600)\, dx \\ &= \int_0^{200} (1200 - 0.03x^2)\, dx \\ &= (1200x - 0.01x^3) \Big|_0^{200} \\ &= 1200(200) - 0.01(200)^3 \\ &\approx \$160{,}000\end{aligned}$$

69. Find the market price at $x = 120$.

$$B = d(120) = 300e^{-0.2\sqrt{120}} \approx 33.545$$

$$\begin{aligned}(\text{Consumers' surplus}) &= \int_0^{120} (300e^{-0.2\sqrt{x}} - 33.545)\, dx \\ &\approx \$5623\end{aligned}$$

70. Find the market price at $x = 100$.

$$B = d(100) = \frac{100}{1+\sqrt{100}} = \frac{100}{11}$$

$$\begin{aligned}(\text{Consumers' surplus}) &= \int_0^{100} \left(\frac{100}{1+\sqrt{x}} - \frac{100}{11}\right) dx \\ &\approx \$611\end{aligned}$$

71. $$\begin{aligned}(\text{Gini Index}) &= 2\int_0^1 [x - L(x)]\, dx \\ &= 2\int_0^1 (x - x^{3.5})\, dx \\ &= 2\left(\frac{x^2}{2} - \frac{x^{4.5}}{4.5}\right) \Big|_0^1 \\ &= 2\left[\left(\frac{1}{2} - \frac{1}{4.5}\right) - 0\right] \approx 0.56\end{aligned}$$

72. $$\begin{aligned}(\text{Gini Index}) &= 2\int_0^1 (x - x^{2.5})\, dx \\ &= 2\left(\frac{x^2}{2} - \frac{x^{3.5}}{3.5}\right) \Big|_0^1 \\ &= 2\left(\frac{1}{2} - \frac{1}{3.5}\right) \approx 0.43\end{aligned}$$

73. $$(\text{Gini Index}) = 2\int_0^1 \left(x - \frac{x}{2-x}\right) dx \approx 0.23$$

74. $$(\text{Gini Index}) = 2\int_0^1 \left(x - \frac{e^{x^4} - 1}{e - 1}\right) dx \approx 0.68$$

75. $\int x^2\sqrt[3]{x^3-1}\,dx = \frac{1}{3}\int u^{1/3}\,dx$ Let $u = x^3 - 1$, $du = 3x^2\,dx$

$= \frac{1}{3}\left(\frac{3}{4}u^{4/3}\right)+C$

$= \frac{1}{4}(x^3-1)^{4/3}+C$

76. $\int x^3\sqrt{x^4-1}\,dx = \frac{1}{4}\int u^{1/2}\,dx$ Let $u = x^4 - 1$, $du = 4x^3\,dx$

$= \frac{1}{4}\left(\frac{2}{3}u^{3/2}\right)+C$

$= \frac{1}{6}(x^4-1)^{3/2}+C$

77. Cannot be found using substitution formulas.

78. Cannot be found using substitution formulas.

79. $\int \frac{dx}{9-3x} = -\frac{1}{3}\int \frac{du}{u} = -\frac{1}{3}\ln|u|+C$ Let $u = 9 - 3x$, $du = -3\,dx$

$= -\frac{1}{3}\ln|9-3x|+C$

80. $\int \frac{dx}{1-2x} = -\frac{1}{2}\int \frac{du}{u}$ Let $u = 1 - 2x$, $du = -2\,dx$

$= -\frac{1}{2}\ln|u|+C$

$= -\frac{1}{2}\ln|1-2x|+C$

81. $\int \frac{dx}{(9-3x)^2} = -\frac{1}{3}\int \frac{1}{u^2}\,du$ Let $u = 9 - 3x$, $du = -3\,dx$

$= -\frac{1}{3}(-u^{-1})+C$

$= \frac{1}{3}\left(\frac{1}{9-3x}\right)+C$

82. $\int \frac{dx}{(1-2x)^2} = -\frac{1}{2}\int \frac{1}{u^2}\,du$ Let $u = 1 - 2x$, $du = -2\,dx$

$= -\frac{1}{2}(-u^{-1})+C$

$= \frac{1}{2}\left(\frac{1}{1-2x}\right)+C$

83. $\int \frac{x^2}{\sqrt[3]{8+x^3}}\,dx = \frac{1}{3}\int \frac{1}{u^{1/3}}\,du = \frac{1}{3}\int u^{-1/3}\,du$ Let $u = 8 + x^3$, $du = 3x^2\,dx$

$= \frac{1}{3}\left(\frac{3}{2}u^{2/3}\right)+C$

$= \frac{1}{2}\left(8+x^3\right)^{2/3}+C$

84. $\int \frac{x}{\sqrt{9+x^2}}\,dx = \frac{1}{2}\int u^{-1/2}\,du$ Let $u = 9 + x^2$, $du = 2x\,dx$

$= \frac{1}{2}(2u^{1/2})+C$

$= \left(9+x^2\right)^{1/2}+C$

85. $$\int \frac{w+3}{(w^2+6w-1)^2}\,dw = \frac{1}{2}\int \frac{1}{u^2}\,du = \frac{1}{2}\int u^{-2}\,du$$
$$= \frac{1}{2}(-u^{-1}) + C$$
$$= -\frac{1}{2}\left(\frac{1}{w^2+6w-1}\right) + C$$

Let $u = w^2 + 6w - 1$
$du = (2w+6)\,dw$
$= 2(w+3)\,dw$

86. $$\int \frac{t-2}{(t^2-4t+1)^2}\,dw = \frac{1}{2}\int \frac{1}{u^2}\,du$$
$$= \frac{1}{2}(-u^{-1}) + C$$
$$= -\frac{1}{2}\left(\frac{1}{t^2-4t+1}\right) + C$$

Let $u = t^2 - 4t + 1$
$du = (2t-4)\,dt$
$= 2(t-2)\,dt$

87. $$\int \frac{(1+\sqrt{x})^2}{\sqrt{x}}\,dx = 2\int u^2\,du = 2\cdot\frac{u^3}{3} + C$$
$$= \frac{2}{3}(1+\sqrt{x})^3 + C$$

Let $u = 1 + x^{1/2}$
$du = \frac{1}{2}x^{-1/2}\,dx$

88. $$\int \frac{(1+\sqrt[3]{x})^2}{\sqrt[3]{x^2}}\,dx = 3\int u^2\,du$$
$$= u^3 + C$$
$$= (1+\sqrt[3]{x})^3 + C$$

Let $u = 1 + \sqrt[3]{x}$
$du = \frac{1}{3}x^{-2/3}\,dx$

89. $$\int \frac{e^x}{e^x-1}\,dx = \int \frac{1}{u}\,du$$
$$= \ln|u| + C$$
$$= \ln\left|e^x - 1\right| + C$$

Let $u = e^x - 1$
$du = e^x\,dx$

90. $$\int \frac{1}{x\ln x}\,dx = \int \frac{1}{u}\,du$$
$$= \ln|u| + C$$
$$= \ln|\ln x| + C$$

Let $u = \ln x$
$du = \frac{1}{x}\,dx$

91. **a.** $$\int_0^3 x\sqrt{x^2+16}\,dx = \frac{1}{2}\int_{16}^{25} u^{1/2}\,du$$
$$= \frac{1}{2}\left(\frac{2}{3}u^{3/2}\right)\Bigg|_{16}^{25}$$
$$= \frac{1}{3}(25^{3/2} - 16^{3/2})$$
$$= \frac{125-64}{3} = \frac{61}{3} \approx 20.33$$

Let $u = x^2 + 16$
$du = 2x\,dx$
For $x = 0,\ u = 16$
$x = 3,\ u = 25$

b. Using FnInt, the integral is 20.33.

92. **a.** $$\int_0^4 \frac{dz}{\sqrt{2z+1}} = \frac{1}{2}\int_1^9 u^{-1/2}\,du$$
$$= \frac{1}{2}(2u^{1/2})\Bigg|_1^9$$
$$= 9^{1/2} - 1^{1/2} = 2$$

Let $u = 2z + 1$
$du = 2\,dz$
For $z = 0,\ u = 1$
$z = 4,\ u = 9$

b. Using FnInt, the integral is 2.

93. **a.** $$\int_0^4 \frac{w}{\sqrt{25-w^2}}\,dw = -\frac{1}{2}\int_{25}^{9} u^{-1/2}\,du$$
$$= -\frac{1}{2}(2u^{1/2})\Big|_{25}^{9}$$
$$= -(9^{1/2}-25^{1/2}) = -(-2) = 2$$

Let $u = 25 - w^2$
$du = -2w\,dw$
For $w = 0,\ u = 25$
$w = 4,\ u = 9$

b. Using FnInt, the integral is 2.

94. **a.** $$\int_1^2 \frac{x+1}{(x^2+2x-2)^2}\,dx = \frac{1}{2}\int_1^6 u^{-2}\,du$$
$$= \frac{1}{2}\left(\frac{1}{-1}u^{-1}\right)\Big|_1^6$$
$$= -\frac{1}{2}\left(\frac{1}{6}-1\right) = \frac{5}{12} \approx 0.417$$

Let $u = x^2 + 2x - 2$
$du = 2x + 2\,dx$
$= 2(x+1)\,dx$
For $x = 1,\ u = 1$
$x = 2,\ u = 6$

b. Using FnInt, the integral is 0.417.

95. **a.** $$\int_3^9 \frac{dx}{x-2} = \ln|x-2|\ \Big|_3^9$$
$$= \ln 7 - \ln 1 = \ln 7 \approx 1.95$$

b. Using FnInt, the integral is 1.95.

96. **a.** $$\int_4^5 \frac{dx}{x-6} = \ln|x-6|\ \Big|_4^5$$
$$= \ln 1 - \ln 2 = -\ln 2 \approx -0.69$$

b. Using FnInt, the integral is –0.69.

97. **a.** $$\int_0^1 x^3 e^{x^4}\,dx = \frac{1}{4}\int_0^1 e^u\,du$$
$$= \frac{1}{4}e^u\Big|_0^1$$
$$= \frac{1}{4}e - \frac{1}{4} \approx 0.430$$

Let $u = x^4$
$du = 4x^3\,dx$
For $x = 0,\ u = 0$
$x = 1,\ u = 1$

b. Using FnInt, the integral is 0.430.

98. **a.** $$\int_0^1 x^4 e^{x^5}\,dx = \frac{1}{5}\int_0^1 e^u\,du$$
$$= \frac{1}{5}e^u\Big|_0^1 = \frac{1}{5}e - \frac{1}{5} \approx 0.34$$

Let $u = x^5$
$du = 5x^4\,dx$
For $x = 0,\ u = 0$
$x = 1,\ u = 1$

b. Using FnInt, the integral is 0.34.

99. $$\text{Area} = \int_1^3 \frac{x^2+6x}{\sqrt[3]{x^3+9x^2+17}}\,dx$$
$$= \frac{1}{3}\int_{27}^{125} u^{-1/3}\,du$$
$$= \frac{1}{3}\left(\frac{3}{2}u^{2/3}\right)\Big|_{27}^{125}$$
$$= \frac{1}{2}[(125)^{2/3} - (27)^{2/3}]$$
$$= \frac{1}{2}(25-9) = 8 \text{ square units}$$

Let $u = x^3 + 9x^2 + 17$
$du = (3x^2+18x)\,dx$
$= 3(x^2+6x)\,dx$
For $x = 1,\ u = 27$
$x = 3,\ u = 125$

100. $$\text{Area} = \int_0^3 \frac{x+6}{\sqrt{x^2+12x+4}}\,dx = \frac{1}{2}\int_4^{49} u^{-1/2}\,du$$
$$= \frac{1}{2}\left(2u^{1/2}\right)\Big|_4^{49}$$
$$= 7 - 2 = 5 \text{ square units}$$

Let $u = x^2 + 12x + 4$
$du = (2x+12)\,dx$
$= 2(x+6x)\,dx$
For $x = 0,\ u = 4$
$x = 3,\ u = 49$

101. $(\text{Average Value}) = \frac{1}{2-0}\int_0^2 xe^{-x^2}\,dx$

$= \frac{1}{2}\left(-\frac{1}{2}\right)\int_0^{-4} e^u\,du$

$= -\frac{1}{4}(e^{-4}-1)$

Let $u = -x^2$
$du = -2x\,dx$
For $x = 0,\ u = 0$
$x = 2,\ u = -4$

102. $(\text{Average Value}) = \frac{1}{4-2}\int_2^4 \frac{x}{x^2-3}\,dx$

$= \frac{1}{2}\left(\frac{1}{2}\right)\int_1^{13} \frac{1}{u}\,du$

$= \frac{1}{4}\ln u\,\Big|_1^{13} = \frac{1}{4}\ln 13$

Let $u = x^2 - 3$
$du = 2x\,dx$
For $x = 2,\ u = 1$
$x = 4,\ u = 13$

103. $C(x) = \int MC(x)\,dx = \int \frac{1}{\sqrt{2x+9}}\,dx$

$= \frac{1}{2}\int u^{-1/2}\,du = \frac{1}{2}(2u^{1/2}) + K$

$= (2x+9)^{1/2} + K$

Let $u = 2x + 9$
$du = 2\,dx$

Since fixed costs are 100, $c(0) = 100$. Thus

$100 = [2(0)+9]^{1/2} + K$
$= 3 + K$
$K = 97$

Thus, $C(x) = (2x+9)^{1/2} + 97$.

104. $(\text{Total change in temperature}) = \int_1^3 \frac{3x^2}{x^3+1}\,dx$

$= \int_1^{28} \frac{1}{u}\,du = \ln u\,\Big|_1^{28} = \ln 28$

≈ 3.33 degrees

Let $u = x^3 + 1$
$du = 3x^2\,dx$
For $x = 0,\ u = 1$
$x = 3,\ u = 28$

Chapter 6: Integration Techniques

EXERCISES 6.1

1. $\int e^{2x} dx = \frac{1}{2}e^{2x} + C$

3. $\int (x+2)dx = \frac{x^2}{2} + 2x + C$

5. $\int \sqrt{x}\, dx = \int x^{1/2} dx = \frac{2}{3}x^{3/2} + C$

7. $\int (x+3)^4 dx = \frac{(x+3)^5}{5} + C$

9. $\int xe^{2x} dx$ Let $u = x$ and $dv = e^{2x}dx$. Then $du = dx$ and $v = \int e^{2x}dx = \frac{1}{2}\int e^{2x}\, 2\,dx \frac{1}{2}e^{2x}$. Therefore,

$$\int xe^{2x}\, dx = \frac{1}{2}xe^{2x} - \int \frac{1}{2}e^{2x}\, dx$$
$$= \frac{1}{2}xe^{2x} - \frac{1}{2}\left(\frac{1}{2}\right)\int e^{2x}\, 2dx = \frac{1}{2}xe^{2x} - \frac{1}{4}e^{2x} + C$$

11. $\int x^5 \ln x\, dx$ Let $u = \ln x$ and $dv = x^5 dx$. Then $du = \frac{1}{x}dx$ and $v = \int x^2 dx = \frac{x^6}{6}$.

$$\int x^5 \ln x\, dx = \frac{x^6}{6}\ln x - \int \frac{x^6}{6}\left(\frac{1}{x}\right) dx = \frac{x^6}{6}\ln x - \int \frac{x^5}{6}\, dx$$
$$= \frac{x^6}{6}\ln x - \frac{1}{6}\left(\frac{x^6}{6}\right) + C = \frac{x^6}{6}\ln x - \frac{x^6}{36} + C$$

13. $\int (x+2)e^x dx$ Let $u = x+2$ and $dv = e^x dx$. Then $du = dx$ and $v = \int e^x dx = e^x$.

$$\int (x+2)e^x\, dx = (x+2)e^x - \int e^x\, dx = (x+2)e^x - e^x + C$$

15. $x\int \sqrt{x} \ln x\, dx$ Let $u = \ln x$ and $dv = \sqrt{x}\, dx$. Then $du = \frac{1}{x}dx$ and $v = \int \sqrt{x}\, dx = \frac{2}{3}x^{3/2}$.

$$\int \sqrt{x} \ln x\, dx = \frac{2}{3}x^{3/2}\ln x - \int \frac{2}{3}x^{3/2}\left(\frac{1}{x}\right) dx = \frac{2}{3}x^{3/2}\ln x - \frac{2}{3}\int x^{1/2}\, dx$$
$$= \frac{2}{3}x^{3/2}\ln x - \frac{2}{3}\left(\frac{2}{3}x^{3/2}\right) + C = \frac{2}{3}x^{3/2}\ln x - \frac{4}{9}x^{3/2} + C$$

17. $\int (x-3)(x+4)^5 dx$ Let $u = x-3$ and $dv = (x+4)^5 dx$. Then $du = dx$ and $v = \int (x+4)^5\, dx = \frac{(x+4)^6}{6}$.

$$\int (x-3)(x+4)^5\, dx = (x-3)\frac{(x+4)^6}{6} - \int \frac{(x+4)^6}{6}\, dx$$
$$= (x-3)\frac{(x+4)^6}{6} - \frac{1}{6}\left[\frac{(x+4)^7}{7}\right] + C$$
$$= (x-3)\frac{(x+4)^6}{6} - \frac{1}{42}(x+4)^7 + C$$

19. $\int te^{-0.5t} dt$ Let $u = t$ and $dv = e^{-0.5t}dt$. Then $du = dt$ and $v = \int e^{-0.5t}\, dt = \frac{1}{-0.5}e^{-0.5t} = -2e^{-0.5t}$.

$$\int te^{-0.5t} dx = t(-2e^{-0.5t}) - \int (-2e^{-0.5t})\, dt = -2te^{-0.5t} + 2\int e^{-0.5t} dt$$
$$= -2te^{-0.5t} + 2\left(\frac{1}{-0.5}e^{-0.5t}\right) + C = -2te^{-0.5t} - 4e^{-0.5t} + C$$

21. $\int \frac{\ln t}{t^2}\,dt$ Let $u=\ln t$ and $\frac{1}{t^2}dt = t^{-2}dt$. Then $du=\frac{1}{t}dt$ and $v=\int t^{-2}\,dt=\frac{t^{-1}}{-1}=-t^{-1}$.

$$\int \frac{\ln t}{t^2}\,dt = -t^{-1}\ln t - \int(-t^{-1})\left(\frac{1}{t}\right)dt = -t^{-1}\ln t + \int t^{-2}\,dt$$
$$= -t^{-1}\ln t + \left(\frac{t^{-1}}{-1}\right)+C = -t^{-1}\ln t - t^{-1} + C$$

23. $\int s(2s+1)^4 ds$ Let $u=s$ and $dv=(2s+1)^4 ds$. Then $du=ds$ and $v=\int(2s+1)^4 ds$.

$$v=\int(2s+1)^4 ds = \frac{1}{2}\int w^4 dw = \frac{1}{2}\left(\frac{w^5}{5}\right) \quad \text{Let } w=2s+1,\ dw=2ds$$
$$=\frac{(2s+1)^5}{10}$$

$$\int s(2s+1)^4\,ds = \frac{s(2s+1)^5}{10} - \int\frac{(2s+1)^5}{10}\,ds = \frac{s(2s+1)^5}{10} - \frac{1}{10}\int(2s+1)^5 ds$$
$$=\frac{s(2s+1)^5}{10} - \frac{1}{10}\left[\frac{(2s+1)^6}{12}\right]+C$$
$$=\frac{s(2s+1)^5}{10} - \frac{1}{120}(2s+1)^6 + C$$

25. $\int \frac{x}{e^{2x}}\,dx$ Let $u=x$ and $dv=\frac{1}{e^{2x}}dx=e^{-2x}dx$. Then $du=dx$ and $v=\int e^{-2x}\,dx = -\frac{1}{2}e^{-2x}$.

$$\int\frac{x}{e^{2x}}\,dx = -\frac{1}{2}xe^{-2x} - \int\left(-\frac{1}{2}e^{-2x}\right)dx = -\frac{1}{2}xe^{-2x} + \frac{1}{2}\int e^{-2x}dx$$
$$= -\frac{1}{2}xe^{-2x} + \frac{1}{2}\left(-\frac{1}{2}e^{-2x}\right) = -\frac{1}{2}xe^{-2x} - \frac{1}{4}e^{-2x} + C$$

27. $\int\frac{x}{\sqrt{x+1}}\,dx$ Let $u=x$ and $dv=\frac{1}{\sqrt{x+1}}dx=(x+1)^{-1/2}dx$. Then $du=dx$ and $v=\int(x+1)^{-1/2}\,dx=2(x+1)^{1/2}$.

$$\int\frac{x}{\sqrt{x+1}}\,dx = 2x(x+1)^{1/2} - \int 2(x+1)^{1/2}dx = 2x(x+1)^{1/2} - 2\left[\frac{2}{3}(x+1)^{3/2}\right]+C$$
$$=2x(x+1)^{1/2} - \frac{4}{3}(x+1)^{3/2} + C$$

29. $\int xe^{ax}\,dx$ Let $u=x$ and $dv=e^{ax}\,dx$. Then $du=dx$ and $v=\int e^{ax}\,dx=\frac{1}{a}e^{ax}$.

$$\int xe^{ax}\,dx = \frac{1}{a}xe^{ax} - \int\frac{1}{a}e^{ax}\,dx = \frac{1}{a}xe^{ax} - \frac{1}{a}\int e^{ax}\,dx$$
$$=\frac{1}{a}xe^{ax} - \frac{1}{a}\left(\frac{1}{a}e^{ax}\right)+C = \frac{1}{a}xe^{ax} - \frac{1}{a^2}e^{ax}+C$$

31. $\int x^n \ln ax\,dx$ Let $u=\ln ax=\ln a+\ln x$ and $dv=x^n\,dx$. Then $du=\frac{1}{x}dx$ and $v=\int x^n\,dx=\frac{x^{n+1}}{n+1}$.

$$\int x^n\ln ax\,dx = \frac{x^{n+1}}{n+1} - \int\frac{x^{n+1}}{n+1}\left(\frac{1}{x}dx\right) = \frac{x^{n+1}}{n+1}\ln ax - \frac{1}{n+1}\int x^n\,dx$$
$$=\frac{x^{n+1}}{n+1}\ln ax - \frac{1}{n+1}\left(\frac{x^{n+1}}{n+1}\right)+C$$
$$=\frac{x^{n+1}}{n+1}\ln ax - \frac{1}{(n+1)^2}x^{n+1}+C$$

33. $\int \ln x \, dx$ Let $u = \ln x$ and $dv = dx$. Then $du = \frac{1}{x}dx$ and $v = \int dx = x$.

$$\int \ln x \, dx = x\ln x - \int x\left(\frac{1}{x}\right) dx = x\ln x - \int dx = x\ln x - x + C$$

35. $\int x^3 e^{x^2} \, dx$ Let $u = x^2$ and $dv = xe^{x^2} dx$. Then $du = 2x \, dx$ and

$$v = \int xe^{x^2} \, dx = \frac{1}{2}\int e^w \, dw = \frac{1}{2}e^w \quad \text{Let } w = x^2, \; dw = 2x \, dx$$
$$= \frac{1}{2}e^{x^2}$$

$$\int x^3 e^{x^2} \, dx = \frac{1}{2}x^2 e^{x^2} - \int \frac{1}{2}e^{x^2}(2x \, dx)$$
$$= \frac{1}{2}x^2 e^{x^2} - \frac{1}{2}\int e^{x^2}(2x \, dx)$$
$$= \frac{1}{2}x^2 e^{x^2} - \frac{1}{2}e^{x^2} + C \quad \text{(using the substitution } w = x^2 \text{ as before)}$$

37. **a.** $\int xe^{x^2} \, dx = \frac{1}{2}\int e^u \, du = \frac{1}{2}e^u + C = \frac{1}{2}e^{x^2} + C$ Let $u = x^2$, $du = 2x \, dx$

b. $\int \frac{(\ln x)^3}{x} \, dx = \int u^3 \, du = \frac{u^4}{4} + C = \frac{(\ln x)^4}{4} + C$ Let $u = \ln x$, $du = \frac{1}{x} \, dx$

c. $\int x^2 \ln 2x \, dx$ Let $u = \ln 2x = \ln 2 + \ln x$ and $dv = x^2 dx$. Then $du = \frac{1}{x}dx$ and $v = \int x^2 \, dx = \frac{x^3}{3}$.

$$\int x^2 \ln 2x \, dx = \frac{1}{3}x^3 \ln 2x - \int \frac{x^3}{3}\left(\frac{1}{x} \, dx\right) = \frac{1}{3}x^3 \ln 2x - \frac{1}{3}\int x^2 \, dx$$
$$= \frac{1}{3}x^3 \ln 2x - \frac{1}{9}x^3 + C$$

d. $\int \frac{e^x}{e^x + 4} \, dx = \int \frac{1}{u} \, du = \ln|u| + C = \ln\left|e^x + 4\right| + C$ Let $u = e^x + 4$, $du = e^x \, dx$

39. $\int_0^2 xe^{4x} dx$ Let $u = x$ and $dv = e^x dx$. Then $du = dx$ and $v = \int e^x \, dx = e^x$.

$$\int xe^x \, dx = xe^x - \int e^x \, dx = xe^x - e^x + C$$

$$\int_0^2 xe^{4x} dx = \left(xe^x - e^x\right)\Big|_0^2 = 2e^2 - e^2 - \left(0e^0 - e^0\right) = e^2 + 1$$

41. $\int_1^3 x^2 \ln x \, dx$ Let $u = \ln x$ and $dv = x^2 dx$. Then $du = \frac{1}{x}dx$ and $v = \int x^2 \, dx = \frac{1}{3}x^3$.

$$\int x^2 \ln x \, dx = \frac{1}{3}x^3 \ln x - \int \frac{1}{3}x^3 \left(\frac{dx}{x}\right) = \frac{1}{3}x^3 \ln x - \frac{1}{3}\int x^2 \, dx$$
$$= \frac{1}{3}x^3 \ln x - \frac{1}{9}x^3 + C$$

$$\int_1^3 x^2 \ln x \, dx = \left(\frac{1}{3}x^3 \ln x - \frac{1}{9}x^3\right)\Bigg|_1^3 = \left(\frac{27}{3}\ln 3 - \frac{37}{9}\right) - \left(\frac{1}{3}\ln 3 - \frac{1}{9}\right)$$
$$= 9\ln 3 - \frac{27}{9} - 0 + \frac{1}{9} = 9\ln 3 - \frac{26}{9}$$

43. $\int_0^2 z(z-2)^4\,dz$ Let $u=z$ and $dv=(z-2)^4dz$. Then $du=dz$ and $v=\int(z-2)^4\,dz=\frac{(z-2)^5}{5}$.

$$\int z(z-2)^4\,dz=\frac{z(z-2)^5}{5}-\int\frac{(z-2)^5}{5}\,dz=\frac{z(z-2)^5}{5}-\frac{(z-2)^6}{30}+C$$

$$\int_0^2 z(z-2)^4\,dz=\left[\frac{z(z-2)^5}{5}-\frac{(z-2)^6}{30}\right]\Bigg|_0^2=\frac{2(0)^5}{5}-\frac{(0)^6}{30}-\left[\frac{0(-2)^5}{5}-\frac{(-2)^6}{30}\right]$$

$$=\frac{64}{30}=\frac{32}{15}$$

45. $\int_0^{\ln 4} te^t\,dt$ Let $u=t$ and $dv=e^t\,dt$. Then $du=dt$ and $v=\int e^t\,dt=e^t$.

$$\int te^t\,dt=te^t-\int e^t\,dt=te^t-e^t+C$$

$$\int_0^{\ln 4} te^t\,dt=(te^t-e^t)\Big|_0^{\ln 4}=(e^{\ln 4}\ln 4-e^{\ln 4})-(0e^0-e^0)$$

$$=4\ln 4-4+1=4\ln 4-3$$

47. **a.** $\int x(x-2)^5\,dx$ Let $u=x$ and $dv=(x-2)^5dx$. Then $du=dx$ and $v=\int(x-2)^5\,dx=\frac{(x-2)^6}{6}$

$$\int x(x-2)^5\,dx=\frac{x(x-2)^6}{6}-\int\frac{(x-2)^6}{6}\,dx$$

$$=\frac{x(x-2)^6}{6}-\frac{1}{6}\int(x-2)^6\,dx$$

$$=\frac{x(x-2)^6}{6}-\frac{1}{42}(x-2)^7+C$$

b. $\int x(x-2)^5\,dx=\int(u+2)u^5\,du=\int(u^6+2u^5)\,du$ Let $u=x-2$, $du=dx$

$$=\frac{u^7}{7}+2\left(\frac{u^6}{6}\right)+C$$

$$=\frac{(x-2)^7}{7}-\frac{(x-2)^6}{3}+C$$

49. $\int x^ne^x\,dx$ Let $u=x^n$ and $dv=e^xdx$. Then $du=nx^{n-1}\,dx$ and $v=\int e^x\,dx=e^x$.

$$\int x^ne^x\,dx=x^ne^x-\int nx^{n-1}e^x\,dx=x^ne^x-n\int x^{n-1}e^x\,dx$$

51. Using the formula in Exercise 49 with $n=2$, we obtain

$$\int x^2e^x\,dx=x^2e^x-2\int x^{2-1}e^xdx=x^2e^x-2\int xe^xdx$$

Using the formula on the remaining integral with $n=1$, we obtain

$$\int x^2e^x\,dx=x^2e^x-2(xe^x-\int e^x\,dx)$$

$$=x^2e^x-2xe^x+2e^x+C$$

53. $R(x)=\int MR(x)\,dx=\int xe^{-x/4}dx$ Let $u=x$ and $dv=e^{-x/4}dx$. Then $du=dx$ and $v=-4e^{-x/4}$.

$$R(x)=\int xe^{-x/4}dx=-4xe^{-x/4}+\int 4e^{-x/4}\,dx=-4xe^{-x/4}+4(-4e^{-x/4})+C$$
$$=-4xe^{-x/4}-16e^{-x/4}+C$$

Since $R(0)=0$,

$$R(0)=-4(0)e^{-0/4}-16e^{-0/4}+C=0$$
$$0-16(1)+C=0$$
$$C=16$$

Thus, $R(x)=-4xe^{-x/4}-16e^{-x/4}+16$.

55. The present value of a continuous stream of income over T years is $\int_0^T C(t)e^{-rt}dt$; here $C(t)$ = 4 million dollars per year, T = 10 years, and r = 10% = 0.10.

$\int_0^{10} 4te^{-0.1t}dt$ Let $u=4t$ and $dv=e^{-0.1t}dt$ Then $du=4\,dt$ and $v=\int e^{-0.1t}dt=-10e^{-0.1t}$.

Therefore $\int_0^{10} 4te^{-0.1t}dt=-40te^{-0.1t}-\int -10e^{-0.1t}(4dt)$

$$=-40te^{-0.1t}+40\int e^{-0.1t}dt$$
$$=-40te^{-0.1t}+400e^{-0.1t}+K$$

and $\int_0^{10} 4te^{-0.1t}\,dt=(-40te^{-0.1t}-400e^{-0.1t})\Big|_0^{10}$

$$=\left[-40(10)e^{-1}-400e^{-1}\right]-(0-400e^{0})$$
$$=-800e^{-1}+400\approx 105.7$$

The present value of the continuous stream of income over the first ten years is approximately $105.7 million.

57. **a.** To find the amount absorbed in the first 5 hours, integrate the rate from 0 to 5.

$\int_0^5 te^{-0.5t}\,dt$ Let $u=t$ and $dv=e^{-0.5t}dt$.

Then $du=dt$ and $v=\int e^{-0.5t}dt=\frac{1}{-0.5}e^{-0.5t}=-2e^{-0.5t}$.

$$\int te^{-0.5t}\,dt=-2te^{-0.5t}-\int -2e^{-0.5t}\,dt$$
$$=-2te^{-0.5t}+2(-2e^{-0.5t})+C$$
$$=-2te^{-0.5t}-4e^{-0.5t}+C$$

Since none is absorbed at $t=0$,

$$-2(0)e^{-0.5(0)}-4e^{-0.5t(0)}+C=0$$
$$-4+C=0$$
$$C=4$$

$$\int_0^5 te^{-0.5t}dt=(-2te^{-0.5t}-4e^{-0.5t}+4)\Big|_0^5$$
$$=-2(5)e^{-0.5(5)}-4e^{-0.5(5)}+4-[-2(0e^{-0.5(0)}-4e^{-0.5(0)}+4]$$
$$=-10e^{-2.5}-4e^{-2.5}+4\approx 2.85$$

The total amount of the drug absorbed during the first 5 hours is about 2.85 mg.

b. Using FnInt, the integral is about 2.85.

59. The area under the curve is $\int_1^2 x \ln x \, dx$

$\int x \ln x \, dx$ Let $u = \ln x$ and $dv = x dx$. Then $du = \frac{dx}{x}$ and $v = \int x \, dx = \frac{1}{2}x^2$.

$$\int x \ln x \, dx = \frac{1}{2}x^2 \ln x - \int \frac{1}{2}x^2 \left(\frac{dx}{x}\right) = \frac{1}{2}x^2 \ln x - \frac{1}{2}\int x \, dx$$
$$= \frac{1}{2}x^2 \ln x - \frac{1}{4}x^2 + C$$

and the area is $\left(\frac{1}{2}x^2 \ln x - \frac{1}{4}x^2\right)\Big|_1^2 = 2\ln 2 - 1 - \left(\frac{1}{2}\ln 1 - \frac{1}{4}\right)$

$$= 2\ln 2 - \frac{3}{4} = \ln 4 - \frac{3}{4}$$
$$= 2\ln 2 - \frac{3}{4} \approx 0.64 \text{ square unit}$$

61. Cumulative sales $\int_0^4 11xe^{0.2x} dx = 11\int_0^4 xe^{0.2x} dx$.

Use $u = x$ and $dv = e^{0.2x}dx$. Then $du = dx$ and $v = \int e^{0.2x}dx = 5e^{0.2x}$.

$$\int 11xe^{0.2x} dx = 11\left(5xe^{0.2x} - \int 5e^{0.2x} dx\right)$$
$$= 11\left(5xe^{0.2x} - 25e^{0.2x}\right) + C$$
$$= 55xe^{0.2x} - 275e^{0.2x} + C$$

Evaluate from $t = 0$ to $t = 4$.

$$\int_0^4 11xe^{0.2x} dx = \left(55xe^{0.2x} - 275e^{0.2x}\right)\Big|_0^4$$
$$= \left(55(4)e^{0.2(4)} - 275e^{0.2(4)}\right) - \left(55(0)e^{0.2(0)} - 275e^{0.2(0)}\right)$$
$$= 220e^{0.8} - 275e^{0.8} + 275 \approx 152.6$$

The cumulative sales during the first 4 years are about \$153 million.

63. Total water absorbed $\int_0^5 2xe^{-\frac{1}{2}x} dx$. Use $u = x$ and $dv = e^{-\frac{1}{2}x}dx$. Then $du = dx$ and $v = \int e^{-\frac{1}{2}x}dx = -2e^{-\frac{1}{2}x}$.

$$\int 2xe^{-\frac{1}{2}x} dx = 2\int xe^{-\frac{1}{2}x} dx$$
$$= 2\left(-2xe^{-\frac{1}{2}x} - \int -2e^{-\frac{1}{2}x} dx\right)$$
$$= 2\left(-2xe^{-\frac{1}{2}x} - 4e^{-\frac{1}{2}x}\right) + C$$
$$= -4xe^{-\frac{1}{2}x} - 8e^{-\frac{1}{2}x} + C$$

Evaluate from $t = 0$ to $t = 5$.

$$\int_0^5 2xe^{-\frac{1}{2}x} dx = \left(-4xe^{-\frac{1}{2}x} - 8e^{-\frac{1}{2}x}\right)\Big|_0^5 = \left(-4(5)e^{-\frac{1}{2}(5)} - 8e^{-\frac{1}{2}(5)}\right) - \left(-4(0)e^{-\frac{1}{2}(0)} - 8e^{-\frac{1}{2}(0)}\right)$$
$$= -20e^{-\frac{5}{2}} - 8e^{-\frac{5}{2}} + 8 \approx 5.7$$

The total water absorbed during the first 5 hours of exercise is about 5.7 liters.

65. **b.** $\int u\,dv = uv - \int v\,du$

67. $v = e^x$

69. $v = x$

71. $u = \ln x$

73. **a.** $\int x^{-1}\,dx$ Let $u = x^{-1}$ and $dv = dx$. Then $du = -x^{-2}\,dx$ and $v = \int dx = x$

$\int x^{-1}dx = x^{-1}x - \int (-x^{-2})x\,dx + 1 + \int x^{-1}dx.$

b. Since $\int f(x)\,dx = F(x) + C$, integrating $\int x^{-1}\,dx$ on each side of the equation $\int x^{-1}\,dx = 1 + \int x^{-1}\,dx$ produces constants C_1 and C_2 such that $C_1 = 1 + C_2$.

75. $\int x^2 e^{-x}\,dx$ Let $u = x^2$ and $dv = e^{-x}\,dx$. Then $du = 2x\,dx$ and $v = \int e^{-x}dx = -e^{-x}$.

Therefore, $\int x^2 e^{-x}\,dx = -x^2 e^{-x} - \int (-e^{-x})2x\,dx$

$= -x^2 e^{-x} + \int 2xe^{-x}\,dx$

Let $u = 2x$ and $dv = e^{-x}\,dx$. Then $du = 2\,dx$ and $v = -e^{-x}$

$\int x^2 e^{-x}\,dx = -x^2 e^{-x} + \left(-2xe^{-x} - \int (-e^{-x})2\,dx\right)$

$= -x^2 e^{-x} - 2xe^{-x} + 2\int e^{-x}\,dx$

$= -x^2 e^{-x} - 2xe^{-x} - 2e^{-x} + C$

77. $\int (x+1)^2 e^x\,dx$ Let $u = (x+1)^2$ and $dv = e^x dx$. Then $du = 2(x+1)\,dx$ and $v = \int e^x dx = e^x$.

$\int (x+1)^2 e^x\,dx = (x+1)^2 e^x - \int 2(x+1)e^x\,dx.$

Let $u = 2(x+1)$ and $dv = e^x dx$. Then $du = 2dx$ and $v = \int e^x dx = e^x$

$\int (x+1)^2 e^x\,dx = (x+1)^2 e^x - \left[2(x+1)e^x - \int 2e^x\,dx\right]$

$= (x+1)^2 e^x - 2(x+1)e^x + 2e^x + C$

79. $\int x^2 (\ln x)^2\,dx$ Let $u = (\ln x)^2$ and $dv = x^2\,dx$. Then $du = 2\ln x\left(\frac{dx}{x}\right)$ and $v = \int dx = \frac{1}{3}x^3$.

$\int x^2 (\ln x)^2\,dx = \frac{1}{3}x^3 (\ln x)^2 - \int \frac{2}{3}x^3\left(\frac{\ln x}{x}\right)dx$

$= \frac{1}{3}x^3 (\ln x)^2 - \int \frac{2}{3}x^2 \ln x\,dx$

Let $u = \ln x$ and $dv = \frac{2}{3}x^2\,dx$. Then $du = \frac{dx}{x}$ and $v = \frac{2}{9}x^3$

$\int x^2 (\ln x)^2\,dx = \frac{1}{3}x^3 (\ln x)^2 - \left[\frac{2}{9}x^3 \ln x - \int \frac{2}{9}x^3\left(\frac{dx}{x}\right)\right]$

$= \frac{1}{3}x^3 (\ln x)^2 - \frac{2}{9}x^3 \ln x + \frac{2}{9}\int x^2\,dx$

$= \frac{1}{3}x^3 (\ln x)^2 - \frac{2}{9}x^3 \ln x + \frac{2}{27}x^3 + C$

81. **a.** From the explanation preceding Exercise 65, we know

$$\int x^2 e^x\,dx = x^2e^x - 2(xe^x - e^x) + C$$

Evaluate the definite integral.

$$\int_0^2 x^2e^x dx = \left(x^2e^x - 2xe^x + 2e^x\right)\Big|_0^2$$
$$= 2^2e^2 - 2(2)e^2 + 2e^2 - \left[0^2e^2 - 2(0)e^0 + 2e^2\right]$$
$$= 4e^2 - 4e^2 + 2e^2 - [2(1)]$$
$$= 2e^2 - 2 \approx 12.78$$

b. Using FnInt, the integral is about 12.78.

83. $\int x^2e^{-x}dx$ Let $u = x^2$ and $dv = e^{-x}dx$. We make a table of u and $dv = v'$.

Alternating Signs	$u = x^2$ and Its Derivatives	$v' = e^{-x}$ and Its Antiderivatives
+	x^2	e^{-x}
–	$2x$	$-e^{-x}$
+	2	e^{-x}
–	0	$-e^{-x}$

$$\int x^2e^{-x}dx = -x^2e^{-x} - 2xe^{-x} + 2\left(-e^{-x}\right) + C$$
$$= -x^2e^{-x} - 2xe^{-x} - 2e^{-x} + C$$

85. $\int x^3e^{2x}dx$ Let $u = x^3$ and $dv = v' = e^{2x}\,dx$. We make a table of u and v'.

Alternating Signs	$u = x^3$ and Its Derivatives	$v' = e^{2x}$ and Its Antiderivatives
+	x^3	e^{2x}
–	$3x^2$	$\frac{1}{2}e^{2x}$
+	$6x$	$\frac{1}{4}e^{2x}$
–	6	$\frac{1}{8}e^{2x}$
+	0	$\frac{1}{16}e^{2x}$

$$\int x^3e^{2x}dx = \frac{1}{2}x^3e^{2x} - 3x^2\left(\frac{1}{4}e^{2x}\right) + 6x\left(\frac{1}{8}e^{2x}\right) - 6\left(\frac{1}{16}e^{2x}\right) + C$$
$$= \frac{1}{2}x^3e^{2x} - \frac{3}{4}x^2e^{2x} + \frac{3}{4}xe^{2x} - \frac{3}{8}e^{2x} + C$$

87. $\int (x-1)^3e^{3x}dx$ Let $u = (x-1)^3$ and $dv = v' = e^{3x}\,dx$.

Alternating Signs	$u = (x-1)^3$ and Its Derivatives	$v' = e^{3x}$ and Its Antiderivatives
+	$(x-1)^3$	e^{3x}
–	$3(x-1)^2$	$\frac{1}{3}e^{3x}$
+	$6(x-1)$	$\frac{1}{9}e^{3x}$
–	6	$\frac{1}{27}e^{3x}$
+	0	$\frac{1}{81}e^{3x}$

$$\int (x-1)^3e^{3x}dx = \frac{1}{3}(x-1)^3e^{3x} - \frac{1}{3}(x-1)^2e^{3x} + \frac{6}{27}(x-1)e^{3x} - \frac{6}{81}e^{3x} + C$$
$$= \frac{1}{3}(x-1)^3e^{3x} - \frac{1}{3}(x-1)^2e^{3x} + \frac{2}{9}(x-1)e^{3x} - \frac{2}{27}e^{3x} + C$$

EXERCISES 6.2

1. Formula 12 is $\int \frac{1}{x^2(ax+b)}dx = -\frac{1}{b}\left(\frac{1}{x} + \frac{a}{b}\ln\left|\frac{x}{ax+b}\right|\right) + C.$

Here let $a = 5$ and $b = -1$.

3. Use Formula 14 and let $a = -1$ and $b = 7$.

5. Use Formula 9 with $a = -1$ and $b = 1$.

7. Using Formula 16 with $a = 3$, we obtain

$$\int \frac{1}{9-x^2}dx = \frac{1}{2(3)}\ln\left|\frac{3+x}{3-x}\right| + C$$
$$= \frac{1}{6}\ln\left|\frac{3+x}{3-x}\right| + C$$

9. Use Formula 12 with $a = 2$, $b = 1$.

$$\int \frac{1}{x^2(2x+1)}\,dx = -\frac{1}{1}\left(\frac{1}{x} + \frac{2}{1}\ln\left|\frac{x}{2x+1}\right|\right) + C = -\frac{1}{x} - 2\ln\left|\frac{x}{2x+1}\right| + C$$

11. Use Formula 9 with $a = -1$, $b = 1$.

$$\int \frac{x}{1-x}\,dx = \frac{x}{-1} - \frac{1}{(-1)^2}\ln|1-x| + C = -x - \ln|1-x| + C$$

13. Use Formula 10 with $a = 2$, $b = 1$, $c = 1$, $d = 1$.

$$\int \frac{1}{(2x+1)(x+1)}\,dx = \frac{1}{2-1}\ln\left|\frac{2x+1}{x+1}\right| + C = \ln\left|\frac{2x+1}{x+1}\right| + C$$

15. Use Formula 17 with $a = 2$

$$\int \sqrt{x^2-4}\,dx = \frac{x}{2}\sqrt{x^2-4} - \frac{4}{2}\ln\left|x + \sqrt{x^2-4}\right| + C$$
$$= \frac{x}{2}\sqrt{x^2-4} - 2\ln\left|x + \sqrt{x^2-4}\right| + C$$

17. Use Formula 20 with $x = z$ and $a = 1$.

$$\int \frac{1}{z\sqrt{1-z^2}}\,dz = -\ln\left|\frac{1+\sqrt{1-z^2}}{z}\right| + C$$

19. Use Formula 21 with $n = 3$ and $a = 2$

$$\int x^3e^{2x}\,dx = \frac{1}{2}x^3e^{2x} - \frac{3}{2}\int x^{3-1}e^{2x}dx = \frac{1}{2}x^3e^{2x} - \frac{3}{2}\int x^2e^{2x}\,dx$$

Use Formula 21 again with $n = 2$ and $a = 2$

$$\int x^3e^{2x}\,dx = \frac{1}{2}x^3e^{2x} - \frac{3}{2}\left(\frac{1}{2}x^2e^{2x} - \int xe^{2x}\,dx\right)$$

Use Formula 21 again with $n = 1$ and $a = 2$

$$\int x^3e^{2x}\,dx = \frac{1}{2}x^3e^{2x} - \frac{3}{2}\left[\frac{1}{2}x^2e^{2x} - \left(\frac{1}{2}xe^{2x} - \frac{1}{2}\int x^0e^{2x}\,dx\right)\right]$$
$$= \frac{1}{2}x^3e^{2x} - \frac{3}{4}x^2e^{2x} + \frac{3}{4}xe^{2x} - \frac{3}{4}\int e^{2x}\,dx$$
$$= \frac{1}{2}x^3e^{2x} - \frac{3}{4}x^2e^{2x} + \frac{3}{4}xe^{2x} - \frac{3}{4}\left(\frac{1}{2}e^{2x}\right) + C$$
$$= \frac{1}{2}x^3e^{2x} - \frac{3}{4}x^2e^{2x} + \frac{3}{4}xe^{2x} - \frac{3}{8}e^{2x} + C$$

21. Use Formula 23 with $n = -101$

$$\int x^{-101}\ln x\,dx = \frac{1}{-100}x^{-100}\ln x - \frac{1}{(-100)^2}x^{-100} + C$$
$$= -\frac{1}{100}x^{-100}\ln x - \frac{1}{10{,}000}x^{-100} + C$$

23. Use Formula 10 with $a=1,\ b=0,\ c=1,\ d=3$.

$$\int \frac{1}{x(x+3)}\,dx = \frac{1}{1(3)-(0)(1)}\ln\left|\frac{x}{x+3}\right|+C$$
$$=\frac{1}{3}\ln\left|\frac{x}{x+3}\right|+C$$

25. Use Formula 15 with $x=z^2$ and $a=2$. Then $dx = 2z\,dz$.

$$\int \frac{z}{z^4-4}\,dz = \int \frac{\frac{1}{2}dx}{x^2-4} = \frac{1}{2}\left[\frac{1}{2(2)}\right]\ln\left|\frac{x-2}{x+2}\right|+C$$
$$=\frac{1}{8}\ln\left|\frac{x-2}{x+2}\right|+C=\frac{1}{8}\ln\left|\frac{z^2-2}{z^2+2}\right|+C$$

27. Use Formula 17 with $u=3x$ and $a=4$. Then $du=3\,dx$.

$$\int\sqrt{9x^2+16}\,dx = \int\sqrt{u^2+16}\,\frac{du}{3}=\frac{1}{3}\int\sqrt{u^2+16}\,du$$
$$=\frac{1}{3}\left(\frac{u}{2}\sqrt{u^2+16}+\frac{16}{2}\ln\left|u+\sqrt{u^2+16}\right|\right)+C$$
$$=\frac{1}{3}\left(\frac{3x}{2}\sqrt{9x^2+16}+8\ln\left|3x+\sqrt{9x^2+16}\right|\right)+C$$
$$=\frac{1}{2}x\sqrt{9x^2+16}+\frac{8}{3}\ln\left|3x+\sqrt{9x^2+16}\right|+C$$

29. First, we write the integral in a form such that we can use a formula involving $\frac{1}{\sqrt{a^2-x^2}}$.

$$\int\frac{1}{\sqrt{4-e^{2t}}}\,dt = \int\frac{e^t}{e^t\sqrt{4-e^{2t}}}\,dt$$

Now we let $x=e^t$ and use Formula 20 with $a=2$.

$$\int\frac{1}{\sqrt{4-e^{2t}}}\,dt = \int\frac{e^t}{e^t\sqrt{4-e^{2t}}}\,dt = \int\frac{1}{x\sqrt{4-x^2}}\,dt$$
$$=-\frac{1}{2}\ln\left|\frac{2+\sqrt{4-x^2}}{x}\right|+C=-\frac{1}{2}\ln\left|\frac{2+\sqrt{4-e^{2t}}}{e^t}\right|+C$$

31. Use Formula 15 with $x=e^t$ and $a=1$. Then $dx=e^t\,dt$.

$$\int\frac{e^t}{e^{2t}-1}\,dt = \int\frac{1}{x^2-1}\,dx=\frac{1}{2}\ln\left|\frac{x-1}{x+1}\right|+C=\frac{1}{2}\ln\left|\frac{e^t-1}{e^t+1}\right|+C$$

33. Use Formula 18 with $u=x^4,\ a=1$. Then $du=4x^3\,dx$.

$$\int\frac{x^3}{\sqrt{x^8-1}}\,dx = \int\frac{\frac{1}{4}du}{\sqrt{u^2-1}} = \frac{1}{4}\ln\left|u+\sqrt{u^2-1}\right|+C$$
$$=\frac{1}{4}\ln\left|x^4+\sqrt{x^8-1}\right|+C$$

35. First, we write the integral so that we can use a formula involving $\frac{1}{u\sqrt{u+1}}$.

$$\int\frac{1}{x\sqrt{x^3+1}}\,dx = \int\frac{x^2}{x^3\sqrt{x^3+1}}\,dx$$

Let $u=x^3$ and $a=1$. Then $du=3x^2\,dx$. Use Formula 14 with $b=1$.

$$\int\frac{1}{x\sqrt{x^3+1}}\,dx = \int\frac{1}{u\sqrt{u+1}}\,\frac{du}{3}=\frac{1}{3}\left(\frac{1}{\sqrt{1}}\ln\left|\frac{\sqrt{u+1}-\sqrt{1}}{\sqrt{u+1}+\sqrt{1}}\right|\right)+C$$
$$=\frac{1}{3}\ln\left|\frac{\sqrt{x^3+1}-1}{\sqrt{x^3+1}+1}\right|+C$$

37. Use Formula 15 with $x = e^t$ and $a = 1$. Then $dx = e^t\ dt$.

$$\int \frac{e^t}{(e^t-1)(e^t+1)}\,dt = \int \frac{e^t}{e^{2t}-1}\,dt = \int \frac{1}{x^2-1}\,dx = \frac{1}{2}\ln\left|\frac{x-1}{x+1}\right| + C$$
$$= \frac{1}{2}\ln\left|\frac{e^t-1}{e^t+1}\right| + C$$

39. Use Formula 21 with $n = 1$ and $a = \frac{1}{2}$.

$$\int xe^{x/2} = \frac{1}{\frac{1}{2}}xe^{x/2} - \frac{1}{\frac{1}{2}}\int x^{1-1}e^{x/2}dx$$
$$= 2xe^{x/2} - 2\int e^{x/2}\,dx$$
$$= 2xe^{x/2} - 2\left(\frac{1}{\frac{1}{2}}e^{x/2}\right) + C$$
$$= 2xe^{x/2} - 4e^{x/2} + C$$

41. First, rewrite the integral as

$$\int \frac{1}{e^{-x}+4}\,dx = \int \frac{e^x}{1+4e^x}\,dx$$

Use Formula 6 with $u = 1+4e^x$, $du = 4e^x dx$.

$$\int \frac{1}{e^{-x}+4}\,dx = \int \frac{1}{u}\frac{du}{4} = \frac{1}{4}\ln\left|1+4e^x\right| + C$$

43. **a.** Use Formula 17 with $a = 4$.

$$\int_4^5 \sqrt{x^2-16}\,dx = \left(\frac{x}{2}\sqrt{x^2-16} - \frac{16}{2}\ln\left|x+\sqrt{x^2-16}\right|\right)\Bigg|_4^5$$
$$= \frac{5}{2}\sqrt{25-16} - 8\ln\left|5+\sqrt{25-16}\right| - \left(\frac{4}{2}\sqrt{16-16} - 8\ln\left|4+\sqrt{16-16}\right|\right)$$
$$= \frac{5}{2}(3) - 8\ln(5+3) - 2(0) + 8\ln(4+0)$$
$$= \frac{15}{2} - 8\ln 8 + 8\ln 4 \approx 1.95$$

b. Using FnInt, the integral is about 1.95.

45. **a.** Use Formula 15 with $a = 1$.

$$\int_2^3 \frac{1}{x^2-1}\,dx = \frac{1}{2}\ln\left|\frac{x-1}{x+1}\right|\Bigg|_2^3 = \frac{1}{2}\left(\ln\frac{2}{4} - \ln\frac{1}{3}\right)$$
$$= \frac{1}{2}\left(\ln\frac{1}{2} - \ln\frac{1}{3}\right) = \frac{1}{2}\ln\frac{\frac{1}{2}}{\frac{1}{3}} = \frac{1}{2}\ln\frac{3}{2} \approx 0.203$$

b. Using FnInt, the integral is about 0.203.

47. **a.** Use Formula 19 with $a = 5$.

$$\int_3^5 \frac{\sqrt{25-x^2}}{x}\,dx = \sqrt{25-x^2} - 5\ln\left|\frac{5+\sqrt{25-x^2}}{x}\right|\Bigg|_3^5$$
$$= \sqrt{25-25} - 5\ln\frac{5+\sqrt{25-25}}{5} - \left(\sqrt{25-9} - 5\ln\frac{5+\sqrt{25-9}}{3}\right)$$
$$= 0 - 5\ln 1 - \sqrt{16} + 5\ln\frac{5+\sqrt{16}}{3}$$
$$= -4 + 5\ln 3 \approx 1.49$$

b. Using FnInt, the integral is about 1.49.

49. $\int \frac{1}{2x+6}\,dx$ Let $u = 2x+6$. Then $du = 2\,dx$.

$$\int \frac{1}{2x+6}\,dx = \frac{1}{2}\int \frac{du}{u} = \frac{1}{2}\ln|u| + C$$
$$= \frac{1}{2}\ln|2x+6| + C$$

51. Use Formula 9 with $a = 2,\ b = 6$.

$$\int \frac{x}{2x+6}\,dx = \frac{x}{2} - \frac{6}{(2)^2}\ln|2x+6| + C = \frac{x}{2} - \frac{3}{2}\ln|2x+6| + C$$

53. $\int x\sqrt{1-x^2}\,dx$ Let $u = x^2$. Then $du = 2x\,dx$.

$$\int x\sqrt{1-x^2}\,dx = \int (1-u)^{1/2}\left(\frac{1}{2}du\right) = -\frac{1}{2}\int (1-u)^{1/2}(-du)$$
$$= -\frac{1}{2}\left(\frac{2}{3}\right)(1-u)^{3/2} + C = -\frac{1}{3}(1-u)^{3/2} + C = -\frac{1}{3}(1-x^2)^{3/2} + C$$

55. Use Formula 19 with $a = 1$.

$$\int \frac{\sqrt{1-x^2}}{x}\,dx = \sqrt{1-x^2} - 1\ln\left|\frac{1+\sqrt{1-x^2}}{x}\right| + C = \sqrt{1-x^2} - \ln\left|\frac{1+\sqrt{1-x^2}}{x}\right| + C$$

57. $\int \frac{x-1}{(3x+1)(x+1)}\,dx = \int \frac{x}{(3x+1)(x+1)}\,dx - \int \frac{1}{(3x+1)(x+1)}\,dx$

Formula 11 is $\int \frac{x}{(ax+b)(cx+d)}\,dx = \frac{1}{ad-bc}\left(\frac{d}{c}\ln|cx+d| - \frac{b}{a}\ln|ax+b|\right) + C$

Here $a = 3,\ b = 1,\ c = 1,\ d = 1$.

So, $\int \frac{x}{(3x+1)(x+1)}\,dx = \frac{1}{3-1}\left(\ln|x+1| - \frac{1}{3}\ln|3x+1|\right) + C_1$

$$= \frac{1}{2}\left(\ln|x+1| - \frac{1}{3}\ln|3x+1|\right) + C_1$$

Formula 10 is $\int \frac{1}{(ax+b)(cx+d)}\,dx = \frac{1}{ad-bc}\ln\left|\frac{ax+b}{cx+d}\right| + C$.

So, $\int \frac{1}{(3x+1)(x+1)}\,dx = \frac{1}{3-1}\ln\left|\frac{3x+1}{x+1}\right| + C_2$

$$= \frac{1}{2}\ln\left|\frac{3x+1}{x+1}\right| + C_2$$

and $\int \frac{x-1}{(3x+1)(x+1)}\,dx = \frac{1}{2}\left(\ln|x+1| - \frac{1}{3}\ln|3x+1|\right) - \frac{1}{2}\ln\left|\frac{3x+1}{x+1}\right| + C$

$$= \frac{1}{2}\ln|x+1| - \frac{1}{6}\ln|3x+1| - \frac{1}{2}\ln|3x+1| + \frac{1}{2}\ln|x+1| + C$$
$$= -\frac{2}{3}\ln|3x+1| + \ln|x+1| + C = \ln\left|\frac{x+1}{(3x+1)^{2/3}}\right| + C$$

59. $\int \frac{x+1}{x\sqrt{1+x^2}}\,dx = \int \frac{x}{x\sqrt{1+x^2}}\,dx + \int \frac{1}{x\sqrt{1+x^2}}\,dx = \int \frac{1}{\sqrt{1+x^2}}\,dx + \int \frac{1}{x\sqrt{1+x^2}}\,dx$

For the first integral, use Formula 18 with $a = 1$.
For the second integral, use Formula 20 with $a = 1$.

$$\int \frac{x+1}{x\sqrt{1+x^2}}\,dx = \ln\left|x + \sqrt{1+x^2}\right| + \left(-\frac{1}{1}\ln\left|\frac{1+\sqrt{1+x^2}}{x}\right|\right) + C$$
$$= \ln\left|x + \sqrt{1+x^2}\right| - \ln\left|\frac{1+\sqrt{1+x^2}}{x}\right| + C$$

61. $\int \frac{x+1}{x-1}\,dx = \int \frac{x}{x-1}\,dx + \int \frac{1}{x-1}\,dx$

Formula 9 is $\int \frac{x}{ax+b}\,dx = \frac{x}{a} - \frac{b}{a^2}\ln|ax+b| + C$.

Here $a = 1$ and $b = -1$, so $\int \frac{x}{x-1}dx = x\ln|x-1| + C_1$.

For $\int \frac{1}{x-1}\,dx$, Let $u = x-1$
$du = dx$

$$\int \frac{1}{x-1}dx = \int \frac{du}{u} = \ln|u| + C = \ln|x-1| + C_2$$

So $\int \frac{x+1}{x-1}dx = x + \ln|x-1| + \ln|x-1| + C$

$= x + 2\ln|x-1| + C = x + \ln(x-1)^2 + C$

63. To find the total sales, integrate the rate. Use Formula 21 with $n = 2$ and $a = -1$.

$$\int x^2e^{-x}dx = \frac{1}{-1}x^2e^{-x} - \frac{2}{-1}\int x^{2-1}e^{-x}\,dx = -x^2e^{-x} + 2\int xe^{-x}\,dx$$

Now, use Formula 21 again with $a = -1$ and $n = 1$.

$$\int x^2e^{-x}dx = -x^2e^{-x} + 2\int xe^{-x}\,dx = -x^2e^{-x} + 2\left(\frac{1}{-1}xe^{-x} - \frac{1}{-1}\int x^{1-1}e^{-x}\,dx\right)$$
$$= -x^2e^{-x} + 2\left(-xe^{-x} + \int e^{-x}\,dx\right)$$
$$= -x^2e^{-x} + 2\left(-xe^{-x} - e^{-x}\right) + C$$
$$= -x^2e^{-x} - 2xe^{-x} - 2e^{-x} + C$$

Since total sales are 0 at $x = 0$,

$-(0)^2e^0 - 2(0)e^0 - 2e^0 + C = 0$
$-2 + C = 0$
$C = 2$

The total sales in the first x months are $-x^2e^{-x} - 2xe^{-x} - 2e^{-x} + 2$ million sales.

65. $n = 3\int_{0.1}^{0.3} \frac{1}{q^2(1-q)}\,dq$ generations

Formula 12 is $\int \frac{x}{x^2(ax+b)}\,dx = -\frac{1}{b}\left(\frac{1}{x} + \frac{a}{b}\ln\left|\frac{x}{ax+b}\right|\right) + C$.

Here $a = -1$ and $b = 1$.

So, $\int \frac{1}{q^2(1-q)}dq = -\left(\frac{1}{q} - \ln\left|\frac{q}{1-q}\right|\right) + C$

and $3\int_{0.1}^{0.3} \frac{1}{q^2(1-q)}\,dq = -3\left[\frac{1}{q} - \ln\left(\frac{q}{1-q}\right)\right]\Bigg|_{0.1}^{0.3}$

$= -3\left\{\left[\frac{1}{0.3} - \ln\left(\frac{0.3}{0.7}\right)\right] - \left[\frac{1}{0.1} - \ln\left(\frac{0.1}{0.9}\right)\right]\right\}$

≈ 24

It takes about 24 generations to increase the frequency of a gene from 0.1 to 0.3.

67. $C(x) = \int MC(x)\,dx = \int \frac{1}{\sqrt{x^2+1}}\,dx$

Use Formula 18 with $a = 1$.

$C(x) = \int \frac{1}{\sqrt{x^2+1}}\,dx = \ln\left|x+\sqrt{x^2+1}\right| + K$

Since fixed costs are \$2000, $C(0) = 2000$. Thus,

$$C(0) = \ln\left|0+\sqrt{(0)^2+1}\right| + K = 2000$$
$$\ln 1 + K = 2000$$
$$K = 2000$$

The cost function is $C(x) = \ln\left|x+\sqrt{x^2+1}\right| + 2000$

69. Formula 9, with $x = e^{50t}$

71. Formula 18, with $x = t^{50}$

73. 476 uses

75. By a formula (either 22 or 23) and by integration by parts (with $u = \ln x$ and $dv = dx$)

77. Formula 10 (with denominator $(x+1)(x-1) = x^2 - 1$) and formula 15

EXERCISES 6.3

1. $\lim_{x\to\infty} \frac{1}{x^2} = 0$

3. $\lim_{b\to\infty} \left(1 - 2e^{-5b}\right) = 1 - 0 = 1$

5. $\lim_{x\to\infty} (2 - e^{x/2}) = -\infty$ (does not exist)

7. $\lim_{b\to\infty} (3 + \ln b) = \infty$ (does not exist)

9. $\lim_{x\to-\infty} \frac{1}{x^3} = 0$

11. $\lim_{a\to-\infty} e^{2a} = 0$

13. $\lim_{a\to-\infty} e^{-2a} = \infty$ (does not exist)

15. $\lim_{x\to-\infty} \frac{1}{\left(x^2+1\right)^2} = 0$

17. First integrate over the interval from $x = 1$ to $x = b$.

$$\int_1^b \frac{1}{x^3}\,dx = \int_1^b x^{-3}\,dx = \left.\frac{x^{-2}}{2}\right|_1^b = \left.-\frac{1}{2x^2}\right|_1^b$$
$$= -\frac{1}{2b^2} - \left(-\frac{1}{2}\right)$$
$$= -\frac{1}{2b^2} + \frac{1}{2}$$

Now take the limit as $b \to \infty$.

$$\lim_{b\to\infty}\left(-\frac{1}{2b^2} + \frac{1}{2}\right) = 0 + \frac{1}{2} = \frac{1}{2}$$

Thus, $\int_1^\infty \frac{1}{x^3}\,dx = \frac{1}{2}$

19. Integrate from $x = 2$ to $x = b$.

$$\int_2^b 3x^{-4}\,dx = 3\left.\left(\frac{x^{-3}}{-3}\right)\right|_2^b = \left.-\frac{1}{x^3}\right|_2^b$$
$$= -\frac{1}{b^3} - \left(-\frac{1}{2^3}\right) = -\frac{1}{b^3} + \frac{1}{8}$$

Now take the limit as $b \to \infty$.

$$\int_2^\infty 3x^{-4}\,dx = \lim_{b\to\infty}\left(-\frac{1}{b^3} + \frac{1}{2^3}\right) = \left(0 + \frac{1}{8}\right)$$
$$= \frac{1}{8}$$

21. $\int_2^\infty \frac{1}{x}dx = \lim_{b\to\infty}\int_2^b \frac{1}{x}dx = \lim_{b\to\infty} \ln x \Big|_2^b$

$\lim_{b\to\infty} \ln x$ does not exist, so the integral is divergent.

23. $\int_1^\infty \frac{1}{x^{1.01}}\,dx == \lim_{b\to\infty}\int_1^b x^{-1.01}\,dx$

$$= \lim_{b\to\infty} \frac{x^{-1.01}}{-0.01}\Big|_1^b = \lim_{b\to\infty}\left[\frac{b^{-1.01}}{-0.01} - \frac{(1)^{-0.01}}{-0.01}\right]$$

$$= 0 + 100 = 100$$

25. $\int_0^\infty e^{-0.05t}dt = \lim_{b\to\infty}\int_0^b e^{-0.05t}dt = \lim_{b\to\infty} -\frac{1}{0.05}e^{-0.05t}\Big|_0^b$

$$= \lim_{b\to\infty}[-20e^{-0.05b} + (20e^{-0.05(0)})] = 0 + 20(1) = 20$$

27. $\int_5^\infty \frac{1}{(x-4)^3}\,dx = \int_5^b \frac{1}{(x-4)^3}\,dx = \lim_{b\to\infty} -\frac{1}{2}(x-4)^{-2}\Big|_5^b$

$$= \lim_{b\to\infty} -\frac{1}{2}\left[\frac{1}{(b-4)^2} - \frac{1}{(5-4)^2}\right] = -\frac{1}{2}(0-1) = \frac{1}{2}$$

29. $\int_0^\infty \frac{x}{x^2+1}\,dx = \lim_{b\to\infty}\int_0^b \frac{x}{x^2+1}\,dx$ Let $u = x^2+1$

$du = 2x\,dx$

$$\int \frac{x}{x^2+1}\,dx = \int \frac{1}{u}\left(\frac{du}{2}\right) = \frac{1}{2}\ln|u| + C$$

$$= \frac{1}{2}\ln\left|x^2+1\right| + C$$

Thus, $\lim_{b\to\infty}\int_0^b \frac{x}{x^2+1}\,dx = \lim_{b\to\infty}\left[\frac{1}{2}\ln(x^2+1)\right]\Big|_0^b$

$$= \lim_{b\to\infty}\left[\frac{1}{2}\ln(b^2+1) - \frac{1}{2}(1)^2\right]$$

$$= \lim_{b\to\infty}\left[\frac{1}{2}\ln(b^2+1) - 0\right]$$

Since $\lim_{b\to\infty}\ln(b^2+1)$ does not exist, the integral is divergent.

31. $\int_0^\infty x^2e^{-x^3}\,dx = \lim_{b\to\infty}\int_0^b x^2e^{-x^3}\,dx$ Let $u = -x^3$

$du = -3x^2\,dx$

Then $\int x^2e^{-x^3}\,dx = \int -\frac{1}{3}e^u\,du = -\frac{1}{3}e^u + C = -\frac{1}{3}e^{-x^3} + C$

and $\lim_{b\to\infty}\int_0^b x^2e^{-x^3}\,dx = \lim_{b\to\infty} -\frac{1}{3}e^{-x^3}\Big|_0^b = \lim_{b\to\infty} -\frac{1}{3}\left(\frac{1}{e^{b^3}} - \frac{1}{e^0}\right)$

$$= -\frac{1}{3}(0-1) = \frac{1}{3}$$

33. $\int_{-\infty}^0 e^{3x}\,dx = \lim_{b\to-\infty}\int_b^0 e^{3x}\,dx = \lim_{b\to-\infty}\frac{1}{3}e^{3x}\Big|_b^0$

$$= \lim_{b\to-\infty}\frac{1}{3}e^0 - \frac{1}{3}e^{3b} = \frac{1}{3} - 0 = \frac{1}{3}$$

35. $\int_{-\infty}^{1} \frac{1}{2-x}\,dx = \lim_{a\to-\infty} \int_{a}^{1} \frac{1}{2-x}\,dx$ Let $u = 2-x$
$du = -dx$

Then $\int \frac{1}{2-x}\,dx = \int -\frac{du}{u} = -\ln|u| + C = -\ln|2-x| + C$

$$\lim_{a\to-\infty} \int_{a}^{1} \frac{1}{2-x}\,dx = \lim_{a\to-\infty} -\ln|2-x| \Big|_{a}^{1} = \lim_{a\to-\infty} -(\ln 1 - \ln|2-a|)$$

$\lim_{a\to-\infty} \ln|2-a|$ does not exist, so the integral is divergent.

37. $\int_{-\infty}^{\infty} \frac{e^x}{(1+e^x)^2}\,dx = \lim_{b\to-\infty} \int_{b}^{0} \frac{e^x}{(1+e^x)^2}\,dx + \lim_{a\to\infty} \int_{0}^{a} \frac{e^x}{(1+e^x)^2}\,dx$

$$\begin{aligned}\int \frac{e^x}{(1+e^x)^2}\,dx &= \int \frac{1}{u^2}\,du = -u^{-1} + C \\ &= -(1+e^x)^{-1} + C \\ &= (-e^x - 1)^{-1} + C\end{aligned}$$

Thus,

$$\begin{aligned}&\lim_{b\to-\infty} \int_{b}^{0} \frac{e^x}{(1+e^x)^2}\,dx + \lim_{a\to\infty} \int_{0}^{a} \frac{e^x}{(1+e^x)^2}\,dx \\ &= \lim_{b\to-\infty} (-e^x - 1)^{-1} \Big|_{b}^{0} + \lim_{a\to\infty} (-e^x - 1)^{-1} \Big|_{0}^{a} \\ &= \lim_{b\to-\infty} \left[(-e^0 - 1)^{-1} + (e^b + 1)^{-1}\right] + \lim_{a\to\infty} \left[(-e^a - 1)^{-1} + (e^0 + 1)^{-1}\right] \\ &= \lim_{b\to-\infty} \left(-\frac{1}{2} + \frac{1}{e^b + 1}\right) + \lim_{a\to\infty} \frac{1}{-e^a - 1} + \frac{1}{2} \\ &= -\frac{1}{2} + \frac{1}{0+1} + 0 + \frac{1}{2} \qquad \left(\text{Since } \lim_{x\to-\infty} e^x = 0\right) \\ &= 1\end{aligned}$$

39. $\int_{-\infty}^{\infty} \frac{e^x}{1+e^x}\,dx = \lim_{a\to-\infty} \int_{a}^{0} \frac{e^x}{1+e^x}\,dx + \lim_{b\to\infty} \int_{0}^{b} \frac{e^x}{1+e^x}\,dx$

To evaluate the integral, let $u = 1 + e^x$
$du = e^x\,dx$

Then $\int \frac{e^x}{1+e^x}\,dx = \int \frac{du}{u} = \ln|u| + C = \ln|1+e^x| + C$

$$\begin{aligned}\lim_{a\to-\infty} \int_{a}^{0} \frac{e^x}{1+e^x}\,dx &= \lim_{a\to-\infty} \ln(1+e^x) \Big|_{a}^{0} = \lim_{a\to-\infty} \left[\ln 2 - \ln(1+e^a)\right] \\ &= \ln 2 - \ln 1 = \ln 2\end{aligned}$$

$$\lim_{b\to\infty} \int_{0}^{b} \frac{e^x}{1+e^x}\,dx = \lim_{b\to\infty} \ln(1+e^x) \Big|_{0}^{b} = \lim_{b\to\infty} \left[\ln(1+e^b) - \ln(1+e^0)\right]$$

$\lim_{b\to\infty} \ln(1+e^b)$ does not exist, so $\int_{0}^{\infty} \frac{e^x}{1+e^x}\,dx$ and $\int_{-\infty}^{\infty} \frac{e^x}{1+e^x}\,dx$ are divergent.

41. **c.**

x	y_1	y_2
1	2	0.74682
10	104.16	0.88623
100	396,478	0.88623
1000	3.3×10^{15}	1.4×10^{-7}

$\int_0^\infty e^{\sqrt{x}}\,dx$ diverges and

$\int_0^\infty e^{-x^2}\,dx$ converges to 0.88623.

Note: The calculator gives misleading results for large values of x (870 or above).

43. The size of the permanent endowment needed to generate an annual \$12,000 forever at 6% is

$$\begin{aligned}\int_0^\infty 12{,}000e^{-0.06t}\,dt &= \lim_{b\to\infty}\int_0^b 12{,}000e^{-0.06t}\,dt\\ &= \lim_{b\to\infty}\left(\frac{12{,}000}{-0.06}e^{-0.06t}\right)\Bigg|_0^b\\ &= \lim_{b\to\infty}(-200{,}000e^{-0.06b}+200{,}000e^{-0.06(0)})\\ &= \lim_{b\to\infty}(-200{,}000e^{-0.06b}+200{,}000)\\ &= \$200{,}000\end{aligned}$$

since $\lim_{b\to\infty} e^{-0.06b} = 0$

45. **a.** The size of the permanent endowment needed to generate an annual \$1000 forever at 10% is

$\int_0^\infty 1000e^{-0.1t}\,dt$, since $r = 10\% = 0.1$.

$$\lim_{b\to\infty}\int_0^b 1000e^{-0.1t}\,dt = \lim_{b\to\infty} -10{,}000\,e^{-0.1t}\Big|_0^b$$

$$\lim_{b\to\infty} -10{,}000\left(\frac{1}{e^{0.1b}}-\frac{1}{e^0}\right) = -10{,}000(0-1) = 10{,}000.$$

The amount needed is \$10,000.

b. $$\int_0^{100} 1000e^{-0.1t}\,dt = -10{,}000e^{-0.1t}\Big|_0^{100}$$

$$= -10{,}000\left(\frac{1}{e^{10}}-\frac{1}{e^0}\right) \approx 9999.55$$

The fund needed to provide an endowment for 100 years is \$9999.55, which is practically the same as one needed to provide an endowment forever.

47. $C(t) = 50\sqrt{t}$ thousand dollars and $r = 0.05$.

$$(\text{Capital value}) = \int_0^\infty 50\sqrt{t}e^{-0.05t}\,dt = \lim_{b\to\infty} 50\sqrt{t}e^{-0.05t}\,dt$$
$$\approx 3963$$

The capital value is about 3963 thousand dollars, or \$3,963,000.

49. To find the total output, integrate from 0 to ∞.

$$\int_0^\infty 50e^{-0.05t}\,dt = \lim_{b\to\infty}\int_0^b 50e^{-0.05t}\,dt = \lim_{b\to\infty}\left(-\frac{50}{0.05}e^{-0.05t}\right)\Bigg|_0^b$$
$$= \lim_{b\to\infty}(-1000e^{-0.05b} + 1000e^{-0.05(0)})$$
$$= 0 + 1000 = 1000$$

The total lifetime output of the well is 1,000,000 barrels.

51. The area under the curve $y = \frac{1}{x^{3/2}}$ from $x = 1$ to ∞ (above the x-axis) is

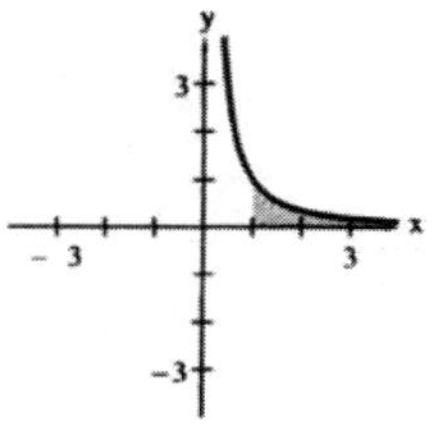

$$\int_1^\infty \frac{1}{x^{3/2}} = \lim_{b\to\infty}\int_1^b x^{-3/2}dx$$
$$= \lim_{b\to\infty}\frac{x^{-1/2}}{-\frac{1}{2}}\Bigg|_1^b = \lim_{b\to\infty} -2x^{-1/2}\Big|_1^b$$
$$= \lim_{b\to\infty} -2\left(\frac{1}{b^{1/2}} - 1\right)$$
$$= -2(0-1) = 2 \text{ square units}$$

53. $$\text{Area} = \int_0^\infty e^{-ax}dx = \lim_{b\to\infty}\int_0^b e^{-ax}dx = \lim_{b\to\infty}\left(-\frac{1}{a}e^{-ax}\right)\Bigg|_0^b$$
$$= \lim_{b\to\infty}\left(-\frac{1}{a}e^{-ab} + \frac{1}{a}e^0\right) = 0 + \frac{1}{a} = \frac{1}{a} \text{ square units}$$

55. The proportion of rats who needed more than 10 seconds to reach the end of the maze is

$$\int_{10}^\infty 0.05e^{-0.05s}ds = \lim_{b\to\infty}\int_{10}^b 0.05e^{-0.05s}ds = \lim_{b\to\infty} -e^{-0.05s}\Big|_{10}^b$$
$$= \lim_{b\to\infty} -\left(\frac{1}{e^{0.05b}} - \frac{1}{e^{0.5}}\right) = -\left(0 - \frac{1}{e^{0.5}}\right)$$
$$\approx 0.607 = 60.7\%$$

57. The proportion of light bulbs that last longer than 1200 hours is

$$\int_{1200}^\infty 0.001e^{-0.001s}\,ds = \lim_{b\to\infty}\int_{1200}^b 0.001e^{-0.001s}\,ds$$
$$= \lim_{b\to\infty}\left(-\frac{0.001}{0.001}e^{-0.001s}\right)\Bigg|_{1200}^b$$
$$= \lim_{b\to\infty}\left(-e^{-0.001b} + e^{-0.001(1200)}\right)$$
$$= 0 + e^{-1.2} \approx 0.30 = 30\%$$

59. $$\int_0^\infty 16{,}000e^{-0.8t}dt = \lim_{b\to\infty}\int_0^b 16{,}000e^{-0.8t}dt$$
$$= \lim_{b\to\infty}\frac{16{,}000}{-0.8}e^{-0.8t}\Big|_0^b$$
$$= \lim_{b\to\infty} -20{,}000\left(\frac{1}{e^{0.8b}} - \frac{1}{e^0}\right)$$
$$= -20{,}000(0-1) = 20{,}000$$

The total number of books sold is 20,000 books.

61. **b.** $\lim_{x\to\infty}\frac{1}{x^3}$ exists (and equals zero)

63. **a.** $\lim_{x\to-\infty} e^{3x}$ exists (and equals zero)

65. False (see example 4)

67. Diverges (Hint: think of the graph)

69. True (Hint: think of the graph)

71. $$\lim_{t\to\infty} D\left(\frac{1-(1+r)^{-t}}{r}\right)=\frac{D}{r}\left(\lim_{t\to\infty}1-\lim_{t\to\infty}(1+r)^{-t}\right)=\frac{D}{r}\left(1-\lim_{t\to\infty}\frac{1}{(1+r)^t}\right)=\frac{D}{r}$$

↗ Approaches zero as t approaches infinity.

73. $$\begin{aligned}\int_0^\infty 2000\cdot 1.05^{-x}dx &= \lim_{b\to\infty}\left(2000\int_0^b 1.05^{-x}\,dx\right)\\ &= \lim_{b\to\infty}\left[2000\left(\frac{1}{(-1)\ln 1.05}\right)1.05^{-x}\Big|_0^b\right]\\ &\approx \lim_{b\to\infty}\left(-40,992\cdot 1.05^{-b}+40,992\cdot 1.05^{0}\right)\\ &= \$40,992\end{aligned}$$

EXERCISES 6.4

1. $\int_1^3 x^2\,dx$

a. $n=4$ trapezoids, so $\Delta x=\frac{b-a}{n}=\frac{3-1}{4}=\frac{1}{2}=0.5$

x	$f(x)=x^2$
1	$1\to 0.5$
1.5	2.25
2	4
2.5	6.25
3	$9\to 4.5$

Adding, $0.5+2.25+4+6.25+4.5=17.5$

and $\int_1^3 x^2\,dx\approx 17.5(\Delta x)=17.5(0.5)=8.75$

b. $\int_1^3 x^2\,dx=\frac{1}{3}x^3\Big|_1^3=\frac{1}{3}(27-1)=\frac{26}{3}\approx 8.667$

c. Actual error $=\left|8.75-\frac{26}{3}\right|\approx 0.083$

d. Relative error $=\frac{0.083}{\frac{26}{3}}(100\%)\approx 0.958\%\approx 1\%$

3. $\int_2^4 \frac{1}{x}\,dx$

a. $n = 4$ trapezoids, so $\Delta x = \frac{4-2}{4} = \frac{1}{2} = 0.5$

x	$f(x) = \frac{1}{x}$
2.0	$\frac{1}{2} \to \frac{1}{4}$
2.5	0.4
3.0	0.333
3.5	0.286
4.0	$0.25 \to 0.125$

Adding
$0.25 + 0.4 + 0.333 + 0.286 + 0.125 = 1.394$

and $\int_2^4 \frac{1}{x}\,dx \approx 1.394(\Delta x) = 1.394(0.5) = 0.697$

b. $\int_2^4 \frac{1}{x}\,dx = \ln x \,\Big|_2^4 = \ln 4 - \ln 2 \approx 0.693$

c. Actual error $= |0.697 - 0.693| = 0.004$

d. Relative error $= \frac{0.004}{0.693} \approx 0.006 \approx 0.6\%$

5. $\int_0^1 \sqrt{1+x^2}\,dx;\ n = 3$

$\Delta x = \frac{b-a}{n} = \frac{1-0}{3} = \frac{1}{3}$

x	$f(x) = \sqrt{1+x^2}$
0	$1 \to 0.5$
$\frac{1}{3}$	1.054
$\frac{2}{3}$	1.202
1	$1.414 \to 0.707$

Adding, $0.5 + 1.054 + 1.202 + 0.707 = 3.463$

and $\int_0^1 \sqrt{1+x^2}\,dx \approx 3.463\left(\frac{1}{3}\right) = 1.154$

7. $\int_0^1 e^{-x^2}\,dx \qquad \Delta x = \frac{1-0}{4} = 0.25$

x	$f(x) = e^{-x^2}$
0	$1 \to 0.5$
0.25	0.939
0.5	0.779
0.75	0.570
1.0	$0.368 \to 0.184$

Adding, $0.5 + 0.939 + 0.779 + 0.570 + 0.184 = 2.972$

$\int_0^1 e^{-x^2}\,dx \approx 2.972(\Delta x) = 2.972(0.25) = 0.743$

9.

n	$\int_1^2 \sqrt{\ln x}\,dx$
10	0.586287
50	0.592013
100	0.592385
200	0.592518
500	0.592572

$\int_1^2 \sqrt{\ln x}\,dx \approx 0.593$

11.

n	$\int_{-1}^1 \sqrt{16+9x^2}\,dx$
10	8.70879
50	8.69726
100	8.69690
200	8.69681
500	8.69679

$\int_{-1}^1 \sqrt{16+9x^2}\,dx \approx 8.6968$

13.

n	$\int_{-1}^1 e^{x^2}\,dx$
10	2.96131
50	2.92675
100	2.92567
200	2.92539
500	2.92532

$\int_{-1}^1 e^{x^2}\,dx \approx 2.925$

15. The proportion of people in the United States with IQs between A and B is

$$\int_{\frac{A-100}{15}}^{\frac{B-100}{15}} \frac{1}{\sqrt{2\pi}} e^{-\frac{1}{2}x^2}\,dx;$$

here $A = 100$ and $B = 130$.

$$\int_{\frac{100-100}{15}}^{\frac{130-100}{15}} \frac{1}{\sqrt{2\pi}} e^{-\frac{1}{2}x^2}\,dx = \int_0^2 \frac{1}{\sqrt{2\pi}} e^{-\frac{1}{2}x^2}\,dx$$

n	$\int_0^2 \frac{1}{\sqrt{2\pi}} e^{-\frac{1}{2}x^2}\,dx$
10	0.4769
100	0.4772
500	0.4772

Approximately 0.4772 or 47.22% of the U.S. population have IQs between 100 and 130.

17. The total oil consumed is the integral from $t = 10$ to $t = 20$.

$$\int_{10}^{20} f(x)\,dx = \int_{10}^{20} \sqrt{12+0.025x^2}\,dx$$

n	$\int_{10}^{20} \sqrt{12+0.025x^2}\,dx$
10	42.16
20	42.15
30	42.15

The total amount of oil consumed from 2010 to 2020 is about 42 billion tons.

19. $\int_1^3 x^2 dx;\ n = 4$

$$\Delta x = \frac{3-1}{4} = \frac{1}{2}$$

x	$f(x) = x^2$	Weight	$f(x) \cdot$ weight
1	1	1	1
1.5	2.25	4	9
2.0	4.0	2	8
2.5	6.25	4	25
3.0	9.0	1	9
			52

$$\int_1^3 x^2 dx \approx 52\left(\frac{\Delta x}{3}\right) \approx 52\left(\frac{1}{6}\right) \approx 8.667$$

Simpson's rule is exact for quadratics; the answer obtained in Exercise 1 using 4 trapezoids had about a 1% error.

21. $\int_2^4 \frac{1}{x} dx;\ n = 4$

$$\Delta x = \frac{4-2}{4} = \frac{1}{2} = 0.5$$

x	$f(x) = \frac{1}{x}$	Weight	$f(x)$ • weight
2	0.5	1	0.5
2.5	0.4	4	1.6
3.0	0.333	2	0.666
3.5	0.286	4	1.144
4	0.25	1	0.25

$0.5 + 1.6 + 0.666 + 1.144 + 0.25 = 4.16$

$$\int_2^4 \frac{1}{x} dx \approx 4.16\left(\frac{\Delta x}{3}\right) = 4.16\left(\frac{0.5}{3}\right) \approx 0.693$$

Simpson's rule produces an estimate that is within 0.001 of the exact answer. The trapezoidal rule had about a 0.6% error (within 0.004).

23. $\int_0^1 \sqrt{1+x^2}\,dx;\ n = 4 \qquad \Delta x = \frac{1-0}{4} = \frac{1}{4}$

x	$f(x) = \sqrt{1+x^2}$	Weight	$f(x)$ • weight
0	1	1	1
0.25	1.031	4	4.124
0.50	1.118	2	2.236
0.75	1.250	4	5.0
1.0	1.414	1	1.414

$1 + 4.124 + 2.236 + 5.0 + 1.414 = 13.774$

$$\int_0^1 \sqrt{1+x^2}\,dx \approx 13.774\left(\frac{\Delta x}{3}\right) = 13.774\left(\frac{1}{12}\right) \approx 1.148$$

25. $\int_0^1 e^{-x^2} dx;\ n = 4 \qquad \Delta x = \frac{1-0}{4} = 0.25$

x	$f(x)$	Weight	$f(x)$ • weight
0	1	1	1
0.25	0.939	4	3.56
0.5	0.779	2	1.558
0.75	0.570	4	2.280
1	0.368	1	0.368

$1 + 3.756 + 1.558 + 2.280 + 0.368 = 8.962$

$$\int_0^1 e^{-x^2} dx \approx 8.962\left(\frac{\Delta x}{3}\right) = 8.962\left(\frac{0.25}{3}\right) \approx 0.747$$

27.

n	$\int_1^2 \sqrt{\ln x}\,dx$
10	0.59001145
20	0.59168072
50	0.59236060
100	0.59250920
200	0.59256171

$$\int_1^2 \sqrt{\ln x}\,dx \approx 0.593$$

29.

n	$\int_{-1}^1 \sqrt{16+9x^2}\,dx$
10	8.6967628
20	8.6967836
50	8.6967849
100	8.6967850
200	8.6967850

$$\int_{-1}^1 \sqrt{16+9x^2}\,dx \approx 8.6967850$$

31.

n	$\int_{-1}^1 e^{x^2}\,dx$
10	2.9262039
20	2.9253628
50	2.9253050
100	2.9253036
200	2.9253035

$$\int_{-1}^1 e^{x^2}\,dx \approx 2.92530$$

33. Let $x = \frac{1}{t}$. Then $dx = -\frac{dt}{t^2}$.

$$\int_1^\infty \frac{1}{x^3+1}\,dx = \int_1^0 \frac{1}{\left(\frac{1}{t}\right)^3+1}\left(-\frac{dt}{t^2}\right) = \int_1^0 \frac{-t}{1+t^3}\,dt = \int_0^1 \frac{t}{1+t^3}\,dt$$

t	$f(t) = \frac{t}{1+t^3}$	Weight	$f(t)\cdot$ weight
0	0	1	0
0.25	0.2462	4	0.9848
0.5	0.4444	2	0.8888
0.75	0.5275	4	2.1100
1	0.5	1	0.5

$0 + 0.9848 + 0.8888 + 2.1100 + 0.5 = 4.4836$

$$\int_1^\infty \frac{1}{x^3+1}\,dx \approx 4.4836\left(\frac{\Delta x}{3}\right) = 4.4836\left(\frac{0.25}{3}\right) \approx 0.374$$

35. The length of the cable is

$$\int_{-400}^{400} \sqrt{1+\left(\frac{x}{1000}\right)^2}\,dx$$

For $n = 10$, the integral is about 820.8.
For $n = 50$, the integral is about 820.8.

$$\int_{-400}^{400} \sqrt{1+\left(\frac{x}{1000}\right)^2}\,dx \approx 821$$

The length of the cable is about 821 feet.

37.

n	$\int_0^{10} \sqrt{4096+100x^2}\,dx$
10	845.3
20	845.3

$$\int_0^{10} \sqrt{4096+100x^2}\,dx \approx 845$$

The total GWP is approximately \$845 trillion.

39. The second derivative of a linear function is zero, so the error formula gives zero error, making the approximation exact. (Alternatively, the graph of a linear function *is* a trapezoid, so approximating by a trapezoid will give the exact answer.)

41. 9

43. $\frac{1}{2}\cdot 0+4+5+3+\frac{1}{2}\cdot 0=12$ square units

45. $n = 2$ (Since the method is exact for *any n*, use the smallest number possible.)

47. **i.** For the 3 points on the parabola, the following equations hold:

$$a(-d)^2+b(-d)+c=y_1$$
$$a(0)^2+b(0)+c=y_2$$
$$a(d)^2+b(d)+c=y_3$$

ii. Simplifying gives

$$ad^2-bd+c=y_1$$
$$c=y_2$$
$$ad^2+bd+c=y_3$$

iii. Add the first and last equations and 4 times the middle equation.

$$ad^2-bd+c+\left(ad^2+bd+c\right)+4c=y_1+y_3+4y_2$$
$$2ad^2+6c=y_1+4y_2+y_3$$

iv. Now the area under the parabola is

$$\text{Area}=\int_{-d}^{d}\left(ax^2+bx+c\right)dx=\left(\tfrac{ax^3}{3}+\tfrac{bx^2}{2}+cx\right)\Big|_{-d}^{d}=\tfrac{ad^3}{3}+\tfrac{bd^2}{2}+cd-\left[\frac{a(-d)^3}{3}+\frac{b(-d)^2}{2}+c(-d)\right]$$
$$=\tfrac{ad^3}{3}+\tfrac{bd^2}{2}+cd+\tfrac{ad^3}{3}-\tfrac{bd^2}{2}+cd=\tfrac{2}{3}ad^3+2cd=\left(2ad^2+6c\right)\tfrac{d}{3}$$

v. From step **iii**: $2ad^2+6c=y_1+4y_2+y_3$

Thus, $\text{Area}=\left(2ad^2+6c\right)\tfrac{d}{3}=(y_1+4y_2+y_3)\tfrac{d}{3}$

vi. Since $y_1=f(-d), y_2=f(0), y_3=f(d)$, and Δx is equal to d,

$\text{Area}=\left[f(-d)+4f(0)+f(d)\right]\frac{\Delta x}{3}$

which is Simpson's rule using one parabola ($n = 2$).

ATTENTION *BRIEF APPLIED CALCULUS* USERS: SECTIONS 6.5 AND 6.6 ARE EQUIVALENT TO SECTIONS 9.1 AND 9.2 OF *APPLIED CALCULUS*, WHICH BEGINS ON PAGE 454 OF THIS MANUAL.

REVIEW EXERCISES AND CHAPTER TEST FOR CHAPTER 6

1. $\int xe^{2x}dx$ Let $u = x$ and $dv = e^{2x}dx$. Then $du = dx$ and $v = \int e^{2x}dx = \frac{1}{2}e^{2x}$.

$$\int xe^{2x}dx = \frac{1}{2}xe^{2x} - \int \frac{1}{2}e^{2x}dx = \frac{1}{2}xe^{2x} - \frac{1}{2}\left(\frac{1}{2}e^{2x}\right) + C = \frac{1}{2}xe^{2x} - \frac{1}{4}e^{2x} + C$$

2. $\int xe^{-x}dx$ Let $u = x$ and $dv = e^{-x}dx$. Then $du = dx$ and $v = \int e^{-x}dx = -e^{-x}$.

$$\int xe^{-x}dx = -xe^{-x} - \int\left(-e^{-x}\right)dx = -xe^{-x} + \int e^{-x}dx = -xe^{-x} - e^{-x} + C$$

3. $\int x^8 \ln x\, dx$ Let $u = \ln x$ and $dv = x^8dx$. Then $du = \frac{1}{x}dx$ and $v = \int x^8dx = \frac{x^9}{9}$.

$$\int x^8 \ln x\, dx = \frac{x^9}{9}\ln x - \int \frac{x^9}{9}\frac{1}{x}dx = \frac{x^9}{9}\ln x - \frac{1}{9}\int x^8dx = \frac{x^9}{9}\ln x - \frac{1}{81}x^9 + C$$

4. $\int \sqrt[4]{x}\ln x\, dx$ Let $u = \ln x$ and $dv = \sqrt[4]{x}dx$. Then $du = \frac{1}{x}dx$ and $v = \int \sqrt[4]{x}dx = \int x^{1/4}dx = \frac{4}{5}x^{5/4}$.

$$\int \sqrt[4]{x}dx = \frac{4}{5}x^{5/4}\ln x - \int \frac{4}{5}x^{5/4}\left(\frac{1}{x}\right)dx = \frac{4}{5}x^{5/4}\ln x - \frac{4}{5}\int x^{1/4}dx = \frac{4}{5}x^{5/4}\ln x - \frac{16}{25}x^{5/4} + C$$

5. Let $u = (x-2)$ and $dv = (x+1)^5dx$. Then $du = dx$ and $v = \int (x+1)^5dx = \frac{(x+1)^6}{6}$.

$$\int (x-2)(x+1)^5dx = \frac{1}{6}(x-2)(x+1)^6 - \int \frac{(x+1)^6}{6}dx = \frac{1}{6}(x-2)(x+1)^6 - \frac{1}{42}(x+1)^7 + C$$

6. Let $u = (x+3)$ and $dv = (x-1)^4dx$. Then $du = dx$ and $v = \int (x-1)^4dx = \frac{(x-1)^5}{5}$.

$$\int (x+3)(x-1)^4dx = \frac{1}{5}(x+3)(x-1)^5 - \int \frac{(x-1)^5}{5}dx = \frac{1}{5}(x+3)(x-1)^5 - \frac{1}{30}(x-1)^6 + C$$

7. Let $u = \ln t$ and $dv = \frac{1}{\sqrt{t}}dt$. Then $du = \frac{1}{t}dt$ and $v = \int \frac{1}{\sqrt{t}}dt = \int t^{-1/2}dt = 2t^{1/2}$.

$$\int \frac{\ln t}{\sqrt{t}}dt = 2t^{1/2}\ln t - \int 2t^{1/2}\left(\frac{1}{t}\right)dt = 2t^{1/2}\ln t - 2\int t^{-1/2}dt$$
$$= 2t^{1/2}\ln t - 2\left(2t^{1/2}\right) + C = 2t^{1/2}\ln t - 4t^{1/2} + C$$

8. Let $u = x^4$. Then $du = 4x^3du$.

$$\int x^7e^{x^4}dx = \int x^4e^{x^4}\left(x^3dx\right) = \int ue^u\left(\frac{du}{4}\right) = \frac{1}{4}\int ue^u du$$

Now, let $w = u$ and $dv = e^u du$. Then $dw = du$ and $v = \int e^u du = e^u$.

$$\frac{1}{4}\int ue^u du = \frac{1}{4}\left(ue^u - \int e^u du\right) = \frac{1}{4}ue^u - \frac{1}{4}e^u + C$$

Since $u = x^4$, $\int x^7e^{x^4}dx = \frac{1}{4}x^4e^{x^4} - \frac{1}{4}e^{x^4} + C$

9. Let $u = x^2$ and $dv = e^x dx$. Then $du = 2x\,dx$ and $v = \int e^x dx = e^x$.

$$\int x^2 e^x dx = x^2 e^x - \int 2xe^x dx = x^2 e^x - 2\int xe^x dx$$

Now, do another integration by parts with $u = x$ and $dv = e^x dx$. Then $du = dx$ and $v = \int e^x dx = e^x$.

$$\int x^2 e^x dx = x^2 e^x - 2\int xe^x dx = x^2 e^x - 2\left(xe^x - \int e^x dx\right) = x^2 e^x - 2xe^x + 2e^x + C$$

10. Let $u = (\ln x)^2$ and $dv = dx$. Then $du = 2\ln x \frac{1}{x} dx$ and $v = \int dx = x$.

$$\int (\ln x)^2 dx = x(\ln x)^2 - \int x\left(\frac{1}{x}\right)(2\ln x)dx = x(\ln x)^2 - 2\int \ln x\, dx$$

Now let $u = \ln x$ and $dv = dx$. Then $du = \frac{1}{x} dx$ and $v = \int dx = x$.

$$\int (\ln x)^2 dx = x(\ln x)^2 - 2\int \ln x\, dx = x(\ln x)^2 - 2\left[x\ln x - \int \frac{1}{x}(x)dx\right]$$

$$= x(\ln x)^2 - 2x\ln x + 2x + C$$

11. Let $u = x$ and $dv = (x+a)^n$. Then $du = dx$ and $v = \int (x+a)^n dx = \frac{(x+a)^{n+1}}{n+1}$.

$$\int x(x+a)^n dx = x\frac{(x+a)^{n+1}}{n+1} - \int \frac{(x+a)^{n+1}}{n+1} dx = \frac{1}{n+1}x(x+a)^{n+1} - \frac{1}{(n+1)(n+2)}(x+a)^{n+2} + C$$

12. Let $u = x$ and $dv = (1-x)^n dx$. Then $du = dx$ and $v = \int (1-x)^n dx = -\frac{(1-x)^{n+1}}{n+1}$.

$$\int x(1-x)^n dx = -x\frac{(1-x)^{n+1}}{n+1} - \int -\frac{(1-x)^{n+1}}{n+1} dx$$

$$= -\frac{1}{n+1}x(1-x)^{n+1} + \frac{1}{(n+1)(n+2)}(1-x)^{n+2}(-1) + C$$

$$= -\frac{1}{n+1}x(1-x)^{n+1} - \frac{1}{(n+1)(n+2)}(1-x)^{n+2} + C$$

13. $\int_0^5 xe^x dx$ Let $u = x$ and $dv = e^x dx$. Then $du = dx$ and $v = e^x$.

$$\int xe^x dx = xe^x - \int e^x dx = xe^x - e^x + C$$

$$\int_0^5 xe^x dx = \left(xe^x - e^x\right)\Big|_0^5 = \left(5e^5 - e^5\right) - \left(0 - e^0\right) = 4e^5 + 1$$

14. Let $u = \ln x$ and $dv = x\,dx$. Then $du = \frac{1}{x} dx$ and $v = \int x\,dx = \frac{x^2}{2}$.

$$\int x\ln x\, dx = \frac{x^2}{2}\ln x - \int \frac{x^2}{2}\left(\frac{1}{x}\right)dx = \frac{x^2}{2}\ln x - \frac{1}{2}\left(\frac{x^2}{2}\right) + C = \frac{x^2}{2}\ln x - \frac{x^2}{4} + C$$

$$\int_1^e x\ln x\, dx = \left(\frac{x^2}{2}\ln x - \frac{x^2}{4}\right)\Big|_1^e = \frac{e^2}{2}\ln e - \frac{e^2}{4} - \left[\frac{(1)^2}{2}\ln 1 - \frac{(1)^2}{4}\right] = \frac{e^2}{2} - \frac{e^2}{4} + \frac{1}{4} = \frac{e^2}{4} + \frac{1}{4}$$

15. Let $u = 1 - x$. Then $du = -dx$.

$$\int \frac{dx}{1-x} = \int \frac{1}{u}(-du) = -\int \frac{1}{u} du = -\ln|u| + C = -\ln|1-x| + C$$

16. Let $u=-x^2$. Then $du=-2x\,dx$.

$$\int xe^{-x^2}dx=\int e^u\left(-\tfrac{du}{2}\right)=-\tfrac{1}{2}e^u+C=-\tfrac{1}{2}e^{-x^2}+C$$

17. $\int x^3\ln 2x\,dx$ Let $u=\ln 2x$ and $dv=x^3dx$. Then $du=\frac{2dx}{2x}=\frac{dx}{x}$ and $v=\frac{1}{4}x^4$.

$$\int x^3\ln 2x\,dx=\tfrac{1}{4}x^4\ln 2x-\int\tfrac{1}{4}x^4\left(\tfrac{dx}{x}\right)=\tfrac{1}{4}x^4\ln 2x-\tfrac{1}{4}\int x^3dx$$
$$=\tfrac{1}{4}x^4\ln 2x-\tfrac{1}{16}x^4+C=\tfrac{1}{4}x^4\left(\ln 2x-\tfrac{1}{4}\right)+C$$

18. Let $u=1-x$. Then $du=-dx$.

$$\int\frac{dx}{(1-x)^2}=\int\frac{1}{u^2}(-du)=-\int u^{-2}du=-\left(\frac{u^{-1}}{-1}\right)+C=u^{-1}+C=\frac{1}{1-x}+C$$

19. Let $u=\ln x$. Then $du=\frac{1}{x}dx$.

$$\int\frac{\ln x}{x}dx=\int u\,du=\frac{u^2}{2}+C=\frac{(\ln x)^2}{2}+C$$

20. Let $u=e^{2x}+1$. Then $du=2e^{2x}dx$.

$$\int\frac{e^{2x}}{e^{2x}+1}dx=\int\frac{1}{u}\left(\frac{du}{2}\right)=\frac{1}{2}\ln|u|+C=\frac{1}{2}\ln\left|e^{2x}+1\right|+C$$

21. $\int\frac{e^{\sqrt{x}}}{\sqrt{x}}dx$ Let $u=\sqrt{x}$. Then $du=\frac{dx}{2\sqrt{x}}$.

$$\int\frac{e^{\sqrt{x}}}{\sqrt{x}}dx=\int e^u 2\,du=2e^u+C=2e^{\sqrt{x}}+C$$

22. Let $u=e^{2x}+1$. Then $du=2e^{2x}dx$.

$$\int\left(e^{2x}+1\right)^3e^{2x}dx=\int u^3\left(\tfrac{du}{2}\right)=\tfrac{1}{2}\left(\tfrac{u^4}{4}\right)+C=\tfrac{1}{8}\left(e^{2x}+1\right)^4+C$$

23. **a.** The present value of a continuous stream of income over 10 years is

$$\int_0^{10}C(t)e^{-rt}dt,\ C(t)=25t,\ r=0.05$$

$$\int_0^{10}25te^{-0.05t}dt$$

Let $u=25t$ and $dv=e^{-0.05t}dt$. Then $du=25\,dt$ and $v=\int e^{-0.05t}dt=\frac{1}{-0.05}e^{-0.05t}=-20e^{-0.05t}$.

$$\int 25te^{-0.05t}dt=25t\left(-20e^{-0.05t}\right)-\int 25\left(-20e^{-0.05t}\right)dt=-500te^{-0.05t}+500\int e^{-0.05t}dt$$
$$=-500te^{-0.05}+500\left(-20e^{-0.05t}\right)+K=-500te^{-0.05t}-10{,}000e^{-0.05t}+K$$

$$\int_0^{10}25te^{-0.05t}dt=\left(-500te^{-0.05t}-10{,}000e^{-0.05t}\right)\Big|_0^{10}$$
$$=-500(10)e^{-0.05(10)}-10{,}000e^{-0.05(10)}-\left[500(0)e^{-0.05(0)}-10{,}000e^{-0.05(0)}\right]$$
$$=-5000e^{-0.5}-10{,}000e^{-0.5}+10{,}000=-15{,}000e^{-0.5}+10{,}000\approx 902$$

The present value is about \$902 million

b. Using FnInt, the integral is about 902.

24. **a.** To find the total leakage during the first 3 months, integrate $te^{0.2t}$ from 0 to 3.

Let $u = t$ and $dv = e^{0.2t}dt$. Then $du = dt$ and $v = \int e^{0.2t}dt = \frac{1}{0.2}e^{0.2t} = 5e^{0.2t}$.

$$\int te^{0.2t}dt = 5te^{0.2t} - 5\int e^{0.2t}dt = 5te^{0.2t} - 25e^{0.2t} + C$$

$$\int_0^3 te^{0.2t}dt = \left(5te^{0.2t} - 25e^{0.2t}\right)\Big|_0^3$$
$$= 5(3)e^{0.2(3)} - 25e^{0.2(3)} - \left[5(0)e^{0.2(0)} - 25e^{0.2(0)}\right]$$
$$= 15e^{0.6} - 25e^{0.6} - 0 + 25e^{0} = -10e^{0.6} + 25 \approx 6.78$$

The total leakage is about 678 gallons.

b. Using FnInt, the integral is approximately 6.78.

25. Use Formula 16 with $a = 5$.

$$\int \frac{1}{25-x^2}dx = \frac{1}{2(5)}\ln\left|\frac{5+x}{5-x}\right| + C = \frac{1}{10}\ln\left|\frac{5+x}{5-x}\right| + C$$

26. Use Formula 15 with $a = 2$.

$$\int \frac{1}{x^2-4}dx = \frac{1}{4}\ln\left|\frac{x-2}{x+2}\right| + C$$

27. Use Formula 11 with $a = 1$, $b = -1$, $c = 1$, $d = -2$.

$$\int \frac{x}{(x-1)(x-2)}dx = \frac{1}{(1)(-2)-(-1)(1)}\left(\frac{-2}{1}\ln|x-2| - \frac{-1}{1}\ln|x-1|\right) + C$$
$$= \frac{1}{-1}\left(-2\ln|x-2| + \ln|x-1|\right) + C = 2\ln|x-2| - \ln|x-1| + C$$

28. Use Formula 10 with $a = 1$, $b = -1$, $c = 1$, $d = -2$.

$$\int \frac{1}{(x-1)(x-2)}dx = \frac{1}{(1)(-2)-(-1)(1)}\ln\left|\frac{x-1}{x-2}\right| + C = -\ln\left|\frac{x-1}{x-2}\right| + C$$

29. Use Formula 14 with $b = 1$, $a = 1$.

$$\int \frac{1}{x\sqrt{x+1}}dx = \frac{1}{\sqrt{1}}\ln\left|\frac{\sqrt{x+1}-\sqrt{1}}{\sqrt{x+1}+\sqrt{1}}\right| + C = \ln\left|\frac{\sqrt{x+1}-1}{\sqrt{x+1}+1}\right| + C$$

30. Use Formula 13 with $a = 1$, $b = 1$.

$$\int \frac{x}{\sqrt{x+1}}dx = \frac{2(1)x-4(1)}{3(1)^2}\sqrt{x+1} + C = \frac{2x-4}{3}\sqrt{x+1} + C$$

31. Use Formula 18 with $a = 3$.

$$\int \frac{1}{\sqrt{x^2+9}}dx = \ln\left|x + \sqrt{x^2+9}\right| + C$$

32. Use Formula 18 with $a = 4$.

$$\int \frac{1}{\sqrt{x^2+16}}dx = \ln\left|x + \sqrt{x^2+16}\right| + C$$

33. Let $x = z^2$. Then $dx = 2z\,dz$; $dz = \frac{dx}{2z} = \frac{dx}{2\sqrt{x}}$. Also $z^3 = x^{3/2}$.

So $\int \frac{z^3}{\sqrt{z^2+1}}\,dz = \int \frac{x^{3/2}}{\sqrt{x+1}}\left(\frac{dx}{2x^{1/2}}\right) = \frac{1}{2}\int \frac{x}{\sqrt{x+1}}\,dx$

Now, use Formula 13 with $a = 1$, $b = 1$.

$\frac{1}{2}\int \frac{x}{\sqrt{x+1}}\,dx = \frac{1}{2}\frac{2x-4}{3}\sqrt{x+1} + C = \frac{1}{3}(x-2)\sqrt{x+1} + C$

So, $\int \frac{z^3}{\sqrt{z^2+1}}\,dz = \frac{1}{3}\left(z^2-2\right)\sqrt{z^2+1} + C$.

34. Let $u = e^t$. Then $du = e^t dt$.

$\int \frac{e^{2t}}{e^t+2}\,dt = \int \frac{u}{u+2}\,du$

Now, use Formula 9 with $a = 1$, $b = 2$.

$\int \frac{u}{u+2}\,du = \frac{u}{1} - \frac{2}{(1)^2}\ln|u+2| + C = u - 2\ln|u+2| + C$

So, $\int \frac{e^{2t}}{e^t+2}\,dt = e^t - 2\ln\left|e^t+2\right| + C$

35. Use Formula 21 with $n = 2$, $a = 2$.

$\int x^2 e^{2x}\,dx = \frac{1}{2}x^2 e^{2x} - \frac{2}{2}\int x^{2-1}e^{2x}\,dx = \frac{1}{2}x^2 e^{2x} - \int xe^{2x}\,dx$

Use Formula 21 again with $n = 1$, $a = 2$.

$$\begin{aligned}\int x^2 e^{2x}\,dx &= \frac{1}{2}x^2 e^{2x} - \int xe^{2x}\,dx = \frac{1}{2}x^2 e^{2x} - \left(\frac{1}{2}xe^{2x} - \frac{1}{2}\int x^{1-1}e^{2x}\,dx\right)\\ &= \frac{1}{2}x^2 e^{2x} - \frac{1}{2}xe^{2x} + \frac{1}{2}\int e^{2x}\,dx = \frac{1}{2}x^2 e^{2x} - \frac{1}{2}xe^{2x} + \frac{1}{4}e^{2x} + C\end{aligned}$$

36. Use Formula 22 with $n = 4$.

$\int (\ln x)^4\,dx = x(\ln x)^4 - 4\int (\ln x)^{4-1}\,dx = x(\ln x)^4 - 4\int (\ln x)^3\,dx$

Use Formula 22 with $n = 3$.

$\int (\ln x)^4\,dx = x(\ln x)^4 - 4\left[x(\ln x)^3 - 3\int (\ln x)^2\,dx\right]$

Use Formula 22 with $n = 2$.

$$\begin{aligned}\int (\ln x)^4\,dx &= x(\ln x)^4 - 4x(\ln x)^3 + 12\left[x(\ln x)^2 - 2\int (\ln x)\,dx\right]\\ &= x(\ln x)^4 - 4x(\ln x)^3 + 12x(\ln x)^2 - 24\int \ln x\,dx\end{aligned}$$

Use Formula 22 with $n = 1$.

$$\begin{aligned}\int (\ln x)^4\,dx &= x(\ln x)^4 - 4x(\ln x)^3 + 12x(\ln x)^2 - 24\left[x\ln x - 1\int (\ln x)^0\,dx\right]\\ &= x(\ln x)^4 - 4x(\ln x)^3 + 12x(\ln x)^2 - 24x\ln x + 24\int dx\\ &= x(\ln x)^4 - 4x(\ln x)^3 + 12x(\ln x)^2 - 24x\ln x + 24x + C\end{aligned}$$

37. To find the cost function, integrate $MC(x)$.

$$\int MC(x)\,dx = \int \frac{1}{(2x+1)(x+1)}\,dx$$

Use Formula 10 with $a = 2$, $b = 1$, $c = 1$, $d = 1$.

$$C(x) = \int \frac{1}{(2x+1)(x+1)}\,dx = \frac{1}{2-1}\ln\left|\frac{2x+1}{x+1}\right| + K = \ln\left|\frac{2x+1}{x+1}\right| + K$$

Since fixed costs are 1000, $C(0) = 1000$. So

$$1000 = \ln\left|\frac{0+1}{0+1}\right| + K$$
$$K = 1000$$

The company's cost function is $C(x) = \ln\left|\frac{2x+1}{x+1}\right| + 1000$.

38. **a.** The total increase in population is

$$\int_0^{30} \sqrt{t^2+1600}\,dt$$

Use formula 17 with $a = 40$.

$$\int_0^{30} \sqrt{t^2+1600}\,dt = \left(\frac{t}{2}\sqrt{t^2+1600} + \frac{1600}{2}\ln\left|t+\sqrt{t^2+1600}\right|\right)\Bigg|_0^{30}$$
$$= \frac{30}{2}\sqrt{(30)^2+1600} + 800\ln\left|30+\sqrt{(30)^2+1600}\right| - \left(\frac{0}{2}\sqrt{(0)^2+1600} + 800\ln\left|0+\sqrt{(0)^2+1600}\right|\right)$$
$$= 15\sqrt{2500} + 800\ln\left|30+\sqrt{2500}\right| - (0 + 800\ln 40)$$
$$= 750 + 800\ln 80 - 800\ln 40 \approx 1305$$

The town's population will increase by a total of 1305 people.

b. Using FnInt, the integral is about 1305.

39. $$\int_1^\infty \frac{1}{x^5}\,dx = \lim_{b\to\infty}\int_1^b \frac{1}{x^5}\,dx = \lim_{b\to\infty}\frac{x^{-4}}{-4}\Bigg|_1^b = \lim_{b\to\infty}\left\{-\frac{1}{4b^4} - \left[-\frac{1}{4(1)^4}\right]\right\}$$
$$= 0 + \frac{1}{4} = \frac{1}{4}$$

40. $$\int_1^\infty \frac{1}{x^6}\,dx = \lim_{b\to\infty}\int_1^b x^{-6}\,dx = \lim_{b\to\infty}\frac{x^{-5}}{-5}\Bigg|_1^b = \lim_{b\to\infty}\left\{-\frac{1}{5b^5} - \left[-\frac{1}{5(1)^5}\right]\right\} = 0 + \frac{1}{5} = \frac{1}{5}$$

41. $$\int_1^\infty \frac{1}{\sqrt[5]{x}}\,dx = \int_1^\infty x^{-1/5}\,dx = \lim_{b\to\infty}\int_1^b x^{-1/5}\,dx = \lim_{b\to\infty}\frac{5}{4}x^{4/5}\Bigg|_1^b = \lim_{b\to\infty}\frac{5}{4}\left(b^{4/5} - 1\right)$$

Since $\lim_{b\to\infty}\frac{5}{4}b^{4/5}$ does not exist, the integral is divergent.

42. $$\int_1^\infty \frac{1}{\sqrt[6]{x}}\,dx = \lim_{b\to\infty}\int_1^b x^{-1/6}\,dx = \lim_{b\to\infty}\frac{6}{5}x^{5/6}\Bigg|_1^b = \lim_{b\to\infty}\left[\frac{6}{5}\left(b^{5/6} - 1\right)\right]$$

Since $\lim_{b\to\infty}\frac{6}{5}b^{5/6}$ does not exist, the integral is divergent.

43. $$\int_0^\infty e^{-2x}\,dx = \lim_{b\to\infty}\int_0^b e^{-2x}\,dx = \lim_{b\to\infty} -\frac{1}{2}e^{-2x}\Bigg|_0^b = \lim_{b\to\infty}\left[-\frac{1}{2}e^{-2b} - \left(-\frac{1}{2}e^0\right)\right]$$
$$= 0 + \frac{1}{2} = \frac{1}{2}$$

44. $\int_4^{\infty} e^{-0.5x}\,dx = \lim_{b\to\infty}\int_4^b e^{-0.5x}\,dx = \lim_{b\to\infty} -2e^{-0.5x}\Big|_4^b$
$= \lim_{b\to\infty}\left[-2e^{-0.5b} - \left(-2e^{-2}\right)\right] = 0 + 2e^{-2} = 2e^{-2}$

45. $\int_0^{\infty} e^{2x}\,dx = \lim_{b\to\infty}\int_0^b e^{2x}\,dx = \lim_{b\to\infty}\frac{1}{2}e^{2x}\Big|_0^b = \lim_{b\to\infty}\frac{1}{2}\left(e^{2b}-1\right)$

Since $\lim_{b\to\infty}\frac{1}{2}e^{2b}$ does not exist, the integral is divergent.

46. $\int_4^{\infty} e^{0.5x}\,dx = \lim_{b\to\infty}\int_4^b e^{0.5x}\,dx = \lim_{b\to\infty} 2e^{0.5x}\Big|_4^b = \lim_{b\to\infty}\left(2e^{0.5b}-2e^{2}\right)$

Since $\lim_{b\to\infty} 2e^{0.5b}$ does not exist, the integral is divergent.

47. $\int_0^{\infty} e^{-t/5}\,dt = \lim_{b\to\infty}\int_0^b e^{-t/5}\,dt = \lim_{b\to\infty} -5e^{-t/5}\Big|_0^b$
$= \lim_{b\to\infty}\left[-5e^{-b/5} - \left(-5e^{0}\right)\right] = 0 + 5 = 5$

48. $\int_{100}^{\infty} e^{-t/10}\,dt = \lim_{b\to\infty}\int_{100}^b e^{-t/10}\,dt = \lim_{b\to\infty} -10e^{-t/10}\Big|_{100}^b$
$= \lim_{b\to\infty}\left[-10e^{-b/10} - \left(-10e^{-10}\right)\right] = 10e^{-10}$

49. $\int_0^{\infty}\frac{x^3}{\left(x^4+1\right)^2}\,dx = \lim_{b\to\infty}\int_0^b\frac{x^3}{\left(x^4+1\right)^2}\,dx = \lim_{b\to\infty}\frac{1}{4}\int_0^b\frac{4x^3}{\left(x^4+1\right)^2}\,dx$
$= \lim_{b\to\infty} -\frac{1}{4}\left(x^4+1\right)^{-1}\Big|_0^b = \lim_{b\to\infty} -\frac{1}{4}\left(\frac{1}{b^4+1} - \frac{1}{0+1}\right) = -\frac{1}{4}(0-1) = \frac{1}{4}$

50. $\int_0^{\infty}\frac{x^4}{\left(x^5+1\right)^2}\,dx = \lim_{b\to\infty}\int_0^b\frac{x^4}{\left(x^5+1\right)^2}\,dx = \lim_{b\to\infty}\frac{1}{5}\int_0^b\frac{5x^4}{\left(x^5+1\right)^2}\,dx$
$= \lim_{b\to\infty} -\frac{1}{5}\left(x^5+1\right)^{-1}\Big|_0^b = \lim_{b\to\infty}\left\{-\frac{1}{5}\left(b^5+1\right)^{-1} - \left[-\frac{1}{5}(0+1)^{-1}\right]\right\} = 0 + \frac{1}{5}(1) = \frac{1}{5}$

51. $\int_{-\infty}^{0} e^{2t}\,dt = \lim_{b\to-\infty}\int_b^0 e^{2t}\,dt = \lim_{b\to-\infty}\frac{1}{2}e^{2t}\Big|_b^0 = \lim_{b\to-\infty}\left(\frac{1}{2}e^{0} - \frac{1}{2}e^{2b}\right) = \frac{1}{2} - 0 = \frac{1}{2}$

52. $\int_{-\infty}^{0} e^{4t}\,dt = \lim_{b\to-\infty}\int_b^0 e^{4t}\,dt = \lim_{b\to-\infty}\frac{1}{4}e^{4t}\Big|_b^0 = \lim_{b\to-\infty}\left(\frac{1}{4}e^{0} - \frac{1}{4}e^{4b}\right) = \frac{1}{4} - 0 = \frac{1}{4}$

53. $\int_{-\infty}^{4}\frac{1}{(5-x)^2}\,dx = \lim_{b\to-\infty}\int_b^4\frac{1}{(5-x)^2}\,dx = \lim_{b\to-\infty}(5-x)^{-1}\Big|_b^4$
$= \lim_{b\to-\infty}\left(\frac{1}{5-4} - \frac{1}{5-b}\right) = 1 - 0 = 1$

54. $$\int_{-\infty}^{8} \frac{1}{(9-x)^2}\,dx = \lim_{b\to-\infty} \int_{b}^{8} \frac{1}{(9-x)^2}\,dx = \lim_{b\to-\infty} (9-x)^{-1}\Big|_b^8$$
$$= \lim_{b\to-\infty} \left(\tfrac{1}{9-8} - \tfrac{1}{9-b}\right) = 1 - 0 = 1$$

55. $$\int_{-\infty}^{\infty} \frac{e^{-x}}{\left(1+e^{-x}\right)^4}\,dx = \lim_{a\to-\infty} \int_{a}^{0} \frac{e^{-x}}{\left(1+e^{-x}\right)^4}\,dx + \lim_{b\to\infty} \int_{0}^{b} \frac{e^{-x}}{\left(1+e^{-x}\right)^4}\,dx$$
$$= \lim_{a\to-\infty} \tfrac{1}{3}\left(1+e^{-x}\right)^{-3}\Big|_a^0 + \lim_{b\to\infty} \tfrac{1}{3}\left(1+e^{-x}\right)^{-3}\Big|_0^b$$
$$= \lim_{a\to-\infty} \left[\tfrac{1}{3}\left(1+e^{-0}\right)^{-3} - \tfrac{1}{3}\left(1+e^{-a}\right)^{-3}\right] + \lim_{b\to\infty} \left[\tfrac{1}{3}\left(1+e^{-b}\right)^{-3} - \tfrac{1}{3}\left(1+e^{-0}\right)^{-3}\right]$$
$$= \tfrac{1}{24} - 0 + \tfrac{1}{3} - \tfrac{1}{24} = \tfrac{1}{3}$$

56. $$\int_{-\infty}^{\infty} \frac{e^{-x}}{\left(1+e^{-x}\right)^3}\,dx = \lim_{a\to-\infty} \int_{a}^{0} \frac{e^{-x}}{\left(1+e^{-x}\right)^3}\,dx + \lim_{b\to\infty} \int_{0}^{b} \frac{e^{-x}}{\left(1+e^{-x}\right)^3}\,dx$$
$$= \lim_{a\to-\infty} \tfrac{1}{2}\left(1+e^{-x}\right)^{-2}\Big|_a^0 + \lim_{b\to\infty} \tfrac{1}{2}\left(1+e^{-x}\right)^{-2}\Big|_0^b$$
$$= \lim_{a\to-\infty} \left[\tfrac{1}{2}\left(1+e^{-0}\right)^{-2} - \tfrac{1}{2}\left(1+e^{-a}\right)^{-2}\right] + \lim_{b\to\infty} \left[\tfrac{1}{2}\left(1+e^{-b}\right)^{-2} - \tfrac{1}{2}\left(1+e^{-0}\right)^{-2}\right]$$
$$= \tfrac{1}{8} - 0 + \tfrac{1}{2} - \tfrac{1}{8} = \tfrac{1}{2}$$

57. $$\int_{0}^{\infty} 6000e^{-rt}\,dt = \int_{0}^{\infty} 6000e^{-0.1t}\,dt = \lim_{b\to\infty} \int_{0}^{b} 6000e^{-0.1t}\,dt$$
$$= \lim_{b\to\infty} -10(6000)e^{-0.1t}\Big|_0^b = \lim_{b\to\infty} -60{,}000\left(e^{-0.1b} - e^0\right)$$
$$= -60{,}000(0-1) = 60{,}000$$

The permanent endowment needed to generate an annual \$6000 forever at 10% continuously compounded interest is \$60,000.

58. The proportion of cars on the road more than 5 years old is
$$\int_{5}^{\infty} 0.21e^{-0.21t}\,dt = \lim_{b\to\infty} \int_{5}^{b} 0.21e^{-0.21t}\,dt$$
$$= \lim_{b\to\infty} -e^{-0.21t}\Big|_5^b = \lim_{b\to\infty} \left[-e^{-0.21b} - \left(-e^{-1.05}\right)\right]$$
$$= 0 + e^{-1.05} \approx 0.35 = 35\%$$

59. The total number of books sold is
$$\int_{0}^{\infty} 12e^{-0.05t}\,dt = \lim_{b\to\infty} \int_{0}^{b} 12e^{-0.05t}\,dt = \lim_{b\to\infty} -240e^{-0.05t}\Big|_0^b$$
$$= \lim_{b\to\infty} \left[-240e^{-0.05b} - \left(-240e^0\right)\right]$$
$$= 0 + 240 = 240 \text{ thousand}$$

The total number of books sold will be about 240,000.

60. The total amount of mercury used is
$$\int_{0}^{\infty} 270e^{-0.02t}\,dt = \lim_{b\to\infty} \int_{0}^{b} 270e^{-0.02t}\,dt$$
$$= \lim_{b\to\infty} -13{,}500e^{-0.02t}\Big|_0^b$$
$$= \lim_{b\to\infty} \left[-13{,}500e^{-0.02b} - \left(-13{,}500e^0\right)\right]$$
$$= 0 + 13{,}500 = 13{,}500 \text{ metric tons}$$

61. b.

b	$\int_1^b \frac{1}{x^3}\,dx$
1	0
10	0.495
100	0.49995
1000	0.4999995

The integral converges to 0.5.

c.

$$\int_1^\infty \frac{1}{x^3}\,dx = \lim_{b\to\infty}\int_1^b x^{-3}\,dx = \lim_{b\to\infty} -\tfrac{1}{2}x^{-2}\Big|_1^b$$
$$= \lim_{b\to\infty} -\tfrac{1}{2}(b)^{-2} - \left[-\tfrac{1}{2}(1)\right] = \tfrac{1}{2}$$

62. b.

b	$\int_1^b \frac{1}{\sqrt[3]{x}}\,dx$
1	0
10	5.424
100	30.817
1000	148.5

The integral diverges.

c.

$$\int_1^\infty \frac{1}{\sqrt[3]{x}}\,dx = \lim_{b\to\infty}\int_1^b x^{-1/3}\,dx = \lim_{b\to\infty} \tfrac{3}{2}x^{2/3}\Big|_1^b$$
$$= \lim_{b\to\infty}\left[\tfrac{3}{2}b^{2/3} - \tfrac{3}{2}(1)^{2/3}\right]$$

Since $\lim_{b\to\infty} \frac{3}{2}b^{2/3}$ does not exist, the integral diverges.

63. $\int_0^1 \sqrt{1+x^4}\,dx$, $n = 3$ $\Delta x = \frac{1-0}{3} = \frac{1}{3} = 0.333$

x	$f(x) = \sqrt{1+x^4}$
0	$1 \to 0.5$
0.333	1.006
0.667	1.094
1	$1.414 \to 0.707$

0.5 + 1.006 + 1.094 + 0.707 = 3.307

$$\int_0^1 \sqrt{1+x^4}\,dx \approx 3.307\left(\tfrac{1}{3}\right) \approx 1.102$$

64. $\int_0^1 \sqrt{1+x^5}\,dx$, $n = 3$ $\Delta x = \frac{1-0}{3} = \frac{1}{3} = 0.333$

x	$f(x) = \sqrt{1+x^5}$
0	$1 \to 0.5$
0.333	1.002
0.667	1.064
1	$1.414 \to 0.707$

0.5 + 1.002 + 1.064 + 0.707 = 3.273

$$\int_0^1 \sqrt{1+x^5}\,dx \approx 3.273\left(\tfrac{1}{3}\right) \approx 1.091$$

65. $\int_0^1 e^{(1/2)x^2}\,dx$, $n = 4$ $\Delta x = \frac{1-0}{4} = \frac{1}{4} = 0.25$

x	$f(x) = e^{(1/2)x^2}$
0	$1 \to 0.5$
0.25	1.032
0.50	1.133
0.75	1.325
1	$1.648 \to 0.824$

0.5 + 1.032 + 1.133 + 1.325 + 0.824 = 4.814

$$\int_0^1 e^{(1/2)x^2}\,dx \approx 4.814\left(\tfrac{1}{4}\right) \approx 1.204$$

66. $\int_0^1 e^{-(1/2)x^2}\,dx$, $n = 4$ $\Delta x = \frac{1-0}{4} = \frac{1}{4} = 0.25$

x	$f(x) = e^{(1/2)x^2}$
0	$1 \to 0.5$
0.25	0.969
0.50	0.882
0.75	0.755
1	$0.607 \to 0.304$

0.5 + 0.969 + 0.882 + 0.755 + 0.304 = 3.410

$$\int_0^1 e^{-(1/2)x^2}\,dx \approx 3.41\left(\tfrac{1}{4}\right) \approx 0.853$$

67. $\int_{-1}^{1} \ln(1+x^2)\,dx$, $n=4$ $\quad \Delta x = \frac{1-(-1)}{4} = \frac{1}{2} = 0.5$

x	$f(x)=\ln(1+x^2)$
–1	$0.693 \to 0.347$
–0.5	0.223
0	0
0.5	0.223
1	$0.693 \to 0.347$

0.347 + 0.223 + 0 + 0.223 + 0.347 = 1.140

$$\int_{-1}^{1} \ln(1+x^2)\,dx \approx (1.140)\left(\tfrac{1}{2}\right) = 0.570$$

68. $\int_{-1}^{1} \ln(x^3+2)\,dx$, $n=4$ $\quad \Delta x = \frac{1-(-1)}{4} = \frac{1}{2} = 0.5$

x	$f(x)=\ln(x^3+2)$
–1	$0 \to 0$
–0.5	0.629
0	0.693
0.5	0.754
1	$1.099 \to 0.550$

0 + 0.629 + 0.693 + 0.754 + 0.550 = 2.626

$$\int_{-1}^{1} \ln(x^3+2)\,dx \approx (2.626)\left(\tfrac{1}{2}\right) = 1.313$$

69.

n	$\int_0^1 \sqrt{1+x^4}\,dx$
10	1.0906
50	1.0895
100	1.0894

$$\int_0^1 \sqrt{1+x^4}\,dx \approx 1.089$$

70.

n	$\int_0^1 \sqrt{1+x^5}\,dx$
50	1.0747
100	1.0747

$$\int_0^1 \sqrt{1+x^5}\,dx \approx 1.075$$

71.

n	$\int_0^1 e^{(1/2)x^2}\,dx$
50	1.1950
100	1.1950

$$\int_0^1 e^{(1/2)x^2}\,dx \approx 1.195$$

72.

n	$\int_0^1 e^{-(1/2)x^2}\,dx$
50	0.8556
100	0.8556

$$\int_0^1 e^{-(1/2)x^2}\,dx \approx 0.856$$

73.

n	$\int_{-1}^{1} \ln(1+x^2)\,dx$
50	0.5282
100	0.5280
200	0.5279

$$\int_{-1}^{1} \ln(1+x^2)\,dx \approx 0.528$$

74.

n	$\int_{-1}^{1} \ln(x^3+2)\,dx$
50	1.3476
100	1.3478

$$\int_{-1}^{1} \ln(x^3+2)\,dx \approx 1.348$$

75. $\int_0^1 \sqrt{1+x^4}\,dx$, $n=4$ $\quad \Delta x = \frac{1-0}{4} = \frac{1}{4} = 0.25$

x	$f(x)$	Weight	$f(x)$ • weight
0	1	1	1
0.25	1.002	4	4.008
0.5	1.0308	2	2.0616
0.75	1.1473	4	4.5892
1	1.4142	1	1.4142

1 + 4.008 + 2.0616 + 4.5892 + 1.4142 = 13.0730

$$\int_0^1 \sqrt{1+x^4}\,dx \approx 13.0730\left(\tfrac{0.25}{3}\right) \approx 1.0894$$

76. $\int_0^1 \sqrt{1+x^5}\,dx$, $n=4$ $\quad \Delta x = \frac{1-0}{4} = \frac{1}{4} = 0.25$

x	$f(x)$	Weight	$f(x)$ • weight
0	1	1	1
0.25	1.0005	4	4.002
0.5	1.0155	2	2.0310
0.75	1.1123	4	4.4492
1	1.4142	1	1.4142

1 + 4.002 + 2.0310 + 4.4492 + 1.4142 = 12.8964

$$\int_0^1 \sqrt{1+x^5}\,dx \approx 12.8964\left(\tfrac{0.25}{3}\right) \approx 1.0747$$

77. $\int_0^1 e^{x^2/2}dx$, $n = 4$ $\Delta x = \frac{1-0}{4} = \frac{1}{4} = 0.25$

x	$f(x)$	Weight	$f(x) \cdot$ weight
0	1	1	1
0.25	1.0317	4	4.1268
0.5	1.1331	2	2.2662
0.75	1.3248	4	5.2992
1	1.6487	1	1.6487

$1 + 4.1268 + 2.2662 + 5.2992 + 1.6487 = 14.3409$

$$\int_0^1 e^{x^2/2}dx \approx 14.3409\left(\tfrac{0.25}{3}\right) \approx 1.1951$$

78. $\int_0^1 e^{-x^2/2}dx$, $n = 4$ $\Delta x = \frac{1-0}{4} = \frac{1}{4} = 0.25$

x	$f(x)$	Weight	$f(x) \cdot$ weight
0	1	1	1
0.25	0.9692	4	3.8768
0.5	0.8825	2	1.7650
0.75	0.7548	4	3.0192
1	0.6065	1	0.6065

$1 + 3.8768 + 1.765 + 3.0192 + 0.6065 = 10.2675$

$$\int_0^1 e^{-x^2/2}dx \approx 10.2675\left(\tfrac{0.25}{3}\right) \approx 0.8556$$

79. $\int_{-1}^1 \ln\left(1+x^2\right)dx$, $n = 4$ $\Delta x = \frac{1-(-1)}{4} = \frac{1}{2} = 0.5$

x	$f(x)$	Weight	$f(x) \cdot$ weight
−1	0.6931	1	0.6931
−0.5	0.2231	4	0.8924
0	0	2	0
0.5	0.2231	4	0.8924
1	0.6932	1	0.6931

$0.6931 + 0.8924 + 0 + 0.8924 + 0.6931 = 3.1710$

$$\int_{-1}^1 \ln\left(1+x^2\right)dx \approx 3.1710\left(\tfrac{0.5}{3}\right) \approx 0.5285$$

80. $\int_{-1}^1 \ln\left(x^3+2\right)dx$, $n = 4$ $\Delta x = \frac{1-(-1)}{4} = \frac{1}{2} = 0.5$

x	$f(x)$	Weight	$f(x) \cdot$ weight
−1	0	1	0
−0.5	0.6286	4	2.5144
0	0.6931	2	1.3862
0.5	0.7538	4	3.0152
1	1.0986	1	1.0986

$0 + 2.5144 + 1.3862 + 3.0152 + 1.0986 = 8.0144$

$$\int_{-1}^1 \ln\left(x^3+2\right)dx \approx 8.0144\left(\tfrac{0.5}{3}\right) \approx 1.3357$$

81.

n	$\int_0^1 \sqrt{1+x^4}\,dx$
200	1.0894294
300	1.0894294

$$\int_0^1 \sqrt{1+x^4}\,dx \approx 1.089429$$

82.

n	$\int_0^1 \sqrt{1+x^5}\,dx$
200	1.0746692
300	1.0746692

$$\int_0^1 \sqrt{1+x^5}\,dx \approx 1.074669$$

83.

n	$\int_0^1 e^{x^2/2}dx$
100	1.1949577
200	1.1949577

$$\int_0^1 e^{x^2/2}dx \approx 1.194958$$

84.

n	$\int_0^1 e^{-x^2/2}dx$
100	0.8556244
200	0.8556244

$$\int_0^1 e^{-x^2/2}dx \approx 0.855624$$

85.

n	$\int_{-1}^1 \ln\left(1+x^2\right)dx$
100	0.5278870
200	0.5278870

$$\int_{-1}^1 \ln\left(1+x^2\right)dx \approx 0.527887$$

86.

n	$\int_{-1}^1 \ln\left(x^3+2\right)dx$
100	1.3478553
200	1.3478554

$$\int_{-1}^1 \ln\left(x^3+2\right)dx \approx 1.347855$$

87. **a.** $\int_1^\infty \frac{1}{x^2+1}dx$ Let $x = \frac{1}{t} = t^{-1}$. Then $dx = -t^{-2}dt$. When $x = 1$, $t = 1$; when $x \to \infty$, $t \to 0$.

$$\int_1^\infty \frac{1}{x^2+1}dx = \int_1^0 \frac{1}{\left(\frac{1}{t}\right)^2+1}\left(\frac{-1}{t^2}\right)dt$$
$$= \int_1^0 \frac{1}{\frac{1+t^2}{t^2}}\left(\frac{-1}{t^2}\right)dt$$
$$= -\int_1^0 \frac{1}{1+t^2}dt = \int_0^1 \frac{1}{1+t^2}dt$$

b. $\Delta t = \frac{1-0}{4} = \frac{1}{4}$

t	$f(t) = \frac{1}{1+t^2}$
0	$1 \to 0.5$
0.25	0.941
0.50	0.8
0.75	0.64
1.0	$0.5 \to 0.25$

$0.5 + 0.941 + 0.8 + 0.64 + 0.25 = 3.131$

$$\int_0^1 \frac{1}{1+t^2}dt \approx 3.131(\Delta t) = 3.131\left(\tfrac{1}{4}\right) \approx 0.783$$

88. **a.** $\int_1^\infty \frac{x^2}{x^4+1}dx$ Let $x = \frac{1}{t} = t^{-1}$. Then $dx = -t^{-2}dt$. When $x = 1$, $t = 1$; when $x \to \infty$, $t \to 0$.

$$\int_1^\infty \frac{x^2}{x^4+1}dx = \int_1^0 \frac{\left(\frac{1}{t}\right)^2}{\left(\frac{1}{t}\right)^4+1}\left(-\frac{1}{t^2}dt\right)$$
$$= \int_1^0 \frac{1}{\frac{1}{t^4}+1}\left(-\frac{1}{t^4}\right)dt$$
$$= \int_0^1 \frac{1}{t^4+1}dt$$

b. $\Delta t = \frac{1-0}{4} = \frac{1}{4}$

t	$f(t) = \frac{1}{t^4+1}$
0	$1 \to 0.5$
0.25	0.996
0.50	0.941
0.75	0.760
1.0	$0.5 \to 0.25$

$0.5 + 0.996 + 0.941 + 0.760 + 0.25 = 3.447$

$$\int_0^1 \frac{1}{t^4+1}dt \approx 3.447(\Delta t) = 3.447\left(\tfrac{1}{4}\right) \approx 0.862$$

Attention *Brief Applied Calculus* users:

For the solutions to Review Exercises #89–115, please see Chapter 9 Review Exercises #1–27 on page 304 of this manual.

Chapter 7: Calculus of Several Variables

EXERCISES 7.1

1. The domain is $\{(x,y)|x \neq 0, y \neq 0\}$.

3. The domain is $\{(x,y)|x \neq y\}$.

5. Since the logarithm is defined only for positive values, the domain is $\{(x,y)|x > 0, y \neq 0\}$.

7. Since the logarithm is defined only for positive values, the domain is $\{(x,y,z)|x \neq 0, y \neq 0, z > 0\}$.

9. $f(3,-9) = \sqrt{99-(3)^2-(-9)^2} = \sqrt{99-9-81}$
$= \sqrt{9} = 3$

11. $g(0,e) = \ln\left[(0)^2 + (e)^4\right] = \ln e^4 = 4$

13. $w(-1,1) = \dfrac{1+2(-1)+3(1)}{(-1)(1)} = \dfrac{1-2+3}{-1} = -2$

15. $h(1,-2) = e^{(1)(-2)+(-2)^2-2} = e^{-2+4-2} = e^0 = 1$

17. $f(1,-1) = (1)e^{-1} - (-1)e^1 = e^{-1} + e$

19. $f(1,-1,1) = (1)e^{-1} + (-1)e^1 + 1e^1$
$= e^{-1} - e + e = e^{-1}$

21. $f(-1,-1,5) = 5\ln\sqrt{(-1)(-1)}$
$= 5\ln\sqrt{1} = 5\ln 1 = 0$

23. Yield $= Y(d,p) = \dfrac{d}{p}$. For $p = 140$ and $d = 2.2$,
$Y(2.2,140) = \dfrac{2.2}{140} \approx 0.0157$

25. $T(90,33) = \frac{33(90)}{33+33} = \frac{33}{66}(90) = \frac{1}{2}(90)$
$= 45$ minutes

27. $P(320,150) = 2(320)^{0.6}(150)^{0.4} \approx 472.7$

29. $P(400,\ 8100) = 1.5(400)^{1/2}(8100)^{1/2} = 2700$
$P(900,\ 3600) = 1.5(900)^{1/2}(3600)^{1/2} = 2700$
$P(3600,\ 900) = 1.5(3600)^{1/2}(900)^{1/2} = 2700$
$P(8100,\ 400) = 1.5(8100)^{1/2}(400)^{1/2} = 2700$
The production level, 2700 is the same for all four pairs.

31. $P(2L,2K) = a(2L)^b(2K)^{1-b} = a2^bL^b2^{1-b}K^{1-b}$
$= a2^b2^{1-b}L^bK^{1-b} = a2^{b+1-b}L^bK^{1-b}$
$= a2^1L^bK^{1-b} = 2\left(aL^bK^{1-b}\right) = 2P(L,K)$

33. If the cities have population 40,000 and 60,000 respectively, and if they are 600 miles apart,
$x = 40{,}000$,
$y = 60{,}000$, and $d = 600$.

$$f(40{,}000,\ 60{,}000,\ 600) = \frac{3(40{,}000)(60{,}000)}{600^2} = \frac{7{,}200{,}000{,}000}{600^2} = 20{,}000 \text{ calls}$$

35. Let x = the number of washers and y = the number of dryers. The cost of making x washers is $210x$, and the cost of making y dryers is $180y$. If fixed costs are \$4000, then
$C(x,\ y) = 210x + 180y + 4000$

37. **a.** Volume = length times width times height. Thus, $V = xyz$.

b. Bottom area $= xy$
2 side areas $= 2yz$
Front and back areas $= 2xz$
Total area $= xy + 2yz + 2xz$

39. **a.**

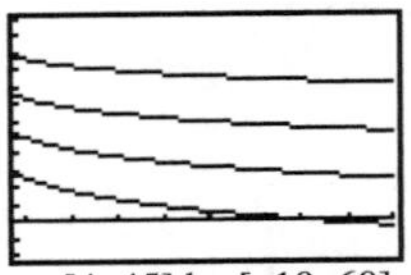

on [4, 45] by [–10, 60]

b. A given wind speed will lower the wind chill further on a cold day than on a warmer day.

c. For the lowest curve, *dy/dx* • –0.63 at $x = 10$, meaning that at 20 degrees and 10 mph of wind, wind chill drops by about 0.63 degrees for each additional 1 mph of wind.
For the highest curve, *dy/dx* • –0.33 at $x = 10$, meaning that at 50 degrees and 10 mph of wind, wind chill drops by only about 0.33 degrees for each additional 1 mph of wind.

d. Yes, the effect of wind on the wind chill index is greater on a colder day.

41. $P = \dfrac{(400)^2}{(400)^2 + (300)^2} = 0.64$ or 64%

43. b

45. There are five relative extreme points: three relative maximum points and two relative minimum points, one hidden behind the front-most peak.

47. **a.** has a relative minimum point

b. has a relative maximum point

49. False: The order is backwards.

51. a

53. b

EXERCISES 7.2

1. **a.** $f(x,y) = x^3 + 3x^2y^2 - 2y^3 - x + y$

$f_x(x,y) = 3x^2 + 2\cdot 3xy^2 - 1 = 3x^2 + 6xy^2 - 1$

b. $f_y(x,y) = 2\cdot 3x^2y - 3\cdot 2y^2 + 1$

$= 6x^2y - 6y^2 + 1$

3. $f(x,y) = 12x^{1/2}y^{1/3} + 8$

a. $f_x(x,y) = \frac{1}{2}(12)x^{-1/2}y^{1/3}$

$= 6x^{-1/2}y^{1/3}$

b. $f_y(x,y) = \frac{1}{3}(12)x^{1/2}y^{-2/3}$

$= 4x^{1/2}y^{-2/3}$

5. $f(x,y) = 100x^{0.05}y^{0.02}$

a. $f_x(x,y) = 0.05\cdot 100x^{-.95}y^{0.02}$

$= 5x^{-.95}y^{0.02}$

b. $f_y(x,y) = 0.02\cdot 100x^{0.05}y^{-0.98}$

$= 2x^{0.05}y^{-0.98}$

7. $f(x,y) = (x+y)^{-1}$

a. $f_x(x,y) = -(x+y)^{-2}$

b. $f_y(x,y) = -(x+y)^{-2}$

9. $f(x,y)=\ln(x^3+y^3)$

a. $f_x(x,y)=\dfrac{1}{x^3+y^3}(3x^2)$
$=\dfrac{3x^2}{x^3+y^3}$

b. $f_y(x,y)=\dfrac{1}{x^3+y^3}(3y^2)$
$=\dfrac{3y^2}{x^3+y^3}$

11. $f(x,y)=2x^3e^{-5y}$

a. $f_x(x,y)=3\cdot 2x^2e^{-5y}=6x^2e^{-5y}$

b. $f_y(x,y)=2x^3e^{-5y}(-5)=-10x^3e^{-5y}$

13. $f(x,y)=e^{xy}$

a. $f_x(x,y)=e^{xy}(y)=ye^{xy}$

b. $f_y(x,y)=e^{xy}(x)=xe^{xy}$

15. $f(x,y)=\ln\sqrt{x^2+y^2}$

a. $f_x(x,y)=\dfrac{1}{\sqrt{x^2+y^2}}\left[\dfrac{1}{2}(x^2+y^2)^{-1/2}(2x)\right]=\dfrac{x}{\sqrt{x^2+y^2}\cdot\sqrt{x^2+y^2}}=\dfrac{x}{x^2+y^2}$

b. $f_y(x,y)=\dfrac{1}{\sqrt{x^2+y^2}}\left[\dfrac{1}{2}(x^2+y^2)^{-1/2}(2y)\right]=\dfrac{y}{\sqrt{x^2+y^2}\cdot\sqrt{x^2+y^2}}=\dfrac{y}{x^2+y^2}$

17. $w=(uv-1)^3$

a. $\dfrac{\partial w}{\partial u}=3(uv-1)^2(v)=3v(uv-1)^2$

b. $\dfrac{\partial w}{\partial v}=3(uv-1)^2(u)=3u(uv-1)^2$

19. $w=e^{\frac{1}{2}(u^2-v^2)}$

a. $\dfrac{\partial w}{\partial u}=e^{\frac{1}{2}(u^2-v^2)}\left(\dfrac{1}{2}\cdot 2u\right)=ue^{\frac{1}{2}(u^2-v^2)}$

b. $\dfrac{\partial w}{\partial v}=e^{\frac{1}{2}(u^2-v^2)}\left[\dfrac{1}{2}(-2v)\right]=-ve^{\frac{1}{2}(u^2-v^2)}$

21. $f(x,y)=4x^3-3x^2y^2-2y^2$

$f_x(x,y)=12x^2-6xy^2$

$f_x(-1,1)=12(-1)^2-6(-1)(1)^2$
$=12+6=18$

$f_y(x,y)=-6x^2y-4y$

$f_y(-1,1)=-6(-1)^2(1)-4(1)$
$=-6-4=-10$

23. $f(x,y)=e^{x^2+y^2}$

$f_x(x,y)=e^{x^2+y^2}(2x)=2xe^{x^2+y^2}$

$f_x(0,1)=2(0)e^{(0)^2+(1)^2}=0$

$f_y(x,y)=e^{x^2+y^2}(2y)=2ye^{x^2+y^2}$

$f_y(0,1)=2(1)e^{(0)^2+(1)^2}=2e$

25. $h(x,y)=x^2y-\ln(x+y)$

$h_x(x,y)=2xy-\dfrac{1}{x+y}$

$h_x(1,1)=2(1)(1)-\dfrac{1}{1+1}$
$=2-\dfrac{1}{2}=\dfrac{3}{2}$

27. $f(x,y)=5x^3-2x^2y^3+3y^4$

$f_x(x,y)=15x^2-4xy^3$

$f_y(x,y)=-6x^2y^2+12y^3$

a. $f_{xx}(x,y)=30x-4y^3$

b. $f_{xy}(x,y)=-12xy^2$

c. $f_{yx}(x,y)=-12xy^2$

d. $f_{yy}(x,y)=-12x^2y+36y^2$

29. $f(x,y)=9x^{1/3}y^{2/3}-4xy^3$

$f_x(x,y)=9\cdot\frac{1}{3}x^{-2/3}y^{2/3}-4y^3=3x^{-2/3}y^{2/3}-4y^3$

$f_y(x,y)=\frac{2}{3}\cdot 9x^{1/3}y^{-1/3}-3\cdot 4xy^2$
$=6x^{1/3}y^{-1/3}-12xy^2$

a. $f_{xx}(x,y)=3\cdot-\frac{2}{3}x^{-5/3}y^{2/3}-0$
$=-2x^{-5/3}y^{2/3}$

b. $f_{xy}(x,y)=\frac{2}{3}\cdot 3x^{-2/3}y^{-1/3}-4\cdot 3y^2$
$=2x^{-2/3}y^{-1/3}-12y^2$

c. $f_{yx}(x,y)=6\cdot\frac{1}{3}x^{-2/3}y^{-1/3}-12y^2$
$=2x^{-2/3}y^{-1/3}-12y^2$

d. $f_{yy}(x,y)=-\frac{1}{3}\cdot 6x^{1/3}y^{-4/3}-2\cdot 12xy$
$=-2x^{1/3}y^{-4/3}-24xy$

31. $f(x,y)=ye^x-x\ln y$

$f_x(x,y)=ye^x-\ln y$

$f_y(x,y)=e^x-\frac{x}{y}$

a. $f_{xx}(x,y)=ye^x$

b. $f_{xy}(x,y)=e^x-\frac{1}{y}$

c. $f_{yx}(x,y)=e^x-\frac{1}{y}$

d. $f_{yy}(x,y)=\frac{x}{y^2}$

33. $f(x,y)=x^4y^3-e^{2x}$

$f_x(x,y)=4x^3y^3-2e^{2x}$

$f_y(x,y)=3x^4y^2-0=3x^4y^2$

a. $f_{xx}(x,y)=4\cdot 3x^2y^3-2\cdot 2e^{2x}=12x^2y^3-4e^{2x}$
$f_{xxy}(x,y)=3\cdot 12x^2y^2-0=36x^2y^2$

b. $f_{xy}(x,y)=3\cdot 4x^3y^2-0=12x^3y^2$
$f_{xyx}(x,y)=12\cdot 3x^2y^2=36x^2y^2$

c. $f_{yx}(x,y)=3\cdot 4x^3y^2=12x^3y^2$
$f_{yxx}(x,y)=12\cdot 3x^2y^2=36x^2y^2$

35. $f=xy^2z^3$

a. $f_x=y^2z^3$

b. $f_y=2xyz^3$

c. $f_z=3xy^2z^2$

37. $f=\left(x^2+y^2+z^2\right)^4$

a. $f_x=4\left(x^2+y^2+z^2\right)^3(2x)=8x\left(x^2+y^2+z^2\right)^3$

b. $f_y=4\left(x^2+y^2+z^2\right)^3(2y)=8y\left(x^2+y^2+z^2\right)^3$

c. $f_z=4\left(x^2+y^2+z^2\right)^3(2z)=8z\left(x^2+y^2+z^2\right)^3$

39. $f=e^{x^2+y^2-z^2}$

a. $f_x=e^{x^2+y^2-z^2}(2x)=2xe^{x^2+y^2-z^2}$

b. $f_y=e^{x^2+y^2-z^2}(2y)=2ye^{x^2+y^2-z^2}$

c. $f_z=e^{x^2+y^2-z^2}(-2z)=-2ze^{x^2+y^2-z^2}$

41. $f=3x^2y-2xz^2$

$f_x(x,y,z)=3\cdot 2xy-2z^2=6xy-2z^2$

So $f_x(2,-1,1)=6(2)(-1)-2(1)^2=-12-2=-14$

43. $f=e^{x^2+2y^2+3z^2}$

$f_y=4ye^{x^2+2y^2+3z^2}$

$f_y(-1,1,-1)=4(1)e^{(-1)^2+2(1)^2+3(-1)^2}$
$=4e^{1+2+3}=4e^6$

45. $P(x,y)=2x^2-3xy+3y^2+150x+75y+200$ (in dollars)

a. The marginal profit function for DVD is

$P_x(x,y)=2\cdot 2x-3y+0+150+0+0=4x-3y+150$

b. $P_x(200,300)=4(200)-3(300)+150=800-900+150=50$

Profit increases by about \$50 per additional DVD when 200 DVD and 300 CD players are produced each day.

c. The marginal profit function for CD players is

$P_y(x,y)=0-3x+2\cdot 3y+0+75+0=-3x+6y+75$

d. $P_y(200,100)=-3(200)+6(100)+75=-600+600+75=75$

Profit increases by about \$75 per additional CD player when 200 DVD and 100 CD players are produced each day.

47. $P(L,K)=270L^{1/3}K^{2/3}$

a. $P_L(L,K)=90L^{-2/3}K^{2/3}=90\left(\frac{K}{L}\right)^{2/3}$

$P_L(27,125)=90\left(\frac{27}{125}\right)^{2/3}=90\left(\frac{5}{3}\right)^2=90\left(\frac{25}{9}\right)=250$

Production increases by 250 for each additional unit of labor.

b. $P_K(L,K)=180L^{1/3}K^{-1/3}=180\left(\frac{L}{K}\right)^{1/3}$

$P_K(27,125)=180\left(\frac{27}{125}\right)^{1/3}=180\left(\frac{3}{5}\right)=108$

Production increases by 108 for each additional unit of capital.

c. An additional unit of labor increases production more.

49. $S(x,y)=200-0.1x+0.2y^2$

$S_x=0-0.1+0=-0.1$

Sales decrease by 0.1 unit for each dollar increase in the price of televisions.

$S_y=0-0+2\cdot 0.2y=0.4y$

Sales increase by $0.4y$ units for each y dollars spent on advertising.

51. **a.** $S(x,y)=7x^{1/3}y^{1/2}$

$S_x(x,y)=\frac{7}{3}x^{-2/3}y^{1/2}$

$S_x(27,4)=\frac{7}{3}(27)^{-2/3}(4)^{1/2}=\frac{7}{3\cdot 9}(2)\approx 0.52$

Status increases by about 0.52 unit for each additional \$1000 of income.

b. $S_y(x,y)=\frac{7}{2}x^{1/3}y^{-1/2}$

$S_y(27,4)=\frac{7}{2}(27)^{1/3}(4)^{-1/2}=\frac{7}{2}(3)\left(\frac{1}{2}\right)=5.25$

Status increases by 5.25 units for each additional year of education.

53. $S(w,v)=0.027wv^2$

a. $S_w(4,60)=0.027v^2\Big|_{(4,60)}=0.027(60)^2=97.2$

For a 4-ton truck traveling at 60 miles per hour, increasing the weight by 1 ton will increase the skid length by 97.2 units.

b. $S_v(4,60)=2(0.027)wv\Big|_{(4,60)}=0.054(4)(60)=12.96$

For a 4-ton truck traveling at 60 miles per hour, increasing the speed by 1 mile per hour will increase the skid length by about 13 units.

55. $f(x,y)=\dfrac{x^2}{x^2+y^2}$

a. $f_x=\dfrac{(x^2+y^2)(2x)-x^2(2x)}{(x^2+y^2)^2}=\dfrac{2xy^2}{(x^2+y^2)^2}$

$f_x(400,\ 300)=0.001152$

b. $0.001152(20)=3.6864$

So about 4 more games.

57. 0

59. $\dfrac{\partial f}{\partial x}=3$, $\dfrac{\partial f}{\partial y}=-2$

61. $\dfrac{\partial f}{\partial x}$ is negative, $\dfrac{\partial f}{\partial y}$ is positive.

63. $f(x,y)=-x+y$

65. $f(x,y)=\pi x+ey+\ln 2$

67. $D=\dfrac{E\cdot V}{T\cdot P}$, $\dfrac{\partial D}{\partial E}>0$, $\dfrac{\partial D}{\partial T}<0$

Your desire to complete the task should increase with the expectation of success and decrease with the time needed to complete the task.

69. a. $B_b(b,m)$ is the rate at which the sales of butter change for each price increase of butter.

b. $B_b(b,m)$ is negative because sales will decrease as the price of butter increases.

c. $B_m(b,m)$ is the rate at which the sales of butter change for each price increase of margarine.

d. $B_m(b,m)$ is positive because sales of butter will increase as the price of margarine increases.

71. $f_{xx}+f_{yy}=2-2=0$, harmonic

73. $f_{xx}+f_{yy}=2+2=4\neq 0$, not harmonic

75. Taking the limit as $h\to 0$:

$\dfrac{\partial f}{\partial x}=4x-3$, $\dfrac{\partial f}{\partial y}=6y+7$

77. Taking the limit as $h\to 0$:

$\dfrac{\partial f}{\partial x}=10x-3$, $\dfrac{\partial f}{\partial y}=7$

EXERCISES 7.3

1. $f(x,y)=x^2+2y^2+2xy+2x+4y+7$

$f_x(x,y)=2x+0+2y+2+0+0=2x+2y+2=0$

$f_y(x,y)=0+4y+2x+0+4+0=2x+4y+4=0$

Solving these equations simultaneously gives $x=0$, $y=-1$.

$f_{xx}=2$, $f_{yy}=4$, and $f_{xy}=2$.

$D=f_{xx}(a,b)\cdot f_{yy}(a,b)-[f_{xy}(a,b)]^2=(2)(4)-2^2$
$=4>0$

and $f_{xx}>0$. Therefore f has a relative minimum at $(0,-1)$. This minimum is

$0^2+2(-1)^2+2(0)(-1)+2(0)+4(-1)+7$
$=2-4+7=5$

3. $f(x,y)=2x^2+3y^2+2xy+4x-8y$

$f_x(x,y)=4x+2y+4=0$

$f_y(x,y)=6y+2x-8=0$

Solving this system of equations gives $x=-2$, $y=2$.

$f_{xx}=4$, $f_{yy}=6$, and $f_{xy}=2$. Thus,

$D=4\cdot 6-(2)^2=24-4=20>0$

and $f_{xx}>0$. Therefore f has a relative minimum at $(-2,2)$ and

$f(-2,2)$
$=2(-2)^2+3(2)^2+2(-2)(2)+4(-2)-8(2)$
$=8+12-8-8-16=-12$

5. $f(x,y)=3xy-2x^2-2y^2+14x-7y-5$
$f_x(x,y)=3y-4x-0+14-0-0$
$=-4x+3y+14=0$
$f_y(x,y)=3x-0-4y+0-7-0=3x-4y-7=0$
Solving these equations simultaneously gives $x=5$, $y=2$.
$f_{xx}=-4, f_{yy}=-4$, and $f_{xy}=3$. Thus,
$D=(-4)(-4)-3^2=7>0$
and $f_{xx}<0$. Therefore f has a relative maximum at $(5,2)$. This maximum is
$3(5)(2)-2(5)^2-2(2)^2+14(5)-7(2)-5=23$

7. $f(x,y)=xy+4x-2y+1$
$f_x(x,y)=y+4=0 \quad y=-4$
$f_y(x,y)=x-2=0 \quad x=2$
$f_{xx}=0, f_{yy}=0$, and $f_{xy}=1$. Thus,
$D=(0)(0)-(1)^2=-1<0$
Since $D<0$, f has a saddle point at $(2,-4)$. There are no relative extrema.

9. $f(x,y)=3x-2y-6$
$f_x=3$ and $f_y=-2$, so there are no relative extrema, since $f_x\neq 0$ and $f_y\neq 0$. Note that this is the equation of a plane.

11. $f(x,y)=e^{\frac{1}{2}(x^2+y^2)}$
$f_x(x,y)=xe^{\frac{1}{2}(x^2+y^2)}=0$
$f_y(x,y)=ye^{\frac{1}{2}(x^2+y^2)}=0$,
Only $(0,0)$ is a solution of these equations.
$f_{xx}=xe^{\frac{1}{2}(x^2+y^2)}(x)+e^{\frac{1}{2}(x^2+y^2)}=x^2e^{\frac{1}{2}(x^2+y^2)}+e^{\frac{1}{2}(x^2+y^2)} \qquad f_{xx}(0,0)=(0)^2e^0+1=1$
$f_{yy}=ye^{\frac{1}{2}(x^2+y^2)}(y)+e^{\frac{1}{2}(x^2+y^2)}=y^2e^{\frac{1}{2}(x^2+y^2)}+e^{\frac{1}{2}(x^2+y^2)} \qquad f_{yy}(0,0)=(0)^2e^0+1=1$
$f_{xy}=xe^{\frac{1}{2}(x^2+y^2)}(y)=xye^{\frac{1}{2}(x^2+y^2)} \qquad f_{xy}(0,0)=(0)(0)e^0=0$
$D=(1)(1)-(0)^2=1>0$ and $f_{xx}>0$.
Thus, f has a relative minimum at $(0,0)$ and
$f(0,0)=e^{\frac{1}{2}((0)^2+(0)^2)}=e^0=1$

13. $f(x,y)=\ln(x^2+y^2+1)$

$f_x=\frac{2x}{x^2+y^2+1}=0$

$f_y=\frac{2y}{x^2+y^2+1}=0$

The only solution to the equation is $x = 0$ and $y = 0$. At $(0,0)$, $f=\ln(0+0+1)=\ln 1=0$. This must be a relative minimum, since $x^2+y^2+1\ge 1$ for all x and y, and so $\ln(x^2+y^2+1)$ must be $= 0$ for all x and y.

15. $f(x,y)=-x^3-y^2+3x-2y$

$f_x(x,y)=-3x^2+3=0$

$-3(x^2-1)=0$ so $x=-1,1$

$f_y(x,y)=-2y-2=0$

$y=-1$

$f_{xx}(x,y)=-6x, f_{yy}(x,y)=-2, f_{xy}(x,y)=0.$

At $(-1,-1)$, $D=-6(-1)(-2)-(0)^2=-12$ so there is a saddle point.

At $(1,-1)$, $D=-6(1)(-2)-(0)^2=12$ and $f_{xx}(1,-1)=-6(1)=-6$ so $(1,-1)$ is a relative maximum and

$f(1,-1)=-(1)^3-(-1)^2+3(1)-2(-1)$
$=-1-1+3+2=3$

17. $f(x,y)=y^3-x^2-2x-12y$

$f_x=0-2x-2-0=-2x-2=0$

$f_y=3y^2-0-0-12=3y^2-12=0$

The solutions are $x=-1$ and $y=\pm 2$.

$f_{xx}=-2, f_{yy}=6y,$ and $f_{xy}=0$.

At $(-1, 2)$

$D=(-2)(6)(2)-0=-24<0$

so there is a saddle point. At $(-1,-2)$,

$D=(-2)(6)(-2)-0=24>0$

and $f_{xx}<0$, so there is a relative maximum of

$(-2)^3-(-1)^2-2(-1)-12(-2)=17$

19. $f(x,y)=x^3-2xy+4y$

$f_x(x,y)=3x^2-2y=0$

$f_y(x,y)=-2x+4=0$

$x=2$

Substituting $x = 2$ into the first equation gives

$3(2)^2-2y=0$

$2y=12$

$y=6$

$f_{xx}(x,y)=6x,\ f_{yy}(x,y)=0,\ f_{xy}(x,y)=-2.$

Thus

$D=6(2)(0)-(-2)^2=-4$ at $(2, 6)$

There is a saddle point at $(2, 6)$. There are no relative extreme values.

21. Product A: $p=12-\frac{1}{2}x \quad (x\le 24)$

Product B: $q=20-y \quad (y\le 20)$

$C(x,y)=9x+16y-xy+7$

The profit function for the two products x and y is

$$P(x,y)=(12-\tfrac{1}{2}x)x+(20-y)y-(9x+16y-xy+7)$$
$$=12x-\tfrac{1}{2}x^2+20y-y^2-9x-16y+xy-7$$
$$=-\tfrac{1}{2}x^2+xy-y^2+3x+4y-7$$

Profit will be maximized only when $P_x=P_y=0$.

$P_x=-x+y+3=0 \quad P_y=x-2y+4=0$

Solving these equations simultaneously gives $x = 10$ and $y = 7$; thus, to maximize profit, 10 units of x at $\left[12-\frac{1}{2}(10)\right]=7$ thousand dollars each and 7 units of y at $(20 - 7) = 13$ thousand dollars each should be produced.

The maximum profit is

$12(10)-\frac{1}{2}(100)+20(7)-49-9(10)-16(7)+(10)(7)-7=22$ thousand dollars

The D test will show that this is indeed a maximum.

23. **a.** America: $p = 20 - 0.2x \quad (x < 100)$
Europe: $q = 16 - 0.1y \quad (y \le 160)$
$C(x, y) = 20 + 4(x + y)$
The profit function for the company is

$$\begin{aligned} P(x,y) &= (20-0.2x)x + (16-0.1y)y - [20+4(x+y)] \\ &= 20x - 0.2x^2 + 16y - 0.1y^2 - 20 - 4x - 4y \\ &= 16x - 0.2x^2 + 12y - 0.1y^2 - 20 \end{aligned}$$

b. To maximize profit, maximize $P(x, y)$.

$$P_x = 16 - 0.4x = 0$$
$$x = 40$$
$$P_y = 12 - 0.2y = 0$$
$$y = 60$$

To maximize profit, 40 cars should be sold in America at 20 – 0.2(40) = 12 thousand dollars and 60 cars should be sold in Europe at 16 – 0.1(60) = 10 thousand dollars.
The D test shows that this is a relative maximum.

25. $$f(x,y) = xy - x^2 - y^2 + 11x - 4y + 120 \quad (x \le 10,\ y \le 4)$$
$$f_x(x,y) = y - 2x + 11 = 0$$
$$f_y(x,y) = x - 2y - 4 = 0$$

Solving the equations simultaneously gives $x = 6$ and $y = 1$; thus to maximize the score, the subject should practice for 6 hours and rest for 1 hour. The D test shows that $(x, y) = (6, 1)$ does maximize f.

27. **a.** To maximize revenue, find the maximum of the revenue function $R(x) = x \bullet p(x)$, where $p = 12 - 0.005x \ (x \le 2400)$.

$$R(x) = x(12 - 0.005x) = 12x - 0.005x^2$$
$$R'(x) = 12 - 0.005(2x) = 12 - 0.01x$$
$$x = \frac{12}{0.01} = 1200$$

$R''(x) = -0.01$; thus R has a maximum at 1200. The price is 12 – 0.005(1200) = \$6 at $x = 1200$, and the revenue = 1200(6) = \$7200.

b. If the first duopolist (d_1) sells x units and the second duopolist (d_2) sells y units, the price at which they must sell their products is
$p = 12 - 0.005(x + y) = 12 - 0.005x - 0.005y$
Now calculate the revenues of d_1 and d_2.

$$\begin{aligned} \text{Revenue of } d_1 &= x(12 - 0.005x - 0.005y) \\ &= 12x - 0.005x^2 - 0.005xy \end{aligned}$$
$$\begin{aligned} \text{Revenue of } d_2 &= y(12 - 0.005x - 0.005y) \\ &= 12y - 0.005xy - 0.005y^2 \end{aligned}$$

To maximize revenue, set the partial of the revenue of d_1 with respect to x equal to zero and set the partial of the revenue of d_2 with respect to y equal to zero.
$12 - 0.01x - 0.005y = 0$
$12 - 0.005x - 0.01y = 0$
Solving the equations simultaneously gives
$x = 800$ and $y = 800$. Each duopolist sells 800 units to maximize their revenues. The price at $x = 800$ is 12 – 0.005(800 + 800) = \$4 and

$$\text{Revenue of } d_1 = 12(800) - 0.005(800)^2 - 0.005(800)(800) = \$3200$$

$$\text{Revenue of } d_2 = 12(800) - 0.005(800)(800) - 0.005(800)^2 = \$3200$$

The D test shows that this is a maximum.

c. More goods are produced under a duopoly.

d. The price is lower under a duopoly.

29. Let x = the number of \$400 price decreases for sedans.
Let y = the number of \$600 price decreases for SUVs. Then the profit is

$$P(x,y) = \underbrace{\underbrace{\underbrace{(200-4x)}_{\text{price}}\underbrace{(8+2x)}_{\text{quantity}}}_{\text{for sedans}} + \underbrace{\underbrace{(250-6y)}_{\text{price}}\underbrace{(4+y)}_{\text{quantity}}}_{\text{for SUVs}}}_{\text{REVENUE}} - \underbrace{\underbrace{\underbrace{(200-4x)}_{\text{unit cost}}\underbrace{(8+2x)}_{\text{quantity}}}_{\text{for sedans}} + \underbrace{\underbrace{(250-6y)}_{\text{unit cost}}\underbrace{(4+y)}_{\text{quantity}}}_{\text{for SUVs}}}_{\text{COST}}$$

$$= 496 + 32x - 8x^2 + 36y - 6y^2$$

$$P_x = 32 - 16x$$

$$P_y = 36 - 12y$$

Set the partials equal to zero and solve to find the critical values: $\begin{aligned} 32-16x=0 \Rightarrow 16(2-x)=0 \Rightarrow x=2 \\ 36-12y=0 \Rightarrow 12(3-y)=0 \Rightarrow y=3 \end{aligned}$

These two values give one point. CP:(2,3).

For the D-test, calculate the second partials. $\begin{aligned} &P_{xx}=-16 \quad P_{xy}=0 \quad P_{yy}=-12 \\ &D=P_{xx}P_{yy}-(P_{xy})^2=(-16)(-12)-0^2=192>0 \end{aligned}$

D is positive and P_{xx} is negative so profit is indeed maximized at $x = 2$ and $y = 3$.
To find the prices, subtract the appropriate number of price decreases from the original price.
price of sedan = 20,000 – 2(400) = \$19,200 price of SUV = 25,000 – 3(600) = \$23,200
To find the quantity sold, add the additional sales to the original sales.
quantity of sedans sold = 8 + 2(2) = 12 per day quantity of SUVs sold= 4 + 3(1) = 7 per day
To maximize profit, the dealer should sell 12 sedans per day for \$19,200 each and 7 SUVs per day for \$23,200 each.

31. America: $p = 20 - 0.2x$ $(x \le 100)$
Europe: $q = 16 - 0.1y$ $(y \le 160)$
Asia: $r = 12 - 0.1z$ $(z \le 120)$

a. The revenue from America is $(20 - 0.2x)x$, that from Europe is $(16 - 0.1y)y$, and that from Asia is $(12 - 0.1z)z$; the cost function is $22 + 4(x + y + z)$, all in thousands of dollars. The profit function is thus

$$\begin{aligned} P(x,y,z) &= (20-0.2x)x + (16-0.1y)y + (12-0.1z)z - [22+4(x+y+z)] \\ &= 20x - 0.2x^2 + 16y - 0.1y^2 + 12z - 0.1z^2 - 22 - 4x - 4y - 4z \\ &= -0.2x^2 - 0.1y^2 - 0.1z^2 + 16x + 12y + 8z - 22 \end{aligned}$$

b. Profit will be maximized when $P_x = P_y = P_z = 0$:

$$P_x = -0.4x + 16 = 0 \qquad P_y = -0.2y + 12 = 0 \qquad P_z = -0.2z + 8 = 0$$
$$x = 40 \qquad y = 60 \qquad z = 40$$

Thus for maximum profit 40 cars should be sold in America, 60 in Europe, and 40 in Asia.

33. $f(x,y)=x^3+y^3-3xy$

$f_x(x,y)=3x^2-3y=0$

$f_y(x,y)=3y^2-3x=0$

$$x=\frac{3y^2}{3}=y^2$$

Substituting in the first equation,

$3\left(y^2\right)^2-3y=0$

$3y^4-3y=0$

$3y\left(y^3-1\right)=0$

$y=0,1$

At $y=0,\ x=0^2=0$. At $y=1,\ x=1^2=1$.

$f_{xx}=6x, f_{yy}=6y, f_{xy}=-3.$

At (0, 0), $D=6(0)\cdot 6(0)-(-3)^2=-9.$ Saddle point

At (1, 1), $D=6(1)\cdot 6(1)-(-3)^2=27.$ Since $D \bullet 0$ and $f_{xx}(1,1)=6>0,$ there is a relative minimum at (1, 1) and $f(1,1)=(1)^3+(1)^3-3(1)(1)=-1$

35. $f(x,y)=12xy-x^3-6y^2$

$f_x=12y-3x^2=0$

$f_y=12x-12y=0$

Solving the second equation gives $x=y$; substituting in the first equation gives $12x=3x^2$; $x=0$ or 4. Thus relative extrema will possibly be found at (0, 0) or (4, 4).

$f_{xx}=-6x, f_{yy}=-12, f_{xy}=12.$

At (0, 0), $D=-6(0)(-12)-12^2=-144.$ Saddle point. At (4, 4),

$D=-6(4)(-12)-12^2=144>0$, and $f_{xx}<0$, so there is a relative (and absolute) maximum at this point of $12(4)(4)-4^3-6(4)^2=32$.

37. $f(x,y)=2x^4+y^2-12xy$

$f_x(x,y)=8x^3-12y=0$

$f_y(x,y)=2y-12x=0$

$$x=\frac{1}{6}y$$

Substituting into the first equation gives

$8\left(\frac{1}{6}y\right)^3-12y=0$

$\frac{y^3}{27}-12y=0$

$y^3-324y=0$

$y\left(y^2-324\right)=0$

$y=0,\ 18,\ -18$

At $y=0,\ x=\frac{1}{6}(0)=0.$ At $y=18,\ x=\frac{1}{6}(18)=3$. At $y=-18,\ x=\frac{1}{6}(-18)=-3$.

$f_{xx}=24x^2, f_{yy}=2, f_{xy}=-12$.

At (0, 0), $D=24(0)^2(2)-(-12)^2=-144.$ Saddle point.

At (3, 18), $D=24(3)^2(2)-(-12)^2=288.$ Since $f_{xx}(3,18)=24(3)^2=432$, (3, 18) is a relative minimum.

At (–3, –18), $D=24(-3)^2(2)-(-12)^2=288.$ Since $f_{xx}(-3,-18)=24(-3)^2=432$, (–3, –18) is a relative minimum.

$f(3,18)=2(3)^4+(18)^2-12(3)(18)=-162$

$f(-3,-18)=2(-3)^4+(-18)^2-12(-3)(-18)=-162$

39. Saddle point

41. True

43. The function has a relative maximum or minimum at the critical point.

45. Very little. Depending on the sign of D, there may be a relative minimum point or a saddle point.

47. False: No conclusion can be drawn. For example, at (0, 0), $x^4 + y^4$ has a relative minimum, $-x^4 - y^4$ has a relative maximum, and $x^4 - y^4$ has a saddle point, and in each case $D = 0$.

48. True: $0 > f_{xx} f_{yy} \geq f_{xx} f_{yy} - (f_{xy}) = D$

EXERCISES 7.4

1.

x	y	xy	x^2
1	2	2	1
2	5	10	4
3	9	27	9
$\Sigma x = 6$	$\Sigma y = 16$	$\Sigma xy = 39$	$\Sigma x^2 = 14$

$$a = \frac{n\Sigma xy - (\Sigma x)(\Sigma y)}{n\Sigma x^2 - (\Sigma x)^2} = \frac{3(39)-(6)(16)}{3(14)-(6)^2} = 3.5$$

$$b = \frac{1}{n}(\Sigma y - a\Sigma x) = \frac{1}{3}[16-(3.5)(6)] \approx -1.67$$

The least squares line is $y = 3.5x - 1.67$.

3.

x	y	xy	x^2
1	6	6	1
3	4	12	9
6	2	12	36
$\Sigma x =$ 10	$\Sigma y = 12$	$\Sigma xy = 30$	$\Sigma x^2 = 46$

$$a = \frac{n\Sigma xy - (\Sigma x)(\Sigma y)}{n\Sigma x^2 - (\Sigma x)^2} = \frac{3(30)-(10)(12)}{3(46)-(10)^2} \approx -0.79$$

$$b = \frac{1}{n}(\Sigma y - a\Sigma x) = \frac{1}{3}[12+0.79(10)] \approx 6.63$$

The least squares line is $y = -0.79x + 6.63$.

5.

x	y	xy	x^2
0	7	0	0
1	10	10	1
2	10	20	4
3	15	45	9
$\Sigma x = 6$	$\Sigma y = 42$	$\Sigma xy = 75$	$\Sigma x^2 = 14$

$$a = \frac{n\Sigma xy - (\Sigma x)(\Sigma y)}{n\Sigma x^2 - (\Sigma x)^2} = \frac{4(75)-(6)(42)}{4(14)-(6)^2} = 2.4$$

$$b = \frac{1}{n}(\Sigma y - a\Sigma x) = \frac{1}{4}[42-(2.4)(6)] = 6.9$$

The least squares line is $y = 2.4x + 6.9$.

7.

x	y	xy	x^2
−1	10	−10	1
0	8	0	0
1	5	5	1
3	0	0	9
5	−2	−10	25
$\Sigma x = 8$	$\Sigma y = 21$	$\Sigma xy = -15$	$\Sigma x^2 = 36$

$$a = \frac{n\Sigma xy - (\Sigma x)(\Sigma y)}{n\Sigma x^2 - (\Sigma x)^2} = \frac{5(-15)-(8)(21)}{5(36)-(8)^2} \approx -2.1$$

$$b = \frac{1}{n}(\Sigma y - a\Sigma x) = \frac{1}{5}[21+(2.1)(8)] \approx 7.6$$

The least squares line is $y = -2.1x + 7.6$.

9. Let x = year and let y = sales (in millions).

x	y	xy	x^2
1	7	7	1
2	10	20	4
3	11	33	9
4	14	56	16
$\Sigma x = 10$	$\Sigma y = 42$	$\Sigma xy = 116$	$\Sigma x^2 = 30$

$$a = \frac{n\Sigma xy - (\Sigma x)(\Sigma y)}{n\Sigma x^2 - (\Sigma x)^2} = \frac{4(116)-(10)(42)}{4(30)-(10)^2} = 2.2$$

$$b = \frac{1}{n}(\Sigma y - a\Sigma x) = \frac{1}{4}[42-(2.2)(10)] = 5$$

The least squares line is $y = 2.2x + 5$. There will be about $(2.2)(5) + 5 = 16$ million sales in the fifth year.

11. Let x = year and let y = number of arrests.

x	y	xy	x^2
1	120	120	1
2	110	220	4
3	90	270	9
4	100	400	16
$\Sigma x = 10$	$\Sigma y = 420$	$\Sigma xy = 1010$	$\Sigma x^2 = 30$

$$a = \frac{n\Sigma xy - (\Sigma x)(\Sigma y)}{n\Sigma x^2 - (\Sigma x)^2} = \frac{4(1010)-(10)(420)}{4(30)-(10)^2} = -8$$

$$b = \frac{1}{n}(\Sigma y - a\Sigma x) = \frac{1}{4}[420+8(10)] = 125$$

The least squares line is $y = -8x + 125$.
For year 5, $x = 5$ and $y = -8(5) + 125 = 85$ arrests

13. Let x = the time period (a span of 20 years) and y = the high average.

x	y		x
1	.422		1
2	.424		2
3	.406		3
4	.390		4
5	.394		5
$\Sigma x = 15$	$\Sigma y = 2.036$		$\Sigma x = 15$

$$a = \frac{n\Sigma xy - (\Sigma x)(\Sigma y)}{n\Sigma x^2 - (\Sigma x)^2} = \frac{5(6.018) - (15)(2.036)}{5(55) - (15)^2} = -0.009$$

$$b = \tfrac{1}{n}(\Sigma y - a\Sigma x) = \tfrac{1}{5}[2.036 - (-0.009)(15)] = 0.4342$$

The least squares line is $y = -0.009x + 0.4342$.
For $x = 6$ (the period 2001–2010), the high average should be about $-0.009(6) + 0.4342 \approx .380$.

15. Let y = percent of males who smoke

x	y	xy	x^2
1	32	32	1
2	28	56	4
3	27	81	9
4	25	100	16
5	23	115	25
$\Sigma x = 15$	$\Sigma y = 135$	$\Sigma xy = 384$	$\Sigma x^2 = 55$

$$a = \frac{n\Sigma xy - (\Sigma x)(\Sigma y)}{n\Sigma x^2 - (\Sigma x)^2} = \frac{5(364) - (15)(126)}{5(55) - (15)^2} = -1.4$$

$$b = \tfrac{1}{n}(\Sigma y - a\Sigma x) = \tfrac{1}{5}[126 - (-1.4)(15)] = 29.4$$

The least squares line is $y = -1.4x + 29.4$.
In the year 2020 ($x = 7$), $-1.4(7) + 29.4 = 19.6\%$ or about 20% of males will smoke.

17. Let x be the number of cigarettes smoked daily and y the resultant life expectancy.

x	y	xy	x^2
0	73.6	0	0
5	69.0	345.0	25
15	68.1	1021.5	225
30	67.4	2022.0	900
40	65.3	2612.0	1600
$\Sigma x = 90$	$\Sigma y = 343.4$	$\Sigma xy = 6000.5$	$\Sigma x^2 = 2750$

$$a = \frac{n\Sigma xy - (\Sigma x)(\Sigma y)}{n\Sigma x^2 - (\Sigma x)^2} = \frac{5(6000.5) - (90)(343.4)}{5(2750) - (90)^2} \approx -0.160$$

$$b = \tfrac{1}{n}(\Sigma y - a\Sigma x) = \tfrac{1}{5}[343.4 + (0.160)(90)] \approx 71.6$$

The least squares line is $y = -0.16x + 71.6$.
For each cigarette smoked per day, life expectancy is decreased by about 0.16 year, or a little over 8 weeks.

19.

x	y	$Y = \ln y$	xY	x^2
1	2	0.69	0.69	1
2	4	1.39	2.78	4
3	7	1.95	5.85	9
$\Sigma x = 6$		$\Sigma Y = 4.03$	$\Sigma xY = 9.32$	$\Sigma x^2 = 14$

$$A = \frac{n\Sigma xY - (\Sigma x)(\Sigma Y)}{n\Sigma x^2 - (\Sigma x)^2} = \frac{3(9.32) - (6)(4.03)}{3(14) - (6)^2} \approx 0.63$$

$$b = \tfrac{1}{n}(\Sigma Y - A\Sigma x) = \tfrac{1}{3}[4.03 - (0.63)(6)] \approx 0.08$$

$B = e^b = e^{0.08} \approx 1.08$, and the least squares curve is $y = 1.08e^{0.63x}$.

21.

x	y	$Y = \ln y$	xY	x^2
1	10	2.30	2.30	1
3	5	1.61	4.83	9
6	1	0	0	36
$\Sigma x = 10$		$\Sigma Y = 3.91$	$\Sigma xY = 7.13$	$\Sigma x^2 = 46$

$$A = \frac{n\Sigma xY - (\Sigma x)(\Sigma Y)}{n\Sigma x^2 - (\Sigma x)^2} = \frac{3(7.13) - (10)(3.91)}{3(46) - (10)^2} \approx -0.466$$

$$b = \tfrac{1}{n}(\Sigma Y - A\Sigma x) = \tfrac{1}{3}[3.91 + (0.466)(10)] \approx 2.86$$

$B = e^b = e^{2.86} \approx 17.46$, and the least squares curve is $y = 17.46e^{-0.47x}$.

23.

x	y	$Y = \ln y$	xY	x^2
0	1	0	0	0
1	2	0.69	0.69	1
2	5	1.61	3.22	4
3	10	2.30	6.90	9
$\Sigma x = 6$		$\Sigma Y = 4.60$	$\Sigma xY = 10.81$	$\Sigma x^2 = 14$

$$A = \frac{n\Sigma xY - (\Sigma x)(\Sigma Y)}{n\Sigma x^2 - (\Sigma x)^2} = \frac{4(10.81) - (6)(4.6)}{4(14) - (6)^2} \approx 0.78$$

$$b = \tfrac{1}{n}(\Sigma Y - A\Sigma x) = \tfrac{1}{3}[4.6 - (0.78)(6)] = -0.02$$

$B = e^b = e^{-0.02} \approx 0.98$, and the least squares curve is $y = 0.98e^{0.78x}$.

25.

x	y	$Y = \ln y$	xY	x^2
−1	20	3.00	−3.00	1
0	18	2.89	0	0
1	15	2.71	2.71	1
3	4	1.39	4.17	9
5	1	0	0	25
$\Sigma x = 8$		$\Sigma Y = 9.99$	$\Sigma xY = 3.88$	$\Sigma x^2 = 36$

$$A = \frac{n\Sigma xY - (\Sigma x)(\Sigma Y)}{n\Sigma x^2 - (\Sigma x)^2} = \frac{5(3.88) - (8)(9.99)}{5(36) - (8)^2} \approx -0.521$$

$$b = \tfrac{1}{n}(\Sigma Y - A\Sigma x) = \tfrac{1}{5}\left[9.99 + (0.521)(8)\right] = 2.83$$

$B = e^b = e^{2.83} \approx 16.95$, and the least squares curve is $y = 16.95e^{-0.52x}$.

27. Let y = cost in millions of dollars

x	y	$Y = \ln y$	xY	x^2
1	1.373	0.3170	0.3170	1
2	1.440	0.3646	0.7293	4
3	1.567	0.4492	1.3475	9
$\Sigma x = 6$		$\Sigma Y = 1.1308$	$\Sigma xY = 2.3938$	$\Sigma x^2 = 14$

$$A = \frac{n\Sigma xY - (\Sigma x)(\Sigma Y)}{n\Sigma x^2 - (\Sigma x)^2} = \frac{3(2.3938) - 6(1.1308)}{3(14) - 6^2} \approx 0.066$$

$$b = \frac{1}{n}(\Sigma Y - A\Sigma x) = \frac{1}{3}\left[1.1308 - (0.066)(6)\right] \approx 0.245$$

$B = e^b = e^{0.245} \approx 1.28$. The least squares curve is $y = 1.28e^{0.066x}$.

In the year 2020 ($x = 7$), $y = 1.28e^{0.066(7)} \approx \$2.03$ million.

29. Let y = cost of a Super Bowl ticket.

x	y	$Y = \ln y$	xY	x^2
1	175	5.1647	5.1647	1
2	500	6.2146	12.4292	4
3	2600	7.8633	23.5898	9
$\Sigma x = 6$		$\Sigma Y =$ 19.2426	$\Sigma xY =$ 41.1837	$\Sigma x^2 = 14$

$$A = \frac{n\Sigma xY - (\Sigma x)(\Sigma Y)}{n\Sigma x^2 - (\Sigma x)^2} = \frac{3(41.1837) - (6)(19.2426)}{3(14) - (6)^2} \approx 1.349$$

$$b = \tfrac{1}{n}(\Sigma Y - A\Sigma x) = \tfrac{1}{3}[19.2426 - (1.349)(6)] \approx 3.716$$

$B = e^b = e^{3.716} \approx 41.1$, and the least squares curve is $y = 41.1e^{1.35x}$. In the year 2024 ($x = 4$), $y = 41.1e^{1.35(4)} \approx 9100$, the cost of a Super Bowl ticket will be about \$9,100.

31. Let y = the number of elderly.

x	y	$Y = \ln y$	xY	x^2
1	0.4	–0.9162	–0.9162	1
2	0.7	–0.3567	–0.7133	4
3	1.2	0.1823	0.5470	9
$\Sigma x = 6$		$\Sigma Y =$ –1.0906	$\Sigma xY =$ –1.0825	$\Sigma x^2 = 14$

$$A = \frac{n\Sigma xY-(\Sigma x)(\Sigma Y)}{n\Sigma x^2-(\Sigma x)^2} = \frac{3(-1.0825)-(6)(-1.0906)}{3(14)-(6)^2} \approx 0.549$$

$$b=\tfrac{1}{n}(\Sigma Y - A\Sigma x)=\tfrac{1}{3}[-1.0906-(0.549)(6)]\approx -1.462$$

$B = e^b = e^{-1.462} \approx 0.232$, and the least squares curve is $y = 0.232e^{0.549x}$. For $x = 4$, which is the year 2045, $y = 0.232e^{0.549(4)} \approx 2.085$.

Let y = number of children.

x	y	xy	x^2
1	1.6	1.6	1
2	1.8	3.6	4
3	1.9	5.7	9
$\Sigma x = 6$	$\Sigma y = 5.3$	$\Sigma xy = 10.9$	$\Sigma x^2 = 14$

$$a = \frac{n\Sigma xy-(\Sigma x)(\Sigma y)}{n\Sigma x^2-(\Sigma x)^2} = \frac{3(10.9)-(6)(5.3)}{3(14)-(6)^2} \approx 0.15$$

$$b = \tfrac{1}{n}(\Sigma y - a\Sigma x) = \tfrac{1}{3}[5.3-0.15(6)] \approx 1.47$$

The least squares line is $y = 0.15x + 1.47$. For $x = 4$, which is the year 2045, $y = 0.15(4) + 1.47 = 2.07$.

Since 2.085 > 2.07, the elderly population is larger.

33. False (see, for example, the graph on p 491)

35. True

37. $y = ax + b + 5$ (as can be verified using formulas on p492).

39. The formula gives $a = \frac{0}{0}$, which is undefined.

41. The third column requires the logarithm of that number, and log of 0 or negative number are *undefined*. Therefore, the y's must be positive.

EXERCISES 7.5

1. Maximize $f(x, y) = 3xy$ subject to $x + 3y = 12$.

$$F(x,y,\lambda) = 3xy + \lambda(x+3y-12)$$
$$F_x = 3y + \lambda = 0$$
$$F_y = 3x + 3\lambda = 0$$
$$F_\lambda = x + 3y - 12 = 0$$

From the first two equations, $\lambda = -3y = -x$; thus $x = 3y$. Substituting in the third equation,

$$3y + 3y - 12 = 0$$
$$y = 2$$
$$x = 3(2) = 6$$

The maximum constrained value of f occurs at (6, 2), and its value is 3(6)(2) = 36.

3. Maximize $f(x, y) = 6xy$ subject to $2x + 3y = 24$.

$$F(x,y,\lambda) = 6xy + \lambda(2x+3y-24)$$
$$F_x = 6y + 2\lambda = 0 \quad \text{or} \quad y = -\tfrac{1}{3}\lambda$$
$$F_y = 6x + 3\lambda = 0 \quad \text{or} \quad x = -\tfrac{1}{2}\lambda$$

Since $y = -\frac{1}{3}\lambda$ and $x = -\frac{1}{2}\lambda$,

$\frac{3}{2}y = -\frac{1}{3}\left(\frac{3}{2}\right)\lambda = -\frac{1}{2}\lambda = x$ Thus, $F_\lambda = 0$ gives

$$2x + 3y - 24 = 0$$
$$2\left(\tfrac{3}{2}y\right) + 3y - 24 = 0$$
$$6y = 24$$
$$y = 4$$
$$x = \tfrac{3}{2}(4) = 6$$

Thus, the maximum constrained value of f occurs at (6, 4), and its value is 6(6)(4) = 144.

5. Maximize $f(x,y)=xy-2x^2-y^2$ subject to $x+y=8$.

$F(x,y,\lambda)=xy-2x^2-y^2+\lambda(x+y-8)$
$F_x=y-4x+\lambda=0$
$F_y=x-2y+\lambda=0$
$F_\lambda=x+y-8=0$

The first two equations give $\lambda=4x-y=2y-x$. So, $5x=3y$, or $x=\frac{3}{5}y$. Substituting in the third equation gives

$\frac{3}{5}y+y=8$
$\frac{8}{5}y=8$
$y=5$
$x=\frac{3}{5}(5)=3$

Thus, the maximum constrained value of f occurs at (3, 5), and its value is

$(3)(5)-2(3)^2-(5)^2=-28$.

7. Maximize $f(x,y)=x^2-y^2+3$ subject to $2x+y=3$.

$F(x,y,\lambda)=x^2-y^2+3+\lambda(2x+y-3)$
$F_x=2x+2\lambda=0$ or $x=-\lambda$
$F_y=-2y+\lambda=0$ or $y=\frac{1}{2}\lambda$

Thus, $y=\frac{1}{2}\lambda=\frac{1}{2}(-x)=-\frac{1}{2}x$. $F_\lambda=0$ gives

$2x+y-3=0$
$2x+\left(-\frac{1}{2}x\right)-3=0$
$\frac{3}{2}x=3$
$x=2$
$y=-\frac{1}{2}(2)=-1$

Thus, the maximum constrained value of f occurs at (2, –1), and its value is

$(2)^2-(-1)^2+3=6$.

9. Maximize $f(x, y) = \ln(xy)$ subject to $x + y = 2e$.

$F(x,y,\lambda)=\ln(xy)+\lambda(x+y-2e)$
$F_x=\frac{y}{xy}+\lambda=\frac{1}{x}+\lambda=0$
$F_y=\frac{x}{xy}+\lambda=\frac{1}{y}+\lambda=0$
$F_\lambda=x+y-2e=0$

From the first two equations, $\lambda=-\frac{1}{x}=-\frac{1}{y}$; thus $x=y$. Substituting in the third equation,

$x+x=2e$
$2x=2e$
$x=e$

Since $x=y$, $y=e$. The maximum constrained value of f occurs at (e, e), and its value is

$\ln(e\cdot e)=\ln e^2=2$.

11. Minimize $f(x,y)=x^2+y^2$ subject to $2x+y=15$.

$F(x,y,\lambda)=x^2+y^2+\lambda(2x+y-15)$
$F_x=2x+2\lambda=0$ or $x=-\lambda$
$F_y=2y+\lambda=0$ or $y=-\frac{1}{2}\lambda$

Thus, $y=-\frac{1}{2}\lambda=\frac{1}{2}(-\lambda)=\frac{1}{2}x$. $F_\lambda=0$ gives

$2x+y-15=0$
$2x+\left(\frac{1}{2}x\right)-15=0$
$\frac{5}{2}x=15$
$x=6$
$y=\frac{1}{2}(6)=3$

Thus, the minimum constrained value of f occurs at (6, 3), and its value is $(6)^2+(3)^2=45$.

13. Minimize $f(x, y) = xy$ subject to $y = x + 8$.

$F(x,y,\lambda)=xy+\lambda(x-y+8)$
$F_x=y+\lambda=0$ or $y=-\lambda$
$F_y=x-\lambda=0$ or $x=\lambda$

Thus, $\lambda=-y=x$. $F_\lambda=0$ gives

$x-y+8=0$
$(-y)-y+8=0$
$-2y=-8$
$y=4$
$x=-(4)=-4$

Thus, the minimum constrained value of f occurs at (–4, 4), and its value is (–4)(4) = –16.

15. Minimize $f(x,y)=x^2+y^2$ subject to $2x+3y=26$.

$F(x,y,\lambda)=x^2+y^2+\lambda(2x+3y-26)$
$F_x=2x+2\lambda=0$ or $x=-\lambda$
$F_y=2y+3\lambda=0$ or $y=-\frac{3}{2}\lambda$

Thus, $y=-\frac{3}{2}\lambda=-\frac{3}{2}(-x)=\frac{3}{2}x$. $F_\lambda=0$ gives

$2x+3y-26=0$
$2x+3\left(\frac{3}{2}x\right)=26$
$\frac{13}{2}x=26$
$x=4$
$y=\frac{3}{2}(4)=6$

Thus, the minimum constrained value of f occurs at (4, 6), and its value is $(4)^2+(6)^2=52$.

17. Minimize $f(x,y)=\ln\left(x^2+y^2\right)$ subject to $2x+y=25$.

$F(x,y,\lambda)=\ln\left(x^2+y^2\right)+\lambda(2x+y-25)$

$F_x=\frac{2x}{x^2+y^2}+2\lambda=0$ or $\lambda=-\frac{x}{x^2+y^2}$

$F_y=\frac{2y}{x^2+y^2}+\lambda=0$ or $\lambda=-\frac{2y}{x^2+y^2}$

Equating expressions for λ gives

$$\frac{-x}{x^2+y^2}=\frac{-2y}{x^2+y^2}$$
$$x=2y$$

$F_\lambda=0$ gives

$$2x+y-25=0$$
$$2(2y)+y-25=0$$
$$5y=25$$
$$y=5$$
$$x=2(5)=10$$

Thus, the minimum constrained value of f occurs at (10, 5), and its value is $\ln\left[(10)^2+(5)^2\right]=\ln 125$.

19. Minimize $f(x,y)=e^{x^2+y^2}$ subject to $x+2y=10$.

$F(x,y,\lambda)=e^{x^2+y^2}+\lambda(x+2y-10)$

$F_x=2xe^{x^2+y^2}+\lambda=0$ or $\lambda=-2xe^{x^2+y^2}$

$F_y=2ye^{x^2+y^2}+2\lambda=0$ or $\lambda=-ye^{x^2+y^2}$

Equating expressions for λ gives

$$-2xe^{x^2+y^2}=-ye^{x^2+y^2}$$
$$2x=y$$

$F_\lambda=0$ gives

$$x+2y-10=0$$
$$x+2(2x)-10=0$$
$$5x=10$$
$$x=2$$
$$y=2(2)=4$$

Thus, the minimum constrained value of f occurs at (2, 4), and its value is $e^{4+16}=e^{20}$.

21. Maximize $f(x, y) = 2xy$ subject to $x^2+y^2=8$.

$F(x,y,\lambda)=2xy+\lambda\left(x^2+y^2-8\right)$

$F_x=2y+2x\lambda=0$

$F_y=2x+2y\lambda=0$

$F_\lambda=x^2+y^2-8=0$

From the first two equations, $\lambda=-\frac{2y}{2x}=-\frac{2x}{2y}$; so $\frac{y}{x}=\frac{x}{y}$, and $x^2=y^2$; $x=\pm y$.

Substituting $x = y$ in the third equation,

$$x^2+y^2=8$$
$$2x^2=8$$
$$x=\pm 2$$
$$y=\pm 2$$

Substituting $x = -y$ in the third equation again gives $x = \pm 2$, but now $y = \pm 2$; thus there are 4 points at which constrained extrema may occur, (2, 2), (2, –2), (–2, 2) and (–2, –2). At (2, 2) and (–2, –2), $f = 8$, and at (2, –2) and (–2, 2), $f = -8$; thus the constrained maximum is 8 at (2, 2) and (–2, –2), and the constrained minimum is –8 at (2, –2) and (–2, 2).

23. Maximize $f(x, y) = x + 2y$ subject to $2x^2+y^2=72$.

$F(x,y,\lambda)=x+2y+\lambda\left(2x^2+y^2-72\right)$

$F_x=1+4\lambda x=0$ or $\lambda=-\frac{1}{4x}$

$F_y=2+2\lambda y=0$ or $\lambda=-\frac{1}{y}$

Equating expressions for λ gives

$$-\frac{1}{4x}=-\frac{1}{y}$$
$$y=4x$$

$F_\lambda=0$ gives

$$2x^2+y^2-72=0$$
$$2x^2+(4x)^2-72=0$$
$$2x^2+16x^2=72$$
$$x^2=4$$
$$x=\pm 2$$
$$y=\pm 8$$

$f(2, 8) = 2 + 2(8) = 18$

$f(-2, -8) = -2 + 2(-8) = -18$

Thus, the maximum constrained value of f occurs at (2, 8) where $f = 18$. The minimum constrained value of f occurs at (–2, –8), where $f = -18$.

25. **a.** Call the horizontal strip of fence x and the three vertical strips of fence y. Then maximize the area xy subject to $x + 3y = 6000$.

$$F(x, y, \lambda) = xy + \lambda(x + 3y - 6000)$$
$$F_x = y + \lambda = 0 \quad \text{or} \quad \lambda = -y$$
$$F_y = x + 3\lambda = 0 \quad \text{or} \quad \lambda = -\tfrac{1}{3}x$$

Thus, $-y = -\frac{1}{3}x$ or $x = 3y$. $F_\lambda = 0$ gives

$$x + 3y - 6000 = 0$$
$$3y + 3y = 6000$$
$$6y = 6000$$
$$y = 1000$$
$$x = 3(1000) = 3000$$

The largest area that can be enclosed under the given conditions is (3000)(1000) = 3 million square feet. The dimensions are 1000 feet perpendicular to the building and 3000 feet parallel to the building.

b. Since $\lambda = -y$, $\lambda = -1000$. $|\lambda| = 1000$, which is the number of additional objective units per additional constraint unit; thus each extra foot of fence will enclose about 1000 additional square feet in area.

27. Let x = radius of the bottom and let h = height of the cylinder. We must minimize surface area = $\pi x^2 + 2\pi xh$ subject to volume = $\pi x^2 h = 160$.

$$F(x, y, \lambda) = \pi x^2 + 2\pi xh + \lambda(\pi x^2 h - 160)$$
$$F_x = 2\pi x + 2\pi h + 2\pi\lambda xh = 0$$
$$x + h + \lambda hx = 0$$
$$\lambda = -\tfrac{x+h}{hx}$$
$$F_h = 2\pi x + \lambda\pi x^2 = 0$$
$$2x + \lambda x^2 = 0$$
$$\lambda = -\tfrac{2}{x}$$

Equating expressions for λ gives

$$-\tfrac{x+h}{hx} = -\tfrac{2}{x}$$
$$-\tfrac{x+h}{h} = -2$$
$$-x - h = -2h$$
$$h = x$$

$F_\lambda = 0$ gives

$$\pi x^2 h - 160 = 0$$
$$\pi x^3 = 160$$
$$x = \sqrt[3]{\tfrac{160}{\pi}} \approx 3.7 = h$$

The least amount of materials is used when the radius of the bottom is about 3.7 feet and the height is 3.7 feet.

29. Call the sides of the square base x and the length of the box y. Maximize $x^2 y$ subject to $4x + y = 84$ (note that the full allowed 84 inches of length plus girth needs to be used, since, for a given square base, increasing the length of the box so that the full 84 inches are used will always increase the volume of the box).

$$F(x, y, \lambda) = x^2 y + \lambda(4x + y - 84)$$
$$F_x = 2xy + 4\lambda = 0$$
$$F_y = x^2 + \lambda = 0$$
$$F_\lambda = 4x + y - 84 = 0$$

From the first two equations,

$\lambda = -\frac{2xy}{4} = -\frac{xy}{2} = -x^2$. Since x • 0, $\frac{y}{2} = x$, or

$y = 2x$. Substituting into the third equation gives,

$$4x + 2x = 84$$
$$6x = 84$$
$$x = 14$$
$$y = 2(14) = 28$$

The largest box with a square base whose length plus girth does not exceed 84 inches has dimensions 14 inches by 14 inches by 28 inches and volume (28)(14)(14) = 5488 cubic inches.

31. a. Maximize $P = 280L^{3/4}K^{1/4}$ subject to $525L + 0.28K = 16{,}800$.

$$F(L,K,\lambda) = 280L^{3/4}K^{1/4} + \lambda(525L + 0.28K - 16{,}800)$$

$$F_L = 210L^{-1/4}K^{1/4} + 525\lambda = 0$$

Then $\lambda = -0.4L^{-1/4}K^{1/4} = -0.4\frac{K^{1/4}}{L^{1/4}}$

$$F_K = 70L^{3/4}K^{-3/4} + 0.28\lambda = 0$$

Then $\lambda = -250L^{3/4}K^{-3/4} = -250\frac{L^{3/4}}{K^{3/4}}$

$$F_\lambda = 525L + 0.28K - 16{,}800$$

Equation expressions for λ gives

$$-0.4\frac{K^{1/4}}{L^{1/4}} = -250\frac{L^{3/4}}{K^{3/4}}$$

$$-0.4K^{1/4+3/4} = -250L^{1/4+3/4}$$

$$K = 625L$$

Set $F_\lambda = 0$ and substitute $625L$ for K

$$525L + 0.28K - 16{,}800 = 0$$

$$525L + 0.28(625L) - 16{,}800 = 0$$

$$700L = 16{,}800$$

$$L = 24$$

Substitute 24 for L and solve for K and P.

$$K = 625(24) = 15{,}000$$

$$P = 280(24)^{3/4}(15{,}000)^{1/4} = 33{,}600$$

Therefore, $L = 24$, $K = 15{,}000$, and $P = 33{,}600$.

b. To find $|\lambda|$, we can use either equation for λ from above and substitute 15,000 for K and 24 for L.

$$|\lambda| = \left|-0.4\frac{K^{1/4}}{L^{1/4}}\right| = \left|-0.4\frac{15{,}000^{1/4}}{24^{1/4}}\right| = 2$$

Each additional dollar of budget increases production by about 2 units.

c. By increasing the budget by \$300, we have

$$|\lambda|\cdot 300 = 2\cdot 300 = 600$$

So the increase in production is about 600 units.

33. a. Maximize $P = 180L^{1/2}K^{1/2}$ subject to $45L + 20K = 2520$.

$$F(L,K,\lambda) = 180L^{1/2}K^{1/2} + \lambda(45L + 20K - 2520)$$

$$F_L = 90L^{-1/2}K^{1/2} + 45\lambda = 0$$

Then $\lambda = -2L^{-1/2}K^{1/2} = -2\frac{K^{1/2}}{L^{1/2}}$

$$F_K = 90L^{1/2}K^{-1/2} + 20\lambda = 0$$

Then $\lambda = -4.5L^{1/2}K^{-1/2} = -4.5\frac{L^{1/2}}{K^{1/2}}$

$$F_\lambda = 45L + 20K - 2520$$

Equation expressions for λ gives

$$-2\frac{K^{1/2}}{L^{1/2}} = -4.5\frac{L^{1/2}}{K^{1/2}}$$

$$-2K^{1/2+1/2} = -4.5L^{1/2+1/2}$$

$$K = 2.25L$$

Set $F_\lambda = 0$ and substitute $2.25L$ for K

$$45L + 20K - 2520 = 0$$

$$45L + 20(2.25L) - 2520 = 0$$

$$90L = 2520$$

$$L = 28$$

Substitute 28 for L and solve for K and P.

$$K = 2.25(28) = 63$$

$$P = 180(28)^{1/2}(63)^{1/2} = 7560$$

Therefore, $L = 28$, $K = 63$, and $P = 7560$.

b. To find $|\lambda|$, we can use either equation for λ from above and substitute 63 for K and 28 for L.

$$|\lambda| = \left|-2\frac{K^{1/2}}{L^{1/2}}\right| = \left|-2\frac{63^{1/2}}{28^{1/2}}\right| = 3$$

Each additional dollar of budget increases production by about 3 units.

c. By increasing the budget by \$50, we have

$$|\lambda|\cdot 50 = 3\cdot 50 = 150$$

So the increase in production is about 150 units.

35. To minimize the cost, minimize $(50)(2)x^2+(30)(4)xy=100x^2+120xy$ subject to $x^2y=45$.

$$F(x,y,\lambda)=100x^2+120xy+\lambda\left(x^2y-45\right)$$
$$F_x=200x+120y+2xy\lambda=0$$
$$F_y=120x+x^2\lambda=0$$
$$F_\lambda=x^2y-45=0$$

From the first two equations,

$\lambda=-\frac{200x+120y}{2xy}=-\frac{100x+60y}{xy}$ and

$\lambda=-\frac{120x}{x^2}=-\frac{120}{x}$, since x and y • 0. Thus,

$$\frac{100x+60y}{xy}=\frac{120}{x}$$
$$100x+60y=120y$$
$$y=\frac{100}{60}x=\frac{5}{3}x$$

Substituting in the third equation gives

$$x^2\left(\frac{5}{3}x\right)=45$$
$$x^3=27$$
$$x=3$$
$$y=\frac{5}{3}(3)=5$$

The box of minimum cost subject to the conditions given will have dimensions 3 inches by 3 inches by 5 inches.

37. Minimize $f(x,y,z)=x^2+y^2+z^2$ subject to $2x+y-z=12$.

$$F(x,y,z,\lambda)=x^2+y^2+z^2+\lambda(2x+y-z-12)$$
$$F_x=2x+2\lambda=0 \quad \text{or} \quad \lambda=-x$$
$$F_y=2y+\lambda=0 \quad \text{or} \quad \lambda=-2y$$
$$F_z=2z-\lambda=0 \quad \text{or} \quad \lambda=2z$$

Equating the first and third equations gives $-x=2z$ or $x=-2z$. Equating the second and third equations gives $-2y=2z$ or $y=-z$. $F_\lambda=0$ gives

$$2x+y-z-12=0$$
$$2(-2z)+(-z)-z-12=0$$
$$-6z=12$$
$$z=-2$$
$$y=-(-2)=2$$
$$x=-2(-2)=4$$

Thus, the minimum constrained value of f occurs at (4, 2, –2), where $f=4^2+2^2+(-2)^2=24$.

39. Maximize $f(x, y, z) = x + y + z$, subject to $x^2+y^2+z^2=12$.

$$F(x,y,z,\lambda)=x+y+z+\lambda\left(x^2+y^2+z^2-12\right)$$
$$F_x=1+2x\lambda=0$$
$$F_y=1+2y\lambda=0$$
$$F_z=1+2z\lambda=0$$
$$F_\lambda=x^2+y^2+z^2-12=0$$

So, $\lambda=-\frac{1}{2x}=-\frac{1}{2y}=-\frac{1}{2z}$ and $x=y=z$.

Substituting in the fourth equation,

$$x^2+x^2+x^2=12$$
$$3x^2=12$$
$$x=\pm 2 \quad \text{so } y=z=\pm 2$$

The maximum constrained value of f occurs at (2, 2, 2), where $f = 2 + 2 + 2 = 6$. Note that $f(-2, -2, -2) = -6$ which is less than $f(2, 2, 2)$.

41. Area of roof = xy, and cost of roof is $32xy$. Area of each side wall = yz, and each costs $10yz$. Area of front and back walls = xz, and each costs $10xz$. Area of floor = xy, and cost of floor is $8xy$.

$$\text{Cost of building} = 32xy + 2(10yz) + 2(10xz) + 8xy$$
$$= 40xy + 20yz + 20xz$$

Maximize cost function subject to $xyz = 2000$.

$$F(x, y, z, \lambda) = 40xy + 20yz + 20xz + \lambda(xyz - 2000)$$
$$F_x = 40y + 20z + \lambda yz = 0$$
$$F_y = 40x + 20z + \lambda xz = 0$$
$$F_z = 20y + 20x + \lambda xy = 0$$

Multiplying F_x by x and F_y by y gives

$$40xy + 20xz + \lambda xyz = 0$$
$$-(40xy + 20yz + \lambda xyz) = 0$$
$$20xz - 20yz = 0$$
$$20xz = 20yz$$
$$x = y$$

Substituting into F_y and F_z gives

$$F_y = 40y + 20z + \lambda yz = 0$$
$$F_z = 20y + 20y + \lambda y^2 = 0$$
$$\lambda = -\frac{40y}{y^2} = -\frac{40}{y}$$

Thus, $40y + 20z - \left(\frac{40}{y}\right)yz = 0$

$$40y + 20z - 40z = 0$$
$$z = 2y$$

Thus, using $x = y$, $z = 2y$, and $F_\lambda = 0$ gives

$$xyz - 2000 = 0$$
$$(y)y(2y) = 2000$$
$$y^3 = 1000$$
$$y = 10$$
$$x = 10$$
$$z = 2(10) = 20$$

The dimensions that minimize cost are $x = 10$ feet, $y = 10$ feet, and $z = 20$ feet.

43. False: It should be $F(x, y, \lambda) = f(x, y) + \lambda g(x, y)$

45. True: A solution can occur only at a critical point.

47. The maximum occurs at one critical point, and the minimum occurs at the other.

49. $a \le b$

51. Maximize $P = aL^bK^{1-b}$ subject to $wL + rK = C^*$.

$F(L, K, \lambda) = aL^bK^{1-b} + \lambda(wL + rK - C^*)$

$F_L = abL^{b-1}K^{1-b} + w\lambda = 0$

Then $\lambda = -\frac{ab}{w}L^{b-1}K^{1-b}$

$F_K = a(1-b)L^bK^{-b} + r\lambda = 0$

Then $\lambda = -\frac{a(1-b)}{r}L^bK^{-b}$

$F_\lambda = wL + rK - C^*$

Equation expressions for λ gives

$$-\frac{ab}{w}L^{b-1}K^{1-b} = -\frac{a(1-b)}{r}L^bK^{-b}$$

$$\frac{b}{w}K = \frac{(1-b)}{r}L$$

$$K = \frac{w(1-b)}{br}L$$

Set $F_\lambda = 0$ and substitute $\frac{w(1-b)}{br}L$ for K

$$wL + rK - C^* = 0$$

$$wL + r\left(\frac{w(1-b)}{br}L\right) - C^* = 0$$

$$wL + \frac{w(1-b)}{b}L = C^*$$

$$wL + \frac{w}{b}L - wL = C^*$$

$$L^* = \left(\frac{b}{w}\right)C^*$$

Substitute $\left(\frac{b}{w}\right)C^*$ for L and solve for K, P, and λ.

$$K^* = \frac{w(1-b)}{br}\left(\left(\frac{b}{w}\right)C^*\right) = \left(\frac{1-b}{r}\right)C^*$$

$$P^* = a\left(\left(\frac{b}{w}\right)C^*\right)^b\left(\left(\frac{1-b}{r}\right)C^*\right)^{1-b}$$

$$= a\left(\frac{b}{w}\right)^b\left(\frac{1-b}{r}\right)^{1-b}C^*$$

$$\lambda = -\frac{ab}{w}\left(\left(\frac{b}{w}C^*\right)\right)^{b-1}\left(\left(\frac{1-b}{r}\right)C^*\right)^{1-b}$$

$$= -a\left(\frac{b}{w}\right)^b\left(\frac{1-b}{r}\right)^{1-b}$$

53. Solving the equation in Exercise 51 for C^*

$$a\left(\frac{b}{w}\right)^b\left(\frac{1-b}{r}\right)^{1-b}C^* = P^*$$

$$C^* = \frac{1}{a}\left(\frac{b}{w}\right)^{-b}\left(\frac{1-b}{r}\right)^{-(1-b)}P^*$$

which is the equation in Exercise 51.

Start with the equation in Exercise 52 for K^* and substitute for P^*.

$$K^* = \frac{1}{a}\left(\frac{1-b}{r}\right)^b\left(\frac{w}{b}\right)^bP^*$$

$$= \frac{1}{a}\left(\frac{1-b}{r}\right)^b\left(\frac{w}{b}\right)^b a\left(\frac{b}{w}\right)^b\left(\frac{1-b}{r}\right)^{1-b}C^*$$

$$= \left(\frac{1-b}{r}\right)C^*$$

which is the equation in Exercise 51.

Start with the equation in Exercise 52 for K^* and substitute for P^*.

$$K^* = \frac{1}{a}\left(\frac{1-b}{r}\right)^b\left(\frac{w}{b}\right)^bP^*$$

$$= \frac{1}{a}\left(\frac{1-b}{r}\right)^b\left(\frac{w}{b}\right)^b a\left(\frac{b}{w}\right)^b\left(\frac{1-b}{r}\right)^{1-b}C^*$$

$$= \left(\frac{1-b}{r}\right)C^*$$

which is the equation in Exercise 51.

Start with the equation in Exercise 52 for K^* and substitute for P^*.

$$L = \frac{1}{a}\left(\frac{1-b}{r}\right)^{b-1}\left(\frac{w}{b}\right)^{b-1}P^*$$

$$= \frac{1}{a}\left(\frac{1-b}{r}\right)^{b-1}\left(\frac{w}{b}\right)^{b-1}a\left(\frac{b}{w}\right)^b\left(\frac{1-b}{r}\right)^{1-b}C^*$$

$$= \left(\frac{b}{w}\right)C^*$$

which is the equation in Exercise 51.

From Exercise 51 we have $\lambda = -a\left(\frac{b}{w}\right)^b\left(\frac{1-b}{r}\right)^{1-b}$. From Exercise 52 we have $\lambda = -\frac{1}{a}\left(\frac{b}{w}\right)^{-b}\left(\frac{1-b}{r}\right)^{-(1-b)}$. If increasing C by 1 increases P by λ, then increasing P by 1 must increase C by $1/\lambda$.

55. From Exercise 50, in Section 3.6 on page 250, we know $MRST = \frac{b}{1-b}\frac{K}{L}$. From Example 3 on page 520, we know that $b = \frac{2}{3}$, $K = 6$, and $L = 48$.

$$MRST = \frac{b}{1-b}\frac{K}{L} = \frac{\frac{2}{3}}{1-\frac{2}{3}}\frac{6}{48} = \frac{1}{4}$$

From Exercise 54, we have $MRST = \frac{w}{r}$. From Example 3 on page 520, we know that $w = 35$ and $r = 140$.

$$MRST = \frac{w}{r} = \frac{35}{140} = \frac{1}{4}.$$

57. From Example 3 on page 520, we have

$P = 840L^{2/3}K^{1/3}$

$P_L = 560L^{-1/3}K^{1/3}$

$P_K = 280L^{2/3}K^{-2/3}$

Then the ratio of P_L to P_K is with $K = 6$ and $L = 48$

$$\frac{P_L}{P_K} = \frac{560L^{-1/3}K^{1/3}}{280L^{2/3}K^{-2/3}} = \frac{2K}{L} = \frac{2(6)}{48} = \frac{1}{4}$$

From Example 3 on page 520, we know that $w = 35$ and $r = 140$. From Exercise 56, we have

$$\frac{P_L}{P_K} = \frac{w}{r} = \frac{35}{140} = \frac{1}{4}.$$

EXERCISES 7.6

1.
$f(x,y) = x^2y^3$

$f_x = 2xy^3$

$f_y = 3x^2y^2$

$df = 2xy^3 \cdot dx + 3x^2y^2 \cdot dy$

3.
$f(x,y) = 6x^{1/2}y^{1/3} + 8$

$f_x = 3x^{-1/2}y^{1/3}$

$f_y = 2x^{1/2}y^{-2/3}$

$df = 3x^{-1/2}y^{1/3} \cdot dx + 2x^{1/2}y^{-2/3} \cdot dy$

5.
$g(x,y) = \frac{x}{y}$

$g_x = \frac{1}{y}$

$g_y = -\frac{x}{y^2}$

$dg = \frac{1}{y} \cdot dx - \frac{x}{y^2} \cdot dy$

7.
$g(x,y) = (x-y)^{-1}$

$g_x = -(x-y)^{-2}$

$g_y = -(x-y)^{-2}(-1) = (x-y)^{-2}$

$dg = -(x-y)^{-2} \cdot dx + (x-y)^{-2} \cdot dy$

9.
$z = \ln(x^3 - y^2)$

$z_x = \frac{3x^2}{x^3-y^2} \qquad z_y = \frac{-2y}{x^3-y^2}$

$dz = \frac{3x^2}{x^3-y^2} \cdot dx - \frac{2y}{x^3-y^2} \cdot dy$

11.
$z = xe^{2y}$

$z_x = e^{2y} \qquad z_y = 2xe^{2y}$

$dz = e^{2y} \cdot dx + 2xe^{2y} \cdot dy$

13.
$w = 2x^3 + xy + y^2$

$w_x = 6x^2 + y \qquad w_y = x + 2y$

$dw = (6x^2 + y) \cdot dx + (x + 2y) \cdot dy$

15.
$f(x,y,z) = 2x^2y^3z^4$

$f_x = 4xy^3z^4 \qquad f_y = 6x^2y^2z^4 \qquad f_z = 8x^2y^3z^3$

$df = 4xy^3z^4 \cdot dx + 6x^2y^2z^4 \cdot dy + 8x^2y^3z^3 \cdot dz$

17.
$f(x,y,z) = \ln(xyz)$

$f_x = \frac{yz}{xyz} = \frac{1}{x} \qquad f_y = \frac{xz}{xyz} = \frac{1}{y} \qquad f_z = \frac{xy}{xyz} = \frac{1}{z}$

$df = \frac{1}{x} \cdot dx + \frac{1}{y} \cdot dy + \frac{1}{z} \cdot dz$

19.
$f(x,y,z) = e^{xyz}$

$df = yze^{xyz} \cdot dx + xze^{xyz} \cdot dy + xye^{xyz} \cdot dz$

21. **a.** $x+\Delta x=4+0.2=4.2 \qquad y+\Delta y=2+(-0.1)=1.9$

$$\Delta f=f(4.2,1.9)-f(4,2)=(4.2)^2+(4.2)(1.9)+(1.9)^3-\left[4^2+(4)(2)+2^3\right]$$
$$=32.479-32=0.479$$

b. $df=(2x+y)\cdot dx+\left(x+3y^2\right)\cdot dy$. Evaluate df at $x=4$, $y=2$, $dx=0.2$, $dy=-0.1$.

$$df=\left[2(4)+2\right](0.2)+\left[4+3(2)^2\right](-0.1)=2+(-0.1)(16)=2-1.6=0.4$$

23. **a.** $x+\Delta x=0+0.05=0.05 \qquad y+\Delta y=1+0.01=1.01$

$$\Delta f=f(0.05,1.01)-f(0,1)=e^{0.05}+(0.05)(1.01)+\ln 1.01-\left[e^0+(0)1+\ln 1\right]$$
$$\approx 1.112-1=0.112$$

b. $df=\left(e^x+y\right)\cdot dx+\left(x+\frac{1}{y}\right)\cdot dy$. Evaluate df at $x=0$, $y=1$, $dx=0.05$, $dy=0.01$.

$$df=\left(e^0+1\right)(0.05)+\left(0+\tfrac{1}{1}\right)(0.01)=0.1+0.01=0.11$$

25. **a.** $x+\Delta x=3+0.03=3.03 \qquad y+\Delta y=2+0.02=2.02 \qquad z+\Delta z=1+0.01=1.01$

$$\Delta f=f(3.03,2.02,1.01)-f(3,2,1)=3.03(2.02)+(1.01)^2-\left[3(2)+1^2\right]$$
$$=7.1407-7=0.1407$$

b. $df=y\cdot dx+x\cdot dy+2z\cdot dz$. Evaluate df at $x=3$, $y=2$, $z=1$, $dx=0.03$, $dy=0.02$, $dz=0.01$.

$$df=2(0.03)+3(0.02)+2(1)(0.01)=0.14$$

27. Let x = length of rectangle and let y = width of rectangle.
$x=150$, $y=100$, $\Delta x=0.5$, $\Delta y=0.5$
$A=xy$
$$dA=y\cdot dx+x\cdot dy$$
$$=100(0.5)+150(0.5)=50+75$$
$$=125 \text{ square feet}$$
$\Delta x=1,\ \Delta y=1$
$dA=100\cdot 1+150\cdot 1=250$ square feet

29. Let $x=200$ DVD and $y=300$ Blu-ray players, and $\Delta x=5$, $\Delta y=4$. To estimate the change in profit $P(x, y)$, find dP.
$$dP=(4x-3y)\cdot dx+(-3x+6y)\cdot dy$$
$$=(4x-3y)\cdot dx-(3x-6y)\cdot dy$$
Evaluate at $x=200$, $y=300$, $\Delta x=5$, $\Delta y=4$.
$$dP=\left[4(200)-3(300)\right](5)-\left[3(200)-6(300)\right](4)$$
$$=-500+4800=4300$$
The extra profit is \$4300.

31. For a truck weighing 4 tons and usually driven at 60 mph, $w=4$, and $v=60$. Find dS for $dw=0.5$ and $dv=5$.
$$S=0.027wv^2$$
$$dS=0.027v^2\cdot dw+0.054wv\cdot dv$$
Evaluate dS at $w=4$, $v=60$, $dw=0.5$, $dv=5$.
$$dS=0.027(60)^2(0.5)+0.054(4)(60)(5)\approx 113 \text{ feet}$$
The extra stopping distance is about 113 feet.

33. Since the error of each measurement is 1%,
$\Delta x=0.01x$
$\Delta y=0.01y$
Area $A=xy$ and the differential is
$$dA=y\cdot dx+x\cdot dy$$
$$=y(0.01x)+x(0.01y)$$
$$=0.01xy+0.01xy=0.02xy=0.02A$$
The percentage error in calculating the area is 0.02 or 2%.

35. $x=250$, $y=160$, $z=150$, $\Delta x=\Delta y=\Delta z=5$.
The differential dC is
$$dC=\frac{1}{y-z}\cdot dx-\frac{x}{(y-z)^2}\cdot dy+\frac{x}{(y-z)^2}\cdot dz=\frac{1}{160-150}(5)-\frac{250}{(160-150)^2}(5)+\frac{250}{(160-150)^2}(5)$$
$$=\frac{5}{10}-2.5(5)+2.5(5)=0.5 \text{ liter per minute}$$
The error in calculating C is 0.5 liter per minute.

37. $V(l,d)=\pi\left(\frac{d}{2}\right)^2 l=\frac{\pi}{4}d^2 l$

$V(12,0.8)=\frac{\pi}{4}(0.8)^2(12)=1.92\pi\approx 6.03\text{ cm}^3$

The estimate of the error is found using the differential.

$$\begin{aligned}dV&=\frac{\pi}{2}(d)(l)\cdot dd+\frac{\pi}{4}d^2\cdot dl\\&=\frac{\pi}{2}(0.8)(12)(0.1)+\frac{\pi}{4}(0.8)^2(0.1)\\&=0.496\pi\approx 1.56\end{aligned}$$

The error may be as much as 1.56 cm^3.

39. 0

41. $f(x,y)=x+y+C$, where C is a constant.

43. It is independent of y (since $f_y=0$)

45. **a.** Since $f(x, y) = c$,

$f(x+\Delta x,F(x+\Delta x))-f(x,F(x))=c-c=0$

b. Adding $-F(x) + F(x) = 0$ gives

$f(x+\Delta x,F(x+\Delta x))-F(x)+F(x)-f(x,F(x))=0$

c. Using the differential approximation formula we know that

$f(x+\Delta x,F+\Delta F)-f(x,F)\approx f_x\cdot\Delta x+f_y\cdot\Delta F$

Thus, $f_x\cdot\Delta x+f_y\cdot\Delta F\approx 0$

d. $f_x\cdot\Delta x\approx -f_y\cdot\Delta F$

$\frac{\Delta F}{\Delta x}\approx -\frac{f_x}{f_y}$

e. As Δx and ΔF approach 0, $f(x+\Delta x,F+\Delta F)-f(x,F)$ approaches $f_x\cdot dx+f_y\cdot dF$. Thus,

$f_x\cdot dx+f_y\cdot dF=0$

$\frac{dF}{dx}=-\frac{f_x}{f_y}$

EXERCISES 7.7

1. $$\begin{aligned}\int_1^{x^2}8xy^3dy&=8x\left(\frac{y^4}{4}\right)\Big|_1^{x^2}=2xy^4\Big|_1^{x^2}\\&=2x\left[\left(x^2\right)^4-1\right]\\&=2x\left(x^8-1\right)\\&=2x^9-2x\end{aligned}$$

3. $$\begin{aligned}\int_{-y}^{y}9x^2y\,dx&=9y\left(\frac{x^3}{3}\right)\Big|_{-y}^{y}\\&=3yx^3\Big|_{-y}^{y}=3y\left(y^3\right)-3y(-y)^3\\&=3y^4+3y^4=6y^4\end{aligned}$$

5. $$\begin{aligned}\int_0^x(6y-x)dy&=\left[6\left(\frac{y^2}{2}\right)-xy\right]\Big|_0^x\\&=\left(3y^2-xy\right)\Big|_0^x\\&=3x^2-x(x)-\left[3(0)^2-x(0)\right]\\&=3x^2-x^2=2x^2\end{aligned}$$

7. $$\begin{aligned}\int_0^2\int_0^1 4xy\,dx\,dy&=\int_0^2 4y\left(\frac{x^2}{2}\right)\Big|_0^1dy=\int_0^2 2yx^2\Big|_0^1dy\\&=\int_0^2\left[2y(1)^2-2y(0)^2\right]dy=\int_0^2 2y\,dy\\&=2\left(\frac{y^2}{2}\right)\Big|_0^2=y^2\Big|_0^2\\&=(2)^2-(0)^2=4\end{aligned}$$

9. $$\int_0^2\int_0^1 x\,dy\,dx = \int_0^2 x(y)\Big|_0^1 dx = \int_0^2 x(1-0)\,dx = \frac{x^2}{2}\Big|_0^2 = \frac{2^2}{2} - 0 = 2$$

11. $$\int_0^1\int_0^2 x^3y^7dx\,dy = \int_0^1 y^7\left(\frac{x^4}{4}\right)\Big|_0^2 dy = \int_0^1 y^7\left(\frac{2^4}{4}-\frac{0^4}{4}\right)dy = \int_0^1 4y^7dy = 4\left(\frac{y^8}{8}\right)\Big|_0^1 = 4\left(\frac{1^8}{8}-\frac{0^8}{8}\right) = \frac{1}{2}$$

13. $$\int_1^3\int_0^2 (x+y)\,dy\,dx = \int_1^3\left(xy+\frac{y^2}{2}\right)\Big|_0^2 dx = \int_1^3\left[(2x+2)-(0+0)\right]dx = \left(x^2+2x\right)\Big|_1^3 = (9+6)-(1+2) = 12$$

15. $$\int_{-1}^1\int_0^3\left(x^2-2y^2\right)dx\,dy = \int_{-1}^1\left(\frac{x^3}{3}-2y^2x\right)\Big|_0^3 dy = \int_{-1}^1\left[\frac{3^3}{3}-2y^2(3)-(0)\right]dy$$
$$= \int_{-1}^1\left(9-6y^2\right)dy = \left[9y-6\left(\frac{y^3}{3}\right)\right]\Big|_{-1}^1 = \left(9y-2y^3\right)\Big|_{-1}^1$$
$$= 9(1)-2(1)^3-\left[9(-1)-2(-1)^3\right] = 9-2-(-9+2) = 7+7 = 14$$

17. $$\int_{-3}^3\int_0^3 y^2e^{-x}dy\,dx = \int_{-3}^3\left(\frac{y^3}{3}e^{-x}\right)\Big|_0^3 dx = \int_{-3}^3\left(9e^{-x}-0\right)dx = -9e^{-x}\Big|_{-3}^3 = -9\left(e^{-3}-e^3\right) = 9\left(e^3-e^{-3}\right)$$

19. $$\int_{-2}^2\int_{-1}^1 ye^{xy}dx\,dy = \int_{-2}^2 y\left(\frac{1}{y}e^{xy}\right)\Big|_{-1}^1 dy = \int_{-2}^2 e^{xy}\Big|_{-1}^1 dy = \int_{-2}^2\left(e^y-e^{-y}\right)dy = \left(e^y+e^{-y}\right)\Big|_{-2}^2 = e^2+e^{-2}-\left(e^{-2}+e^2\right) = 0$$

21. $$\int_0^2\int_x^1 12xy\,dy\,dx = \int_0^2\left[12x\left(\frac{y^2}{2}\right)\right]\Big|_x^1 dx = \int_0^2 6xy^2\Big|_x^1 dx = \int_0^2 6x\left(1-x^2\right)dx$$
$$= \int_0^2\left(6x-6x^3\right)dx = \left(3x^2-\tfrac{3}{2}x^4\right)\Big|_0^2 = \left[3(4)-\tfrac{3}{2}(16)\right]-(0-0) = 12-24 = -12$$

23. $$\int_3^5\int_0^y (2x-y)\,dx\,dy = \int_3^5\left(x^2-yx\right)\Big|_0^y dy = \int_3^5\left(y^2-y^2\right)dy = \int_3^5 0\,dy = 0$$

25. $$\int_{-3}^{3}\int_{0}^{4x}(y-x)\,dy\,dx=\int_{-3}^{3}\left(\frac{y^2}{2}-xy\right)\Bigg|_0^{4x}dx=\int_{-3}^{3}\left[\frac{(4x)^2}{2}-x(4x)-(0-0)\right]dx$$
$$=\int_{-3}^{3}\left(8x^2-4x^2\right)dx=\int_{-3}^{3}4x^2dx=\frac{4}{3}x^3\Big|_{-3}^{3}=\frac{4}{3}[27-(-27)]$$
$$=\frac{4}{3}(54)=72$$

27. $$\int_0^1\int_{-y}^{y}\left(x+y^2\right)dx\,dy=\int_0^1\left(\frac{x^2}{2}+y^2x\right)\Bigg|_{-y}^{y}dy=\int_0^1\left\{\frac{y^2}{2}+y^3-\left[\frac{(-y)^2}{2}+y^2(-y)\right]\right\}dy$$
$$=\int_0^1\left[\frac{y^2}{2}+y^3-\left(\frac{y^2}{2}-y^3\right)\right]dy=\int_0^1 2y^3dy$$
$$=\frac{1}{2}y^4\Big|_0^1=\frac{1}{2}(1)^4-\frac{1}{2}(0)^4=\frac{1}{2}$$

29. **a.** $\int_0^2\int_1^3 3xy^2dy\,dx$ and $\int_1^3\int_0^2 3xy^2dx\,dy$

b. $$\int_0^2\int_1^3 3xy^2dy\,dx=\int_0^2\left[3x\left(\frac{y^3}{3}\right)\right]\Bigg|_1^3dx$$
$$=\int_0^2 xy^3\Big|_1^3dx$$
$$=\int_0^2 x(27-1)\,dx=\int_0^2 26x\,dx$$
$$=13x^2\Big|_0^2=13(4-0)=52$$

$$\int_1^3\int_0^2 3xy^2dx\,dy=\int_1^3\left(\frac{3}{2}x^2y^2\right)\Bigg|_0^2dy$$
$$=\int_1^3\frac{3}{2}y^2(4-0)\,dy$$
$$=\int_1^3 6y^2dy=2y^3\Big|_1^3$$
$$=2(27-1)=2(26)=52$$

31. **a.** $\int_{-1}^{1}\int_0^2 ye^xdy\,dx$ and $\int_0^2\int_{-1}^{1} ye^xdx\,dy$

b. $$\int_{-1}^{1}\int_0^2 ye^xdy\,dx=\int_{-1}^{1}\frac{e^xy^2}{2}\Bigg|_0^2dx=\int_{-1}^{1}2e^xdx$$
$$=2e^x\Big|_{-1}^{1}=2e-2e^{-1}$$

$$\int_0^2\int_{-1}^{1} ye^xdx\,dy=\int_0^2 ye^x\Big|_{-1}^{1}dy$$
$$=\int_0^2\left(ey-e^{-1}y\right)dy$$
$$=e\left(\frac{y^2}{2}\right)-e^{-1}\left(\frac{y^2}{2}\right)\Bigg|_0^2$$
$$=2e-2e^{-1}$$

33. $f(x,y)=x+y;R=\{(x,y)|0\le x\le 2,0\le y\le 2\}$

The volume under the surface f and above the region R is

$$\iint_R(x+y)\,dx\,dy=\int_0^2\int_0^2(x+y)\,dx\,dy$$
$$=\int_0^2\left(\frac{x^2}{2}+xy\right)\Bigg|_0^2dy$$
$$=\int_0^2[(2+2y)-(0+0)]\,dy$$
$$=\left(2y+y^2\right)\Big|_0^2$$
$$=2(2)+4-0=8\text{ cubic units}$$

35. $f(x,y)=2-x^2-y^2; R=\{(x,y)|0\le x\le 1, 0\le y\le 1\}$

The volume under the surface *f* and above the region *R* is

$$\iint_R (2-x^2-y^2)\,dx\,dy = \int_0^1\int_0^1 (2-x^2-y^2)\,dx\,dy$$
$$= \int_0^1 \left(2x-\frac{x^3}{3}-y^2x\right)\Bigg|_0^1 dy$$
$$= \int_0^1 \left[2-\frac{(1)^3}{3}-y^2\right]dy = \int_0^1\left(\frac{5}{3}-y^2\right)dy$$
$$= \left(\frac{5}{3}y-\frac{y^3}{3}\right)\Bigg|_0^1 = \frac{5}{3}-\frac{1}{3}=\frac{4}{3}\text{ cubic units}$$

37. $f(x,y)=24xy; R=\{(x,y)|0\le x\le 1, x^2\le y\le \sqrt{x}\}$

The volume under the surface *f* and above the region *R* is

$$\iint_R 24xy\,dx\,dy = \int_0^1\int_{x^2}^{\sqrt{x}} 24xy\,dy\,dx = \int_0^1 (12xy^2)\Big|_{x^2}^{\sqrt{x}} dx = \int_0^1 12x\left[(\sqrt{x})^2-(x^2)^2\right]dx$$
$$=12\int_0^1 x(x-x^4)\,dx = 12\int_0^1 (x^2-x^5)\,dx$$
$$=12\left(\frac{x^3}{3}-\frac{x^6}{6}\right)\Bigg|_0^1 = 12\left(\frac{1}{3}-\frac{1}{6}\right)-0 = 2\text{ cubic units}$$

39. $f(x,y)=e^y; R=\{(x,y)|0\le x\le 1, x\le y\le 2x\}$

The volume under the surface *f* and above the region *R* is

$$\iint_R e^y\,dx\,dy = \int_0^1\int_x^{2x} e^y\,dy\,dx = \int_0^1 e^y\Big|_x^{2x} dx = \int_0^1 (e^{2x}-e^x)\,dx$$
$$=\left(\frac{1}{2}e^{2x}-e^x\right)\Bigg|_0^1 = \frac{1}{2}e^{2(1)}-e^1-\left(\frac{1}{2}e^{2(0)}-e^0\right)=\frac{1}{2}e^2-e+\frac{1}{2}\text{ cubic units}$$

41. $f(x,y)=48+4x-2y$

The average temperature over the region is $\frac{1}{\text{area of }R}\iint_R (48+4x-2y)\,dx\,dy$, where $R=\{(x,y)|-2\le x\le 2, 0\le y\le 3\}$.

$$(\text{Average temperature}) = \frac{1}{(4)(3)}\int_{-2}^{2}\int_0^3 (48+4x-2y)\,dy\,dx$$
$$=\frac{1}{12}\int_{-2}^{2}\left(48y+4xy-y^2\right)\Big|_0^3 dx = \frac{1}{12}\int_{-2}^{2}\left[48(3)+4x(3)-9\right]dx = \frac{1}{12}\int_{-2}^{2}(135+12x)\,dx$$
$$=\frac{1}{12}\left(135x+6x^2\right)\Big|_{-2}^{2} = \frac{1}{12}\left[135(2)+6(4)-135(-2)-6(4)\right] = \frac{1}{12}(540) = 45^\circ$$

43. The total population is equal to the integral of $P(x,y)=12000e^{x-y}$ over R for $R=\{(x,y)|0\le x\le 2,-1\le y\le 1\}$

$$\begin{aligned}(\text{Total population}) &= \int_{-1}^{1}\int_{0}^{2}12000e^{x-y}\,dx\,dy\\ &=12000\left(\int_{-1}^{1}e^{x-y}\Big|_0^2\,dy\right)=12000\left(\int_{-1}^{1}e^{2-y}-e^{-y}dy\right)\\ &=12000\left(-e^{2-y}+e^{-y}\Big|_{-1}^{1}\right)=12000(-e^{1}+e^{-1}-(e^{3}+e^{1}))\\ &=12000(-e^{3}-2e+e^{-1})\approx 180{,}200 \text{ people}\end{aligned}$$

45. The volume of the building is the volume under the surface $f(x,y)=40-0.006x^2+0.003y^2$ and above the region $R=\{(x,y)|-50\le x\le 50,-100\le y\le 100\}$.

$$\begin{aligned}\iint_R\left(40-0.006x^2+0.003y^2\right)dx\,dy &= \int_{-50}^{50}\int_{-100}^{100}\left(40-0.006x^2+0.003y^2\right)dy\,dx\\ &=\int_{-50}^{50}\left(40y-0.006x^2y+0.001y^3\right)\Big|_{-100}^{100}dx\\ &=\int_{-50}^{50}\left[40(100)-0.006x^2(100)+0.001(100)^3-40(-100)+0.006x^2(-100)-0.001(-100)^3\right]dx\\ &=\int_{-50}^{50}\left(4000-0.6x^2+1000+4000-0.6x^2+1000\right)dx\\ &=\int_{-50}^{50}\left(10{,}000-1.2x^2\right)dx\\ &=\left(10{,}000x-0.4x^3\right)\Big|_{-50}^{50}\\ &=(10{,}000)(50)-0.4(50)^3-(10{,}000)(-50)+(0.4)(-50)^3=900{,}000 \text{ cubic units}\end{aligned}$$

47.
$$\begin{aligned}\int_1^2\int_0^3\int_0^1\left(2x+4y-z^2\right)dx\,dy\,dz &= \int_1^2\int_0^3\left(x^2+4yx-z^2x\right)\Big|_0^1 dy\,dz\\ &=\int_1^2\int_0^3\left[(1)^2+4y(1)-z^2(1)\right]dy\,dz=\int_1^2\int_0^3\left(1+4y-z^2\right)dy\,dz\\ &=\int_1^2\left(y+2y^2-z^2y\right)\Big|_0^3 dz=\int_1^2\left[3+2(3)^2-3z^2\right]dz\\ &=\int_1^2\left(21-3z^2\right)dz=\left(21z-z^3\right)\Big|_1^2=(42-8)-(21-1)\\ &=34-20=14\end{aligned}$$

49.
$$\begin{aligned}\int_1^2\int_0^2\int_0^1 2xy^2z^3dx\,dy\,dz &= \int_1^2\int_0^2\left(x^2y^2z^3\right)\Big|_0^1 dy\,dz\\ &=\int_1^2\int_0^2 y^2z^3(1-0)\,dy\,dz=\int_1^2\int_0^2 y^2z^3dy\,dz\\ &=\int_1^2\left(\frac{y^3}{3}z^3\right)\Big|_0^2 dz=\int_1^2 z^3\left(\frac{8}{3}-0\right)dz\\ &=\int_1^2\frac{8}{3}z^3dz=\frac{8}{3}\left(\frac{z^4}{4}\right)\Big|_1^2=\frac{2}{3}z^4\Big|_1^2=\frac{2}{3}(16-1)=\frac{2}{3}(15)=10\end{aligned}$$

51. **a** is a double integral and **b** is an iterated integral.

53. 6

55. Yes, $\int_c^d e^{f(x)}\,dx \cdot \int_a^b e^{g(y)}\,dy$

57. True

59. widgets

REVIEW EXERCISES AND CHAPTER TEST FOR CHAPTER 7

1. The domain of f is $\{(x, y) \mid x \geq 0, y \neq 0\}$.

2. The domain of f is $\{(x, y) \mid y > 0\}$.

3. The domain of f is $\{(x, y) \mid x \neq 0, y > 0\}$.

4. The domain of f is $\{(x, y) \mid x \neq 0, y > 0\}$.

5. $f(x, y) = 2x^5 - 3x^2y^3 + y^4 - 3x + 2y + 7$

a. $f_x = 10x^4 - 6xy^3 - 3$

b. $f_y = -3x^2(3y^2) + 4y^3 + 2$
$= -9x^2y^2 + 4y^3 + 2$

c-d. $f_{xy} = f_{yx} = -18xy^2$

6. $f(x, y) = 3x^4 + 5x^3y^2 - y^6 - 6x + y - 9$

a. $f_x = 12x^3 + 15x^2y^2 - 6$

b. $f_y = 10x^3y - 6y^5 + 1$

c-d. $f_{xy} = f_{yx} = 30x^2y$

7. $f(x, y) = 18x^{2/3}y^{1/3}$

a. $f_x = 12x^{-1/3}y^{1/3}$

b. $f_y = 6x^{2/3}y^{-2/3}$

c-d. $f_{xy} = f_{yx} = 4x^{-1/3}y^{-2/3}$

8. $f(x, y) = \ln(x^2 + y^3)$

a. $f_x = \frac{2x}{x^2+y^3}$

b. $f_y = \frac{3y^2}{x^2+y^3}$

c-d. $f_{xy} = f_{yx} = -\frac{6xy^2}{(x^2+y^3)^2}$

9. $f(x, y) = e^{x^3-2y^3}$

a. $f_x = 3x^2e^{x^3-2y^3}$

b. $f_y = -6y^2e^{x^3-2y^3}$

c-d. $f_{xy} = f_{yx} = -18x^2y^2e^{x^3-2y^3}$

10. $f(x, y) = 3x^2e^{-5y}$

a. $f_x = 6xe^{-5y}$

b. $f_y = -15x^2e^{-5y}$

c-d. $f_{xy} = f_{yx} = -30xe^{-5y}$

11. $f(x, y) = ye^{-x} - x\ln y$

a. $f_x = -ye^{-x} - \ln y$

b. $f_y = e^{-x} - \frac{x}{y}$

c-d. $f_{xy} = f_{yx} = -e^{-x} - \frac{1}{y}$

12. $f(x, y) = x^2e^y + y\ln x$

a. $f_x = 2xe^y + \frac{y}{x}$

b. $f_y = x^2e^y + \ln x$

c-d. $f_{xy} = f_{yx} = 2xe^y + \frac{1}{x}$

13. $f(x,y)=\dfrac{x+y}{x-y}$

a. $f_x=\dfrac{x-y-(x+y)}{(x-y)^2}=\dfrac{-2y}{(x-y)^2}$

$f_x(1,-1)=\dfrac{2}{(2)^2}=\dfrac{1}{2}$

b. $f_y=\dfrac{x-y-(-1)(x+y)}{(x-y)^2}=\dfrac{2x}{(x-y)^2}$

$f_y(1,-1)=\dfrac{2}{(2)^2}=\dfrac{1}{2}$

14. $f(x,y)=\dfrac{x}{x^2+y^2}$

a. $f_x=\dfrac{x^2+y^2-2x(x)}{(x^2+y^2)^2}=\dfrac{y^2-x^2}{(x^2+y^2)^2}$

$f_x(1,-1)=\dfrac{1-1}{(2)^2}=0$

b. $f_y=\dfrac{-2y(x)}{(x^2+y^2)^2}=-\dfrac{2xy}{(x^2+y^2)^2}$

$f_y(1,-1)=-\dfrac{2(1)(-1)}{2^2}=\dfrac{1}{2}$

15. $f(x,y)=(x^3+y^2)^3$

a. $f_x=3(x^3+y^2)^2(3x^2)=9x^2(x^3+y^2)^2$

$f_x(1,-1)=9(1+1)^2=36$

b. $f_y=6y(x^3+y^2)^2$

$f_y(1,-1)=6(-1)(1+1)^2=-24$

16. $f(x,y)=(2xy-1)^4$

a. $f_x=4(2xy-1)^3(2y)=8y(2xy-1)^3$

$f_x(1,-1)=8(-1)[2(1)(-1)-1]^3$
$=-8(-3)^3=216$

b. $f_y=8x(2xy-1)^3$

$f_y(1,-1)=8(1)[2(1)(-1)-1]^3$
$=8(-3)^3=-216$

17. $P(L,K)=160L^{3/4}K^{1/4}$

a. $P_L(L,K)=120L^{-1/4}K^{1/4}$

$P_L(81,16)=120(81)^{-1/4}(16)^{1/4}$
$=120\left(\frac{1}{3}\right)(2)=80$

Production increases by 80 for each additional unit of labor.

b. $P_K(L,K)=40L^{3/4}K^{-3/4}$

$P_K(81,16)=40(81)^{3/4}(16)^{-3/4}$
$=40\left(\frac{27}{8}\right)=135$

Production increases by 135 for each additional unit of capital.

c. An additional unit of capital is more effective.

18. $S(x,y)=60x^2+90y^2-6xy+200$
$S_x=120x-6y$
$S_x(2,3)=120(2)-6(3)=222$
$S_y=180y-6x$
$S_y(2,3)=180(3)-6(2)=528$

$S_x=222$ is the rate at which sales increase for each additional \$1000 spent on TV ads.
$S_y=528$ is the rate at which sales increase for each additional \$1000 spent on print ads.

19. $f(x,y)=2x^2-2xy+y^2-4x+6y-3$
$f_x=4x-2y-4=0$
$f_y=-2x+2y+6=0$
Solving these equations simultaneously gives $x=-1, y=-4$.
$f_{xx}=4, f_{yy}=2, f_{xy}=-2. D=2(4)-(-2)^2=4>0$
and $f_{xx}>0$. Thus, f has a relative minimum at $(-1,-4)$ and $f(-1,-4)=2-(2)(-1)(-4)+(-4)^2-4(-1)+6(-4)-3=2-8+16+4-24-3=-13$

20. $f(x,y)=x^2-2xy+2y^2-6x+4y+2$
$f_x=2x-2y-6=0$
$f_y=-2x+4y+4=0$
Solving these equations simultaneously gives $x=4, y=1$. $f_{xx}=2,\ f_{yy}=4,\ f_{xy}=-2$.
$D=2(4)-(-2)^2=4>0$
and $f_{xx}>0$. Thus, f has a relative minimum at (4, 1) and
$f(4,1)=16-8+2-24+4+2=-8$

21. $f(x,y)=2xy-x^2-5y^2+2x-10y+3$
$f_x=2y-2x+2=0$
$f_y=2x-10y-10=0$
Solving these equations simultaneously gives $x=0, y=-1. f_{xx}=-2, f_{yy}=-10, f_{xy}=2$.
$D=(-2)(-10)-2^2=16>0$ and $f_{xx}<0$.
Therefore f has a relative (and absolute) maximum at (0, –1) and
$f(0,-1)=-5(-1)^2+-10(-1)+3=8$

22. $f(x,y)=2xy-5x^2-y^2+10x-2y+1$
$f_x=2y-10x+10=0$
$f_y=2x-2y-2=0$
Solving these equations simultaneously gives $x=1, y=0. f_{xx}=-10, f_{yy}=-2, f_{xy}=2$.
$D=-10(-2)-2^2=20-4=16$ and $f_{xx}<0$.
Thus, f has a relative maximum at (1, 0) and $f(1,0)=-5+10+1=6$

23. $f(x,y)=2xy+6x-y+1$
$f_x=2y+6=0$ or $y=-3$
$f_y=2x-1=0$ or $x=\frac{1}{2}$
$f_{xx}=0,\ f_{yy}=0,\ f_{xy}=2.$
$D=(0)(0)-2^2=-4<0$
f has no relative extrema. The point $\left(\frac{1}{2},-3\right)$ is a saddle point.

24. $f(x,y)=4xy-4x+2y-4$
$f_x=4y-4=0$ or $y=1$
$f_y=4x+2$ or $x=-\frac{1}{2}$
$f_{xx}=0,\ f_{yy}=0,\ f_{xy}=4.$
$D=(0)(0)-4^2=-16<0$
f has no relative extrema. The point $\left(-\frac{1}{2},1\right)$ is a saddle point.

25. $f(x,y)=e^{-\left(x^2+y^2\right)}$
$f_x=-2xe^{-\left(x^2+y^2\right)}=0$
$f_y=-2ye^{-\left(x^2+y^2\right)}=0$
$e^{-(x^2+y^2)}>0$ for all x and y, so the only values that satisfy these equations are $x=0$ and $y=0$. By inspection it can be seen that f must have a relative maximum of $e^0=1$ at (0, 0). For any other (x, y), $x^2+y^2>0$, so $e^{-\left(x^2+y^2\right)}=\frac{1}{e^k}$. with k positive, for any x and y not both equal to zero. $\frac{1}{e^k}<1$ if $k>0$, so $e^0=1$ must be a maximum of f.

26. $f(x,y)=e^{2\left(x^2+y^2\right)}$
$f_x=4xe^{2\left(x^2+y^2\right)}=0$
$f_y=4ye^{2\left(x^2+y^2\right)}=0$
Only $x=0$ and $y=0$ satisfy these equations. (0, 0) must be a relative minimum because $e^{2(x^2+y^2)}$ is larger than $e^0=1$ because $x^2+y^2>0$. At (0, 0),
$f(0,0)=e^{2\left(0^2+0^2\right)}=e^0=1$.

27. $f(x,y)=\ln(5x^2+2y^2+1)$

$f_x=\dfrac{10x}{5x^2+2y^2+1}=0$

$f_y=\dfrac{4y}{5x^2+2y^2+1}=0$

Only (0, 0) satisfies these equations because $5x^2+2y^2+1>0$. f has a minimum at (0, 0) because $f(0, 0) = \ln(0 + 0 + 1) = 0$, $5x^2+2y^2+1\ge 1$ and ln is a strictly increasing function.

28. $f(x,y)=\ln(4x^2+3y^2+10)$

$f_x=\dfrac{8x}{4x^2+3y^2+10}=0$

$f_y=\dfrac{6y}{4x^2+3y^2+10}=0$

Only (0, 0) satisfies these equations. f has a minimum at (0, 0) because $f(0, 0) = \ln 10$ and for any other x, y, $\ln(4x^2+3y^2+10)>\ln 10$ because $4x^2+3y^2>0$.

29. $f(x,y)=x^3-y^2-12x-6y$

$f_x=3x^2-12=0$

$f_y=-2y-6=0$

These equations give $x^2=4$

$x=\pm 2$

and $2y=-6$

$y=-3$

so extrema may occur at (2, –3) or (–2, –3).

$f_{xx}=6x, f_{yy}=-2, f_{xy}=0.$

At (2, –3)

$D = 6(2)(-2) - 0 = -24 < 0$, so there is a saddle point.

At (–2, –3)

$D = 6(-2)(-2) - 0 = 24 > 0$ and $f_{xx}=-12<0$, so f has a relative maximum.

$f(-2,-3)=(-2)^3-(-3)^2-12(-2)-6(-3)=25$

30. $f(x,y)=y^2-x^3+12x-4y$

$f_x=-3x^2+12=0$ or $x=\pm 2$

$f_y=2y-4=0$ or $y=2$

$f_{xx}=-6x, f_{yy}=2, f_{xy}=0.$

For (2, 2), $f_{xx}=-6(2)=-12$ and

$D=-12(2)-0^2=-24<0$

(2, 2) is a saddle point.

For (–2, 2), $f_{xx}=-6(-2)=12$ and

$D=12(2)-0^2=24>0$

Thus, at (–2, 2), f has a relative minimum and

$f(-2,2)=(2)^2-(-2)^3+12(-2)-4(2)=-20$

31. **a.** 18-foot boat costs: \$3000

22-foot boat costs: \$5000

Let x = number of 18-foot boats and y = number of 22-foot boats

$C(x, y) = 3000x + 5000y + 6000$

b.
$$\begin{aligned}R(x,y)&=x\cdot p+y\cdot q\\&=x(7000-20x)+y(8000-30y)\\&=7000x-20x^2+8000y-30y^2\end{aligned}$$

c.
$$\begin{aligned}P(x,y)&=R(x,y)-C(x,y)\\&=7000x-20x^2+8000y-30y^2-(3000x+5000y+6000)\\&=4000x-20x^2+3000y-30y^2-6000\end{aligned}$$

d. Profit is maximized when $P_x=0$ and $P_y=0$.

$P_x=4000-40x=0$ or $x=100$

$P_y=3000-60y=0$ or $y=50$

Maximum profit is

$$\begin{aligned}P(100,50)&=4000(100)-20(100)^2+3000(50)-30(50)^2-6000\\&=400{,}000-200{,}000+150{,}000-75{,}000-6000\\&=\$269{,}000\end{aligned}$$

The company maximizes profit if it makes and sells 100 18-foot boats that sell for \$3000 and 50 22-foot boats that sell for \$5000. The maximum profit is \$269,000. The D test shows that this is a maximum.

32. **a.** Let x = number of harvesters sold in America, and let y = number of harvesters sold in Europe

$$R(x, y) = x \cdot p + y \cdot q = 80x - 0.2x^2 + 64y - 0.1y^2$$

$$P(x,y) = R(x,y) - C(x,y)$$
$$= 80x - 0.2x^2 + 64y - 0.1y^2 - (100 + 12x + 12y)$$
$$= 68x - 0.2x^2 + 52y - 0.1y^2 - 100$$

b. $P_x = 68 - 0.4x = 0$ or $x = 170$

$P_y = 52 - 0.2y = 0$ or $y = 260$

Price in America = 80– 0.2(170) = \$46 thousand
Price in Europe = 64 – 0.1(260) = \$38 thousand
To maximize profit, the company should sell 170 harvesters at \$46,000 in America and 260 harvesters at \$38,000 in Europe. The D test shows that this is a maximum.

33. **a.**

x	y	xy	x^2
1	–1	–1	1
3	6	18	9
4	6	24	16
5	10	50	25
$\Sigma x = 13$	$\Sigma y = 21$	$\Sigma xy = 91$	$\Sigma x^2 = 51$

$$a = \frac{n\Sigma xy - (\Sigma x)(\Sigma y)}{n\Sigma x^2 - (\Sigma x)^2} = \frac{4(91)-(13)(21)}{4(51)-13^2} = 2.6$$

$$b = \tfrac{1}{n}(\Sigma y - a\Sigma x) = \tfrac{1}{4}[21 - (2.6)(13)] = -3.2$$

The least squares line is $y = 2.6x - 3.2$.

b. A graphing calculator gives $y = 2.6x - 3.2$.

34. **a.**

x	y	xy	x^2
1	7	7	1
2	4	8	4
4	2	8	16
5	–1	–5	25
$\Sigma x = 12$	$\Sigma y = 12$	$\Sigma xy = 18$	$\Sigma x^2 = 46$

$$a = \frac{n\Sigma xy - (\Sigma x)(\Sigma y)}{n\Sigma x^2 - (\Sigma x)^2} = \frac{4(18)-(12)(12)}{4(46)-12^2} \approx -1.8$$

$$b = \tfrac{1}{n}(\Sigma y - a\Sigma x) = \tfrac{1}{4}[12 + (1.8)(12)] = 8.4$$

The least squares line is $y = -1.8x + 8.4$.

b. A graphing calculator gives $y = -1.8x + 8.4$

35.

x	y	xy	x^2
1	20.1	20.1	1
2	25.5	51.0	4
3	31.2	93.6	9
4	35.0	140	16
5	40.2	201	25
$\Sigma x = 15$	$\Sigma y = 152$	$\Sigma xy = 505.7$	$\Sigma x^2 = 55$

$$a = \frac{n\Sigma xy - (\Sigma x)(\Sigma y)}{n\Sigma x^2 - (\Sigma x)^2} = \frac{5(505.7)-(15)(152)}{5(55)-(15)^2} = 4.97$$

$$b = \tfrac{1}{n}(\Sigma y - a\Sigma x) = \tfrac{1}{5}[152 - (4.97)(15)] = 15.49$$

The least squares line is $y = 4.97x + 15.49$.
When $x = 6$, 4.97(6) + 15.49 = 45.32
The prediction is 45.3 million.

36.

x	y	$Y = \ln y$	xY	x^2
1	5.7	1.7405	1.7405	1
2	4.8	1.5686	3.1372	4
3	4.6	1.5261	4.5782	9
4	3.8	1.3350	5.3400	16
$\Sigma x = 10$		$\Sigma Y =$ 6.1702	$\Sigma xY =$ 14.7959	$\Sigma x^2 = 30$

$$A = \frac{n\Sigma xY - (\Sigma x)(\Sigma Y)}{n\Sigma x^2 - (\Sigma x)^2} = \frac{4(14.7959)-(10)(6.1702)}{4(30)-(10)^2} \approx -0.1259$$

$$b = \tfrac{1}{n}(\Sigma Y - A\Sigma x) = \tfrac{1}{4}[6.1702 - (-0.1259)(10)] \approx 1.857$$

$B = e^b = e^{1.857} \approx 6.4$, and the least squares curve is $y = 6.4e^{-0.1259x}$. In the year 2023 ($x = 6$), $y = 6.4e^{-0.1259(6)} \approx 3$, the number of violent crimes per 1000 people will be about 3.

37. Maximize $f(x,y) = 6x^2 - y^2 + 4$ subject to $3x + y = 12$.

$$F(x,y,\lambda) = 6x^2 - y^2 + 4 + \lambda(3x + y - 12)$$
$$F_x = 12x + 3\lambda = 0$$
$$F_y = -2y + \lambda = 0$$
$$F_\lambda = 3x + y - 12 = 0$$

From the first two equations, $\lambda = \frac{-12x}{3} = 2y$; so $y = -2x$. Substituting in the third equation,

$$3x - 2x - 12 = 0$$
$$x = 12$$
$$y = -2(12) = -24$$

The maximum constrained value of f occurs at (12, –24), and its value is

$$6(12)^2 - (-24)^2 + 4 = 292.$$

38. Maximize $f(x,y) = 4xy - x^2 - y^2$ subject to $x + 2y = 26$.

$$F(x,y,\lambda) = 4xy - x^2 - y^2 + \lambda(x + 2y - 26)$$
$$F_x = 4y - 2x + \lambda = 0 \quad \text{or} \quad \lambda = 2x - 4y$$
$$F_y = 4x - 2y + 2\lambda = 0 \quad \text{or} \quad \lambda = -2x + y$$

Equating expressions for λ gives

$$2x - 4y = -2x + y$$
$$4x = 5y$$
$$x = \tfrac{5}{4}y$$

$F\lambda = 0$ gives

$$x + 2y - 26 = 0$$
$$\tfrac{5}{4}y + 2y - 26 = 0$$
$$\tfrac{13}{4}y = 26$$
$$y = 8$$
$$x = \tfrac{5}{4}(8) = 10$$

The maximum constrained value of f occurs at (10, 8), and its value is

$$4(10)(8) - (10)^2 + (8)^2 \approx 156.$$

39. Minimize $f(x,y) = 2x^2 + 3y^2 - 2xy$ subject to $2x + y = 18$.

$$F(x,y,\lambda) = 2x^2 + 3y^2 - 2xy + \lambda(2x + y - 18)$$
$$F_x = 4x - 2y + 2\lambda = 0 \quad \text{or} \quad \lambda = -2x + y$$
$$F_y = 6y - 2x + \lambda = 0 \quad \text{or} \quad \lambda = 2x - 6y$$

Equating expressions for λ gives

$$-2x + y = 2x - 6y$$
$$7y = 4x$$
$$y = \tfrac{4}{7}x$$

$F_\lambda = 0$ gives

$$2x + y - 18 = 0$$
$$2x + \tfrac{4}{7}x - 18 = 0$$
$$\tfrac{18}{7}x = 18$$
$$x = 7$$
$$y = \tfrac{4}{7}(7) = 4$$

The minimum constrained value of f occurs at (7, 4), and its value is

$$2(7)^2 + 3(4)^2 - 2(7)(4) = 90.$$

40. Minimize $f(x, y) = 12xy - 1$ subject to $y - x = 6$.

$$F(x,y,\lambda) = 12xy - 1 + \lambda(y - x - 6)$$
$$F_x = 12y - \lambda = 0 \quad \text{or} \quad \lambda = 12y$$
$$F_y = 12x + \lambda 0 = \quad \text{or} \quad \lambda = -12x$$

Thus $y = -x$. $F_\lambda = 0$ gives

$$y - x - 6 = 0$$
$$-x - x - 6 = 0$$
$$-2x = 6$$
$$x = -3$$
$$y = 3$$

The minimum constrained value of f occurs at (–3, 3), and its value is
$12(-3)(3) - 1 = -109$.

41. Minimize $f(x,y)=e^{x^2+y^2}$ subject to $x+2y=15$.

$F(x,y,\lambda)=e^{x^2+y^2}+\lambda(x+2y-15)$

$F_x=2xe^{x^2+y^2}+\lambda=0$

$F_y=2ye^{x^2+y^2}+2\lambda=0$

$F_\lambda=x+2y-15=0$

From the first two equations,

$$\lambda=-2xe^{x^2+y^2}=-\frac{2ye^{x^2+y^2}}{2}=-ye^{x^2+y^2}$$
$$-2x=-y$$
$$y=2x$$

Substituting in the third equation,

$$x+2(2x)-15=0$$
$$5x=15$$
$$x=3$$
$$y=2(3)=6$$

The minimum constrained value of f occurs at (3, 6) and its value is $e^{3^2+6^2}=e^{45}\approx 3.49\times10^{19}$.

42. Maximize $f(x,y)=e^{-x^2-y^2}$ subject to $2x+y=5$.

$F(x,y,\lambda)=e^{-x^2-y^2}+\lambda(2x+y-5)$

$F_x=-2xe^{-x^2-y^2}+2\lambda=0$ or $\lambda=xe^{-x^2-y^2}$

$F_y=-2ye^{-x^2-y^2}+\lambda=0$ or $\lambda=2ye^{-x^2-y^2}$

Equating expressions for λ gives

$$xe^{-x^2-y^2}=2ye^{-x^2-y^2}$$
$$x=2y$$

$F_\lambda=0$ gives

$$2x+y-5=0$$
$$2(2y)+y-5=0$$
$$5y=5$$
$$y=1$$
$$x=2(1)=2$$

The maximum constrained value of f occurs at (2, 1), and its value is $e^{-4-1}=e^{-5}\approx 0.0067$

43. Find the extreme values of $f(x, y) = 6x - 18y$ subject to $x^2+y^2=40$.

$F(x,y,\lambda)=6x-18y+\lambda(x^2+y^2-40)$

$F_x=6+2\lambda x=0$ or $\lambda=-\frac{3}{x}$

$F_y=-18+2\lambda y=0$ or $\lambda=\frac{9}{y}$

Equating expressions for λ gives

$$-\frac{3}{x}=\frac{9}{y}$$
$$y=-3x$$

$F_\lambda=0$ gives

$$x^2+y^2-40=0$$
$$x^2+(-3x)^2-40=0$$
$$10x^2=40$$
$$x=\pm2$$

For $x = 2$, $y = -6$, and $f(2, -6) = 12 + 108 = 120$.
For $x = -2$, $y = 6$, and $f(-2, 6) = -12 - 108 = -120$.
The maximum constrained value of f occurs at $f(2, -6)$, and $f(2, -6) = 120$.
The minimum constrained value of f occurs at (2, –6), and $f(-2, 6) = -120$.

44. Find the extreme values of $f(x, y) = 4xy$ subject to $x^2+y^2=32$.

$F(x,y,\lambda)=4xy+\lambda(x^2+y^2-32)$

$F_x=4y+2\lambda x=0$ or $\lambda=-\frac{2y}{x}$

$F_y=4x+2\lambda y=0$ or $\lambda=-\frac{2x}{y}$

Equating expressions for λ gives

$$-\frac{2x}{y}=-\frac{2y}{x}$$
$$-2x^2=-2y^2$$
$$x^2=y^2$$

$F_\lambda=x^2+y^2-32=0$ gives

$$y^2+y^2-32=0$$
$$2y^2=32$$
$$y=\pm4$$
$$x=\pm4$$

$f(4, 4) = 4 \cdot 4 \cdot 4 = 64$
$f(4, -4) = 4 \cdot 4 \cdot (-4) = -64$
$f(-4, 4) = 4(-4)(4) = -64$
$f(-4, -4) = 4(-4)(-4) = 64$
The minimum constrained value of f is –64, which occurs at (4, –4) and (–4, 4).
The maximum constrained value of f is 64, which occurs at (4, 4) and (–4, –4).

45. $P(x,y,\lambda)=300x^{2/3}y^{1/3}+\lambda(x+y-60{,}000)$

a. $P_x=300\cdot\frac{2}{3}x^{-1/3}y^{1/3}+\lambda$
$=200x^{-1/3}y^{1/3}+\lambda=0$
$P_y=\frac{1}{3}\cdot300x^{2/3}y^{-2/3}+\lambda$
$=100x^{2/3}y^{-2/3}+\lambda=0$
$P_\lambda=x+y-60{,}000=0$

Solving the first two equations for λ and equating the results gives

$$\lambda=-200x^{-1/3}y^{1/3}=-100x^{2/3}y^{-2/3}$$
$$2x^{-1/3}y^{1/3}=x^{2/3}y^{-2/3}$$

Multiplying both sides by $x^{1/3}y^{2/3}$, $2y=x$.

Substituting in the third equation,

$$2y+y=60{,}000$$
$$3y=60{,}000$$
$$y=20{,}000$$
$$x=2(20{,}000)=40{,}000$$

When $60,000 can be spent on production and advertising, the maximum profit will occur if $40,000 is spent on production and $20,000 on advertising.

b. $|\lambda|=200(40{,}000)^{-1/3}(20{,}000)^{1/3}\approx159$

When $40,000 is spent on production and $20,000 is spent on advertising, profit increases by about $159 for each additional dollar spent.

46. a. Maximize $4x+2xy+8y$ subject to $2x+y=8$.

$F(x,y,\lambda)=4x+2xy+8y+\lambda(2x+y-8)$
$F_x=4+2y+2\lambda=0$ or $\lambda=-y-2$
$F_y=2x+8+\lambda=0$ or $\lambda=-2x-8$

Equating expressions for λ gives

$$-y-2=-2x-8$$
$$-y=-2x-6$$
$$y=2x+6$$

$F_\lambda=0$ gives

$$2x+y-8=0$$
$$2x+(2x+6)-8=0$$
$$4x-2=0$$
$$x=\frac{1}{2}$$
$$y=2\left(\frac{1}{2}\right)+6=7$$

Giving $\frac{1}{2}$ ounce of the first supplement and 7 ounces of the second supplement maximizes the nutritional value.

b. $\lambda=|-7-2|=9$. Nutritional value increases by 9 units for each additional dollar spent.

47. a. To maximize production, maximize $P=400L^{3/5}K^{2/5}$ subject to the cost $50L+60K=1500$.

$F(L,K,\lambda)=400L^{3/5}K^{2/5}+\lambda(50L+60K-1500)$
$F_L=240L^{-2/5}K^{2/5}+50\lambda=0$ $\quad\lambda=-\frac{24}{5}L^{-2/5}K^{2/5}=-\frac{24K^{2/5}}{5L^{2/5}}$
$F_K=160L^{3/5}K^{-3/5}+60\lambda=0$ $\quad\lambda=-\frac{8}{3}L^{3/5}K^{-3/5}=-\frac{8L^{3/5}}{3K^{3/5}}$

Equation expressions for λ gives

$$-\frac{24K^{2/5}}{5L^{2/5}}=-\frac{8L^{3/5}}{3K^{3/5}}$$
$$9K^{2/5+3/5}=5L^{2/5+3/5}$$
$$K=\frac{5}{9}L$$

$F_\lambda=0$ gives

$$50L+60K-1500=0$$
$$50L+60\left(\frac{5}{9}L\right)-1500=0$$
$$\frac{250}{3}L=1500$$
$$L=18$$
$$K=\frac{5}{9}(18)=10$$

The company should use 18 units of labor and 10 units of capital to maximize production.

b. $|\lambda|=\left|\frac{24}{5}\frac{K^{2/5}}{L^{2/5}}\right|=\left|\frac{24}{5}\left(\frac{K}{L}\right)^{2/5}\right|=\left|\frac{24}{5}\left(\frac{10}{18}\right)^{2/5}\right|\approx3.79$

Output increases by about 3.79 for each additional dollar.

48. Volume = $(x)(x)(y) = x^2y = 576$

Surface area of the box = area of base + areas of the 4 sides and 2 dividers

Area of base = x^2

Area of 4 sides = $4(xy) = 4xy$

Area of 2 dividers = $2xy$

Thus, minimize $x^2 + 6xy$ subject to $x^2y = 576$.

$F(x, y, \lambda) = x^2 + 6xy + \lambda(x^2y - 576)$

$F_x = 2x + 6y + 2\lambda xy = 0$ or $\lambda = \frac{-2x-6y}{2yx} = \frac{-x-3y}{xy}$

$F_y = 6x + \lambda x^2 = 0$ or $\lambda = \frac{-6x}{x^2} = -\frac{6}{x}$

Equating expressions for λ gives

$$\frac{-x-3y}{xy} = -\frac{6}{x}$$
$$-x^2 - 3xy = -6xy$$
$$-x^2 + 3xy = 0$$
$$x(-x + 3y) = 0$$
$$x = 3y$$

$F_\lambda = 0$ gives

$$x^2y - 576 = 0$$
$$(3y)^2 y - 576 = 0$$
$$9y^3 = 576$$
$$y^3 = 64$$
$$y = 4$$
$$x = 4(3) = 12$$

The box should be 12 inches by 12 inches by 4 inches.

49. $f(x, y) = 3x^2 + 2xy + y^2$

$f_x = 6x + 2y$

$f_y = 2x + 2y$

$df = (6x + 2y) \cdot dx + (2x + 2y) \cdot dy$

50. $f(x, y) = x^2 + xy - 3y^2$

$f_x = 2x + y$

$f_y = x - 6y$

$df = (2x + y) \cdot dx + (x - 6y) \cdot dy$

51. $g(x, y) = \ln(xy)$

$g_x = \frac{y}{xy} = \frac{1}{x}$

$g_y = \frac{x}{xy} = \frac{1}{y}$

$dg = \frac{1}{x} \cdot dx + \frac{1}{y} \cdot dy$

52. $g(x, y) = \ln(x^3 + y^3)$

$g_x = \frac{3x^2}{x^3 + y^3}$

$g_y = \frac{3y^2}{x^3 + y^3}$

$dg = \frac{3x^2}{x^3 + y^3} \cdot dx + \frac{3y^2}{x^3 + y^3} \cdot dy$

53. $z = e^{x-y}$

$z_x = e^{x-y}$

$z_y = -e^{x-y}$

$dz = e^{x-y} \cdot dx - e^{x-y} \cdot dy$

54. $z = e^{xy}$

$z_x = ye^{xy}$

$z_y = xe^{xy}$

$dz = ye^{xy} \cdot dx + xe^{xy} \cdot dy$

55. $x = 2$ thousand dollars of television advertising
$y = 3$ thousand dollars of print advertising
Find the total differential dS for $\Delta x = 0.5$ and $\Delta y = -0.5$

$$S(x,y) = 60x^2 - 6xy + 90y^2 + 200$$
$$S_x = 120x - 6y$$
$$S_y = -6x + 180y$$
$$\begin{aligned} dS &= (120x - 6y)\cdot dx + (-6x + 180y)\cdot dy \\ &= (120\cdot 2 - 6\cdot 3)(0.5) + (-6\cdot 2 + 180\cdot 3)(-0.5) \\ &= 111 + (-264) = -153 \end{aligned}$$

Sales would decrease by about \$153,000.
If $\Delta x = 0.25$ and $\Delta y = -0.25$,

$$\begin{aligned} dS &= (120\cdot 2 - 6\cdot 3)(0.25) + (-6\cdot 2 + 180\cdot 3)(-0.25) \\ &= -76.5 \end{aligned}$$

Sales would decrease by \$76,000.

56. Let x = the length of the base of the triangle. Let y = the length of the height of the triangle. Find the differential dA for $A = \frac{1}{2}xy$ with
$\Delta x = 0.01x$ and $\Delta y = 0.01y$.

$$A_x = \frac{1}{2}y \quad \text{and} \quad A_y = \frac{1}{2}x$$
$$\begin{aligned} dA &= \frac{1}{2}y(0.01x) + \frac{1}{2}x(0.01y) \\ &= \frac{1}{2}(0.01xy) + \frac{1}{2}(0.01xy) \\ &= 0.02\left(\frac{1}{2}xy\right) = 0.02A \end{aligned}$$

The error in calculating the area is 2%.

57.
$$\begin{aligned} \int_0^4 \int_{-1}^1 2xe^{2y}\,dy\,dx &= \int_0^4 2x\left(\frac{1}{2}e^{2y}\right)\Big|_{-1}^1 dx \\ &= \int_0^4 xe^{2y}\Big|_{-1}^1 dx \\ &= \int_0^4 x\left(e^2 - e^{-2}\right)dx \\ &= \frac{x^2}{2}\left(e^2 - e^{-2}\right)\Big|_0^4 \\ &= \frac{16}{2}\left(e^2 - e^{-2}\right) = 8\left(e^2 - e^{-2}\right) \end{aligned}$$

58.
$$\begin{aligned} \int_{-1}^1 \int_0^3 \left(x^2 - 4y^2\right)dx\,dy &= \int_{-1}^1 \left(\frac{x^3}{3} - 4xy^2\right)\Big|_0^3 dy \\ &= \int_{-1}^1 \left[\frac{3^3}{3} - 4(3)y\right]dy \\ &= \int_{-1}^1 \left(9 - 12y^2\right)dy \\ &= \left(9y - 4y^3\right)\Big|_{-1}^1 \\ &= 5 - (-5) = 10 \end{aligned}$$

59.
$$\begin{aligned} \int_{-1}^1 \int_{-y}^y (x+y)\,dx\,dy &= \int_{-1}^1 \left(\frac{x^2}{2} + xy\right)\Big|_{-y}^y dy \\ &= \int_{-1}^1 \left[\frac{y^2}{2} + y^2 - \left(\frac{y^2}{2} - y^2\right)\right]dy \\ &= \int_{-1}^1 2y^2\,dy = \frac{2}{3}y^3\Big|_{-1}^1 \\ &= \frac{2}{3}(1) - \frac{2}{3}(-1) = \frac{4}{3} \end{aligned}$$

60.
$$\begin{aligned} \int_{-2}^2 \int_{-x}^x (x+y)\,dy\,dx &= \int_{-2}^2 \left(xy + \frac{y^2}{2}\right)\Big|_{-x}^x dy \\ &= \int_{-2}^2 \left[x^2 + \frac{x^2}{2} - \left(-x^2 + \frac{x^2}{2}\right)\right]dx \\ &= \int_{-2}^2 2x^2\,dx = \frac{2}{3}x^3\Big|_{-2}^2 \\ &= \frac{2}{3}(2)^3 - \frac{2}{3}(-2)^3 = \frac{32}{3} \end{aligned}$$

61. The volume under the surface $f(x, y) = 8 - x - y$ and above the region $R = \{(x, y) \mid 0 \le x \le 2, 0 \le y \le 4\}$ is

$$\begin{aligned} \iint_R (8 - x - y)\,dx\,dy &= \int_0^2 \int_0^4 (8 - x - y)\,dy\,dx = \int_0^2 \left(8y - xy - \frac{y^2}{2}\right)\Big|_0^4 dx \\ &= \int_0^2 \left[(32 - 4x - 8) - (0 - 0 - 0)\right]dx = \int_0^2 (24 - 4x)\,dx \\ &= \left(24x - 2x^2\right)\Big|_0^2 = 24(2) - 2(4) - 0 = 40 \text{ cubic units} \end{aligned}$$

62. $R = \{(x, y) \mid 0 \le x \le 4, 0 \le y \le 2\}$

$$\iint_R (6-x-y)\,dx\,dy = \int_0^4 \int_0^2 (6-x-y)\,dy\,dx$$
$$= \int_0^4 \left(6y - xy - \frac{y^2}{2}\right)\Bigg|_0^2 dx$$
$$= \int_0^4 \left(6\cdot 2 - 2x - \frac{2^2}{2}\right) dx$$
$$= \left(10x - x^2\right)\Big|_0^4 = 40 - 16$$
$$= 24 \text{ cubic units}$$

63. $R = \left\{(x, y) \middle| 0 \le x \le 1,\ x^4 \le y \le \sqrt[4]{x}\right\}$

$$\iint_R 12xy^3\,dx\,dy = \int_0^1 \int_{x^4}^{\sqrt[4]{x}} 12xy^3\,dy\,dx$$
$$= \int_0^1 3xy^4 \Big|_{x^4}^{\sqrt[4]{x}} dx$$
$$= \int_0^1 \left[3x\left(\sqrt[4]{x}\right)^4 - 3x\left(x^4\right)^4\right] dx$$
$$= \int_0^1 \left(3x^2 - 3x^{17}\right) dx$$
$$= \left(x^3 - \frac{3x^{18}}{18}\right)\Bigg|_0^1 = \left[1 - \frac{(1)^{18}}{6}\right]$$
$$= \tfrac{5}{6} \text{ cubic unit}$$

64. $R = \left\{(x, y) \middle| 0 \le x \le 1,\ x^5 \le y \le \sqrt[5]{x}\right\}$

$$\iint_R 15xy^4\,dx\,dy = \int_0^1 \int_{x^5}^{\sqrt[5]{x}} 15xy^4\,dy\,dx = \int_0^1 3xy^5 \Big|_{x^5}^{\sqrt[5]{x}} dx = \int_0^1 \left[3x\left(\sqrt[5]{x}\right)^5 - 3x\left(x^5\right)^5\right] dx = \int_0^1 \left(3x^2 - 3x^{26}\right) dx$$
$$= \left(x^3 - \frac{3x^{27}}{27}\right)\Bigg|_0^1 = 1 - \tfrac{1}{9} = \tfrac{8}{9} \text{ cubic unit}$$

65. The average population is

$$\frac{1}{\text{area of } R} \iint_R P(x, y)\,dx\,dy = \frac{1}{(4)(4)} \int_{-2}^2 \int_{-2}^2 (12{,}000 + 100x - 200y)\,dx\,dy = \tfrac{1}{16} \int_{-2}^2 \left(12{,}000x + 50x^2 - 200xy\right)\Big|_{-2}^2 dy$$
$$= \tfrac{1}{16} \int_{-2}^2 \left[(24{,}000 + 200 - 400y) - (-24{,}000 + 200 + 400y)\right] dy$$
$$= \tfrac{1}{16} \int_{-2}^2 (48{,}000 - 800y)\,dy = \tfrac{1}{16}\left(48{,}000y - 400y^2\right)\Big|_{-2}^2$$
$$= \tfrac{1}{16}\left[(96{,}000 - 1600) - (-96{,}000 - 1600)\right] = \frac{192{,}000}{16} = 12{,}000$$

The average population over the region is 12,000 people per square mile

66. The total value of the land is

$$\int_{-2}^2 \int_{-2}^2 (40 - 4x - 2y)\,dx\,dy = \int_{-2}^2 \left(40x - 2x^2 - 2xy\right)\Big|_{-2}^2 dy$$
$$= \int_{-2}^2 \left\{40(2) - 2(2)^2 - 2(2)y - \left[40(-2) - 2(-2)^2 - 2(-2)y\right]\right\} dy$$
$$= \int_{-2}^2 \left[72 - 4y - (-88 + 4y)\right] dy = \int_{-2}^2 (160 - 8y)\,dy = \left(160y - 4y^2\right)\Big|_{-2}^2$$
$$= 160(2) - 4(2)^2 - \left[160(-2) - 4(-2)^2\right] = 320 - 16 - (-336) = 640$$

The value of the land is \$64,000,000.

CUMULATIVE REVIEW—CHAPTERS 1–7

1.

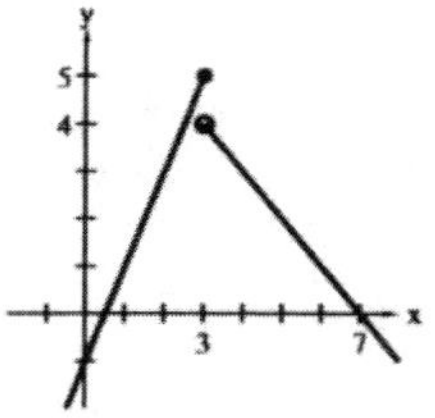

2. $\left(\frac{1}{8}\right)^{-2/3} = 8^{2/3} = \left(8^{1/3}\right)^2 = 2^2 = 4$

3. $f'(x) = \lim\limits_{h\to0} \dfrac{f(x+h)-f(x)}{h} = \lim\limits_{h\to0} \dfrac{\frac{1}{x+h}-\frac{1}{x}}{h}$

$= \lim\limits_{h\to0} \dfrac{\frac{x-(x+h)}{x(x+h)}}{h} = \lim\limits_{h\to0} \dfrac{\frac{-h}{x(x+h)}}{h}$

$= \lim\limits_{h\to0} \dfrac{-h}{hx(x+h)} = \lim\limits_{h\to0} \dfrac{-1}{x(x+h)} = -\dfrac{1}{x^2}$

4. $f(x) = 12\sqrt[3]{x^2} - 4 = 12x^{2/3} - 4$

$f'(x) = 12\left(\frac{2}{3}x^{-1/3}\right) = 8x^{-1/3}$

$f'(8) = 8(8)^{-1/3} = 8\left(\frac{1}{8}\right)^{1/3} = 8\cdot\frac{1}{2} = 4$

5. $S(p) = \frac{800}{p+8} = 800(p+8)^{-1}$

$S'(p) = -800(p+8)^{-2}$

$S'(12) = -800(12+8)^{-2} = -800\left(\frac{1}{20}\right)^2$

$= -800\cdot\frac{1}{400} = -2$

For each 1 dollar increase in the price of the disposable cameras above \$12, sales decrease by 2.

6. $\frac{d}{dx}\left[x^2 + (2x+1)^4\right]^3$

$= 3\left[x^2 + (2x+1)^4\right]^2\left[2x + 4(2x+1)^3(2)\right]$

$= \left[6x + 24(2x+1)^3\right]\left[x^2 + (2x+1)^4\right]^2$

7. $f(x) = x^3 + 9x^2 - 48x - 148$

$f'(x) = 3x^2 + 18x - 48$

$f''(x) = 6x + 18$

$3x^2 + 18x - 48 = 0$

$3\left(x^2 + 6x - 16\right) = 0$

$3(x+8)(x-2) = 0$

$x = 2,\ -8$

Critical values at $x = 2$ and -8.

$f' > 0$ ↗ | $x = -8$ rel max $(-8, 300)$ | $f' < 0$ ↘ | $x = 2$ rel min $(2, -200)$ | $f' > 0$ ↗

$6x + 18 = 0$

$f'' < 0$ con dn | $f'' = 0$ $x = -3$ | $f'' > 0$ con up

$(-3, 50)$ is a point of inflection.

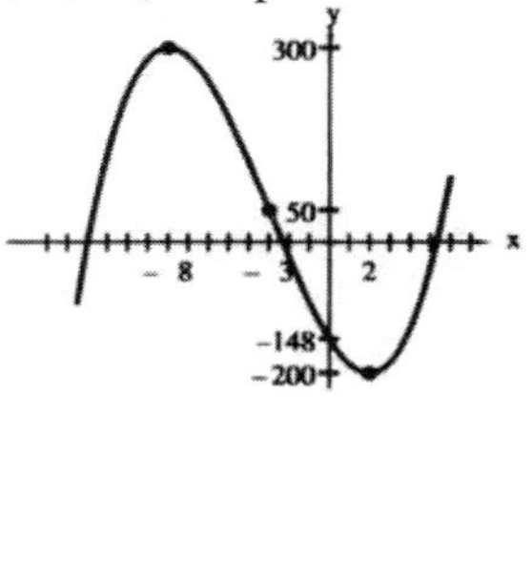

8. $f(x) = \dfrac{1}{x^2-4x} = \left(x^2 - 4x\right)^{-1}$

$f'(x) = -\left(x^2 - 4x\right)^{-2}(2x - 4)$

$= -\dfrac{2x-4}{\left(x^2-4x\right)^2} = 0$

$2x - 4 = 0$

$x = 2$

$f' > 0$ ↗ | f' und $x = 0$ | $f' > 0$ ↗ | $x = 2$ rel max $\left(2, -\frac{1}{4}\right)$ | $f' < 0$ ↘ | f' und $x = 4$ | $f' < 0$ ↘

f' undefined at $x = 0, 4$

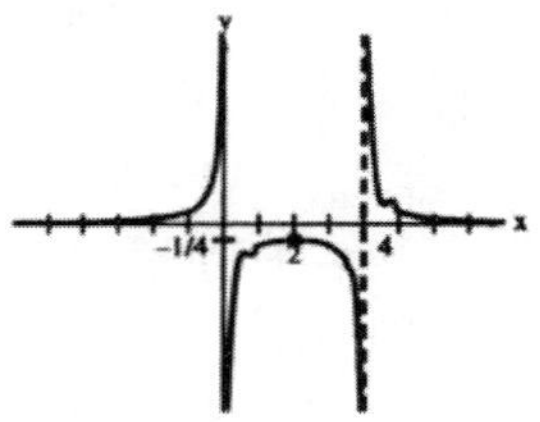

9. Let x = length of side perpendicular to the wall
Let y = length of side parallel to the wall

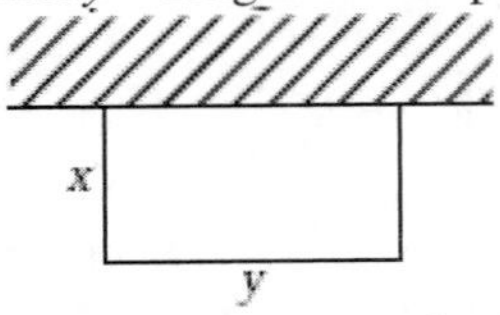

Since the total length of the fence is 80,

$2x + y = 80$

$y = 80 - 2x$

To find the maximum area, maximize
$A = xy = x(80 - 2x)$.

$A = x(80 - 2x)$

$= 80x - 2x^2$

$A' = 80 - 4x = 0$

$-4x = -80$

$x = 20$

$y = 80 - 2(20) = 40$

Thus, to maximize the area, the length of the side parallel to the wall should be 40 feet and the length of the side perpendicular to the wall should be 20 feet.

10. Let x = the length of one side of the square base
Let y = the height

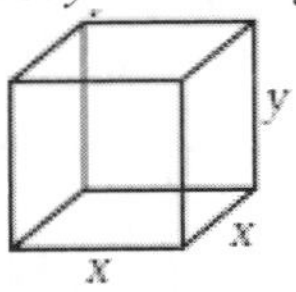

Then $V = x(x)(y) = x^2 y = 108$. Thus, $y = \frac{108}{x^2}$.

The surface area of the box is

Area of base $= x \cdot x = x^2$

Area of 4 sides $= 4(x \cdot y) = 4xy$

Thus to minimize the surface area, minimize
$S = x^2 + 4xy$.

$S = x^2 + 4xy$

$S = x^2 + 4x\left(\frac{108}{x^2}\right) = x^2 + \frac{432}{x}$

$S' = 2x - \frac{432}{x^2} = 0$

$2x^3 - 432 = 0$

$x^3 = 216$

$x = 6$

$y = \frac{108}{6^2} = 3$

The length of the sides of the base should be 6 feet and the height should be 3 feet to minimize the materials.

11. Since the volume of the sphere is increasing at the rate of 128 cubic feet per minute, $V' = 128$, where V is the volume of the sphere. The equation for the volume of a sphere is $V = \frac{4}{3}\pi r^3$.

Since one rate is given and another rate needs to be found, differentiate implicitly:

$\frac{dV}{dt} = 4\pi r^2 \frac{dr}{dt}$

Since $\frac{dV}{dt} = 128$ and $r = 4$,

$128 = 4\pi(4)^2 \frac{dr}{dt}$

$\frac{dr}{dt} = \frac{128}{64\pi} = \frac{2}{\pi} \approx 0.64$

At $r = 4$, the radius is increasing by about 0.64 foot per minute.

12. **a.** For quarterly compounding,
$r = \frac{0.06}{4} = 0.015$ and $n = 4 \bullet 3 = 12$.

$P(1+r)^n = 1000(1+0.015)^{12} \approx \$1195.62$

b. For continuous compounding, $r = 0.06$ and $n = 3$.

$Pe^{rn} = 1000e^{0.06(3)} \approx \$1197.22$

13. To find when the population of the second county will overtake that of the first, solve $P(t) = Q(t)$.

$$12{,}000e^{0.02t} = 9000e^{0.04t}$$
$$\ln\left(12{,}000e^{0.02t}\right) = \ln\left(9000e^{0.04t}\right)$$
$$\ln(12{,}000) + \ln\left(e^{0.02t}\right) = \ln(9000) + \ln\left(e^{0.04t}\right)$$
$$\ln 12{,}000 + 0.02t = \ln 9000 + 0.04t$$
$$0.02t = \ln 12{,}000 - \ln 9000$$
$$t = 50(\ln 12{,}000 - \ln 9000)$$
$$\approx 14.4 \text{ years}$$

14. For monthly compounding, $r = \frac{0.06}{12} = 0.005$. To find when a sum P increases to $1.5P$, solve

$$P(1+0.005)^{12t} = 1.5P$$
$$1.005^{12t} = 1.5$$
$$\ln(1.005)^{12t} = \ln(1.5)$$
$$12t \ln(1.005) = \ln 1.5$$
$$t = \frac{\ln 1.5}{12 \ln 1.005} \approx 6.8 \text{ years}$$

15. $f(x) = e^{-\frac{x^2}{2}}$

$f'(x) = -xe^{-\frac{x^2}{2}} = 0$

Only zero is a critical point.

$f' > 0$ | $f' < 0$

$x = 0$
↗ rel max ↘
(0, 1)

$$f''(x) = -e^{-\frac{1}{2}x^2} + x^2 e^{\frac{1}{2}x^2} = 0$$
$$-e^{-\frac{1}{2}x^2}(1 - x^2) = 0$$
$$x = 1, -1$$

$f'' > 0$ | $f'' < 0$ | $f'' > 0$

con up | -1 | con dn | 1 | con up

Inflection points at $(-1, e^{-1/2})$, $(1, e^{-1/2})$

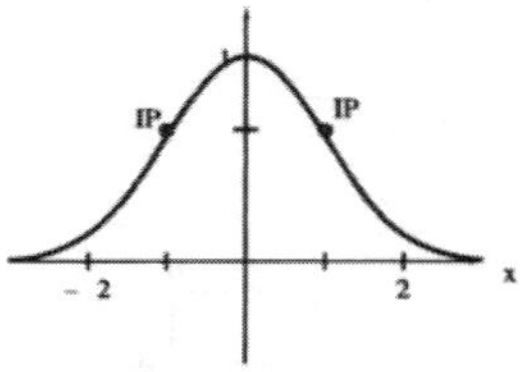

16. $\int (12x^2 - 4x + 1)\,dx$

$$= \int 12x^2 dx - \int 4x\, dx + \int dx$$
$$= 12\left(\frac{x^3}{3}\right) - 4\left(\frac{x^2}{2}\right) + x + C$$
$$= 4x^3 - 2x^2 + x + C$$

17. To find the total amount of pollution, find the integral of $18e^{0.02t}$.

$$\int 18e^{0.02t}\,dt = 18\left(\frac{1}{0.02}e^{0.02t}\right) + C$$
$$= 900e^{0.02t} + C$$

Since the total amount during the next t years is 0 at $t = 0$,

$$900e^{0.02(0)} + C = 0$$
$$900e^{0} + C = 0$$
$$C = -900$$

Thus, the total amount of pollution is $900e^{0.02t} - 900$ million gallons.

18. First, find where the curves intersect by setting the equations equal to each other and solving.

$$20 - x^2 = 8 - 4x$$
$$x^2 - 4x - 12 = 0$$
$$(x-6)(x+2) = 0$$
$$x = -2,\ 6$$

Now, find which curve is the upper curve and which is the lower curve on [–2, 6] by evaluating at $x = 0$.

$$20 - (0)^2 = 20$$
$$8 - 4(0) = 8$$

Thus, $y = 20 - x^2$ is the upper curve and $y = 8 - 4x$ is the lower curve on [–2, 6].

$$\text{Area} = \int_{-2}^{6} \left[(20 - x^2) - (8 - 4x) \right] dx$$
$$= \int_{-2}^{6} (12 + 4x - x^2)\, dx = \left(12x + 2x^2 - \tfrac{x^3}{3} \right)\Big|_{-2}^{6}$$
$$= \left[12(6) + 2(6^2) - \tfrac{(6)^3}{3} \right] - \left[12(-2) + 2(-2)^2 - \tfrac{(-2)^3}{3} \right]$$
$$= 72 - \left(-\tfrac{40}{3} \right) = \tfrac{256}{3}$$

19.
$$\text{Average Value} = \tfrac{1}{b-a} \int_a^b f(x)\, dx$$
$$= \tfrac{1}{4-0} \int_0^4 12\sqrt{x}\, dx$$
$$= \tfrac{1}{4}(12) \int_0^4 x^{1/2}\, dx$$
$$= 3\left(\tfrac{2}{3} x^{3/2} \right)\Big|_0^4$$
$$= 3\left[\tfrac{2}{3}(4)^{3/2} - \tfrac{2}{3}(0)^{3/2} \right]$$
$$= 3\left(\tfrac{16}{3} \right) = 16$$

20. **a.** Let $u = x^3 + 1$. Then $du = 3x^2 dx$.

$$\int \frac{x^2 dx}{x^3+1} = \int \frac{\frac{du}{3}}{u} = \frac{1}{3} \int \frac{du}{u} = \frac{1}{3} \ln|u| + C$$
$$= \tfrac{1}{3} \ln|x^3 + 1| + C$$

b. Let $u = \sqrt{x}$. Then $du = \frac{1}{2} x^{-1/2} dx$ or $2du = \frac{dx}{\sqrt{x}}$.

$$\int \frac{e^{\sqrt{x}} dx}{\sqrt{x}} = \int e^u (2du) = 2 \int e^u du$$
$$= 2e^u + C = 2e^{\sqrt{x}} + C$$

21. Let $u = x$ and $dv = e^{4x} dx$. Then $du = dx$ and $v = \frac{1}{4} e^{4x}$.

$$\int x e^{4x} dx = x\left(\tfrac{1}{4} e^{4x} \right) - \int \tfrac{1}{4} e^{4x} dx$$
$$= \tfrac{1}{4} x e^{4x} - \tfrac{1}{4}\left(\tfrac{1}{4} e^{4x} \right) + C$$
$$= \tfrac{1}{4} x e^{4x} - \tfrac{1}{16} e^{4x} + C$$

22. Use formula 19 with $a = 2$.

$$\int \frac{\sqrt{4-x^2}}{x} dx$$
$$= \sqrt{4 - x^2} - 2 \ln \left| \frac{2 + \sqrt{4-x^2}}{x} \right| + C$$

23.
$$\int_1^{\infty} \frac{1}{x^3} dx = \lim_{a \to \infty} \int_1^a \frac{1}{x^3} dx = \lim_{a \to \infty} -\frac{x^{-2}}{2} \frac{x^{-2}}{2} \Big|_1^a$$
$$= \lim_{a \to \infty} -\frac{1}{2}\left(\frac{1}{a^2} - \frac{1}{(1)^2} \right)$$
$$= -\tfrac{1}{2}(0 - 1) = \tfrac{1}{2}$$

24. $\int_0^1 \sqrt{x^2 + 1} dx \qquad \Delta x = \frac{1-0}{4} = 0.25$

n	$f(x) = \sqrt{x^2 + 1}$
0	$1 \to 0.5$
0.25	1.0308
0.5	1.1180
0.75	1.2500
1	$1.4142 \to 0.707$

0.5 + 1.0308 + 1.118+ 1.25+ 0.707= 4.6058

$$\int_0^1 \sqrt{x^2 + 1} dx = 4.6058(0.25) \approx 1.1515$$

Using FnInt, the integral is about 1.14779

25. $\int_0^1 \sqrt{x^2+1}\,dx \quad \Delta x = \frac{1-0}{4} = 0.25$

x	$f(x)=\sqrt{x^2+1}$	Weight	$f(x)$ • weight
0	1	1	1
0.25	1.0308	4	4.1231
0.5	1.1180	2	2.2360
0.75	1.2500	4	5.0000
1	1.4142	1	1.4142

1 + 4.1231 + 2.236 + 5 + 1.4142 = 13.7733

$\int_0^1 \sqrt{x^2+1}\,dx \approx 13.7733\left(\frac{\Delta x}{3}\right)$

$= 13.7733\left(\frac{0.25}{3}\right) \approx 1.14778$

Using FnInt, the integral is about 1.14779

26. **Please note that there are two #26's for these Cumulative Review Exercises.**

(#26 IN *APPLIED CALCULUS*)

a. $\{(x, y) \mid x \bullet 0, y > 0, x \bullet y\}$

b. $f(4,1) = \frac{3\sqrt{4}+\ln 1}{4-1} = \frac{3\cdot 2+0}{3} = 2$

(#26 IN *BRIEF APPLIED CALCULUS*)

a.

$y' = x^3 y$

$\frac{dy}{dx} = x^3 y$

$dy = x^3 y\, dx$

$\frac{dy}{y} = x^3\, dx,\ y \neq 0$

$y^{-1}\, dy = x^3\, dx$

$\int y^{-1}\, dy = \int x^3\, dx$

$\ln|y| = \frac{x^4}{4} + C$

$e^{\ln|y|} = e^{\frac{x^4}{4}+C}$

$y = e^{\frac{x^4}{4}+C}$

$y = e^{\frac{x^4}{4}} \cdot e^C$

$y = ce^{\frac{x^4}{4}}$

b.

$y = ce^{\frac{x^4}{4}}$

$2 = ce^0$

$2 = c\cdot 1$

$2 = c \rightarrow y = 2e^{\frac{x^4}{4}}$

27.

$f(x,y) = x\ln y + ye^{2x}$

$f_x = \ln y + 2ye^{2x}$

$f_y = \frac{x}{y} + e^{2x}$

28.

$f(x,y) = 2x^2 - 2xy + y^2 + 4x - 6y + 12$

$f_x = 4x - 2y + 4 = 0$

$f_y = -2x + 2y - 6 = 0$

Solving these equations simultaneously gives $x = 1, y = 4$.

$f_{xx} = 4, f_{yy} = 2, f_{xy} = -2.$

$D = (4)(2) - (-2)^2 = 4 > 0.$

Since $f_{xx} > 0$, f has a relative minimum at (1, 4) and

$f(1,4) = 2(1)^2 - 2(1)(4) + (4)^2 + 4(1) - 6(4) + 12$
$= 2.$

29.

x	y	xy	x^2
1	–3	–3	1
3	1	3	9
5	3	15	25
7	8	56	49
$\Sigma x = 16$	$\Sigma y = 9$	$\Sigma xy = 71$	$\Sigma x^2 = 84$

$$a = \frac{n\Sigma xy - (\Sigma x)(\Sigma y)}{n\Sigma x^2 - (\Sigma x)^2} = \frac{4(71)-(16)(9)}{4(84)-(16)^2} = 1.75$$

$$b = \frac{1}{n}(\Sigma y - a\Sigma x) = \frac{1}{4}[9 - 1.75(16)] = -4.75$$

The least squares line is $y = 1.75x - 4.75$.

30. Minimize $f(x,y) = 3x^2 + 2y^2 - 2xy$ subject to $x + 2y = 18$.

$F(x,y,\lambda) = 3x^2 + 2y^2 - 2xy + \lambda(x + 2y - 18)$

$F_x = 6x - 2y + \lambda = 0$ or $\lambda = -6x + 2y$

$F_y = 4y - 2x + 2\lambda = 0$ or $\lambda = x - 2y$

Equating expressions for λ gives

$$-6x + 2y = x - 2y$$
$$4y = 7x$$
$$y = \frac{7}{4}x$$

$F_\lambda = 0$ gives

$$x + 2y - 18 = 0$$
$$x + 2\left(\frac{7}{4}x\right) - 18 = 0$$
$$\frac{9}{2}x = 18$$
$$x = 4$$
$$y = \frac{7}{4}(4) = 7$$

The minimum constrained value of f occurs at (4, 7), and $f(4,7) = 3(4)^2 + 2(7)^2 - 2(4)(7) = 90$.
The D test shows that this is a minimum.

31. $f(x,y) = 2x^2 + xy - 3y^2 + 4$

$f_x = 4x + y$ and $f_y = x - 6y$

$df = f_x \cdot dx + f_y \cdot dy = (4x + y) \cdot dx + (x - 6y) \cdot dy$

32.

$$\text{Volume} = \int_0^2 \int_0^3 (12 - x - 2y)\,dy\,dx$$
$$= \int_0^2 \left(12y - xy - y^2\right)\Big|_0^3 dx$$
$$= \int_0^2 \left\{12(3) - 3x - (3)^2 - \left[12(0) - 0x - (0)^2\right]\right\}$$
$$= \int_0^2 (27 - 3x)\,dx = 27x - \frac{3}{2}x^2\Big|_0^2$$
$$= 27(2) - \frac{3}{2}(2)^2 - (0 - 0) = 54 - 6$$
$$= 48 \text{ cubic units}$$

Chapter 8: Trigonometric Functions

EXERCISES 8.1

1. $90° = \frac{\pi}{2}$

3. $45° = \frac{\pi}{4}$

5. $360° + 180° = 540° = 540 \cdot \frac{\pi}{180} = 3\pi$

7. $-45° = -\frac{\pi}{4}$

9. **a.** $30° = 30 \cdot \frac{\pi}{180} = \frac{\pi}{6}$

b. $225° = 225 \cdot \frac{\pi}{180} = \frac{5\pi}{4}$

c. $-180° = -180 \cdot \frac{\pi}{180} = -\pi$

11. **a.** $60° = 60 \cdot \frac{\pi}{180} = \frac{\pi}{3}$

b. $315° = 315 \cdot \frac{\pi}{180} = \frac{7\pi}{4}$

c. $-120° = -120 \cdot \frac{\pi}{180} = -\frac{2\pi}{3}$

13. **a.** $540° = 540 \cdot \frac{\pi}{180} = 3\pi$

b. $135° = 135 \cdot \frac{\pi}{180} = \frac{3\pi}{4}$

c. $1° = 1 \cdot \frac{\pi}{180} = \frac{\pi}{180}$

15. **a.** $\frac{\pi}{4} = \frac{\pi}{4} \cdot \frac{180}{\pi} = 45°$

b. $\frac{2\pi}{3} = \frac{2\pi}{3} \cdot \frac{180}{\pi} = 120°$

c. $\frac{-5\pi}{6} = \frac{-5\pi}{6} \cdot \frac{180}{\pi} = -150°$

17. **a.** $\frac{\pi}{3} = \frac{\pi}{3} \cdot \frac{180}{\pi} = 60°$

b. $\frac{5\pi}{4} = \frac{5\pi}{4} \cdot \frac{180}{\pi} = 225°$

c. $-\frac{7\pi}{6} = -\frac{7\pi}{6} \cdot \frac{180}{\pi} = -210°$

19. **a.** $\frac{9\pi}{4} = \frac{9\pi}{4} \cdot \frac{180}{\pi} = 405°$

b. $-3\pi = -3\pi \cdot \frac{180}{\pi} = -540°$

c. $5 = 5 \cdot \frac{180}{\pi} = \frac{900°}{\pi}$

21. Let s = the length of the arc. Then the angle is

$$\frac{\pi}{9} = \frac{s}{27}$$

$s = 3\pi \approx 9.42$ inches

23. Let x = measure of the central angle in radians. Then

$$x = \frac{s}{r} = \frac{18}{12} = \frac{3}{2} \text{ radians}$$

25. Let s = the distance between New York City and Paris. Then

$$0.9 = \frac{s}{4000}$$

$$s = 0.9(4000) = 3600 \text{ miles}$$

27. False: The sides are in *proportion* to 1, 2, $\sqrt{3}$.

29. False: The units are reversed.

31. 0

33. 1

35. 0.247

37. **i.** The outer square has sides that each measure $a + b$.

Thus Area of outer square $= (a+b)(a+b) = (a+b)^2$

ii. The inner square has sides that each measure c, and, thus,

Area of inner square $= c \cdot c = c^2$

iii. Area of outer square = Area of inner square + area of 4 triangles

$$(a+b)^2 = c^2 + 2ab$$
$$a^2 + 2ab + b^2 = c^2 + 2ab$$
$$a^2 + b^2 = c^2$$

EXERCISES 8.2

1. $x = 7,\ y = 24,\ r = 25$

$\sin\theta = \dfrac{24}{25} \qquad \cos\theta = \dfrac{7}{25}$

3. $r^2 = 12^2 + 5^2 = 144 + 25 = 169.$
Thus, $r = 13$.

$\sin\theta = \dfrac{y}{r} = \dfrac{5}{13} \qquad \cos\theta = \dfrac{x}{r} = \dfrac{12}{13}$

5. $x = -20,\ y = 21,\ r = 29$

$\sin\theta = \dfrac{21}{29} \qquad \cos\theta = -\dfrac{20}{29}$

7. $r^2 = (-8)^2 + (-6)^2 = 64 + 36 = 100.$
Thus, $r = 10$.

$\cos\theta = \dfrac{-8}{10} = -\dfrac{4}{5} \qquad \sin\theta = \dfrac{-6}{10} = -\dfrac{3}{5}$

9. $x = 96,\ y = -28,\ r = 100$

$\sin\theta = \dfrac{-28}{100} = -\dfrac{7}{25} \qquad \cos\theta = \dfrac{96}{100} = \dfrac{24}{25}$

11. **a.** $\sin\dfrac{\pi}{3} = \dfrac{\sqrt{3}}{2}$

b. $\cos\dfrac{\pi}{3} = \dfrac{1}{2}$

c. $\sin\dfrac{\pi}{4} = \dfrac{1}{\sqrt{2}} = \dfrac{\sqrt{2}}{2}$

d. $\cos\dfrac{\pi}{4} = \dfrac{1}{\sqrt{2}} = \dfrac{\sqrt{2}}{2}$

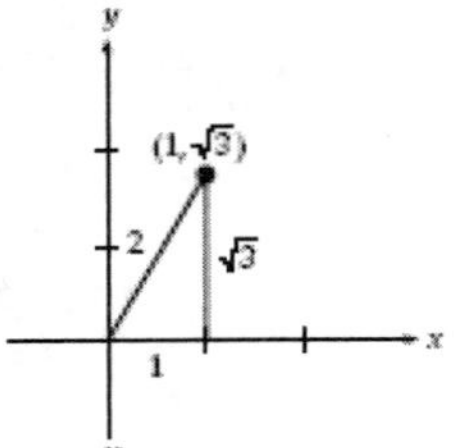

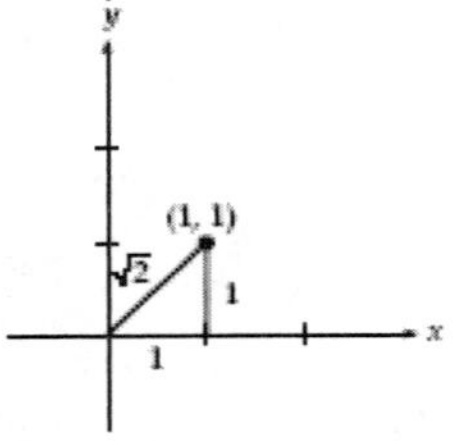

13. **a.** $\sin\dfrac{5\pi}{4} = -\dfrac{1}{\sqrt{2}} = -\dfrac{\sqrt{2}}{2}$

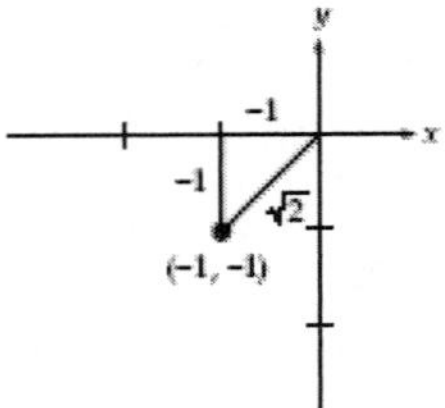

b. $\cos\dfrac{5\pi}{4} = -\dfrac{1}{\sqrt{2}} = -\dfrac{\sqrt{2}}{2}$

c. $\sin\dfrac{5\pi}{6} = \dfrac{1}{2}$

d. $\cos\dfrac{5\pi}{6} = -\dfrac{\sqrt{3}}{2}$

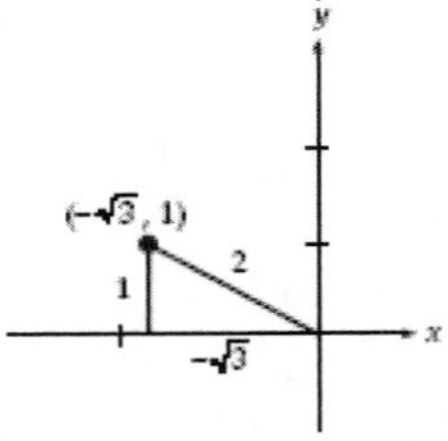

15. **a.** $\sin\left(-\dfrac{\pi}{3}\right) = \dfrac{-\sqrt{3}}{2}$

b. $\cos\left(-\dfrac{\pi}{3}\right) = \dfrac{1}{2}$

c. $\sin\left(-\dfrac{\pi}{4}\right) = \dfrac{-1}{\sqrt{2}} = -\dfrac{\sqrt{2}}{2}$

d. $\cos\left(-\dfrac{\pi}{4}\right) = \dfrac{1}{\sqrt{2}} = \dfrac{\sqrt{2}}{2}$

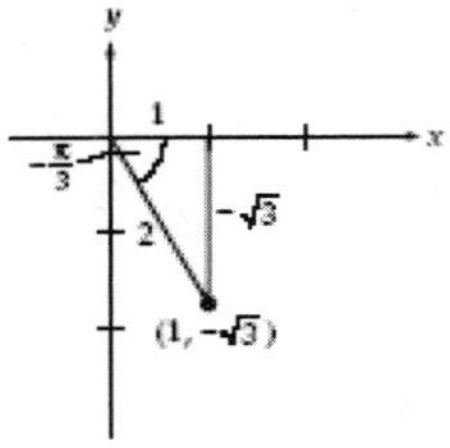

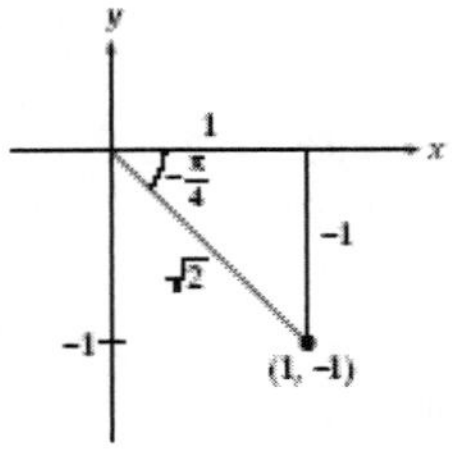

17. **a.** $\sin\dfrac{\pi}{10} \approx 0.31$

b. $\cos\dfrac{\pi}{10} \approx 0.95$

c. $\sin 1 = 0.84$

d. $\cos 1 = 0.54$

19.

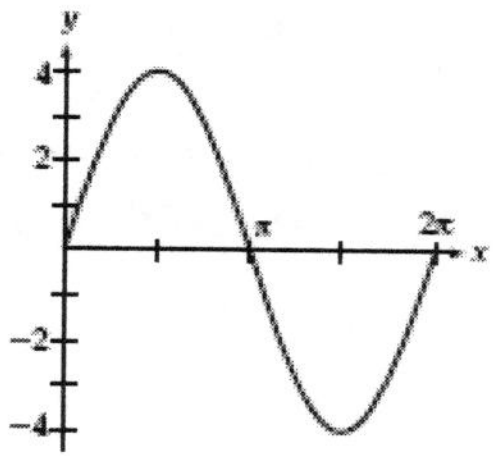

21.

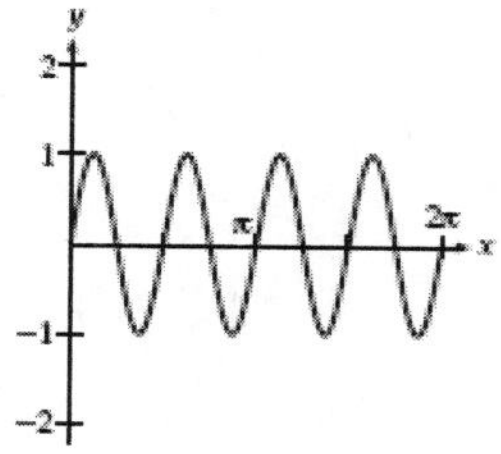

23.

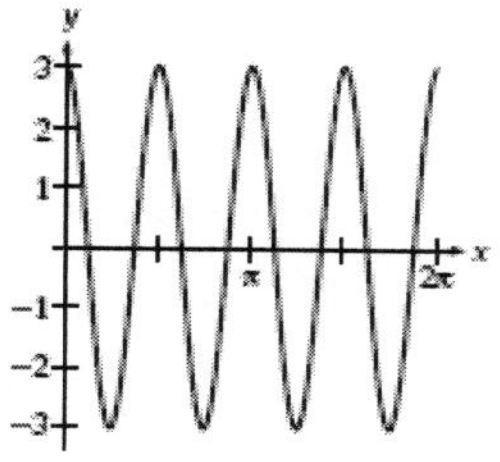

25.

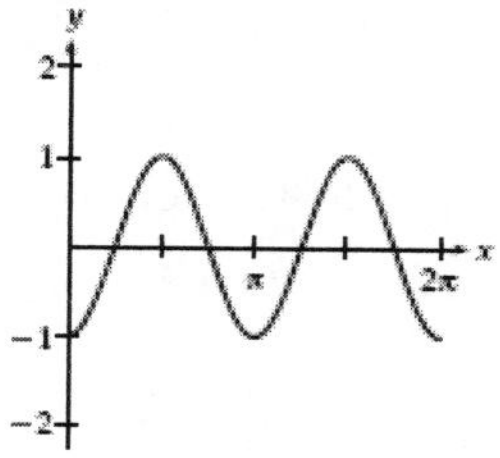

27. **a.** $\sin\frac{5\pi}{12} = \sin\left(\frac{\pi}{4}+\frac{\pi}{6}\right) = \sin\frac{\pi}{4}\cos\frac{\pi}{6} + \cos\frac{\pi}{4}\sin\frac{\pi}{6}$

$= \frac{\sqrt{2}}{2}\cdot\frac{\sqrt{3}}{2} + \frac{\sqrt{2}}{2}\cdot\frac{1}{2} = \frac{\sqrt{6}}{4} + \frac{\sqrt{2}}{4}$

b. $\cos\frac{5\pi}{12} = \cos\left(\frac{\pi}{4}+\frac{\pi}{6}\right) = \cos\frac{\pi}{4}\cos\frac{\pi}{6} - \sin\frac{\pi}{4}\sin\frac{\pi}{6}$

$= \frac{\sqrt{2}}{2}\cdot\frac{\sqrt{3}}{2} - \frac{\sqrt{2}}{2}\cdot\frac{1}{2} = \frac{\sqrt{6}}{4} - \frac{\sqrt{2}}{4}$

29. Let x = height of the wall. Using the definition of the sine, we have

$\sin 70^\circ = \frac{x}{30}$

$x = 30\sin 70^\circ \approx 28.2$ feet

31. Let x = distance between you and the boat. Then

$\cos 48^\circ = \frac{200}{x}$

$x = \frac{200}{\cos 48^\circ} \approx 299$ feet

33. On day 213,

$S(213) = 700 - 600\cos\frac{2\pi(213)}{365} \approx 1219$ cones

35. **a.** $L(0) = 4.9 + 0.4\cos\frac{\pi(0)}{2} = 4.9 + 0.4(1)$

$= 5.3$ liters

b. $L(2) = 4.9 + 0.4\cos\frac{2\pi}{2} = 4.9 + 0.4(-1)$

$= 4.5$ liters

37. **a.**

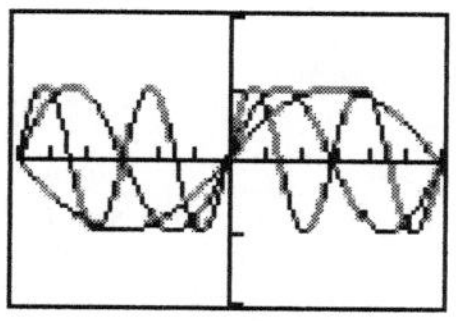

on $[-2\pi, 2\pi]$ by $[-2, 2]$

b.

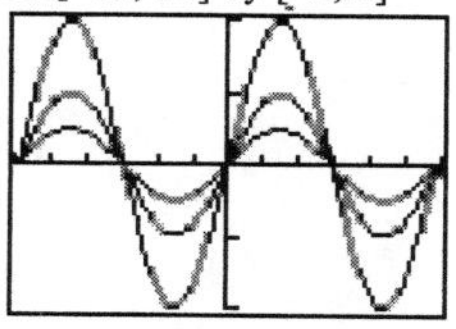

on $[-2\pi, 2\pi]$ by $[-2, 2]$

c. $y = 2\sin 2x$ has amplitude 2 and period π.
$y = -2\sin x$ has period 2π and amplitude 2.
It is an upside down sine graph.
$y = -2\sin 2x$ is an upside down sine graph that has period π and amplitude 2.

39. **a.**

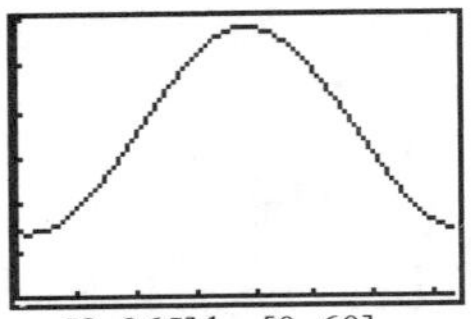

on [0, 365] by [0, 60]

b. $T(10) = 36 + 22 \sin \dfrac{2\pi(10-101)}{365} \approx 14°$

c. $T(193) = 36 + 22 \sin \dfrac{2\pi(193-101)}{365} \approx 58°$

41. **b.** x is the number of years after 1960, so for 2018, $x = 2018 - 1960 = 58$. $C(58) \approx 402.1$, so the carbon dioxide level on January 1, 2018 will be about 402 ppm.

43. Shifting the cosine curve to the right by $\frac{\pi}{2}$ will turn it into a sine curve.

45. Maximum height is 11, minimum height is 5.

47. **a.** 0

b. 1

49. True

51. 1 (since $\sin^2 x + \cos^2 = 1$)

53. **a.** Since the sum of the angles of a triangle is 180° = π radians, and one angle of a right triangle is $\frac{\pi}{2}$ radians, the sum of the other two angles is $\pi - \frac{\pi}{2} = \frac{\pi}{2}$ radians. If one acute angle is of size t radians, then the other must be of size $\frac{\pi}{2} - t$ radians.

b. Since the "opposite" side for angle t is the "adjacent" side for angle $\frac{\pi}{2} - t$, we have

$$\sin t = \frac{\text{opp side for angle } t}{\text{hyp}} = \frac{\text{adj side for angle } (\frac{\pi}{2} - t)}{\text{hyp}} = \cos\left(\frac{\pi}{2} - t\right)$$

$$\cos t = \frac{\text{adj side for angle } t}{\text{hyp}} = \frac{\text{opp side for angle } (\frac{\pi}{2} - t)}{\text{hyp}} = \sin\left(\frac{\pi}{2} - t\right)$$

EXERCISES 8.3

1. $\frac{d}{dt}(t^3 \sin t) = 3t^2 \sin t + t^3 \cos t$ (the Product Rule)

3. $\dfrac{d}{dt}\dfrac{\cos t}{t} = \dfrac{t(-\sin t) - 1 \cdot \cos t}{t^2}$ (the Quotient Rule)

$$= \frac{-t \sin t - \cos t}{t^2}$$

5. $\frac{d}{dt}\sin(t^3 + 1) = \cos(t^3 + 1) \cdot (3t^2) = 3t^2 \cos(t^3 + 1)$

7. $\frac{d}{dt}(\sin 5t) = \cos 5t \cdot (5) = 5 \cos 5t$

9. $\frac{d}{dt}(2 \cos 3t) = -2 \sin 3t \cdot (3) = -6 \sin 3t$

11. $\dfrac{d}{dt}\cos\dfrac{\pi t}{180} = -\left(\sin\dfrac{\pi t}{180}\right) \cdot \left(\dfrac{\pi}{180}\right) = -\dfrac{\pi}{180}\sin\dfrac{\pi t}{180}$

13. **a.** $f'(t) = 2t + \cos \pi t \cdot (\pi) = 2t + \pi \cos \pi t$

b. $f'(0) = 2(0) + \pi \cos \pi(0) = \pi(1) = \pi$

15. **a.** $f'(t) = -4 \sin \frac{t}{2} \cdot \left(\frac{1}{2}\right) = -2 \sin \frac{t}{2}$

b. $f'\left(\frac{\pi}{3}\right) = -2 \sin\left(\frac{1}{2} \cdot \frac{\pi}{3}\right)$
$= -2 \sin \frac{\pi}{6} = -2\left(\frac{1}{2}\right) = -1$

17. **a.** $f'(t) = \cos(\pi - t) \cdot (-1) = -\cos(\pi - t)$

b. $f'\left(\frac{\pi}{2}\right) = -\cos\left(\pi - \frac{\pi}{2}\right) = -\cos\left(\frac{\pi}{2}\right) = 0$

19. $f'(t) = -\sin(t + \pi)^3 \cdot [3(t + \pi)^2]$
$= -3(t + \pi)^2 \sin(t + \pi)^3$

21. $f'(t) = 3\cos^2(t + \pi) \cdot [-\sin(t + \pi)]$
$= -3\cos^2(t + \pi)\sin(t + \pi)$

23. $f'(t) = \cos(\pi - t)^2 \cdot [2(\pi - t)(-1)]$
$= -2(\pi - t)\cos(\pi - t)^2$

25. $f'(t) = 2\sin t^3 \cdot \cos t^3 \cdot (3t^2) = 6t^2 \sin t^3 \cos t^3$

27. $f'(x) = -\sin\sqrt{x+1} \cdot \left[\frac{1}{2}(x+1)^{-1/2}\right]$
$= -\frac{1}{2}(x+1)^{-1/2} \sin\sqrt{x+1}$

29. $f'(x) = 4\cos x + (\cos 4x)(4) + 4\sin^3 x(\cos x)$
$= 4\cos x + 4\cos 4x + 4\sin^3 x \cos x$

31. $f'(x) = \sin e^x \cdot e^x = e^x \sin e^x$

33. $f'(z) = e^{z + \cos z}(1 - \sin z)$

35. $f'(z) = 4(1 - \cos z)^3 \cdot [-(-\sin z)] = 4\sin z \cdot (1 - \cos z)^3$

37. **a.** $f'(z) = \frac{1}{\sin z} \cdot \cos z = \frac{\cos z}{\sin z}$

b. $f'\left(\frac{\pi}{2}\right) = \frac{\cos \frac{\pi}{2}}{\sin \frac{\pi}{2}} = \frac{0}{1} = 0$

39. $f'(z) = \cos(z + \ln z) \cdot \left(1 + \frac{1}{z}\right)$

41. $f'(x) = 6\sin x \cos x - 8\cos x \cdot (-\sin x)$
$= 6\sin x \cos x + 8\cos x \sin x = 14\sin x \cos x$

43. $f'(x) = -\frac{1}{x^2}\cos x^2 + \frac{1}{x}(-\sin x^2) \cdot 2x$
$= -\frac{1}{x^2}\cos x^2 - 2\sin x^2$

45. $f'(x) = -e^{-x}\sin e^x + e^{-x}\cos e^x \cdot e^x$
$= -e^{-x}\sin e^x + \cos e^x$

47. **a.** $$f'(x)=\frac{\sin x\cdot(-\sin x)-\cos x\cdot(\cos x)}{\sin^2 x}=\frac{-\sin^2 x-\cos^2 x}{\sin^2 x}=\frac{-(\sin^2 x+\cos^2 x)}{\sin^2 x}=\frac{-1}{\sin^2 x}$$

b. $$f'\left(\frac{\pi}{3}\right)=-\frac{1}{\left(\sin\frac{\pi}{3}\right)^2}=-\frac{1}{\frac{3}{4}}=-\frac{4}{3}$$

49. $f'(t)=2\sin t\cos t-2\cos t\cdot(-\sin t)=4\sin t\cos t$

51. **a.** $$f'(t)=\frac{\cos t\cdot(\cos t)-(-\sin t)\cdot(1+\sin t)}{\cos^2 t}=\frac{\cos^2 t+\sin^2 t+\sin t}{\cos^2 t}=\frac{1+\sin t}{\cos^2 t}$$

b. $$f'(\pi)=\frac{1+\sin\pi}{(\cos\pi)^2}=\frac{1+0}{(-1)^2}=1$$

53. $f'(x)=\cos(\cos x)\cdot(-\sin x)=-\cos(\cos x)\sin x$

55. $$f'(x)=2\left(\cos\frac{x}{2}-1\right)\cdot\left(-\sin\frac{x}{2}\right)\cdot\left(\frac{1}{2}\right)=-\sin\frac{x}{2}\cdot\left(\cos\frac{x}{2}-1\right)$$

57. $$f'(x)=2x\sin x+x^2\cos x+2\cos x+2x(-\sin x)-2\cos x=x^2\cos x$$

59. $f'(t)=2\sin(t^2+1)\cos(t^2+1)\cdot(2t)$

$4t\sin(t^2+1)\cos(t^2+1)$

61. $f(x)=\sin x+\cos x$

$f(0)=\sin 0+\cos 0=1$ The point is (0, 1).

$f'(x)=\cos x-\sin x$

$f'(0)=\cos 0-\sin 0=1$ Slope at $x=0$ is 1.

Using the point (0, 1) and slope 1, we find the equation of the tangent line at the point.

$y-1=1(x-0)$

$y=x+1$

63. $f(x)=\ln\left(2e^{\sin x}-1\right)$

$f(\pi)=\ln\left(2e^{\sin\pi}-1\right)=0$

The point is $(\pi, 0)$.

$$f'(x)=\frac{2\cos x\cdot e^{\sin x}}{2e^{\sin x}-1}$$

$$f'(\pi)=\frac{2\cos\pi\cdot e^{\sin\pi}}{2e^{\sin\pi}-1}=\frac{2\cdot(-1)\cdot 1}{2-1}=-2$$

Slope at $x=\pi$ is -2.

Using the point $(\pi, 0)$ and slope –2, we find the equation of the tangent line at the point.

$y-0=-2(x-\pi)$

$y=-2x+2\pi$

65. $f'(t)=\sin t+t\cos t$

$f''(t)=\cos t+\cos t+t(-\sin t)=2\cos t-t\sin t$

67. $f'(x) = e^{\sin x} \cos x$
$f''(x) = (e^{\sin x} \cos x) \cos x + e^{\sin x}(-\sin x)$
$= e^{\sin x} \cos^2 x - e^{\sin x} \sin x$

69. $f'(z) = \sin z + z \cos z + (-\sin z) = z \cos z$
$f''(z) = \cos z + z(-\sin z) = \cos z - z \sin z$

71. The rate of change of sales is $S'(x)$

$$S(t) = 500 - 500\cos\frac{\pi t}{26}$$

$$S'(t) = -500\left[-\sin\frac{\pi t}{26}\right]\cdot\left(\frac{\pi}{26}\right) = \frac{500\pi}{26}\sin\frac{\pi t}{26}$$

For $t = 19$, $S'(19) = \frac{500\pi}{26}\sin\frac{19\pi}{26} \approx 45$

In mid May (t = 19), sales are increasing by about 45 air conditioners per week.

73. To find the maximum volume, solve $V'(\theta) = 0$.

$$V(\theta) = 1000\sin\theta$$

$$V'(\theta) = 100\cos\theta = 0$$

$$\cos\theta = 0$$

$$\theta = 90^\circ$$

75. $S(\theta) = \dfrac{\sqrt{3} - \cos\theta}{\sin\theta}$

$$S'(\theta) = \frac{\sin\theta[-(-\sin\theta)] - \cos\theta\cdot(\sqrt{3} - \cos\theta)}{\sin^2\theta} = 0$$

$$\frac{\sin^2\theta - \sqrt{3}\cos\theta + \cos^2\theta}{\sin^2\theta} = 0$$

$$\frac{1 - \sqrt{3}\cos\theta}{\sin^2\theta} = 0$$

$$\cos\theta = \frac{1}{\sqrt{3}}$$

77. **a.** Using NDeriv, $D'(x) = 0$ at $x = 0.7854$.

b. $x = 0.7854 = 0.7854\cdot\dfrac{180}{\pi} = 45^\circ$

79. **a.**

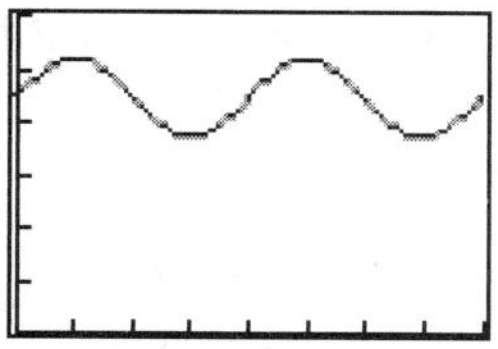

on [0, 1.6] by [0, 120]

b. Using MAXIMUM, blood pressure is greatest after 0.2 seconds and 1 second. The blood pressure then is 105.

c. Using MINIMUM, blood pressure is lowest after 0.6 seconds and 1.4 seconds. The blood pressure then is 75.

81. 0 (since you are differentiating a constant)

83. 0 (since $\sin^2 x + \cos^2 x = 1$)

85. 4

87. $\cos(x + 973)$

89. 1 since $\lim\limits_{h\to 0}\dfrac{\sin h}{h} = 1$

91. **a.**

$$\frac{d}{dx}(\sin^2 x + \cos^2 x) = \frac{d}{dx}(1)$$

$$2\sin x\cos x + 2\cos x\left(\frac{d}{dx}\cos x\right) = 0$$

b.

$$2\cos x\left(\frac{d}{dx}\cos x\right) = -2\sin x\cos x$$

$$\frac{d}{dx}\cos x = -\sin x$$

93. For $f(t) = t$,

$$\frac{d}{dt}\sin f(t) = \cos f(t)\cdot\frac{df}{dt}$$

$$\frac{d}{dt}\sin t = \cos t\cdot\frac{d}{dt}(t)\cdot\cos t\cdot 1 = \cos t$$

and

$$\frac{d}{dt}\cos f(t) = -\sin f(t)\cdot\frac{df}{dt}$$

$$\frac{d}{dt}\cos t = -\sin t\cdot\frac{d}{dt}(t) = -\sin t\cdot 1 = -\sin t$$

95. $\sin(s+t) = (\sin s)(\cos t) + (\cos s)(\sin t)$

$$\frac{d}{ds}\sin(s+t) = (\cos s)(\cos t) - (\sin s)(\sin t)$$

$$\cos(s+t) = (\cos s)(\cot t) - (\sin s)(\sin t)$$

EXERCISES 8.4

1. $$\int(\cos t - \sin t)dt = \int\cos t\,dt + \int(-\sin t)\,dt = \sin t + \cos t + C$$

3. $$\int(1+\sin x)dx = \int 1\,dx + \int\sin x\,dx = x - \cos x + C$$

5. $$\int(t+1)\cos(t^2+2t)\,dt$$

Let $u = t^2 + 2t$. Then $du = (2t+2)\,dt = 2(t+1)\,dt$.

$$\int(t+1)\cos(t^2+2t)\,dt = \int\cos u\left(\frac{du}{2}\right) = \frac{1}{2}\sin u + C = \frac{1}{2}\sin(t^2+2t) + C$$

7. $$\int\sin \pi t\,dt = -\frac{1}{\pi}\cos \pi t + C$$

9. $$\int\cos\frac{\pi}{2}dt = \frac{1}{\frac{\pi}{2}}\sin\frac{\pi}{2}t + C = \frac{2}{\pi}\sin\frac{\pi}{2}t + C$$

11. $$\int\sin(\pi - t)\,dt$$

Let $u = \pi - t$. Then $du = -dt$.

$$\int\sin(\pi-t)\,dt = \int\sin u(-du) = -\int\sin u\,du = -(-\cos u) + C = \cos(\pi - t) + C$$

13. $$\int\cos\frac{2\pi(t+20)}{365}dt$$

Let $u = \frac{2\pi(t+20)}{365} = \frac{2\pi t + 40\pi}{365}$. Then $du = \frac{2\pi}{365}dt$.

$$\int\cos\frac{2\pi(t+20)}{365}dt = \int\cos u\left(\frac{365}{2\pi}du\right) = \frac{365}{2\pi}\int\cos u\,du = \frac{365}{2\pi}\sin u + C = \frac{365}{2\pi}\sin\frac{2\pi(t+20)}{365} + C$$

15. $$\int\sin^2 t\cos t\,dt$$

Let $u = \sin t$. Then $du = \cos t\,dt$.

$$\int\sin^2 t\cos t\,dt = \int u^2\,du = \frac{u^3}{3} + C = \frac{\sin^3 t}{3} + C$$

17. $\displaystyle\int \frac{\sin t}{\cos^3 t}\,dt$

Let $u = \cos t$. Then $du = -\sin t\,dt$.

$$\int \frac{\sin t}{\cos^3 t}\,dt = \int \frac{1}{u^3}(-du) = -\int u^{-3}\,du$$
$$= -\frac{u^{-2}}{-2} = C = \frac{1}{2}(\cos t)^{-2} + C$$
$$= \frac{1}{2\cos^2 t} + C$$

19. $\displaystyle\int e^{1+\sin x}\cos x\,dx$

Let $u = 1 + \sin x$. Then $du = \cos x\,dx$.

$$\int e^{1+\sin x}\cos x\,dx = \int e^u\,du = e^u + C$$
$$= e^{1+\sin x} + C$$

21. $\displaystyle\int \frac{\sin t}{\cos t + 3}\,dt$

Let $u = \cos t + 3$. Then $du = -\sin t\,dt$.

$$\int \frac{\sin t}{\cos t + 3}\,dt = \int \frac{1}{u}(-du) = -\ln|u| + C$$
$$= -\ln|\cos t + 3| + C$$

23. $\displaystyle\int \frac{\sin w}{\sqrt{1-\cos w}}\,dw$

Let $u = 1 - \cos w$.
Then $du = -(-\sin w)dw = \sin w\,dw$.

$$\int \frac{\sin w}{\sqrt{1-\cos w}}\,dw = \int \frac{1}{\sqrt{u}}\,du = \int u^{-1/2}\,du$$
$$= \frac{u^{1/2}}{\frac{1}{2}} + C$$
$$= 2u^{1/2} + C = 2\sqrt{1-\cos w} + C$$

25. $\displaystyle\int \frac{\sin y \cos y}{\sqrt{1+\sin^2 y}}\,dy$

Let $u = 1 + \sin^2 y$. Then $du = 2\sin y\cos y\,dy$.

$$\int \frac{\sin y\cos y}{\sqrt{1+\sin^2 y}}\,dy = \int \frac{1}{\sqrt{u}}\left(\frac{1}{2}du\right)$$
$$= \frac{1}{2}\int u^{-1/2}\,du$$
$$= \frac{1}{2}\left(\frac{u^{1/2}}{\frac{1}{2}}\right) + C = u^{1/2} + C = \sqrt{1+\sin^2 y} + C$$

27. $$\int\left(x^2 - \frac{1}{x} + \cos 2x\right)dx = \int x^2\,dx - \int \frac{1}{x}\,dx + \int \cos 2x\,dx$$
$$= \frac{x^3}{3} - \ln|x| + \frac{1}{2}\sin 2x + C$$

29. **a.** $$\int_0^{1/2} \sin \pi t\,dx = -\frac{1}{\pi}\cos \pi t\Big|_0^{1/2}$$
$$= -\frac{1}{\pi}\left(\cos\frac{\pi}{2} - \cos 0\right)$$
$$= -\frac{1}{\pi}(0-1) = \frac{1}{\pi}$$

b. FnInt gives $0.31831 \approx \frac{1}{\pi}$

31. **a.** $$\int_0^{\pi}(t + \cos t)\,dt = \left(\frac{t^2}{2} + \sin t\right)\Big|_0^{\pi}$$
$$= \frac{\pi^2}{2} + \sin \pi - \left(\frac{0^2}{2} + \sin 0\right)$$
$$= \frac{\pi^2}{2} + 0 - 0 = \frac{\pi^2}{2}$$

b. FnInt gives $4.93480 \approx \frac{\pi^2}{2}$.

33. **a.** $\int_0^1 (t+\sin \pi t)\,dt = \left(\frac{t^2}{2}-\frac{1}{\pi}\cos \pi t\right)\Big|_0^1 = \frac{1^2}{2}-\frac{1}{\pi}\cos \pi(1) - \left[\frac{0^2}{2}-\frac{1}{\pi}\cos \pi(0)\right]$

$= \frac{1}{2}-\frac{1}{\pi}(-1)-\left[0-\frac{1}{\pi}(1)\right] = \frac{1}{2}+\frac{1}{\pi}+\frac{1}{\pi} = \frac{1}{2}+\frac{2}{\pi}$

b. FnInt gives $1.13662 \approx \frac{1}{2}+\frac{2}{\pi}$.

35. **a.** $\int_{\pi/4}^{3\pi/4} \cos\left(t+\frac{\pi}{4}\right)dt = \sin\left(t+\frac{\pi}{4}\right)\Big|_{\pi/4}^{3\pi/4} = \sin\left(\frac{3\pi}{4}+\frac{\pi}{4}\right)-\sin\left(\frac{\pi}{4}+\frac{\pi}{4}\right)$

$= \sin \pi - \sin\frac{\pi}{2} = 0-1 = -1$

b. FnInt gives –1.

37. Area under curve $= \int_0^{\pi} \cos\frac{\pi}{2}\,dx = 2\sin\frac{x}{2}\Big|_0^{\pi}$

$= 2\left(\sin\frac{\pi}{2}-\sin\frac{0}{2}\right) = 2(1-0)$

$= 2$

39. Area under curve $= \int_0^{\pi/2} (\sin x+\cos x)\,dx = (-\cos x+\sin x)\Big|_0^{\pi/2}$

$= -\cos\frac{\pi}{2}+\sin\frac{\pi}{2}-(-\cos 0+\sin 0)$

$= -0+1-(-1+0) = 1+1 = 2$

FnInt gives 2.

41. $\frac{d}{dx}\left(-\frac{1}{a}\cos at+C\right) = -\frac{1}{a}(-\sin at\cdot c)$

$= -\frac{1}{a}(-a)\sin at = \sin at$

$\frac{d}{dx}\left(\frac{1}{a}\sin at+C\right) = \frac{1}{a}\cos at\cdot a$

$= \frac{a}{a}\cos at = \cos at$

43. (Total sales in the first quarter) $= \int_0^{13} S(t)dt = \int_0^{13}\left(10+9\cos\frac{\pi t}{26}\right)dt$

$= \left[10t+9\left(\frac{26}{\pi}\right)\sin\frac{\pi t}{26}\right]\Big|_0^{13}$

$= \left[10(13)+\frac{9\cdot 26}{\pi}\sin\left(\frac{13\pi}{26}\right)\right]-\left[10(0)+\frac{9\cdot 26}{\pi}\sin\left(\frac{0\pi}{26}\right)\right]$

$\approx 130+74.48\sin\frac{\pi}{2}-0-0 = 130+74.48(1) = 204$

Total sales in the first quarter will be about \$204,000.

45. **a.** (Total pollution) $= \int_0^{18} P(t)\,dt = \int_0^{18}\left(5+3\sin\frac{\pi t}{6}\right)dt$

$$= \left[5t+3\left(\frac{6}{\pi}\right)\left(-\cos\frac{\pi t}{6}\right)\right]\Bigg|_0^{18} = \left(5t-\frac{18}{\pi}\cos\frac{\pi t}{6}\right)\Bigg|_0^{18}$$

$$= 18(5)-\frac{18}{\pi}\cos\frac{18\pi}{6}-\left[5(0)-\frac{18}{\pi}\cos\frac{0\pi}{6}\right]$$

$$= 90-\frac{18}{\pi}\cos 3\pi-0+\frac{18}{\pi}\cos 0 = 90-\frac{18}{\pi}(-1)+\frac{18}{\pi}(1)$$

$$= 90+\frac{36}{\pi} \approx 101 \text{ tons}$$

b. (Average pollution) $= \frac{1}{18-0}\int_0^{18} P(t)\,dt = \frac{1}{18}\int_0^{18} P(t)\,dt$

$$= \frac{1}{18}\left(90+\frac{36}{\pi}\right) = 5+\frac{2}{\pi} \approx 5.6 \text{ tons per month}$$

47. $\int_0^{12}\left[300-300\cos\frac{\pi t}{6}\right]dt = \left[300t-300\left(\frac{6}{\pi}\right)\sin\frac{\pi t}{6}\right]\Bigg|_0^{12}$

$$= \left[300(12)-\frac{1800}{\pi}\sin\frac{12\pi}{6}\right]-\left[300(0)-\frac{1800}{\pi}\sin\frac{0\pi}{6}\right]$$

$$= 3600-\frac{1800}{\pi}\sin 2\pi-0$$

$$= 3600 \text{ birds}$$

49. FnInt gives $2\int_0^{122}\left[75-\left(60-30\cos\frac{2\pi t}{365}\right)\right]dt \approx \$6668.$

51. (Average occupancy) $= \frac{1}{10-0}\int_0^{10}\left(200+100\cos\frac{\pi t}{26}\right)dt \approx 277$ guests (using FnInt)

53. **a.** [Average temperature (Jan.–Feb.)] $= \frac{1}{59-0}\int_0^{59}\left[54+24\sin\frac{2\pi(x-101)}{365}\right]dx \approx 32.3$ degrees

b. [Average temperature (July-August)] $= \frac{1}{243-181}\int_{181}^{243}\left[54+24\sin\frac{2\pi(x-101)}{365}\right]dx \approx 75.6$ degrees

55. 0 Integrating over a whole period includes an area above the x-axis and an equal (negative) area *below* the x-axis.

57. -1000 (since $\cos\pi$ is a constant and equal to -1)

59. 0 (since you are integrating over many whole periods)

61. 9 (since, in the long run, $\cos x$ will average out to zero)

63. 54° Fahrenheit

EXERCISES 8.5

1. **a.** $\tan\theta = \frac{y}{x} = \frac{5}{12}$

b. $\cot\theta = \frac{x}{y} = \frac{12}{5}$

c. $\sec\theta = \frac{r}{x} = \frac{13}{12}$

d. $\csc\theta = \frac{r}{y} = \frac{13}{5}$

3. **a.** $\tan\theta = \frac{y}{x} = \frac{-8}{-15} = \frac{8}{15}$

b. $\cot\theta = \frac{x}{y} = \frac{-15}{-8} = \frac{15}{8}$

c. $\sec\theta = \frac{r}{x} = \frac{17}{-15} = -\frac{17}{15}$

d. $\csc\theta = \frac{r}{y} = \frac{17}{-8} = -\frac{17}{8}$

5. **a.** $\tan\frac{\pi}{6} = \frac{y}{x} = \frac{1}{\sqrt{3}} = \frac{\sqrt{3}}{3}$

b. $\csc\frac{\pi}{6} = \frac{r}{y} = \frac{2}{1} = 2$

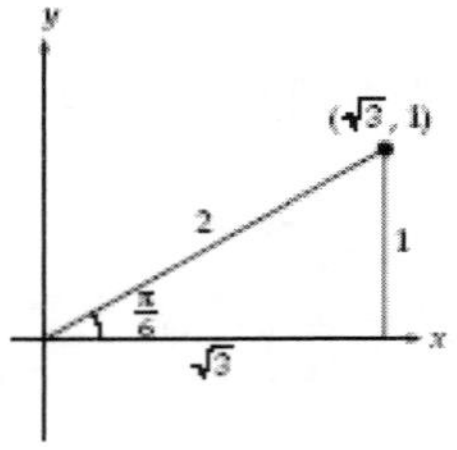

7. **a.** $\tan\frac{3\pi}{2} = \frac{y}{x} = \frac{-1}{0}$ undefined

b. $\csc\frac{3\pi}{2} = \frac{r}{y} = \frac{1}{-1} = -1$

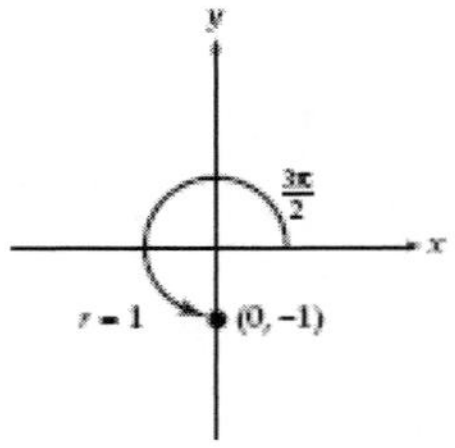

9. **a.** $\tan\frac{\pi}{4} = \frac{y}{x} = \frac{1}{1} = 1$

b. $\cot\frac{\pi}{4} = \frac{x}{y} = \frac{1}{1} = 1$

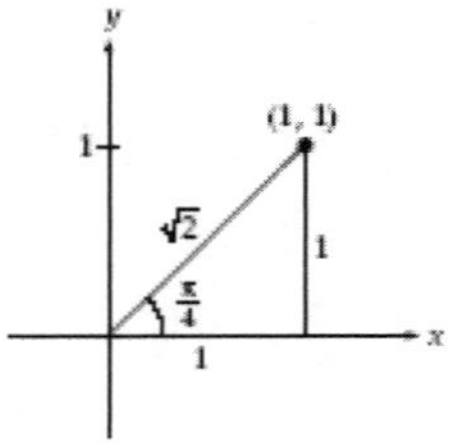

11. **a.** $\tan\frac{7\pi}{6} = \frac{y}{x} = \frac{-1}{-\sqrt{3}} = \frac{\sqrt{3}}{3}$

b. $\csc\frac{7\pi}{6} = \frac{r}{y} = \frac{2}{-1} = -2$

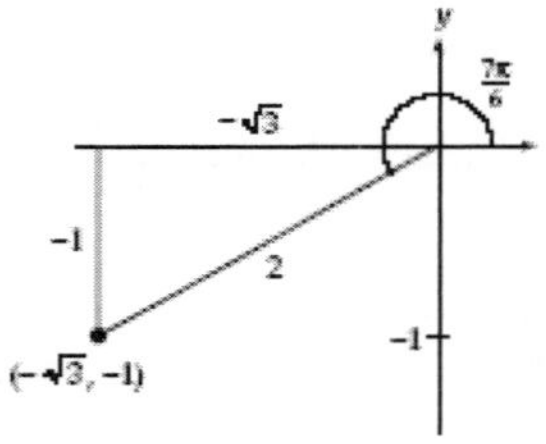

13. $\cot\frac{\pi}{5} = \frac{1}{\tan\frac{\pi}{5}} \approx 1.3764$

15. $\sec\frac{5\pi}{12} = \frac{1}{\cos\frac{5\pi}{12}} \approx 3.8637$

17. Let x = height of the flagpole. Then

$$\tan 62° = \frac{x}{30}$$
$$x = 30\tan 62° \approx 56 \text{ feet}$$

19. Let (x, y) be a point on the line. If $(a, 0)$ is the x-intercept, then $\Delta y = y - 0$ and $\Delta x = x - a$. Thus,

$$\text{Slope of the line} = \frac{\Delta y}{\Delta x} = \tan\theta$$

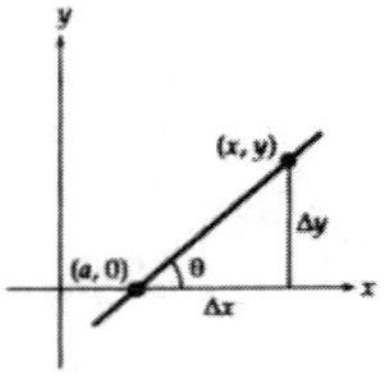

21. Use the formula $y = x\tan\theta - \left(\frac{4x}{v}\sec\theta\right)^2$ with $x = 90$, $v = 60$, and $\theta = 40°$. Then

$$y = x\tan\theta - \left(\frac{4x}{v}\sec\theta\right)^2$$
$$= 90\tan 40° - \left(\frac{4\cdot 90}{60}\sec 40°\right)^2$$
$$\approx 14.2 \text{ feet}$$

Since the height of the football is greater than 10 feet, the ball will clear the goal post.

23. The maximum occurs at about 51.3°. for a height of 20.25 ft.

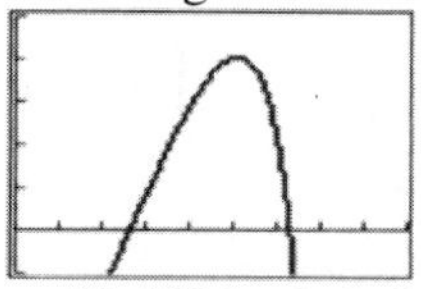

on [0, 90] by [–5, 25]

25.

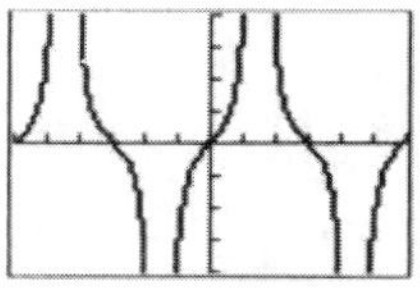

on $[-2\pi, 2\pi]$ by [–4, 4]

The period is 2π, and the function is undefined at odd multiples of $\frac{\pi}{2}$.

27.
$$\sin^2 t + \cos^2 t = 1$$
$$\frac{\sin^2 t}{\cos^2 t} + \frac{\cos^2 t}{\cos^2 t} = \frac{1}{\cos^2 t}$$
$$\tan^2 t + 1 = \sec^2 t$$

29.
$$\frac{d}{dt}(t\cot t) = 1(\cot t) + t\cdot(-\csc^2 t) = \cot t - t\csc^2 t$$

31. $\frac{d}{dt}\tan(t^3+1) = \sec^2(t^3+1)\cdot(3t^2) = 3t^2\sec^2(t^3+1)$

33.
$$\frac{d}{dt}\sec(\pi x+1) = \sec(\pi x+1)\tan(\pi x+1)\cdot\pi = \pi\sec(\pi x+1)\tan(\pi x+1)$$

35. $\frac{d}{d}\cot^2 z^3 = 2\cot z^3(-\csc^2 z^3)\cdot(3z^2) = -6z^2\cot z^3\csc^2 z^3$

37.
$$f(x) = \tan\left(x^2 + \frac{\pi}{4} - 1\right)$$
$$f(1) = \tan\left(1^2 + \frac{\pi}{4} - 1\right) = 1$$
The point is (1, 1).
$$f'(x) = \sec^2\left(x^2 + \frac{\pi}{4} - 1\right)\cdot 2x$$
$$f'(1) = \sec^2\left(1^2 + \frac{\pi}{4} - 1\right)\cdot 2\cdot 1 = 4$$
Slope at $x = 1$ is 4.

Using the point $(1,\ 1)$ and slope 4, we find the equation of the tangent line at the point.
$$y - 1 = 4(x-1)$$
$$y = 4x - 3$$

39.
$$f(x) = e^{\cot x - 1}$$
$$f\left(\frac{\pi}{4}\right) = e^{\cot\frac{\pi}{4} - 1} = e^0 = 1$$
The point is $\left(\frac{\pi}{4},\ 1\right)$.
$$f'(x) = e^{\cot x - 1}(-\csc^2 x)$$
$$f'\left(\frac{\pi}{4}\right) = e^{\cot\frac{\pi}{4} - 1}\left(-\csc^2\frac{\pi}{4}\right) = 1(-2) = -2$$
Slope at $x = \frac{\pi}{4}$ is -2.

Using the point $\left(\frac{\pi}{4},\ 1\right)$ and slope –2, we find the equation of the tangent line at the point.
$$y - 1 = -2\left(x - \frac{\pi}{4}\right)$$
$$y = -2x + 1 + \frac{\pi}{2}$$

41.
$$S(x) = 6\cot\frac{\pi x}{180}$$
$$S'(x) = 6\left(-\csc^2\frac{\pi x}{180}\right)\cdot\frac{\pi}{180} = -\frac{\pi}{30}\csc^2\frac{\pi x}{180}$$
$$S'(10) = -\frac{\pi}{30}\csc^2\frac{10\pi}{180} \approx -3.47$$
The shadow shortens by about 3.47 feet for each additional degree of elevation of the sun.

43.
$$\int t^2\sec^2(t^3+1)\,dt$$
Let $u = t^3 + 1$. Then $du = 3t^2 dt$.
$$\int t^2\sec^2(t^3+1)dt = \int\sec^2 u\left(\frac{du}{3}\right) = \frac{1}{3}\int\sec^2 u\,du = \frac{1}{3}\tan u + C = \frac{1}{3}\tan(t^3+1) + C$$

45. $\int \csc \pi t \cot \pi t \, dt$

Let $u = \pi t$. Then $du = \pi \, dt$.

$$\int \csc \pi t \cot \pi t \, dt = \int \csc u \cot u \left(\frac{du}{\pi}\right) = \frac{1}{\pi}\int \csc u \cot u \, du = \frac{1}{\pi}(-\csc u) + C = -\frac{1}{\pi}\csc \pi t + C$$

47. $\int \tan(1-t)dt$

Let $u = 1 - t$. Then $du = -dt$.

$$\int \tan(1-t)\,dt = \int \tan u(-du) = -\int \tan u \, du = \ln|\cos u| + C = \ln|\cos(1-t)| + C$$

49. $\int x^4 \csc x^5 \, dx$

Let $u = x^5$. Then $du = 5x^4 \, dx$.

$$\int x^4 \csc x^5 \, dx = \int \csc u \left(\frac{du}{5}\right) = \frac{1}{5}\ln|\csc u - \cot u| + C = \frac{1}{5}\ln\left|\csc x^5 - \cot x^5\right| + C$$

51. $\int \cot^4 x \csc^2 x \, dx$

Let $u = \cot x$. Then $du = -\csc^2 x \, dx$.

$$\int \cot^4 x \csc^2 x \, dx = \int u^4(-du) = -\frac{u^5}{5} + C = -\frac{\cot^5 x}{5} + C$$

53. $$\int_{\pi/4}^{\pi/2} \cot t \, dt = \ln|\sin t|\Big|_{\pi/4}^{\pi/2} = \ln\left|\sin\frac{\pi}{2}\right| - \ln\left|\sin\frac{\pi}{4}\right| = \ln 1 - \ln\frac{\sqrt{2}}{2} = 0 - \ln\frac{\sqrt{2}}{2} = -\ln\frac{\sqrt{2}}{2} \approx 0.347$$

55. **a.** $$\int_0^{\pi/4} \sec x \, dx = \ln|\sec x + \tan x|\Big|_0^{\pi/4} = \ln\left|\sec\frac{\pi}{4} + \tan\frac{\pi}{4}\right| - \ln|\sec 0 + \tan 0| = \ln\left|\frac{2}{\sqrt{2}} + 1\right| - \ln|1+0| = \ln\left|\sqrt{2}+1\right| \approx 0.8814$$

b. FnInt gives 0.8814.

57. False: The tangent and cotangent have period π.

59. $\cos\theta$

61. There are none. Does not exist.

63. 0 (Exercise 27 shows that $\csc^2 t - \cot^2 t = 1$, and differentiating 1 gives 0.)

65. The slope is always negative, in fact, always less than or equal to –1 (agreeing with the graph of cot t.)

67. Since $\tan^2 t + 1 = \sec^2 t$,

$$\frac{d}{dt}(\tan^2 t + 1) = \frac{d}{dt}\sec^2 t$$
$$\frac{d}{dt}\tan^2 t + \frac{d}{dt}(1) = \frac{d}{dt}\sec^2 t$$
$$\frac{d}{dt}\tan^2 t = \frac{d}{dt}\sec^2 t$$

69. **a.** $$\frac{d}{dt}\sec t = \frac{d}{dt}(\cos t)^{-1} = -1(\cos t)^{-2}\cdot(-\sin t) = \frac{1}{\cos^2 t} = \sin t = \frac{1}{\cos t}\cdot\frac{\sin t}{\cos t} = \sec t \tan t$$

b.

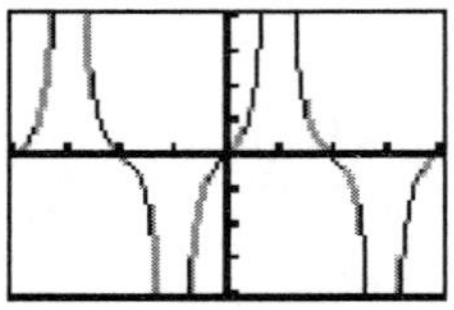

on $[-2\pi, 2\pi]$ by $[-4, 4]$

The graph of the derivative of sec t is $y = \sec t \tan t$.

71. **a.** $\int \tan t \sec^2 t\, dt$

Let $u = \tan t$. Then $du = \sec^2 t\, dt$.

$$\int \tan t \sec^2 t\, dt = \int u\, du = \frac{u^2}{2} + C = \frac{\tan^2 t}{2} + C$$

b. Let $u = \sec t$. Then $du = \sec t \tan t\, dt$.

$$\int \tan t \sec^2 t\, dt = \int \sec t(\sec t \tan t\, dt) = \int u\, du$$

$$= \frac{u^2}{2} + C = \frac{\sec^2 t}{2} + C$$

c. Since $\tan^2 t + 1 = \sec^2 t$, $\frac{1}{2}\tan^2 t$ and $\frac{1}{2}\sec^2 t$ differ by a constant.

Since C represents an arbitrary constant, the two answers are equivalent.

73. $\int \tan t\, dt = \int \frac{\sin t}{\cos t} dt$

Let $u = \cos t$. Then $du = -\sin t\, dt$.

$$\int \tan t\, dt = \int \frac{\sin t}{\cos t} dt = -\int \frac{1}{u} du$$

$$= -\ln|u| + C = -\ln|\cos t| + C$$

75. $\int \tan t\, dt = \int \frac{\sin t}{\cos t} dt$

Let $u = \cos t$. Then $du = -\sin t\, dt$.

$$\int \tan t\, dt = \int \frac{\sin t}{\cos t} dt = -\int \frac{1}{u} du$$

$$= -\ln|u| + C = -\ln|\cos t| + C = \ln|\cos t|^{-1} + C = \ln\left|\frac{1}{\cos t}\right| + C$$

$$= \ln|\sec t| + C$$

$$\int \cot t\, dt = \int \frac{\cos t}{\sin t} dt$$

Let $u = \sin t$. Then $du = \cos t\, dt$.

$$\int \cot t\, dt = \int \frac{\cos t}{\sin t} dt = \int \frac{1}{u} du = \ln|u| + C = \ln|\sin t| + C$$

$$= \ln\left|\frac{1}{\sin t}\right|^{-1} + C = -\ln|\csc t| + C$$

77. $\int \csc t\, dt = \int \csc t\left(\frac{\csc t - \cot t}{\csc t - \cot t}\right) dt = \int \frac{\csc^2 t - \csc t \cot t}{\csc t - \cot t} dt$

Let $u = \csc t - \cot t$. Then $du = (-\csc t \cot t + \csc^2 t)dt$.

$$\int \csc\, dt = \int \frac{\csc^2 t - \csc t \cot t}{\csc t - \cot t} dt = \int \frac{1}{u} du = \ln|u| + C$$

$$= \ln|\csc t - \cot t| + C$$

REVIEW EXERCISES AND CHAPTER TEST FOR CHAPTER 8

1. **a.** $150° = 150 \cdot \dfrac{\pi}{180} = \dfrac{5\pi}{6}$

b. $\dfrac{\pi}{9} = \dfrac{\pi}{9} \cdot \dfrac{180}{\pi} = 20°$

2. **a.** $120° = 120 \cdot \dfrac{\pi}{180} = \dfrac{2\pi}{3}$

b. $\dfrac{\pi}{5} = \dfrac{\pi}{5} \cdot \dfrac{180}{\pi} = 36°$

3. From noon to 12:40 P.M., the minute hand sweeps out an angle that measures $\frac{40}{60}(360°) = 240°$. To find the arc length, use the formula $s = r\theta$ for $r = 26$ and $\theta = 240° = 240 \cdot \frac{\pi}{180} = \frac{4\pi}{3}$. Thus,

$s = 26 \cdot \dfrac{4\pi}{3} = \dfrac{104\pi}{3} \approx 108.853$ feet

4. From 3 P.M. to 3:45 P.M., the minute hand sweeps out an angle that measures $\frac{45}{60}(360) = 270° = 270 \cdot \frac{\pi}{180} = \frac{3\pi}{2}$. Thus, the arc length is

$s = r\theta = 26\left(\dfrac{3\pi}{2}\right) = \dfrac{78\pi}{2} = 39\pi \approx 122.46$ feet

5. **a.** $\sin \dfrac{2\pi}{3} = \dfrac{y}{r} = \dfrac{\sqrt{3}}{2}$

b. $\cos \dfrac{2\pi}{3} = \dfrac{x}{r} = \dfrac{-1}{2} = -\dfrac{1}{2}$

c. $\sin \dfrac{5\pi}{4} = \dfrac{y}{r} = \dfrac{-1}{\sqrt{2}} = -\dfrac{\sqrt{2}}{2}$

d. $\cos \frac{5\pi}{4} = \dfrac{x}{r} = \dfrac{-1}{\sqrt{2}} = -\dfrac{\sqrt{2}}{2}$

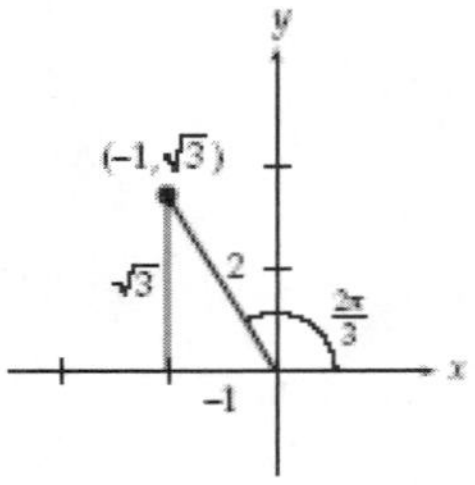

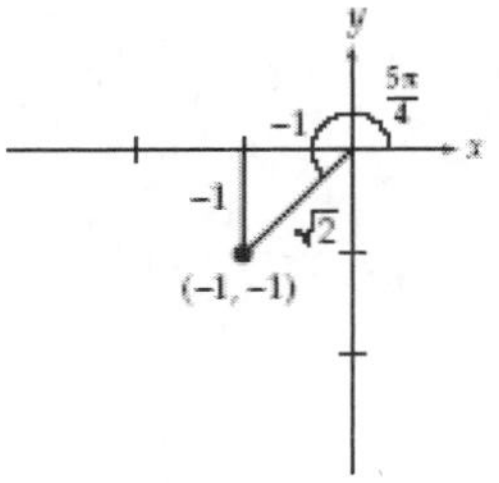

6. **a.** $\sin \dfrac{7\pi}{6} = \dfrac{y}{r} = \dfrac{-1}{2} = -\dfrac{1}{2}$

b. $\cos \dfrac{7\pi}{6} = \dfrac{x}{r} = \dfrac{-\sqrt{3}}{2} = -\dfrac{\sqrt{3}}{2}$

c. $\sin \dfrac{3\pi}{4} = \dfrac{y}{r} = \dfrac{1}{\sqrt{2}} = \dfrac{\sqrt{2}}{2}$

d. $\cos \dfrac{3\pi}{4} = \dfrac{x}{r} = \dfrac{-1}{\sqrt{2}} = -\dfrac{\sqrt{2}}{2}$

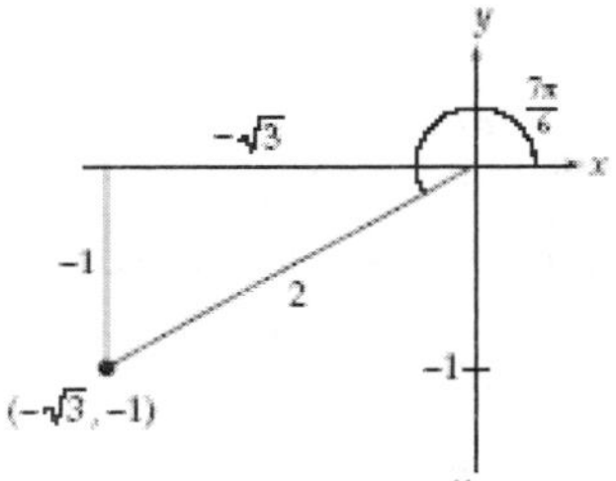

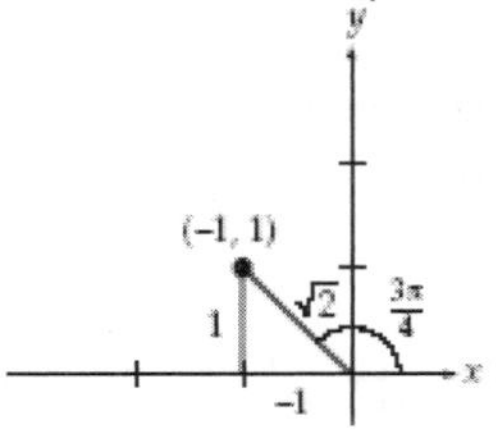

7.

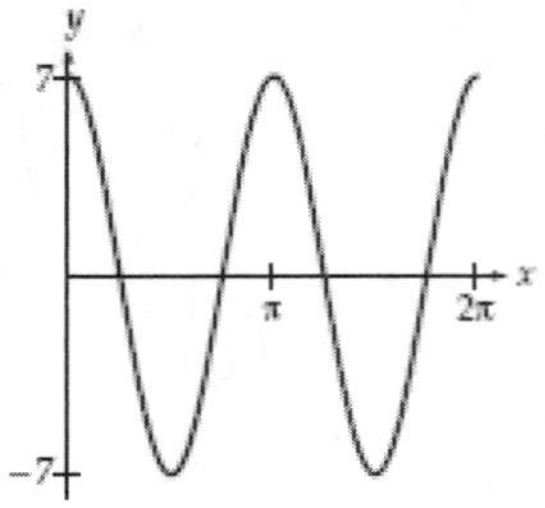

8.

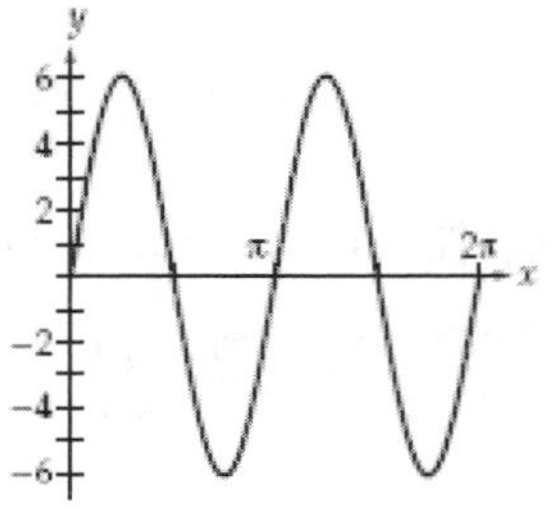

9. $\dfrac{\text{Height in middle}}{\text{Length of rafter}} = \sin 30°$. If height in middle is 12 and we know $\sin 30° = \frac{1}{2}$, then

(Length of rafter) = 2(12) = 24 feet

10. $\sin 30° = \dfrac{\text{depth of water}}{\text{length of anchor line}}$. Since $\sin 30° = \frac{1}{2}$ and the length of the anchor line is 40,

(Depth of water) = $40 \sin 30° = 40\left(\frac{1}{2}\right) = 20$ feet

11. **a.**

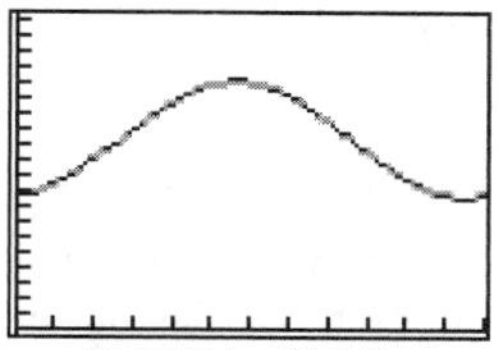

on [0, 365] by [0, 20]

b. $D(121) = 12.25 + 3.75\sin\dfrac{2\pi(121-81)}{365}$
≈ 14.6 hours

c. $D(335) = 12.25 + 3.75\sin\dfrac{2\pi(335-81)}{365}$
≈ 8.7 hours

12. Using the MAX function, we get the maximum sunlight on day 172 (June 21). The MIN function gives the minimum sunlight on day 355 (December 21).

13. $\dfrac{d}{dt}2\cos(4t-1) = -2\sin(4t-1)\cdot 4 = -8\sin(4t-1)$

14. $\dfrac{d}{dt}\sin(2-3t) = \cos(2-3t)\cdot(-3) = -3\cos(2-3t)$

15. $\dfrac{d}{dx}(x^3\sin x + 3x^2\cos x - 6x\sin x - 6\cos x)$
$= 3x^2\sin x + x^3\cos x + 6x\cos x + 3x^2(-\sin x) - 6\sin x - 6x\cos x - 6(-\sin x)$
$= 3x^2\sin x + x^3\cos x + 6x\cos x - 3x^2\sin x - 6\sin x - 6x\cos x + 6\sin x$
$= x^3\cos x$

16. $\dfrac{d}{dt}\left(\dfrac{1+\sin t}{1+\cos t}\right) = \dfrac{(1+\cos t)(\cos t) - (-\sin t)(1+\sin t)}{(1+\cos t)^2}$
$= \dfrac{\cos t + \cos^2 t + \sin t + \sin^2 t}{(1+\cos t)^2} = \dfrac{\cos t + \sin t + 1}{(1+\cos t)^2}$

17. $\dfrac{d}{dt}\left(\sin^3 t^2 - \cos\dfrac{\pi}{3}\right) = 3\sin^2 t^2\cos t^2\cdot(2t) - 0$
$= 6t\sin^2 t^2\cos t^2$

18. $\dfrac{d}{dx}\sin e^{x^2} = \cos e^{x^2}\cdot e^{x^2}\cdot 2x = 2xe^{x^2}\cos e^{x^2}$

19. $\dfrac{d}{dt}\sqrt{\sin t} = \dfrac{d}{dt}(\sin t)^{1/2} = \dfrac{1}{2}(\sin t)^{-1/2}\cos t$

20. $\dfrac{d}{dt}\sin\sqrt{t} = \cos\sqrt{t}\cdot\dfrac{1}{2}t^{-1/2} = \dfrac{1}{2}t^{-1/2}\cos\sqrt{t}$

21. $f(x) = e^{\cos x}$
$f\left(\dfrac{\pi}{2}\right) = e^{\cos\frac{\pi}{2}} = e^0 = 1$ The point is $\left(\dfrac{\pi}{2}, 1\right)$.
$f'(x) = e^{\cos x}\cdot(-\sin x)$
$f'\left(\dfrac{\pi}{2}\right) = e^{\cos\frac{\pi}{2}}\cdot\left(-\sin\dfrac{\pi}{2}\right) = 1\cdot(-1)$
$= -1$ Slope at $x = \dfrac{\pi}{2}$ is -1.

Using the point $\left(\dfrac{\pi}{2}, 1\right)$ and slope –1, we find the equation of the tangent line at the point.

$y - 1 = -1\left(x - \dfrac{\pi}{2}\right)$
$y = -x + 1 + \dfrac{\pi}{2}$

22. **a.**

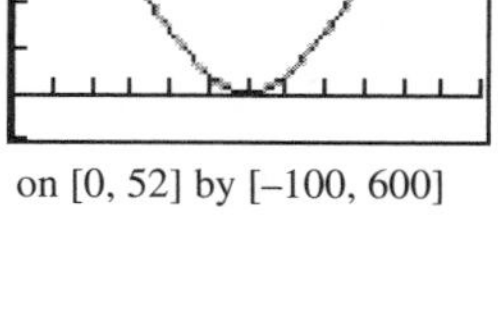

on [0, 52] by [–100, 600]

b. $S(x) = 250 + 250\cos\dfrac{\pi x}{26}$
$S'(x) = -250\sin\dfrac{\pi x}{26}\cdot\dfrac{\pi}{26} = -\dfrac{250\pi}{26}\sin\dfrac{\pi x}{26}$
$S'(37) = -\dfrac{250\pi}{26}\sin\dfrac{37\pi}{26} \approx 29$
Sales are increasing by about 29 sales per week.

23. Using NDERIV and MAX, we find that the rate of increase in sales is greatest in week 39 (sales are increasing at about 30 sales per week.)

24. **a.**

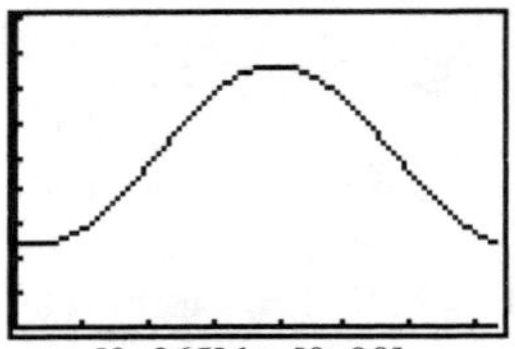

on [0, 365] by [0, 90]

b.
$$T(x)=50-26\cos\frac{2\pi(x-15)}{365}$$
$$T'(x)=-26\left[-\sin\frac{2\pi(x-15)}{365}\right]\left(\frac{2\pi}{365}\right)$$
$$=\frac{52\pi}{365}\sin\frac{2\pi(x-15)}{365}$$
$$T'(288)=\frac{52\pi}{365}\sin\frac{2\pi(288-15)}{365}\approx -0.45$$
The temperature is decreasing by about 0.45 degree per day.

25. Using NDERIV and MAX, we find the rate of increase is the greatest on day 106 (April 16), when the mean temperature is rising by about 0.45 degree per day.

26. Using NDERIV and MAX, we find that the angle that maximizes the volume is $\frac{\pi}{4}\approx 0.7854$.

27.
$$R(\theta)=\frac{r^{-4}-R^{-4}\cos\theta}{\sin\theta}$$
$$R'(\theta)=\frac{\sin\theta[-R^{-4}(-\sin\theta)]-\cos\theta(r^{-4}-R^{-4}\cos\theta)}{\sin^2\theta}$$
$$=\frac{R^{-4}\sin^2\theta-\cos\theta(r^{-4}-R^{-4}\cos\theta)}{\sin^2\theta}=0$$
$$R^{-4}\sin^2\theta-r^{-4}\cos\theta+R^{-4}\cos^2\theta=0$$
$$R^{-4}(1)-r^{-4}\cos\theta=0$$
$$\cos\theta=\frac{R^{-4}}{r^{-4}}=\frac{r^4}{R^4}$$

28. $\int\cos\frac{\pi t}{5}\,dt=\frac{1}{\frac{\pi}{5}}\sin\frac{\pi t}{5}+C=\frac{5}{\pi}\sin\frac{\pi t}{5}+C$

29. $\int t\sin(t^2+1)dt$

Let $u=t^2+1$. Then $du=2t\,dt$.
$$\int t\sin(t^2+1)\,dt=\int\sin u\left(\frac{du}{2}\right)$$
$$=\frac{1}{2}(-\cos u)+C$$
$$=-\frac{1}{2}\cos(t^2+1)+C$$

30. Let $u=\pi-t$. Then $du=-dt$.
$$\int\sin(\pi-t)\,dt=\int\sin u(-du)$$
$$=-(-\cos u)+C$$
$$=\cos(\pi-t)+C$$

31. Let $u=e^{-x}$. Then $du=-e^{-x}\,dx$.
$$\int e^{-x}\sin e^{-x}\,dx=\int\sin u(-du)$$
$$=-(-\cos u)+C$$
$$=\cos e^{-x}+C$$

32. Let $u = \cos x$. Then $du = -\sin x\, dx$.

$$\int e^{\cos x} \sin x\, dx = \int e^u(-du) = -e^u + C$$
$$= -e^{\cos x} + C$$

33. Let $u = \sin t$. Then $du = \cos t\, dt$.

$$\int \frac{\cos t}{\sqrt{\sin t}}\, dt = \int \frac{1}{\sqrt{u}}\, du = 2u^{1/2} + C$$
$$= 2\sqrt{\sin t} + C$$

34. Let $u = 1 + \sin 3t$. Then $du = 3 \cos 3t\, dt$.

$$\int \frac{\cos 3t}{1+\sin 3t}\, dt = \int \frac{1}{u}\left(\frac{1}{3}\, du\right) = \frac{1}{3}\int \frac{1}{u}\, du = \frac{1}{3}\ln |u| + C$$
$$= \frac{1}{3}\ln |1 + \sin 3t| + C$$

35. $$\int (x^3 - e^{-x} - \cos \pi x)\, dx = \frac{x^4}{4} - e^{-x}(-1) - \frac{1}{\pi}\sin \pi x + C$$
$$= \frac{x^4}{4} + e^{-x} - \frac{1}{\pi}\sin \pi x + C$$

36. Let $u = \ln y$. Then $du = \frac{1}{y}dy$.

$$\int \frac{\cos(\ln y)}{y}\, dy = \int \cos u\, du = \sin u + C$$
$$= \sin(\ln y) + C$$

37. Let $u = \frac{1}{z}$. Then $du = -\frac{1}{z^2}dz$.

$$\int \frac{\sin \frac{1}{z}}{z^2}\, dz = \int \sin u(-du) = -\int \sin u\, du$$
$$= -(-\cos u) + C$$
$$= \cos \frac{1}{z} + C$$

38. **a.** $$\int_{\pi/6}^{\pi/2} \sin 2t\, dt = -\frac{1}{2}\cos 2t\Big|_{\pi/6}^{\pi/2} = -\frac{1}{2}\left[\cos 2\left(\frac{\pi}{2}\right) - \cos 2\left(\frac{\pi}{6}\right)\right]$$
$$= -\frac{1}{2}\left(\cos \pi - \cos \frac{\pi}{3}\right) = -\frac{1}{2}\left(-1 - \frac{1}{2}\right) = \frac{3}{4}$$

b. FnInt gives 0.75.

39. **a.** $\int_{\pi/6}^{\pi} \sin^2 t \cos t\, dt$ Let $u = \sin t$. Then $du = \cos t\, dt$.
For $t = \frac{\pi}{6}$, $u = \sin \frac{\pi}{6} = \frac{1}{2}$.
For $t = \pi$, $u = \sin \pi = 0$.

$$\int_{\pi/6}^{\pi} \sin^2 t \cos t\, dt = \int_{1/2}^{0} u^2\, du = \frac{u^3}{3}\Big|_{1/2}^{0} = 0 - \left[\frac{(\frac{1}{2})^3}{3}\right] = -\frac{1}{24}$$

b. FnInt gives $-0.04167 \approx -\frac{1}{24}$.

40. $$(\text{Average value}) = \frac{1}{b-a}\int_a^b f(t)\, dt = \frac{1}{3-0}\int_0^3 \sin \pi t\, dt$$
$$= \frac{1}{3}\left(-\frac{1}{\pi}\cos \pi x\right)\Big|_0^3 = -\frac{1}{3\pi}(\cos 3\pi - \cos 0)$$
$$= -\frac{1}{3\pi}(-1-1) = \frac{2}{3\pi} \approx 0.212$$

41. $$(\text{Average value}) = \frac{1}{3-0}\int_0^3 (200 + 100\cos \pi t)\, dt$$
$$= \frac{1}{3}\left(200t + \frac{100}{\pi}\sin \pi t\right)\Big|_0^3 = \frac{1}{3}\left[200(3) + \frac{100}{\pi}\sin 3\pi - \left(0 + \frac{100}{\pi}\sin 0\right)\right]$$
$$= \frac{1}{3}\left[600 + \frac{100}{\pi}(0) - (0+0)\right] = 200$$

42. Area $= \int_0^{\pi/2} \sin^2 x \cos x\, dx$

$= \int_0^1 u^2\, du = \left.\frac{u^3}{3}\right|_0^1 = \frac{1}{3} - 0 = \frac{1}{3}$ square unit

Let $u = \sin x$. Then $du = \cos x\, dx$.
For $x = 0$, $u = \sin 0 = 0$
For $x = \frac{\pi}{2}$, $u = \sin \frac{\pi}{2} = 1$.

43. Area $= \int_0^{\pi/4} (\cos x - \sin x)\, dx = [\sin x - (-\cos x)]\Big|_0^{\pi/4}$

$= (\sin x + \cos x)\Big|_0^{\pi/4} = \left(\sin \frac{\pi}{4} + \cos \frac{\pi}{4}\right) - (\sin 0 + \cos 0)$

$= \left(\frac{\sqrt{2}}{2} + \frac{\sqrt{2}}{2}\right) - (0+1) = \sqrt{2} - 1 \approx 0.414$ square unit

44. **a.**

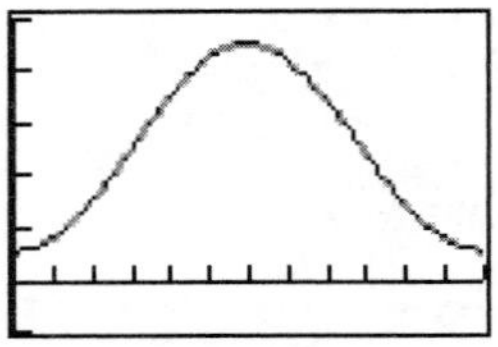

on [0, 52] by [–10, 50]

b. (Total sales during weeks first half of year) $= \int_0^{26} \left(25 - 20\cos \frac{\pi x}{26}\right) dx = \left[25x - \left(20 \sin \frac{\pi x}{26}\right)\left(\frac{26}{\pi}\right)\right]\Big|_0^{26}$

$= \left(25x - \frac{520}{\pi} \sin \frac{\pi x}{26}\right)\Big|_0^{26}$

$= \left[25(26) - \frac{520}{\pi} \sin \frac{26\pi}{26}\right] - \left(25 \cdot 0 - \frac{520}{\pi} \sin \frac{0\pi}{26}\right)$

$= 650 - \frac{520}{\pi}(0) - \left(0 - \frac{520}{\pi} \cdot 0\right) = 650$

Total sales during the first half of the year were $650,000.

45. **a.**

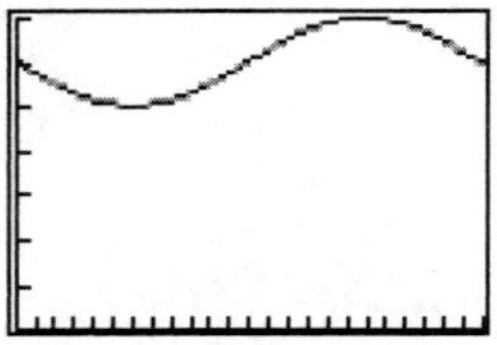

on [0, 24] by [0, 70]

b. (Average temperature during first 6 hours) $= \frac{1}{6-0} \int_0^6 \left(60 - 10\sin \frac{\pi x}{12}\right) dx$

$= \frac{1}{6}\left\{60x - 10\left[\frac{12}{\pi}\left(-\cos \frac{\pi x}{12}\right)\right]\right\}\Big|_0^6$

$= \frac{1}{6}\left(60x + \frac{120}{\pi} \cos \frac{\pi x}{12}\right)\Big|_0^6$

$= \frac{1}{6}\left[60 \cdot 6 + \frac{120}{\pi} \cos \frac{6\pi}{12} - \left(60 \cdot 0 + \frac{120}{\pi} \cos \frac{0 \cdot \pi}{12}\right)\right]$

$= \frac{1}{6}\left[360 + \frac{120}{\pi} \cdot 0 - \left(0 + \frac{120}{\pi} \cdot 1\right)\right]$

$= \frac{1}{6}\left(360 - \frac{120}{\pi}\right) = 60 - \frac{20}{\pi} \approx 53.6$ degrees Fahrenheit

46. **a.** $\tan\frac{\pi}{3} = \frac{y}{x} = \frac{\sqrt{3}}{1} = \sqrt{3}$

b. $\cot\frac{\pi}{3} = \frac{x}{y} = \frac{1}{\sqrt{3}} = \frac{\sqrt{3}}{3}$

c. $\sec\frac{\pi}{3} = \frac{r}{x} = \frac{2}{1} = 2$

d. $\csc\frac{\pi}{3} = \frac{r}{y} = \frac{2}{\sqrt{3}} = \frac{2\sqrt{3}}{3}$

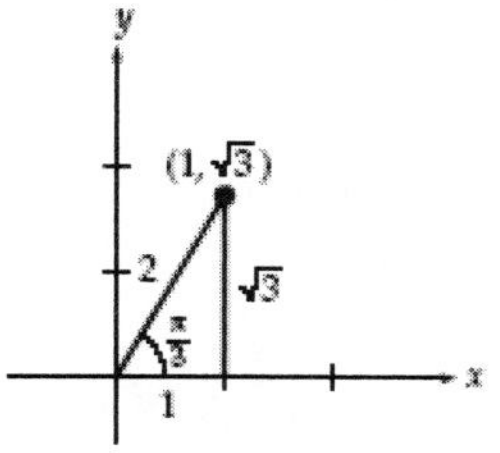

47. **a.** $\tan\frac{3\pi}{4} = \frac{y}{x} = \frac{1}{-1} = -1$

b. $\cot\frac{3\pi}{4} = \frac{x}{y} = \frac{-1}{1} = -1$

c. $\sec\frac{3\pi}{4} = \frac{r}{x} = \frac{\sqrt{2}}{-1} = -\sqrt{2}$

d. $\csc\frac{3\pi}{4} = \frac{r}{y} = \frac{\sqrt{2}}{1} = \sqrt{2}$

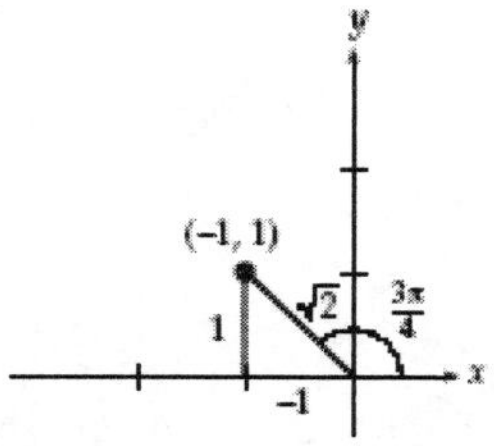

48. Let x = height of monument. Then

$$\tan 40^\circ = \frac{x}{600}$$
$$x = 600\tan 40^\circ \approx 503 \text{ feet}$$

49. Let x = distance from boat to lighthouse. Then

$$\tan 20^\circ = \frac{80}{x}$$
$$x = \frac{80}{\tan 20^\circ} \approx 220 \text{ feet}$$

50. $$\frac{d}{dx}(x\tan x^2) = \tan x^2 + x\sec^2 x^2 \cdot 2x = \tan x^2 + 2x^2\sec^2 x$$

51. $$\frac{d}{dx}(x^2\cot x) = 2x\cot x + x^2(-\csc^2 x) = 2x\cot x - x^2\csc^2 x$$

52. $$\frac{d}{dt}\csc(\pi - t) = -\csc(\pi - t)\cot(\pi - t)\cdot(-1) = \csc(\pi - t)\cot(\pi - t)$$

53. $$\frac{d}{dt}\sec\sqrt{t} = \sec\sqrt{t}\tan\sqrt{t}\cdot\left(\frac{1}{2}t^{-1/2}\right) = \frac{1}{2}t^{-1/2}\sec\sqrt{t}\tan\sqrt{t}$$

54. $$\frac{d}{dt}e^{\tan\pi t} = e^{\tan\pi t}\sec^2\pi t\cdot\pi = \pi e^{\tan\pi t}\sec^2\pi t$$

55. Since $\cot^2 t + 1 = \csc^2 t$,

$$\frac{d}{dt}(\cot^2 t + 1) = \frac{d}{dt}\csc^2 t$$
$$\frac{d}{dt}\cot^2 t + \frac{d}{dt}(1) = \frac{d}{dt}\csc^2 t$$
$$\frac{d}{dt}\cot^2 t = \frac{d}{dt}\csc^2 t$$

56. $$f(x) = \ln(2\tan x - 1)$$

$$f\left(\frac{\pi}{4}\right) = \ln\left(2\tan\frac{\pi}{4} - 1\right) = \ln 1 = 0$$ The point is $\left(\frac{\pi}{4}, 0\right)$.

$$f'(x) = \frac{2\sec^2 x}{2\tan x - 1}$$

$$f'\left(\frac{\pi}{4}\right) = \frac{2\sec^2\frac{\pi}{4}}{2\tan\frac{\pi}{4} - 1} = \frac{2\cdot 2}{2\cdot 1 - 1} = 4$$ Slope at $x = \frac{\pi}{4}$ is 4.

Using the point $\left(\frac{\pi}{4}, 0\right)$ and slope 4, we find the equation of the tangent line at the point.

$$y - 0 = 4\left(x - \frac{\pi}{4}\right)$$
$$y = 4x - \pi$$

57. $\int \sec^2 \frac{\pi t}{2}\, dt$

Let $u = \frac{\pi t}{2}$. Then $du = \frac{\pi}{2}\, dt$.

$$\int \sec^2 \frac{\pi t}{2}\, dt = \int \sec^2 u \left(\frac{2}{\pi} du\right) = \frac{2}{\pi} \tan u + C$$
$$= \frac{2}{\pi} \tan \frac{\pi t}{2} + C$$

58. Let $u = x^2$. Then $du = 2x\, dx$.

$$\int x \tan x^2\, dx = \int \tan u \left(\frac{du}{2}\right)$$
$$= \frac{1}{2}(-\ln |\cos u|) + C$$
$$= -\frac{1}{2} \ln \left|\cos x^2\right| + C$$

59. Let $u = t^3$. Then $du = 3t^2\, dt$.

$$\int t^2 \sec t^3 \tan t^3\, dt = \int \sec u \tan u \left(\frac{du}{3}\right) = \frac{\sec u}{3} + C$$
$$= \frac{1}{3} \sec t^3 + C$$

60. Let $u = \quad - t$. Then $du = -dt$.

$$\int \csc(\pi - t) \cot(\pi - t)\, dt = \int \csc u \cot u(-du)$$
$$= -(-\csc u) + C$$
$$= \csc(\pi - t) + C$$

61. Let $u = \cot x$. Then $du = -\csc^2 x\, dx$.

$$\int \sqrt{\cot x} \csc^2 x\, dx = \int \sqrt{u}\,(-du) = -\frac{2}{3} u^{3/2} + C = -\frac{2}{3}(\cot x)^{3/2} + C$$

62. Let $u = \tan x$. Then $du = \sec^2 x\, dx$.

$$\int \frac{\sec^2 x}{\sqrt{\tan x}}\, dx = \int \frac{du}{\sqrt{u}} = 2u^{1/2} + C = 2\sqrt{\tan x} + C$$

63. **a.** $\displaystyle\int_0^{\pi/4} \sec t\, dt = \ln |\sec t + \tan t| \Big|_0^{\pi/4}$

$$= \ln \left|\sec \frac{\pi}{4} + \tan \frac{\pi}{4}\right| - \ln |\sec 0 + \tan 0|$$
$$= \ln \left|\sqrt{2} + 1\right| - \ln |1 + 0| = \ln \left|\sqrt{2} + 1\right|$$

b. FnInt gives $0.8814 \approx \ln \left|\sqrt{2} + 1\right|$.

64. **a.** $\displaystyle\int_{\pi/4}^{\pi/2} \csc^2 t\, dt = -\cot t \Big|_{\pi/4}^{\pi/2} = -\left(\cot \frac{\pi}{2} - \cot \frac{\pi}{4}\right)$

$$= -(0 - 1) = 1$$

b. FnInt gives 1.

65. **a.** Area under curve $= \displaystyle\int_{\pi/4}^{\pi/2} \cot x\, dx = \ln |\sin x| \Big|_{\pi/4}^{\pi/2}$

$$= \ln \left|\sin \frac{\pi}{2}\right| - \ln \left|\sin \frac{\pi}{4}\right| = \ln 1 - \ln \left|\frac{\sqrt{2}}{2}\right| = -\ln \frac{\sqrt{2}}{2}$$

b. FnInt gives $0.3466 \approx -\ln \frac{\sqrt{2}}{2}$.

66. **a.** Area under curve $= \int_{\pi/4}^{\pi/2} \csc x\,dx = \ln|\csc x - \cot x| \Big|_{\pi/4}^{\pi/2}$

$$= \ln\left|\csc\frac{\pi}{2} - \cot\frac{\pi}{2}\right| - \ln\left|\csc\frac{\pi}{4} - \cot\frac{\pi}{4}\right|$$

$$= \ln|1-0| - \ln\left|\sqrt{2}-1\right| = -\ln(\sqrt{2}-1)$$

b. FnInt gives $0.8814 \approx -\ln(\sqrt{2}-1)$.

67.

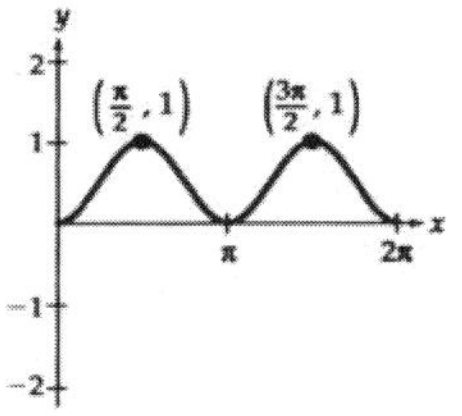

68.

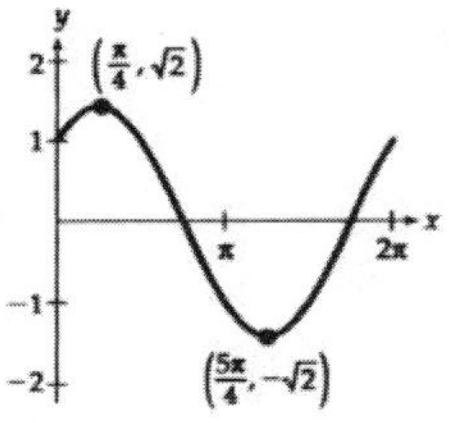

69. **a.** $\sin 2x + \cos y = 0$

$$2\cos 2x + (-\sin y)\frac{dy}{dx} = 0$$

$$-\sin y\frac{dy}{dx} = -2\cos 2x$$

$$\frac{dy}{dx} = \frac{2\cos 2x}{\sin y}$$

b. $$\left.\frac{dy}{dx}\right|_{(\frac{\pi}{2},\frac{\pi}{2})} = \frac{2\cos 2(\frac{\pi}{2})}{\sin\frac{\pi}{2}} = \frac{2\cos\pi}{\sin\frac{\pi}{2}} = \frac{2(-1)}{1} = -2$$

70. **a.** $\sin y - \cos 4x = 0$

$$\cos y\frac{dy}{dx} - (-4\sin 4x) = 0$$

$$\cos y\frac{dy}{dx} = -4\sin 4x$$

$$\frac{dy}{dx} = -\frac{4\sin 4x}{\cos y}$$

b. $$\left.\frac{dy}{dx}\right|_{(\frac{\pi}{8},0)} = -\frac{4\sin 4(\frac{\pi}{8})}{\cos 0} = -\frac{4\sin\frac{\pi}{2}}{\cos 0} = -\frac{4(1)}{1} = -4$$

71. $\frac{dy}{dt} = 10$ and we are asked to find $\frac{d\theta}{dt}$. We know that $\tan\theta = \frac{y}{200}$.

Differentiating implicitly, we get

$$\sec^2\theta\frac{d\theta}{dt} = \frac{1}{200}\frac{dy}{dt}$$

$$\frac{d\theta}{dt} = \frac{\cos^2\theta}{200}\frac{dy}{dt}$$

Since $\frac{dy}{dt} = 10$, $\frac{d\theta}{dt}$ at $\theta = \frac{\pi}{3}$ is

$$\frac{d\theta}{dt} = \frac{\cos^2\frac{\pi}{3}}{200}(10) = \frac{(\frac{1}{2})^2}{20} = \frac{1}{80}$$

The radar angle θ is increasing at $\frac{1}{80}$ radian per second.

72. We wish to find $\frac{d\theta}{dt}$ given $\frac{dx}{dt} = -450$.

$$\cot\theta = \frac{x}{5}$$

$$-\csc^2\theta\frac{d\theta}{dt} = \frac{1}{5}\frac{dx}{dt}$$

$$\frac{d\theta}{dt} = \frac{-\sin^2\theta}{5}\frac{dx}{dt}$$

Since $\frac{dx}{dt} = -450$, $\frac{d\theta}{dt}$ at $\theta = \frac{\pi}{6}$ is

$$\frac{d\theta}{dt} = \frac{-\sin^2\frac{\pi}{6}}{5}(-450) = \frac{\left(\frac{1}{2}\right)^2(450)}{5} = \frac{450}{20}$$

$$= 22.5 \text{ rad/hr} = \frac{22.5 \text{ rad}}{1 \text{ hr}}\cdot\frac{180}{\pi}\cdot\frac{1 \text{ hr}}{60 \text{ min}}$$

$$\approx 21 \text{ deg/min}$$

The angle of the radar beam is increasing by 22.5 radians per hour, or about 21 degrees per minute.

73. $f(x, y) = \sin 2x - y\cos x$

a. $f_x = 2\cos 2x - y(-\sin x) = 2\cos 2x + y\sin x$

b. $f_y = -\cos x$

74. $f(x, y) = \ln \cos(x - y)$

a. $f_x = \dfrac{1}{\cos(x-y)}[-\sin(x-y)] = -\dfrac{\sin(x-y)}{\cos(x-y)}$

b. $f_y = \dfrac{1}{\cos(x-y)}[-\sin(x-y)(-1)]$
$= \dfrac{\sin(x-y)}{\cos(x-y)}$

75. $f(x, y) = (\sin x)e^{\cos y}$

a. $f_x = (\cos x)e^{\cos y}$

b. $f_y = (\sin x)e^{\cos y} \cdot (-\sin y)$
$= (-\sin x \sin y)e^{\cos y}$

76. $f(x, y) = \sin(x^3 y)$

a. $f_x = \cos(x^3 y)(3x^2 y) = 3x^2 y \cos(x^3 y)$

b. $f_y = \cos(x^3 y)(x^3) = x^3 \cos(x^3 y)$

77. Let $u = t$ and $dv = \sin t\, dt$. Then $du = dt$ and $v = -\cos t$.

$$\int t\sin t\, dt = t(-\cos t) - \int(-\cos t)\, dt$$
$$= -t\cos t + \int \cos t\, dt$$
$$= -t\cos t + \sin t + C$$

78. Let $u = t$ and $dv = \cos t\, dt$. Then $du = dt$ and $v = \sin t\, dt$.

$$\int t\cos t\, dt = t\sin t - \int \sin t\, dt$$
$$= t\sin t - (-\cos t) + C$$
$$= t\sin t + \cos t + C$$

79. Let $u = x$ and $dv = \sec^2 x\, dx$. Then $du = dx$ and $v = \tan x$.

$$\int x\sec^2 x\, dx = x\tan x - \int \tan x\, dx$$
$$= x\tan x - (-\ln|\cos x|) + C = x\tan x + \ln|\cos x| + C$$

80. Let $u = x$ and $dv = \sec x \tan x\, dx$. Then $du = dx$ and $v = \sec x$.

$$\int x\sec x\tan x\, dx = x\sec x - \int \sec x\, dx$$
$$= x\sec x - \ln|\sec x + \tan x| + C$$

81. Let $u = e^t$ and $dv = \sin t\, dt$. Then $du = e^t\, dt$ and $v = -\cos t$.

$$\int e^t \sin t\, dt = e^t(-\cos t) - \int -\cos t(e^t\, dt) = -e^t\cos t + \int e^t \cos t\, dt$$

Now let $u = e^t$ and $dv = \cos t\, dt$. Then $du = e^t\, dt$ and $v = \sin t$.

$$\int e^t \sin t\, dt = -e^t\cos t + \int e^t\cos t\, dt$$
$$= -e^t\cos t + e^t\sin t - \int e^t \sin t\, dt$$

$$2\int e^t \sin t\, dt = -e^t\cos t + e^t\sin t$$

$$\int e^t \sin t\, dt = \frac{1}{2}e^t(\sin t - \cos t)$$

82. Let $u = e^t$ and $dv = \cos t\,dt$. Then $du = e^t\,dt$ and $v = \sin t$.

$$\int e^t \cos t\,dt = e^t \sin t - \int \sin t(e^t\,dt) = e^t \sin t - \int e^t \sin t\,dt$$

Let $u = e^t$ and $dv = \sin t\,dt$. Then $du = e^t\,dt$ and $v = -\cos t$.

$$\begin{aligned}\int e^t \cos t\,dt &= e^t \sin t - \int e^t \sin t\,dt \\ &= e^t \sin t - [e^t(-\cos t) - \int e^t(-\cos t)\,dt] \\ &= e^t \sin t - (-e^t \cos t + \int e^t \cos t\,dt) \\ &= e^t \sin t + e^t \cos t - \int e^t \cos t\,dt\end{aligned}$$

$$2\int e^t \cos t\,dt = e^t \sin t + e^t \cos t$$

$$\int e^t \cos t\,dt = \frac{e^t}{2}(\sin t + \cos t)$$

83. For $n = 10$, $\int_0^\pi e^{\sin x}\,dx \approx 6.1923$.
For $n = 100$, $\int_0^\pi e^{\sin x}\,dx \approx 6.2086$.
For $n = 500$, $\int_0^\pi e^{\sin x}\,dx \approx 6.2088$.

84. For $n = 10$, $\int_0^{\pi/2} \sqrt{\sin x}\,dx \approx 1.1852$.
For $n = 100$, $\int_0^{\pi/2} \sqrt{\sin x}\,dx \approx 1.1977$.
For $n = 500$, $\int_0^{\pi/2} \sqrt{\sin x}\,dx \approx 1.1981$.

85. For $n = 10$, $\int_0^\pi e^{\sin x}\,dx \approx 6.2087$.
For $n = 100$, $\int_0^\pi e^{\sin x}\,dx \approx 6.2088$.
For $n = 500$, $\int_0^\pi e^{\sin x}\,dx = 6.2088$.

86. For $n = 10$, $\int_0^{\pi/2} \sqrt{\sin x}\,dx = 1.1931$.
For $n = 100$, $\int_0^{\pi/2} \sqrt{\sin x}\,dx = 1.1977$.
For $n = 500$, $\int_0^{\pi/2} \sqrt{\sin x}\,dx = 1.1981$.

Chapter 9: Differential Equations

EXERCISES 9.1 (Equivalent to Section 6.5 of *Brief Applied Calculus*, 7e.)

1. We calculate y, y', and y'' first.

$y = e^{2x} - 3e^x + 2$

$y' = 2e^{2x} - 3e^x$

$y'' = 4e^{2x} - 3e^x$

Substitute these into the differential equation.

$$y'' - 3y' + 2y = 4$$
$$(4e^{2x} - 3e^x) - 3(2e^{2x} - 3e^x) + 2(e^{2x} - 3e^x + 2) = 4$$
$$4e^{2x} - 3e^x - 6e^{2x} + 9e^x + 2e^{2x} - 6e^x + 4 = 4$$
$$4 = 4$$

The function y satisfies the differential equation.

3. We calculate y and y'.

$y = ke^{ax} - \frac{b}{a}$

$y' = kae^{ax}$

Substitute these into the differential equation.

$y' = ay + b$

$kae^{ax} \stackrel{?}{=} a\left(ke^{ax} - \frac{b}{a}\right) + b$

$kae^{ax} \stackrel{?}{=} ake^{ax} - b + b$

$kae^{ax} = kae^{ax}$

The function y satisfies the differential equation.

5.

$y^2 y' = 4x$

$y^2 \frac{dy}{dx} = 4x$

$y^2 dy = 4x\ dx$

$\int y^2 dy = \int 4x\ dx$

$\frac{y^3}{3} = 2x^2 + C$

$y^3 = 6x^2 + c$

$y = (6x^2 + c)^{1/3}$

7.

$y' = x + y$

$\frac{dy}{dx} = x + y$

Not separable because the equation cannot be written in the form $\frac{dy}{dx} = \frac{f(x)}{g(y)}$.

9.

$y' = 6x^2 y$

$\frac{dy}{dx} = 6x^2 y$

$\frac{dy}{y} = 6x^2 dx$

$\int \frac{dy}{y} = \int 6x^2 dx$

$\ln y = 2x^3 + C$

$y = e^{2x^3 + C}$

$y = e^{2x^3} e^C$

$y = ce^{2x^3}$

To check, find y'.

$y' = ce^{2x^3}(6x^2) = 6x^2 y$

11.
$$y' = \frac{y}{x}$$
$$\frac{dy}{dx} = \frac{y}{x}$$
$$\frac{dy}{y} = \frac{dx}{x}$$
$$\int \frac{dy}{y} = \int \frac{dx}{x}$$
$$\ln y = \ln x + C$$
$$y = e^{\ln x + C} = e^{\ln x} e^{C} = cx$$

To check, $y' = c = \frac{cx}{x} = \frac{y}{x}$.

13.
$$yy' = 4x$$
$$y\frac{dy}{dx} = 4x$$
$$y\ dy = 4x\ dx$$
$$\int y\ dy = \int 4x\ dx$$
$$\frac{y^2}{2} = 2x^2 + C$$
$$y^2 = 4x^2 + c$$
$$y = \pm\sqrt{4x^2 + c}$$

15.
$$y' = e^{xy}$$
$$\frac{dy}{dx} = e^{xy}$$

Not separable because the differential equation cannot be written into the form $\frac{dy}{dx} = \frac{f(x)}{g(y)}$.

17.
$$y' = 9x^2$$
$$\frac{dy}{dx} = 9x^2$$
$$dy = 9x^2 dx$$
$$\int dy = \int 9x^2 dx$$
$$y = 3x^3 + C$$

19.
$$y' = \frac{x}{x^2+1}$$
$$\frac{dy}{dx} = \frac{x}{x^2+1}$$

Let $u = x^2 + 1$. Then $du = 2x\ dx$.
$$\int dy = \int \frac{x}{x^2+1}\, dx = \int \frac{1}{u}\left(\frac{du}{2}\right)$$
$$y = \frac{1}{2}\ln|u| + C$$
$$y = \frac{1}{2}\ln\left|x^2+1\right| + C$$

21.
$$y' = x^3 y$$
$$\frac{dy}{dx} = x^3 y$$
$$\frac{dy}{y} = x^3 dx$$
$$\int \frac{dy}{y} = \int x^3 dx$$
$$\ln y = \frac{x^4}{4} + C$$
$$y = e^{x^4/4 + C}$$
$$y = e^{x^4/4} e^{C}$$
$$y = ce^{x^4/4}$$

23.
$$y' = x^m y^n \qquad (m > 0,\ n \neq 1)$$
$$\frac{dy}{dx} = x^m y^n$$
$$\frac{dy}{y^n} = x^m dx$$
$$\int \frac{dy}{y^n} = \int x^m dx$$
$$\frac{y^{-n+1}}{-n+1} = \frac{x^{m+1}}{m+1} + C$$
$$y^{1-n} = \frac{1-n}{m+1} x^{m+1} + C$$
$$y = \left(\frac{1-n}{m+1} x^{m+1} + c\right)^{1/(1-n)}$$

25.
$$y' = 2\sqrt{y}$$
$$\frac{dy}{dx} = 2\sqrt{y}$$
$$\frac{dy}{\sqrt{y}} = 2\ dx$$
$$\int \frac{dy}{\sqrt{y}} = \int 2\ dx$$
$$\int y^{-1/2} dy = \int 2\ dx$$
$$2y^{1/2} = 2x + C$$
$$y^{1/2} = x + c$$
$$y = (x + c)^2$$

27. $xy' = x^2 + y^2$

$\frac{dy}{dx} = \frac{x^2 + y^2}{x}$

Not separable because the differential equation cannot be written in the form $\frac{dy}{dx} = \frac{f(x)}{g(y)}$.

29. $y' = xy + x$

$\frac{dy}{dx} = x(y+1)$

$\frac{dy}{y+1} = x\ dx$

$\int \frac{dy}{y+1} = \int x\ dx$

$\ln|y+1| = \frac{x^2}{2} + C$

$y + 1 = e^{x^2/2+C}$

$y = e^{x^2/2}e^C - 1 = ce^{x^2/2} - 1$

31. $y' = ye^x - e^x$

$\frac{dy}{dx} = e^x(y-1)$

$\frac{dy}{y-1} = e^x\ dx$

$\int \frac{dy}{y-1} = \int e^x\ dx$

$\ln|y-1| = e^x + C$

$y - 1 = e^{e^x + C} = e^{e^x}e^C = ce^{e^x}$

$y = ce^{e^x} + 1$

33. $y' = ay^2 \quad (a > 0)$

$\frac{dy}{dx} = ay^2$

$\frac{dy}{y^2} = a\ dx$

$\int \frac{dy}{y^2} = \int a\,dx$

$\int y^{-2} dy = \int a\,dx$

$\frac{y^{-1}}{-1} = ax + C$

$y = -\frac{1}{ax + C}$

35. $y' = ay + b$

$\frac{dy}{dx} = ay + b$

$\frac{dy}{ay+b} = dx$

$\frac{1}{a}\int \frac{a\ dy}{ay+b} = \int dx$

$\int \frac{a\ dy}{ay+b} = \int a\ dx$

$\ln|ay+b| = ax + C$

$ay + b = e^{ax+C} = e^{ax}e^C = ce^{ax}$

$y = \frac{ce^{ax} - b}{a} = ke^{ax} - \frac{b}{a}$

since k is arbitrary

37. $y^2 y' = 2x$

$y^2\frac{dy}{dx} = 2x$

$y^2 dy = 2x\ dx$

$\int y^2 dy = \int 2x\,dx$

$\frac{y^3}{3} = x^2 + C$

$y^3 = 3x^2 + c$

$y = \sqrt[3]{3x^2 + c}$

If $y(0) = 2$, then

$2 = \sqrt[3]{3(0)^2 + c}$

$2^3 = 0 + c$

$c = 8$

The particular solution is $y = \sqrt[3]{3x^2 + 8}$.

39.
$$y' = xy$$
$$\frac{dy}{dx} = xy$$
$$\frac{dy}{y} = x\,dx$$
$$\int \frac{dy}{y} = \int x\,dx$$
$$\ln y = \frac{1}{2}x^2 + C$$
$$y = e^{\left(\frac{x^2}{2}+C\right)} = e^{x^2/2}e^C = ce^{x^2/2}$$
If $y(0) = -1$, then
$$-1 = ce^0$$
$$c = -1$$
The particular solution is $y = -e^{x^2/2}$.
To check:
$$y' = -\frac{1}{2}(2x)e^{x^2/2} = -xe^{x^2/2} = x\left(-e^{x^2/2}\right)$$
$$= xy$$
Also $y(0) = -e^0 = -1$.

41.
$$y' = 2xy^2$$
$$\frac{dy}{dx} = 2xy^2$$
$$\frac{dy}{y^2} = 2x\,dx$$
$$\int \frac{dy}{y^2} = \int 2x\,dx$$
$$-\frac{1}{y} = x^2 + C$$
$$y = \frac{1}{-x^2 + c}$$
If $y(0) = 1$
$$1 = \frac{1}{-(0)^2 + C}$$
$$c = 1$$
The particular solution is $y = \frac{1}{1-x^2}$.

To check: $y' = \frac{2x}{(1-x^2)^2} = 2x\left(\frac{1}{(1-x^2)^2}\right)$
$$= 2x\left(\frac{1}{1-x^2}\right)^2 = 2xy^2$$
Also, $y(0) = \frac{1}{1-0^2} = 1$.

43.
$$y' = \frac{y}{x}$$
$$\frac{dy}{dx} = \frac{y}{x}$$
$$\frac{dy}{y} = \frac{dx}{x}$$
$$\int \frac{dy}{y} = \int \frac{dx}{x}$$
$$\ln y = \ln x + C$$
$$y = e^{(\ln x + C)} = e^{\ln x}e^C = cx$$
If $y(1) = 3$
$$3 = c(1)$$
$$c = 3$$
The particular solution is $y = 3x$.
To check:
$$y' = 3 = \frac{3x}{x} = \frac{y}{x}$$ Also $y(1) = 3(1) = 3$.

45.
$$y' = 2\sqrt{y}$$
$$\frac{dy}{dx} = 2\sqrt{y}$$
$$\int \frac{dy}{\sqrt{y}} = \int 2\,dx$$
$$2\sqrt{y} = 2x + C$$
$$\sqrt{y} = x + C$$
$$y = (x+c)^2$$
If $y(1) = 4$, then
$$4 = (1+c)^2$$
$$c^2 + 2c - 3 = 0$$
$$(c+3)(c-1) = 0$$
$$c = -3,\ 1$$
The particular solutions are $y = (x+1)^2$ and $y = (x-3)^2$.

47.
$$y' = y^2 e^x + y^2$$
$$\frac{dy}{dx} = y^2(e^x + 1)$$
$$\frac{1}{y^2}\,dy = (e^x + 1)dx$$
$$\int \frac{1}{y^2}\,dy = \int (e^x + 1)dx$$
$$-\frac{1}{y} = e^x + x + C$$
$$y = \frac{-1}{e^x + x + c}$$
If $y(0) = 1$, then
$$1 = \frac{-1}{e^0 + 0 + c} + = \frac{-1}{1+c}$$
$$1 + c = -1$$
$$c = -2$$
The particular solution is $y = \frac{-1}{e^x + x - 2} = \frac{1}{2 - e^x - x}$.

49.
$$y' = ax^2 y \quad (a > 0)$$
$$\frac{dy}{dx} = ax^2 y$$
$$\frac{dy}{y} = ax^2\,dx$$
$$\int \frac{dy}{y} = \int ax^2 dx$$
$$\ln |y| = \frac{ax^3}{3} + C$$
$$y = e^{\frac{ax^3}{3} + C} = ce^{ax^3/3}$$
If $y(0) = 2$, then
$$2 = ce^{a(0)^3/3}$$
$$c = 2$$
The particular solution is $y = 2e^{ax^3/3}$.

51.
$$-\frac{pD'}{D} = k$$
$$-\frac{p}{D}\frac{dD}{dp} = k$$
$$\frac{dD}{D} = -\frac{k\,dp}{p}$$
$$\int \frac{dD}{D} = -k \int \frac{dp}{p}$$
$$\ln D = -k \ln p + C_1$$
$$D = e^{-k \ln p + C_1} = e^{-k \ln p} e^{C_1} = Cp^{-k}$$
$D(p) = Cp^k$ (for any constant C)

53.
$$y' = d + ry,\ d = 1000,\ r = 0.05$$
$$y' = 1000 + 0.05y$$
$$\frac{dy}{dx} = 1000 + 0.05y$$
$$\frac{dy}{1000 + 0.05y} = dx$$
$$\int \frac{dy}{1000 + 0.05y} = \int dx$$
$$\frac{1}{0.05} \ln |1000 + 0.05y| = x + C$$
$$1000 + 0.05y = e^{0.05x + C} = ce^{0.05x}$$
$$y = ke^{0.05x} - \frac{1000}{0.05}$$
$$y = ke^{0.05x} - 20{,}000$$
If $y(0) = 0$, then
$$0 = ke^{0.05(0)} - 20{,}000$$
$$k = 20{,}000$$
The particular solution is
$y = 20{,}000e^{0.05x} - 20{,}000$.

55. **a.**
$$y' = -0.32(y-70), \quad y(0) = 98.6$$
$$\frac{dy}{dt} = -0.32(y-70)$$
$$\int \frac{dy}{y-70} = \int -0.32\, dt$$
$$\ln(y-70) = -0.32t + C$$
$$y - 70 = e^{(-0.32t+C)} = ce^{-0.32t}$$
$$y = ce^{-0.32t} + 70$$
If $y(0) = 98.6$, then
$$98.6 = c + 70$$
$$c = 28.6$$
So $y = (28.6e^{-0.32t} + 70)$ degrees t hours after the murder.

b. If the temperature of the body is 80 degrees at t hours, then
$$80 = 28.6e^{-0.32t} + 70$$
$$e^{-0.32t} = \frac{10}{28.6}$$
$$t = \frac{\ln\left(\frac{10}{28.6}\right)}{-0.32} \approx 3.28 \text{ hours since the murder}$$

57.
$$y' = 30 - 0.2y$$
$$\frac{dy}{dx} = 30 - 0.2y$$
$$\frac{dy}{30 - 0.2y} = dx$$
$$-\frac{1}{0.2}\ln|30 - 0.2y| = x + C$$
$$\ln|30 - 0.2y| = -0.2x + c$$
$$30 - 0.2y = e^{-0.2x+c}$$
$$y = ke^{-0.2x} + 150$$
If $y(0) = 0$
$$0 = ke^{-0.2(0)} + 150$$
$$0 = k + 150$$
$$k = -150$$
The particular solution is $y = -150e^{-0.2x} + 150$.

59. **a.** Let $y(t)$ = your bank balance. Then y' is the rate of change and $y' = 3 + 0.1y$.

b. Since at $t = 0$ the balance $y = 6$ thousand, $y(0) = 6$.

c.
$$\frac{dy}{dt} = 3 + 0.1y$$
$$\frac{dy}{3 + 0.1y} = dt$$
$$\frac{0.1dy}{3 + 0.1y} = 0.1dt$$
$$\ln(3 + 0.1y) = 0.1t + C$$
$$3 + 0.1y = e^{(0.1t+C)} = ce^{0.1t}$$
$$y = 10(ce^{0.1t} - 3) = ke^{0.1t} - 30$$
If $y(0) = 6$, then
$$6 = k - 30$$
$$k = 36$$
So after t years, you have $36e^{0.1t} - 30$ thousand dollars.

d. For $t = 25$,
$$y = (25) = 36e^{0.1(25)} - 30 = 36e^{2.5} - 30$$
$$\approx \$408.57 \text{ thousand or } \$408,570$$

61. **a.** $y(t)$ = monthly sales after t months. The rate of growth y' is 4 times $y^{0.5}$, or $y' = 4y^{0.5}$.

b. At time $t = 0$, then sales $y(0) = 10{,}000$.

c.
$$y' = 4y^{0.5}$$
$$\frac{dy}{dt} = 4y^{0.5}$$
$$\frac{dy}{y^{0.5}} = 4dt$$
$$\int \frac{dy}{y^{0.5}} = \int 4dt$$
$$2y^{0.5} = 4t + C$$
$$y = (2t + c)^2$$
If $y(0) = 10{,}000$, then
$$10{,}000 = [2(0) + c]^2$$
$$10{,}000 = c^2$$
$$\pm 100 = c$$
The particular solution is $y = (2t + 100)^2$.

d. For $t = 12$,
$$y = [2(12) + 100]^2 = (124)^2 = 15{,}376.$$

63. **a.** Let $y(t)$ = the size of the colony. The rate of change y' is 8 times $y^{3/4}$, or $y' = 8y^{3/4}$.

b. At the time $t = 0$, $y(0) = 10{,}000$

c.
$$y' = \frac{dy}{dt} = 8y^{3/4}$$
$$\frac{dy}{y^{3/4}} = 8\,dt$$
$$\int \frac{dy}{y^{3/4}} = \int 8\,dt$$
$$4y^{1/4} = 8t + C$$
$$y^{1/4} = 2t + c$$
$$y = (2t + c)^4$$
If $y(0) = 10{,}000$, then
$$10{,}000 = [2(0) + c]^4$$
$$10{,}000 = c^4$$
$$10 = c$$
The particular solution is $y = (2t + 10)^4$.

d. At $t = 6$,
$$y(6) = [2(6) + 10]^4 = (22)^4 = 234{,}256$$

65. **a.**

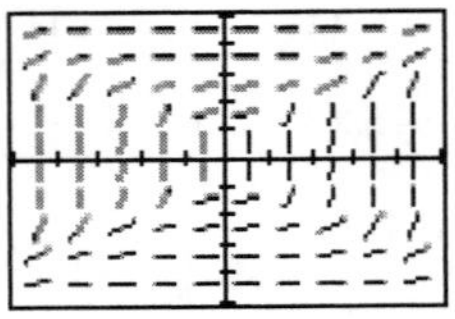

on [–5, 5] by [–5, 5]

b.

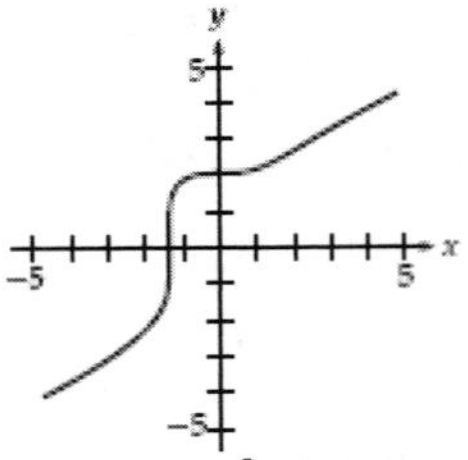

c.
$$\frac{dy}{dx} = \frac{6x^2}{y^4},\ y(0) = 2$$
$$y^4 dy = 6x^2 dx$$
$$\frac{1}{5}y^5 = 2x^3 + C$$
$$y = \sqrt[5]{10x^3 + c}$$
If $y(0) = 2$, then
$$2 = \sqrt[5]{c}$$
$$c = 32$$
The solution is $y = \sqrt[5]{10x^3 + 32}$

d.

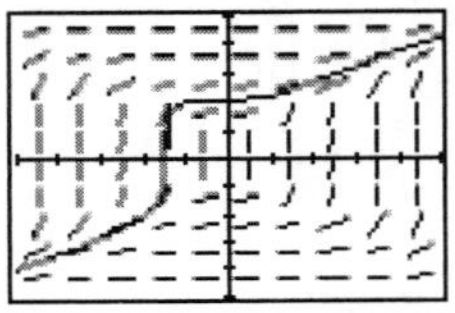

on [–5, 5] by [–5, 5]

67. **a.**

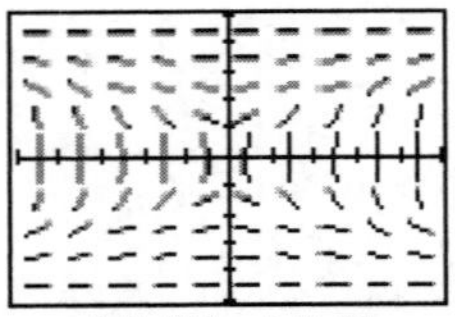

on [–5, 5] by [–5, 5]

b.

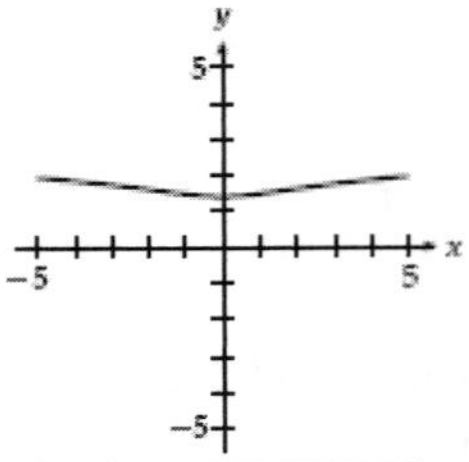

c.
$$\frac{dy}{dx} = \frac{4x}{y^3}$$
$$y^3 dy = 4x\,dx$$
$$\frac{1}{4}y^4 = 2x^2 + c$$
$$y = \pm\sqrt[4]{8x^2 + c}$$
If $y(0) = 2$, then
$$2 = \sqrt[4]{c}$$
$$c = 16$$
The solution is $y = \sqrt[4]{8x^2 + 16}$

d.

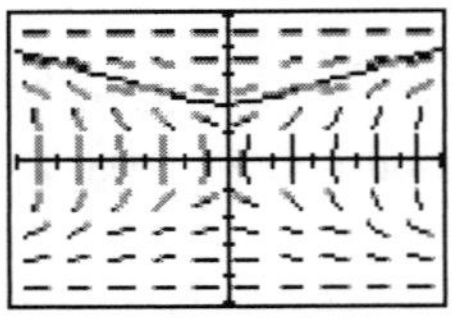

on [–5, 5] by [–5, 5]

69. **a.**

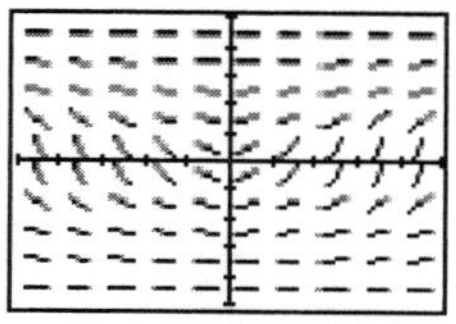

on [–5, 5] by [–5, 5]

b.

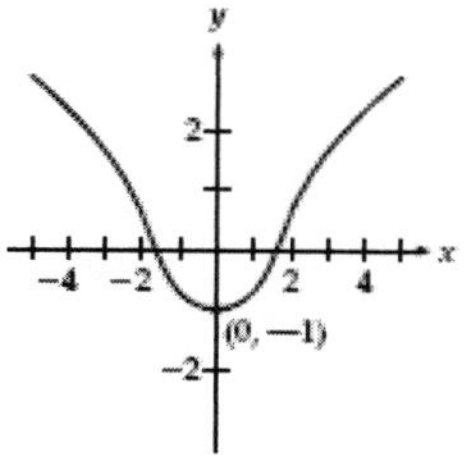

71. **a.** For an example, let $s = 40{,}000$ and $n = 25{,}000$. $r = 0.06$ and $p = 0.75$.

$$W(t) = \tfrac{s-n}{r}(e^{(1-p)rt} - 1)$$
$$= \frac{40{,}000 - 25{,}000}{0.06}(e^{(1-0.75)0.06t} - 1)$$
$$= 250{,}000(e^{0.015t} - 1)$$
$$W(20) = 250{,}000(e^{0.015(20)} - 1)$$
$$\approx 87{,}465$$

Your wealth would be approximately \$87,465 in 20 years.

b. Let $p = 0.70$.

$$\text{Now } W(t) = \tfrac{s-n}{r}(e^{(1-p)rt} - 1)$$
$$= \frac{40{,}000 - 25{,}000}{0.06}(e^{(1-0.70)0.06t} - 1)$$
$$= 250{,}000(e^{0.018t} - 1)$$
$$W(20) = 250{,}000(e^{0.018(20)} - 1)$$
$$\approx 108{,}332$$

Your wealth would be approximately \$108,332 in 20 years. Your wealth would increase by about 108,332 – 87,465 = \$20,867 if you lived more frugally.

c. s = \$40,000. 20% of s = \$80,00.
s' = \$48,000

$$\text{Now } W(t) = \tfrac{s'-n}{r}(e^{(1-p)rt} - 1)$$
$$= \frac{48{,}000 - 25{,}000}{0.06}(e^{(1-0.75)0.06t} - 1)$$
$$= 383{,}333(e^{0.015t} - 1)$$
$$W(20) = 383{,}333(e^{0.015(20)} - 1)$$
$$\approx 134{,}112$$

Your wealth would be approximately \$134,112 in 20 years. Your wealth would increase by about 134,112 – 87,465 = \$46,647 because of the raise.

d. N=\$25,000. 15% of n = \$3,750.
n' = \$28,750

$$W(t) = \tfrac{s-n'}{r}(e^{(1-p)rt} - 1)$$
$$= \frac{40000 - 28750}{0.06}(e^{(1-0.75)0.06t} - 1)$$
$$= 187{,}500(e^{0.015t} - 1)$$
$$W(20) = 187{,}500(e^{0.015(20)} - 1)$$
$$\approx 65{,}599$$

Your wealth would be approximately \$65,559 in 20 years. Your wealth would decrease by about 87,465– 65,599 = \$21,866 because of the mortgage.

73. $y = -2x + C$

75. $y = e^x$

77. Not separable

79. Separable

81. Quadrant I: positive; Quadrant II: negative; Quadrant III: positive; Quadrant IV: negative

83. **a.**
$$\frac{dp}{dt} = -\frac{K}{R}p$$
$$\frac{dp}{p} = -\frac{K}{R}dt$$
$$\int \frac{dp}{p} = -\int \frac{K}{R}dt$$
$$\ln p = -\frac{K}{R}t + C$$
$$p = e^{-\frac{K}{R}t+C}$$
$$p = e^{-\frac{K}{P}t}e^{C}$$
$$p(t) = ce^{-\frac{K}{R}t}$$

b.
$$p_0 = ce^{-\frac{K}{R}t_0}$$
$$\frac{p_0}{e^{-\frac{K}{R}t_0}} = c$$
$$p_0 e^{\frac{K}{R}t_0} = c$$
$$p(t) = p_0 e^{\frac{K}{R}t_0} e^{-\frac{K}{R}t}$$
$$p(t) = p_0 e^{\frac{K}{R}(t_0 - t)}$$

c.
$$\frac{dp}{dt} = KI_0 - \frac{K}{R}P$$
$$\frac{dp}{KI_0 - \frac{K}{R}p} = dt$$
$$\int \frac{dp}{KI_0 - \frac{K}{R}p} = \int dt$$

Using the substitution

$u = KI_0 - \frac{K}{R}p,\ du = -\frac{K}{R}dp$
$$-\frac{R}{K}\int \frac{du}{u} = \int dt$$
$$-\frac{R}{K}\ln u = t + c$$
$$\ln\left(KI_0 - \frac{K}{R}p\right) = -\frac{K}{R}t + c$$
$$KI_0 - \frac{K}{R}p = e^{-\frac{K}{R}t+c} = e^{-\frac{K}{R}t}e^{c} = ce^{-\frac{K}{R}t}$$
$$-\frac{K}{R}p = ce^{-\frac{K}{R}t} - KI_0$$
$$p(t) = RI_0 - \frac{R}{K}ce^{-\frac{K}{R}t} = RI_0 - ce^{-\frac{K}{R}t}$$

d.
$$p_0 = RI_0 - ce^{-\frac{K}{R}t_0}$$
$$p_0 - RI_0 = -ce^{-\frac{K}{R}t_0}$$
$$\frac{RI_0 - p_0}{e^{-\frac{K}{R}t_0}} = c$$
$$(RI_0 - p_0)e^{\frac{K}{R}t_0} = c$$
$$p(t) = RI_0 - (RI_0 - p_0)e^{\frac{K}{R}t_0}e^{-\frac{K}{R}t}$$
$$p(t) = RI_0 - (RI_0 - p_0)e^{\frac{K}{R}(t_0 - t)}$$

e. $p(T) = p(0)$ Using equation from (b) for $p(T)$ and equation from (d) for $p(0)$
$$p_0 e^{\frac{K}{R}(t_0 - T)} = RI_0 - (RI_0 - p_0)e^{\frac{K}{R}(t_0 - 0)}$$
$$p_0 e^{\frac{K}{R}t_0}e^{-\frac{K}{R}T} = RI_0 - RI_0 e^{\frac{K}{R}t_0} + p_0 e^{\frac{K}{R}t_0}$$
$$p_0 e^{\frac{Kt_0}{R}}e^{-\frac{KT}{R}} - p_0 e^{\frac{Kt_0}{R}} = RI_0\left(1 - e^{\frac{Kt_0}{R}}\right)$$
$$\frac{p_0 e^{\frac{Kt_0}{R}}\left(e^{-\frac{KT}{R}} - 1\right)}{1 - e^{\frac{Kt_0}{R}}} = RI_0$$
$$\frac{p_0 e^{\frac{Kt_0}{R}}\left(e^{-\frac{KT}{R}} - 1\right)}{e^{\frac{Kt_0}{R}}\left(e^{-\frac{Kt_0}{R}} - 1\right)} = RI_0$$
$$\frac{p_0}{I_0}\frac{1 - e^{-\frac{KT}{R}}}{1 - e^{-\frac{Kt_0}{R}}} = R$$

EXERCISES 9.2 (Equivalent to Section 6.6 of *Brief Applied Calculus*, 7e.)

1. If $y(t) = ce^{at}$, then
$y' = ace^{at} = ay$ and $y(0) = ce^{a(0)} = ce^0 = c$

3. $y' = 0.02y$
This is the form $y' = ay$ with $a = 0.02$. This is an equation of unlimited growth.

5. $y' = 30(0.5 - y)$
This is of the form $y' = a(M - y)$ with $a = 30$ and $M = 0.5$. This is an equation of limited growth.

7. $y' = 2y^2(0.5 - y)$
This is not an equation of unlimited, limited, or logistic growth.

9. $y' = y(6 - y)$
This is of the form $y' = ay(M - y)$ with $a = 1$ and $M = 6$. This is an equation of logistic growth.

11. $y' = 4y(0.04 - y)$
This is of the form $y' = ay(M - y)$ with $a = 4$ and $M = 0.04$. This is an equation of logistic growth.

13. $y' = 6y$ is an equation of unlimited growth with $a = 6$. The solution is $y = ce^{6t}$. If $y(0) = 1.5$, then
$1.5 = ce^{6(0)}$
$1.5 = c$
The solution is $y = 1.5e^{6t}$.

15. $y' = -y$ is an equation of unlimited growth with $a = -1$. The solution is $y = ce^{-t}$. If $y(0) = 100$, then
$100 = ce^{-0}$
$100 = c$
The solution is $y = 100e^{-t}$.

17. $y' = -0.45y$ is an equation of unlimited growth with $a = -0.45$. The solution is $y = ce^{-0.45t}$. If $y(0) = -1$, then
$-1 = ce^{-0.45(0)}$
$-1 = c$
The solution is $y = -e^{-0.45t}$.

19. $y' = 2(100 - y),\ y(0) = 0$ is an equation of limited growth with $a = 2$ and $M = 100$. The solution is $y = 100(1 - e^{-2t})$.

21. $y' = 0.05(0.25 - y),\ y(0) = 0$ is an equation of limited growth with $a = 0.05$ and $M = 0.25$. The solution is $y = 0.25(1 - e^{-0.05t})$.

23. $y' = 80 - 2y = 2(40 - y),\ y(0) = 0$ is an equation of limited growth with $a = 2$ and $M = 40$. The solution is $y = 40(1 - e^{-2t})$.

25. $y' = 2 - 0.01y = 0.01(200 - y),\ y(0) = 0$ is an equation of limited growth with $a = 0.01$ and $M = 200$. The solution is $200(1 - e^{-0.01t})$.

27. $y' = 5y(100 - y)$ is an equation of logistic growth with $a = 5$ and $M = 100$. The solution is $y = \frac{100}{1+ce^{-500t}}$. If $y(0) = 10$, then
$$10 = \frac{100}{1 + ce^{-500(0)}}$$
$1 + c = 10$
$c = 9$
The solution is $y = \frac{100}{1+9e^{-500t}}$.

29. $y' = 0.25y(0.5 - y)$ is an equation of logistic growth with $a = 0.25$ and $M = 0.5$. The solution is $y = \frac{0.5}{1+ce^{(0.25)(0.5)t}} = \frac{0.5}{1+ce^{0.125t}}$. If $y(0) = 0.1$, then
$$0.1 = \frac{0.5}{1 + ce^{0.125(0)}}$$
$1 + c = 5$
$c = 4$
The solution is $y = \frac{0.5}{1+4e^{0.125t}}$.

31. $y' = 3y(10 - y)$ is an equation of logistic growth with $a = 3$ and $M = 10$. The solution is $y = \frac{10}{1+ce^{-3(10)t}} = \frac{10}{1+ce^{-30t}}$. If $y(0) = 20$, then
$$20 = \frac{10}{1 + ce^{-30(0)}}$$
$1 + c = \frac{1}{2}$
$c = -\frac{1}{2}$
The solution is $y = \frac{10}{1-\frac{1}{2}e^{-30t}} = \frac{20}{2-e^{-30t}}$.

33. $y' = 6y - 2y^2 = 2y(3-y)$ is an equation of logistic growth with $a = 2$ and $M = 3$. The solution is $y = \frac{3}{1+ce^{-3(2)t}} = \frac{3}{1+ce^{-6t}}$. If $y(0) = 1$, then

$$1 = \frac{3}{1+ce^{-6(0)}}$$
$$1 + c = 3$$
$$c = 2$$

The solution is $y = \frac{3}{1+2e^{-6t}}$.

35. The value of the stamp collection increases without limit by 8% per year. Therefore the appropriate equation is $y' = 0.08y$, where y is the value of the collection. $y(0) = 1500$, so the solution of the equation is $y = 1500e^{0.08t}$. This is the value of the collection after t years.

37. Since sales growth at a rate proportional to the distance from the upper limit of 100,000, $y' = a(100,000 - y)$, where y is the sales. The solution is $y = 100{,}000(1 - e^{-at})$. If $y(5) = 10{,}000$, then

$$10,000 = 100,000(1 - e^{-a(5)})$$
$$0.1 = 1 - e^{-5a}$$
$$e^{-5a} = 0.9$$
$$-5a = \ln(0.9)$$
$$a \approx 0.021$$

The solution is $y = 100,000(1 - e^{-0.021t})$.
At the end of the first year,
$y(12) = 100,000(1 - e^{-0.021(12)}) \approx 22,276$

39. If the rate of contributions is proportional to the distance from the goal, then an equation of limited growth is needed. $y' = a(M - y)$, where y is the number of dollars raised. Here $M = 5000$, $y(1) = 1000$, and $y' = a(5000 - y)$. The solution of the equation is $y = M(1 - e^{-at}) = 5000(1 - e^{-at})$. At $t = 1$,

$$1000 = 5000(1 - e^{-a})$$
$$1 - e^{-a} = \frac{1}{5}$$
$$e^{-a} = \frac{4}{5}$$
$$a = -\ln\left(\frac{4}{5}\right) \approx 0.223$$

So the amount y raised in t weeks is $5000(1 - e^{-0.223t})$. When $y = 4000$.

$$1 - e^{-0.223t} = \frac{4}{5}$$
$$e^{-0.223t} = \frac{1}{5}$$
$$t = \frac{\ln\left(\frac{1}{5}\right)}{-0.223} \approx 7.21$$

It will take about 7.21 weeks to raise $4000.

41. Let y = total sales. Since y is proportional both to itself and to the size of the remaining market,
$y' = ay(10,000 - y)$
The solution is $y = \frac{10,000}{1+ce^{-10,000at}} = \frac{10,000}{1+ce^{bt}}$
Since at $t = 0$, $y = 100$,

$$100 = \frac{10,000}{1+ce^{b(0)}}$$
$$1 + c = 100$$
$$c = 99$$

Since at $t = 6$, $y = 2000$,

$$2000 = \frac{10,000}{1+99e^{b(6)}}$$
$$1 + 99e^{6b} = 5$$
$$e^{6b} = \frac{4}{99}$$
$$6b = \ln\left(\frac{4}{99}\right)$$
$$b \approx -0.535$$

The solution is $y = \frac{10,000}{1+99e^{-0.535t}}$.
At the end of the first year ($t = 12$),

$$y = \frac{10,000}{1+9e^{-0.535(12)}} \approx 9856$$

43. The rate of spread of the rumor is proportional to both the number of people who have heard it and the number who haven't. So this is an example of logistic growth, $y = ay(M - y)$, where y is the number of people who have heard the rumor. Here one person starts the rumor, so $y(0) = 1$; also $y(10) = 200$, and $M = 800$. So $y' = ay(800 - y)$.
The solution is

$$y = \frac{M}{1 + ce^{-aMt}} = \frac{800}{1 + ce^{-800at}} = \frac{800}{1 + c^{bt}}$$

At $t = 0$,

$$1 = \frac{800}{1 + c}$$

$$c = 799$$

So $y = \frac{800}{1+799e^{bt}}$. At $t = 10$,

$$200 = \frac{800}{1 + 799e^{10b}}$$

$$799e^{10b} = \left(\frac{800}{200}\right) - 1 = 3$$

$$b = \frac{\ln\left(\frac{3}{799}\right)}{10} \approx -0.558$$

So the number of people y who have heard the rumor after t minutes is $\frac{800}{1+799e^{-0.558t}}$.

After 15 minutes,

$$y = \frac{800}{1 + 799e^{-0.558(15)}} \approx 675 \text{ people}$$

47. Let y = computer sales after t years. Then y is modeled by the logistic equation $y' = ay(360 - y)$

The solution is $y = \frac{360}{1 + ce^{-360at}} = \frac{360}{1 + ce^{bt}}$

Since $y = 289$ at $t = 0$,

$$289 = \frac{360}{1 + ce^{b(0)}}$$

$$1 + c = \frac{360}{289}$$

$$c = \frac{71}{289} \approx 0.246$$

Thus, $y = \frac{360}{1 + 0.246e^{bt}}$. Since $y = 340$ at $t = 5$,

$$340 = \frac{360}{1 + 0.246e^{b(5)}}$$

$$1 + 0.246e^{5b} = \frac{360}{340}$$

$$e^{5b} = \frac{17}{71}$$

$$b = \frac{1}{5}\ln\left(\frac{17}{71}\right) \approx -0.286$$

Thus, the solution is $y = \frac{360}{1 + 0.246e^{-0.286t}}$

To find what the sales will be in 2020, we solve the equation

$$y = \frac{360}{1 + 0.246e^{-0.286(15)}} \approx 358$$

The computer sales in 2020 will reach about 358.

45. Let y = deer population after t years. Then y is modeled by the logistic equation $y' = ay(800 - y)$.
The solution is $y = \frac{800}{1+ce^{-800at}} = \frac{800}{1+ce^{bt}}$.
Since $y = 100$ at $t = 0$,

$$100 = \frac{800}{1 + ce^{b(0)}}$$

$$1 + c = 8$$

$$c = 7$$

Thus, $y = \frac{800}{1+7e^{bt}}$. Since $y = 160$ at $t = 2$,

$$160 = \frac{800}{1 + 7e^{b(2)}}$$

$$1 + 7e^{2b} = 5$$

$$e^{2b} = \frac{4}{7}$$

$$b = \frac{1}{2}\ln\left(\frac{4}{7}\right) \approx -0.28$$

Thus, the solution is $y = \frac{800}{1+7e^{-0.28t}}$.

To find when the deer population will reach 400, we solve the equation

$$400 = \frac{800}{1 + 7e^{-0.28t}}$$

$$1 + 7e^{-0.28t} = 2$$

$$e^{-0.28t} = \frac{1}{7}$$

$$t = -\frac{1}{0.28}\ln\left(\frac{1}{7}\right) \approx 7$$

The population will reach 400 in about 7 years.

49. The equation given, $y' = -0.15y$, is an example of unlimited growth. Since $y(0) = 5$, the solution is $y = 5e^{-0.15t}$. After 2 hours, $y = 5e^{-0.3} \approx 3.70$ units of the drug remain in the blood.

51. **a.** $y' = 20 - 0.1y = 0.1(200 - y)$, $y(0) = 0$ is a limited growth equation with $M = 200$ and $a = 0.1$

b. The solution is
$y = 200(1 - e^{-0.1t})$

c. $0.95M = 0.95(200) = 190$

$$190 = 200(1 - e^{-0.1t})$$
$$0.95 = 1 - e^{-0.1t}$$
$$e^{-0.1t} = 0.05$$
$$t = -\frac{1}{0.1}\ln\ 0.05 \approx 30 \text{ years}$$

The accumulated sediment will reach 190 thousand tons in 30 years.

53. **a.**

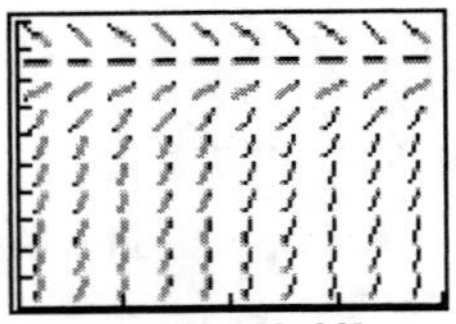

on [0, 3] by [0, 20]
About 16.6 feet per second

b.

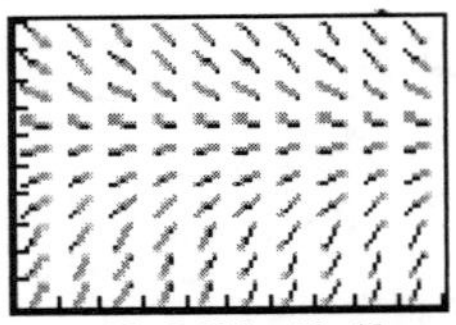

on [0, 0.1] by [0, 1]
About 0.6 foot per second

c.

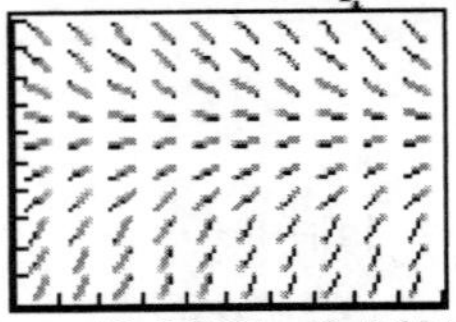

on [0, 0.01] by [0, 0.001]
About 0.006 foot per second

d. About 1/0.006 • 167 seconds, or about 2.8 minutes

55. **a.** Limited
b. Unlimited
c. Logistics

57. **a.** Logistics
b. Unlimited
c. Limited

59. **a.** Approaches 0.03
b. Approaches 2.5
c. Approaches infinity

61.
$$\frac{y'}{y} = ay$$
$$\frac{1}{y}\frac{dy}{dt} = ay$$
$$\frac{dy}{y^2} = a\,dt$$
$$\int\frac{dy}{y^2} = \int a\,dt$$
$$-\frac{1}{y} = at + C$$
$$y = \frac{1}{-at + c}$$

63. $y' = \frac{ay}{x}$; y must be a function of x.
$$\frac{dy}{dx} = \frac{ay}{x}$$
$$\int\frac{dy}{y} = \int a\frac{dx}{x}$$
$$\ln\ y = a\ln\ x + C = \ln x^a + C$$
$$y = e^{(\ln x^a + C)} = x^a e^C = cx^a$$

65. a. $$y' = \frac{dy}{dt} = ay(M-y)$$

$$\frac{dy}{y(M-y)} = a\,dt$$

b. $$\int \frac{dy}{y(M-y)} = \int a\,dt$$

$$\frac{1}{M}\ln\left(\frac{y}{M-y}\right) = at + C$$

c. $$\ln\left(\frac{y}{M-y}\right) = aMt + C$$

$$\frac{y}{M-y} = e^{aMt+C} = ce^{aMt}$$

$$y = ce^{aMt}(M-y)$$

$$y + ce^{aMt}y = cMe^{aMt}$$

$$y(1+ce^{aMt}) = cMe^{aMt}$$

$$y = \frac{cMe^{aMt}}{1+ce^{aMt}}$$

d. $$y = \frac{cMe^{aMt}}{1+ce^{aMt}} \cdot \frac{e^{-aMt}}{e^{-aMt}}$$

$$= \frac{cM}{e^{-aMt}+c}\left(\frac{\frac{1}{c}}{\frac{1}{c}}\right)$$

$$= \frac{M}{ce^{-aMt}+1}$$

EXERCISES 9.3

1. $y' + 2y = 8 \quad p(x) = 2,\ q(x) = 8$

$$I(x) = e^{\int p(x)dx} = e^{\int 2\,dx} = e^{2x}$$

$$e^{2x}y' + 2e^{2x}y = 8e^{2x}$$

$$\frac{d}{dx}(e^{2x}y) = 8e^{2x}$$

$$e^{2x}y = 4e^{2x} + C$$

$$y = 4 + Ce^{-2x}$$

Check:

$y = 4 + Ce^{-2x}$ then $y' = -2Ce^{-2x}$

Substitute these expressions into the differential equation $y' + 2y = 8$.

$$(-2Ce^{-2x}) + 2(4 + Ce^{-2x}) = 8$$

3. $y' - 2y = e^{-2x} \quad p(x) = -2,\ q(x) = e^{-2x}$

$$I(x) = e^{\int -2dx} = e^{-2x}$$

$$e^{-2x}y' - e^{-2x}(2y) = e^{-4x}$$

$$\frac{d}{dx}(e^{-2x}y) = e^{-4x}$$

$$e^{-2x}y = -\frac{1}{4}e^{-4x} + C$$

$$y = -\frac{1}{4}e^{-2x} + Ce^{2x}$$

Check:

$y = -\frac{1}{4}e^{-2x} + Ce^{2x}$ then $y' = \frac{1}{2}e^{-2x} + 2Ce^{2x}$

Substitute these expressions into the differential equation $y' - 2y = e^{-2x}$.

$$\left(\frac{1}{2}e^{-2x} + 2Ce^{2x}\right) - 2\left(-\frac{1}{4}e^{-2x} + Ce^{2x}\right) = e^{-2x}$$

5. $y' + \frac{5}{x}y = 24x^2 \quad p(x) = \frac{5}{x},\ q(x) = 24x^2$

$$I(x) = e^{\int \frac{5}{x}dx} = e^{5\ln x} = e^{\ln x^5} = x^5$$

$$x^5y' + 5x^4y = 24x^7$$

$$\frac{d}{dx}(x^5y) = 24x^7$$

$$x^5y = 3x^8 + C$$

$$y = 3x^3 + Cx^{-5}$$

7. $xy' - y = x^2$

$y' - \frac{y}{x} = x$

Thus, $p(x) = -\frac{1}{x}$, $q(x) = x$.

$I(x) = e^{\int -\frac{1}{x}\,dx} = e^{-\ln x} = \frac{1}{e^{\ln x}} = \frac{1}{x}$

$\frac{1}{x}y' - \frac{y}{x^2} = x\left(\frac{1}{x}\right)$

$\frac{d}{dx}\left(\frac{y}{x}\right) = 1$

$\frac{y}{x} = x + c$

$y = x^2 + Cx$

9. $y' + 3x^2y = 9x^2 \quad p(x) = 3x^2,\ q(x) = 9x^2$

$I(x) = e^{\int 3x^2\,dx} = e^{x^3}$

$e^{x^3}y' + 3e^{x^3}x^2y = 9x^2e^{x^3}$

$\frac{d}{dx}(e^{x^3}y) = 9x^2e^{x^3}$

$e^{x^3}y = 3e^{x^3} + C$

$y = 3 + Ce^{-x^3}$

11. $y' - 2xy = 0 \quad p(x) = -2x,\ q(x) = 0$

$I(x) = e^{\int -2x\,dx} = e^{-x^2}$

$e^{-x^2}y' - 2xe^{-x^2}y = 0$

$\frac{d}{dx}(e^{-x^2}y) = 0$

$e^{-x^2}y = C$

$y = Ce^{x^2}$

13. $(x+1)y' + y = 2x$

$y' + \frac{y}{x+1} = \frac{2x}{x+1}$

Thus, $p(x) = \frac{1}{x+1}$, $q(x) = \frac{2}{x+1}$

$I(x) = e^{\int \frac{1}{x+1}\,dx} = e^{\ln(x+1)} = x+1$

$(x+1)y' + y = 2x$

$\frac{d}{dx}[(x+1)y] = 2x$

$(x+1)y = x^2 + C$

$y = \frac{x^2}{x+1} + \frac{C}{x+1} = \frac{x^2+C}{x+1}$

15. $y' - \frac{2}{x}y = 6x^3 - 9x^2 \quad p(x) = -\frac{2}{x},\ q(x) = 6x^3 - 9x^2$

$I(x) = e^{\int -\frac{2}{x}\,dx} = e^{-2\ln x} = e^{\ln x^{-2}} = x^{-2}$

$x^{-2}y' - 2x^{-3}y = (6x^3 - 9x^2)x^{-2}$

$\frac{d}{dx}(x^{-2}y) = 6x - 9$

$x^{-2}y = 3x^2 - 9x + C$

$y = 3x^4 - 9x^3 + Cx^2$

17. $y' = x + y$

$y' - y = x$

Thus, $p(x) = -1$, $q(x) = x$

$I(x) = e^{\int -1\,dx} = e^{-x}$

$e^{-x}y' - e^{-x}y = xe^{-x}$

$\frac{d}{dx}(-e^{-x}y) = xe^{-x}$

$e^{-x}y = \int xe^{-x}\,dx$

$= -xe^{-x} - \frac{1}{-1}\int e^{-x}\,dx$

$= -xe^{-x} + (-e^{-x}) + C$

$y = -x - 1 + Ce^{x}$

Using integration by parts with $u = x$ and $du = e^{-x}\,dx$.

19. $y' + (\cos x)y = \cos x \quad p(x) = \cos x,\ q(x) = \cos x$

$$I(x) = e^{\int \cos x\,dx} = e^{\sin x}$$
$$e^{\sin x}y' + (\cos x)e^{\sin x}y = (\cos x)e^{\sin x}$$
$$\frac{d}{dx}(e^{\sin x}y) = (\cos x)e^{\sin x}$$
$$e^{\sin x}y = e^{\sin x} + C$$
$$y = 1 + Ce^{-\sin x}$$

21. $y' + 3y = 12e^x,\ y(0) = 5$

$$p(x) = 3,\ q(x) = 12e^x,\ I(x) = e^{\int 3\,dx} = e^{3x}$$
$$e^{3x}y' + 3e^{3x}y = 12e^x(e^{3x})$$
$$\frac{d}{dx}(e^{3x}y) = 12e^{4x}$$
$$e^{3x}y = 3e^{4x} + C$$
$$y = 3e^x + Ce^{-3x}$$

Since $y(0) = 5$,

$5 = e^0 + Ce^{-3(0)}$

$5 = 3 + C$

$C = 2$

Thus, the particular solution is $y = 3e^x + 2e^{-3x}$.

23. $xy' + 2y = 14x^5,\ y(1) = 1$

$$x^2y' + 2xy = 14x^6$$
$$\frac{d}{dx}(x^2y) = 14x^6$$
$$x^2y = 2x^7 + C$$
$$y = 2x^5 + Cx^{-2}$$

Since $y(1) = 1$,

$1 = 2(1)^5 + C(1)^2$

$1 = 2 + C$

$C = -1$

The particular solution is $y = 2x^5 - x^{-2}$.

25. $xy' = 2y + x^2,\ y(1) = 3$

$$xy' - 2y = x^2$$
$$y' - \frac{2}{x}y = x \quad p(x) = -\frac{2}{x},\ q(x) = x$$
$$I(x) = e^{\int -\frac{2}{x}dx} = e^{-2\ln x} = e^{\ln x^{-2}} = x^{-2}$$
$$x^{-2}y' - 2x^{-3}y = x(x^{-2})$$
$$\frac{d}{dx}(x^{-2}y) = \frac{1}{x}$$
$$x^{-2}y = \ln x + C$$
$$y = x^2\ln x + Cx^2$$

If $y(1) = 3$,

$3 = 1^2\ln 1 + C(1)^2$

$3 = 0 + C$

$C = 3$

The particular solution is $y = x^2\ln x + 3x^2$.

27. $y' + 2xy = 4x,\ y(0) = 0$

$$p(x) = 2x,\ q(x) = 4x,\ I(x) = e^{\int 2x\,dx} = e^{x^2}$$
$$e^{x^2}y' + 2xe^{x^2} = 4xe^{x^2}$$
$$\frac{d}{dx}(e^{x^2}y) = 4xe^{x^2}$$
$$e^{x^2}y = 2e^{x^2} + C$$
$$y = 2 + Ce^{-x^2}$$

Since $y(0) = 0$,

$0 = 2 + Ce^{-(0)^2}$

$0 = 2 + C$

$C = -2$

The particular solution is $y = 2 - 2e^{-x^2}$.

29. $y' = y + 1$

$$y' - y = 1 \quad p(x) = -1,\ q(x) = 1$$
$$I(x) = e^{\int -dx} = e^{-x}$$
$$e^{-x}y' - e^{-x}y = e^{-x}$$
$$\frac{d}{dx}(e^{-x}y) = e^{-x}$$
$$e^{-x}y = -e^{-x} + C$$
$$y = -1 + Ce^x$$

Using separation of variables, we have

$$y' = y + 1$$
$$\frac{dy}{dx} = y + 1$$
$$\frac{dy}{y+1} = dx$$
$$\int \frac{dy}{y+1} = \int dx$$
$$\ln(y+1) = x + C$$
$$y + 1 = e^{x+C}$$
$$y = -1 + ce^x$$

31. **a.** Use separation of variables.

$$y' + xy^2 = 0$$
$$y' = \frac{dy}{dx} = -xy^2$$
$$\frac{dy}{y^2} = -x\,dx$$
$$\int \frac{dy}{y^2} = \int -x\,dx$$
$$-\frac{1}{y} = -\frac{x^2}{2} + C$$
$$y = \frac{1}{\frac{x^2}{2} + C} = \frac{2}{c + x^2}$$

b. Use an integrating factor.

$$y' = y + x^2e^x$$
$$y' - y = x^2e^x \quad p(x) = -1,\ q(x) = x^2e^x$$
$$I(x) = e^{\int -dx} = e^{-x}$$
$$e^{-x}y' - e^{-x}y = x^2e^x(e^{-x})$$
$$\frac{d}{dx}(e^{-x}y) = x^2$$
$$e^{-x}y = \frac{x^3}{3} + C$$
$$y = \frac{x^3e^x}{3} + Ce^x$$

33. **a.** $y' = 0.1y + 20e^{0.1t},\ y(0) = 5$

$$y' - 0.1y = 20e^{0.1t}$$

Thus, $p(t) = -0.1$ and $q(t) = 20e^{0.1t}$.

$$I(t) = e^{\int -0.1dt} = e^{-0.1t}$$
$$e^{-0.1t}y' - 0.1e^{-0.1t}y = e^{-0.1t}(20e^{0.1t})$$
$$\frac{d}{dt}(e^{-0.1t}y) = 20$$
$$e^{-0.1t}y = 20t + C$$
$$y = 20te^{0.1t} + Ce^{0.1t}$$

Since $y(0) = 5$,

$$5 = 20(0)e^{0.1(0)} + Ce^{0.1(0)}$$
$$C = 5$$

Thus, $y = 20te^{0.1t} + 5e^{0.1t}$

b. For $t = 2$,

$$y = 20(2)e^{0.1(2)} + 5e^{0.1(2)} \approx \$55 \text{ million}$$

35. **a.** $y' - \frac{3}{t}y = 0,\ y(1) = 125$

$$p(t) = -\frac{3}{t},\ q(t) = 0$$
$$I(t) = e^{\int -\frac{3}{t}dt} = e^{-3\ln t} = e^{\ln t^{-3}} = t^{-3}$$
$$t^{-3}y' - 3t^{-4}y = 0$$
$$\frac{d}{dt}(t^{-3}y) = 0$$
$$t^{-3}y = c$$
$$y = ct^3$$

Since $y(1) = 125$,

$125 = c(1)^3 = C$

Thus, $y = 125t^3$.

b. When $t = 15$,

$y = 125(15)^3 = 421{,}875$ cases

37. **a.** Let $y(t)$ = the total number of investors needed after t years. Thus, $y(0) = 10$.

$$100 \cdot \frac{dy}{dt} = 20y + 800,\ y(0) = 10$$
$$\frac{dy}{dt} - 0.2y = 8$$
$$p(t) = -0.2,\ q(t) = 8,\ I(t) = e^{\int -0.2dt} = e^{-0.2t}$$
$$e^{-0.2t}\frac{dy}{dt} - e^{-0.2t}(0.2y) = e^{-0.2t} \cdot 8$$
$$\frac{d}{dt}\left(ye^{-0.2t}\right) = 8e^{-0.2t}$$
$$ye^{-0.2t} = -40e^{-0.2t} + C$$
$$y = -40 + Ce^{0.2t}$$

Since $y(0) = 10$,

$$10 = -40 + Ce^{0.2(0)}$$
$$50 = Ce^0 = C$$

Thus $y(t) = 50^{0.2t} - 40$

b. Use $y(t) = 50^{0.2t} - 40$.

$$y(10) = 50^{0.2(10)} - 40 \approx 330$$
$$y(20) = 50^{0.2(20)} - 40 \approx 2690$$
$$y(30) = 50^{0.2(30)} - 40 \approx 20{,}131$$
$$y(50) = 50^{0.2(50)} - 40 \approx 1{,}101{,}283$$

After 10, 20, 30 and 50 years, you would need 330, 2690, 20,131, and 1,101,283 investors, respectively.

39. **a.** $y' = ty + t,\ y(0) = 2$

$y' - ty = t$

$p(t) = -t,\ q(t) = t,\ I(t) = e^{\int -t\,dt} = e^{-t^2/2}$

$e^{-t^2/2}y' - te^{-t^2/2}y = te^{-t^2/2}$

$\frac{d}{dt}(e^{-t^2/2}y) = te^{-t^2/2}$

$e^{-t^2/2}y = -e^{-t^2/2} + C$

$y = -1 + Ce^{t^2/2}$

Since $y(0) = 2$,

$2 = -1 + Ce^{0^2/2}$

$3 = Ce^0 = C$

Thus, $y(t) = 3e^{t^2/2} - 1$.

For $t = 2$,

$y(2) = 3e^{(2)^2/2} - 1 \approx 21$ tons

b.

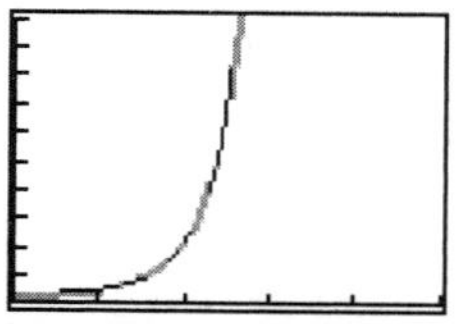

c. The algae bloom will reach 40 tons in about 2.3 weeks (2 weeks and 2 days.)

41. **a.** Let $y(x)$ = the amount of radon after t hours. Thus, $y(0) = 800$ pCi per cubic foot $\times$ 10,000 cubic feet = 8,000,000. The incoming air contains 5 pCi of radon per cubic foot $\times$ 500 cubic feet = 2500 pCi, and $\frac{500}{10,000}y$ is the amount of radon leaving.

Thus,

$y' = 2500 - 0.05y \quad y(0) = 8,000,000$

b. $y' + 0.05y = 2500$

$p(x) = 0.05,\ q(x) = 2500,$

$I(x) = e^{\int 0.05\,dx} = e^{0.05x}$

$e^{0.05x}y' + 0.05e^{0.05x}y = 2500e^{0.05x}$

$\frac{d}{dx}(e^{0.05x}y) = 2500e^{0.05x}$

$e^{0.05x}y = 50,000e^{0.05x} + C$

$y = 50,000 + Ce^{-0.05x}$

Since $y(0) = 8,000,000$

$8,000,000 = 50,000 + Ce^{-0.05(0)}$

$7,950,000 = Ce^0 = C$

Thus, $y(x) = 7,950,000e^{-0.05x} + 50,000.$

c.

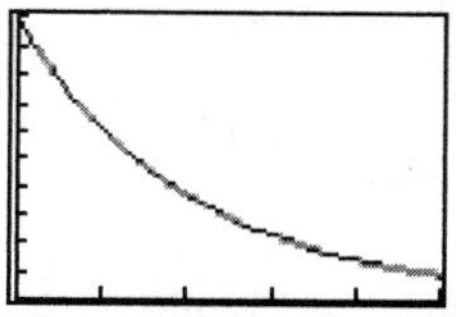

The radon level will fall to 112 pCi per cubic foot in about 40 hours.

43. **a.** Let $y(t)$ = amount of smoke particles in the room. Then $y(0) = 500$ particles per cubic foot $\times$ 12,000 cubic feet = 6,000,000 particles. The incoming air has 0 smoke particles, and $\frac{600}{12,000}y = 0.05y$ leaves the room. And, the smokers add 10,000 particles per minute.

$y' = 0 + 10,000 - 0.05y$

$= 10,000 - 0.05y \quad y(0) = 6,000,000$

b. $y' + 0.05y = 10,000$

$p(t) = 0.05,\ q(t) = 10,000,\ I(t) = e^{\int 0.05\,dt} = e^0$

$e^{0.05t}y' + 0.05e^{0.05t}y = 10,000e^{0.05t}$

$\frac{d}{dt}(e^{0.05t}y) = 10,000e^{0.05t}$

$e^{0.05t}y = 200,000e^{0.05t} + C$

$y = 200,000 + Ce^{-0.05t}$

Since $y(0) = 6,000,000$,

$6,000,000 = 200,000 + Ce^{-0.05(0)}$

$5,800,000 = Ce^0 = C$

Thus, $y(t) = 5,800,000e^{-0.05t} + 200,000.$

c. To find when the level is 100 particles per cubic foot, we must determine when y = 100 particles per cubic foot $\times$ 12,000 cubic feet = 1,200,000 particles Thus,

$1,200,000 = 5,800,000e^{-0.05t} + 200,000$

$1,000,000 = 5,800,000e^{-0.05t}$

$e^{-0.05t} = \frac{1,000,000}{5,80,000}$

$-0.05t = \ln\left(\frac{1,000,000}{5,800,000}\right)$

$t = -\frac{1}{0.05}\ln\left(\frac{1,000,000}{5,800,000}\right) \approx 35$ minutes

45. **a.** Let $y(t)$ = amount of pesticide in the reservoir after t hours. Then $y(0) = 0.1$ gram per gallon × 100,000 gallons = 10,000 grams. Incoming pesticide = 2000 gallons × 0.01 gram per gallon = 20. Outgoing pesticide = $\frac{2000}{100{,}000}y = 0.02y$. Thus,

$$y' = 20 - 0.02y \quad y(0) = 10{,}000$$

b. $y' + 0.02y = 20$

$$p(t) = 0.02,\ q(t) = 20,\ I(t) = e^{0.02\,dt} = e^{0.02t}$$

$$e^{0.02t}y' + 0.02e^{0.02t}y = 20e^{0.02t}$$

$$\frac{d}{dt}(e^{0.02t}y) = 20e^{0.02t}$$

$$e^{0.02t}y = 1000e^{0.02t} + C$$

$$y = 1000 + Ce^{-0.02t}$$

Since $y(0) = 10{,}000$,

$$10{,}000 = 1000 + Ce^{-0.02(0)}$$

$$9000 = Ce^{0} = C$$

Thus, $y(t) = 9000e^{-0.02t} + 1000.$

c.

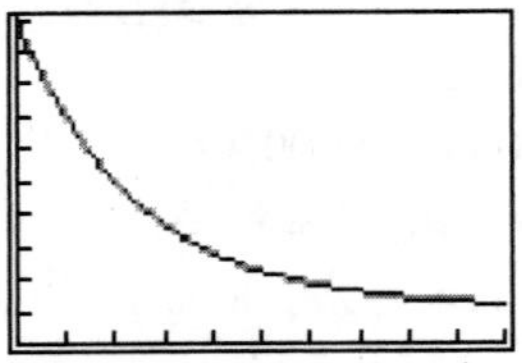

About 110 hours.

d. In the long run, the graph approaches $y = 1000$. Thus, the amount of pesticide in the reservoir will be about 1000 grams.

47. **a.** If a person eats 2100 calories per day, $c = 2100$. Thus,

$$w' + 0.005w = \frac{2100}{3500} = \frac{3}{5}$$

$$p(t) = 0.005,\ q(t) = \frac{3}{5},\ I(t) = e^{\int 0.005\,dt} = e^{0.005t}$$

$$e^{0.005t}w' + 0.005we^{0.005t} = \frac{3}{5}e^{0.005t}$$

$$\frac{d}{dt}(e^{0.005t}w) = \frac{3}{5}e^{0.005t}$$

$$e^{0.005t}w = 120e^{0.005t} + C$$

$$w = 120 + Ce^{-0.005t}$$

If $w(0) = 170$, then

$$170 = 120 + Ce^{-0.005(0)}$$

$$50 = Ce^{0} = C$$

Thus, $w(t) = 50e^{-0.005t} + 120$

b. The person loses 15 pounds when $w = 155$. Thus

$$155 = 50e^{-0.005t} + 120$$

$$35 = 50e^{-0.005t}$$

$$e^{-0.005t} = \frac{35}{50}$$

$$-0.005t = \ln\left(\frac{7}{10}\right)$$

$$t = -\frac{1}{0.005}\ln\left(\frac{7}{10}\right) \approx 71 \text{ days}$$

c. If the person diets indefinitely, the limiting weight is $\lim_{t\to\infty} 50e^{-0.005t} + 120 = 0 + 120 = 120$ pounds.

49. Yes, just divide by $u(x)$ to put it in standard form of $y'+\frac{v(x)}{u(x)}y=\frac{w(x)}{u(x)}$.

51. Unlimited (since it can be written in the form $y'-ay=0$) and limited (since it can be written in the form $y'+ay=aM$).

53. $y'-ay=0$

$I(x)=e^{-\int a\,dx}=e^{-ax}$

55. 800,000 pCi (the volume times the concentration in the incoming air)

57. Consider the differential equation
$y'+p(x)\cdot y=q(x)$.
Then $I(x)=e^{\int p(x)dx+C}=e^{\int p(x)dx}e^{C}=ce^{\int p(x)dx}$
where $c\approx 0$
Multiplying by $I(x)$, we get
$ce^{\int p(x)\,dx}y'+ce^{\int p(x)\,dx}p(x)\cdot y=ce^{\int p(x)\,dx}q(x)$
$e^{\int p(x)dx}y'+e^{\int p(x)dx}p(x)\cdot y=e^{\int p(x)\,dx}q(x)$
which is the result we get when we omit the constant of integration. Thus, we can omit the constant of integration.

59. **a.** $a = 10$ mg, $b = 3$, $c = 0.2$. The differential equation is
$\frac{dy}{dt}=10(3)e^{-3t}-0.2y$ or $\frac{dy}{dt}=30e^{-3t}-0.2y$

b. $\frac{dy}{dt}=y'=30e^{-3t}-0.2y$

$y'+0.2y=30e^{-3t}$

$p(t)=0.2,\ q(t)=30e^{-3t},\ I(t)=e^{\int 0.2\,dt}=e^{0.2t}$

$e^{0.2}y'+0.2e^{0.2t}y=30e^{-3t}(e^{0.2t})$

$\frac{d}{dt}(e^{0.2t}y)=30e^{-2.8t}$

$e^{0.2t}y=-\frac{30}{2.8}e^{-2.8t}+C\approx-10.7e^{-2.8t}+C$

$y=-10.7e^{-3t}+Ce^{-0.2t}$

Since $y(0) = 0$,

$0=-10.7e^{-3(0)}+Ce^{-0.2(0)}=-10.7e^{0}+Ce^{0}$
$C=10.7$
Thus, $y(t)=10.7e^{-0.2t}-10.7e^{-3t}$

c. At $t = 2$, $y=(2)=10.7e^{-0.2(2)}-10.7e^{-3(2)}\approx 7.1$ mg

d.

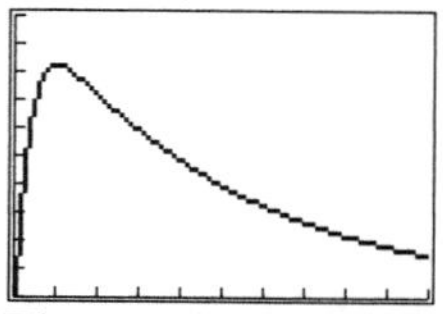

The maximum amount of medication in the bloodstream occurs at about 0.97 hour.

61. The differential equation is $\frac{dy}{dt} = abe^{-bt} - cy$. The initial condition is $y(0) = 0$. Write the equation in standard form:

$$\frac{dy}{dt} + cy = abe^{-bt}$$
$$p(t) = c$$
$$I(t) = e^{\int p(t)dt} = e^{\int cdt} = e^{ct}$$

Use the integrating factor $I(t)$: $e^{ct}\frac{dy}{dt} + e^{ct}cy = e^{ct}abe^{-bt}$

$$\frac{d}{dt}\left[e^{ct}y\right] = abe^{ct-bt}$$

Integrate and solve for y: $e^{ct}y = \int abe^{ct-bt}dt = \frac{abe^{ct-bt}}{c-b} + C$

$$y = \frac{abe^{ct-bt}}{e^{ct}(c-b)} + Ce^{-ct} = \frac{abe^{-bt}}{c-b} + Ce^{-ct}$$

Use the initial condition to solve for C. $0 = \frac{abe^{-b(0)}}{c-b} + Ce^{-c(0)} = \frac{ab}{c-b} + C \Rightarrow C = \frac{-ab}{c-b}$

$$y = \frac{abe^{-bt}}{c-b} - \frac{ab}{c-b}e^{-ct}$$
$$= \frac{ab}{c-b}(e^{-bt} - e^{-ct})$$
$$= -\frac{ab}{c-b}(e^{-ct} - e^{-bt})$$
$$= \frac{ab}{b-c}(e^{-ct} - e^{-bt})$$

EXERCISES 9.4

1. $y' = 5x - 4y$. The slope of the solution at (4, 3) is $g(4, 3) = 5(4) - 4(3) = 20 - 12 = 8$

3. For $y' = 4xy$, $y(1) = 3$, the initial point is (1, 3) and the slope at (1, 3) is $g(1, 3) = 4(1)(3) = 12$

5. $y' = 3x - 2y$, $y(0) = 2$

The initial point is (0, 2), and the slope is $g(x, y) = 3x - 2y$.

The step size is

$\frac{1-0}{4} = 0.25$

Segment 1

Slope $g(0, 2) = 3(0) - 2(2) = -4$

End point $x_1 = 0.25$, $y_1 = 2 + (-4)(0.25) = 1$

Segment 2

Slope $g(0.25, 1) = 3(0.25) - 2(1) = 0.75 - 2 = -1.25$

End point $x_2 = 0.5$, $y_2 = 1 + (-1.25)(0.25) \approx 0.69$

Segment 3

Slope $g(0.5, 0.69) = 3(0.5) - 2(0.69) = 0.12$

End point $x_3 = 0.75$, $y_3 = 0.69 + (0.12)(0.25) = 0.72$

Segment 4

Slope $g(0.75, 0.72) = 3(0.75) - 2(0.72) = 0.81$

End point $x_4 = 1$, $y_4 = 0.72 + 0.81(0.25) \approx 0.92$

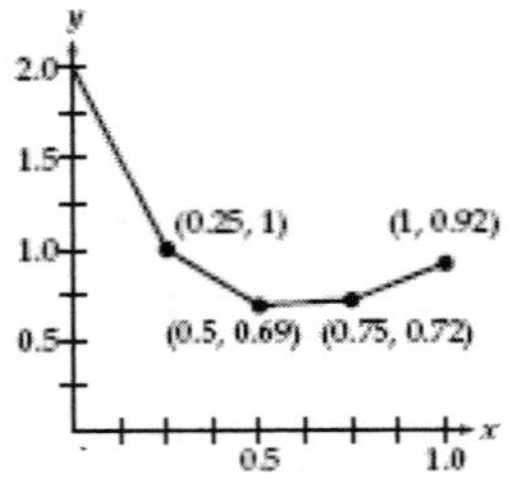

7. $y' = 4xy$, $y(0) = 1$. The initial point is (0, 1), and the slope is $g(x, y) = 4xy$. The step size is $\frac{1-0}{4} = 0.25$

Segment 1
Slope $g(0, 1) = 4(0)(1) = 0$
End point $x_1 = 0.25$, $y_1 = 1 + 0(0.25) = 1$
Segment 2
Slope $g(0.25, 1) = 4(0.25)(1) = 1$
End point $x_2 = 0.50$, $y_2 = 1 + 1(0.25) = 1.25$
Segment 3
Slope $g(0.5, 1.25) = 4(0.5)(1.25) = 2.5$
End point $x_3 = 0.75$, $y_3 = 1.25 + 2.5(0.25) \approx 1.88$
Segment 4
Slope $g(0.75, 1.88) = 4(0.75)(1.88) = 5.64$
End point $x_4 = 1$, $y_4 = 1.88 + 5.64(0.25) = 3.29$

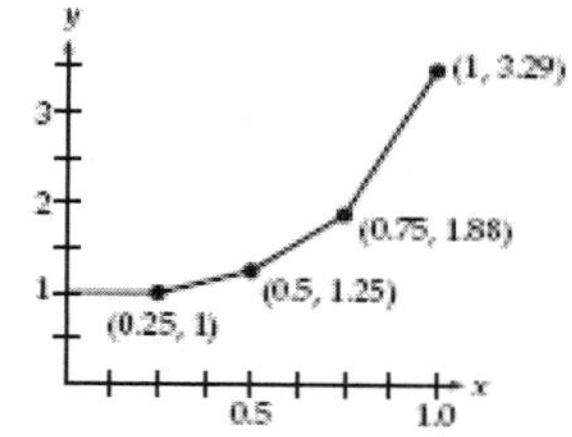

9. $y' + 2y = e^{4x}$, $y(0) = 2$. The initial point is (0, 2), and the slope is $g(x, y) = e^{4x} - 2y$. The step size is $\frac{1-0}{4} = 0.25$

Segment 1
Slope $g(0, 2) = e^{4(0)} - 2(2) = 1 - 4 = -3$
End point $x_1 = 0.25$, $y_1 = 2 + (-3)(0.25) = 1.25$
Segment 2
Slope $g(0.25, 1.25) = e^{4(0.25)} - 2(1.25) = e - 2.5$
End point $x_2 = 0.5$, $y_2 = 1.25 + (e - 2.5)(0.25) \approx 1.30$
Segment 3
Slope $g(0.5, 1.30) = e^{4(0.5)} - 2(1.3) = e^2 - 2.6$
End point $x_3 = 0.75$, $y_3 = 1.30 + (e^2 - 2.6)(0.25) \approx 2.50$
Segment 4
Slope $g(0.75, 2.50) = e^{4(0.75)} - 2(2.50) = e^3 - 5$
End point $x_4 = 1$, $y_4 = 2.50 + (e^3 - 5)(0.25) = 6.27$

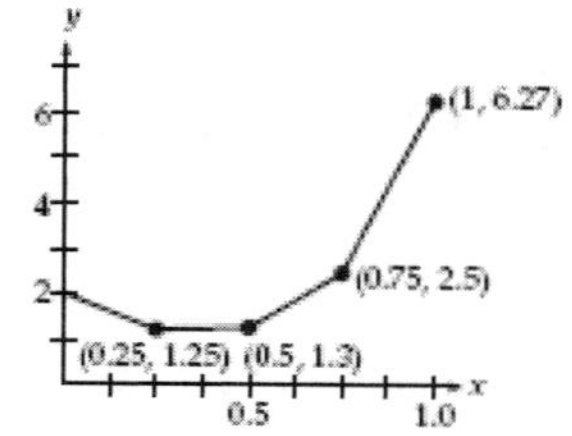

11. **a.** Using a graphing calculator, we get $y(2) \approx 2.2226$.

b. $y' = \frac{x}{y},\ y(0) = 1$

$\frac{dy}{dx} = \frac{x}{y}$

$y\ dy = x\, dx$

$\frac{y^2}{2} = \frac{x^2}{2} + C$

$y = \sqrt{x^2 + c}$

Since $y(0) = 1$,

$1 = \sqrt{0^2 + c}$

$c = 1$

Thus, $y(x) = \sqrt{x^2 + 1}$.

c. $y(2) = \sqrt{2^2 + 1} = \sqrt{5} \approx 2.236$

13. **a.** Using a graphing calculator, we get $y(2) \approx 1.489$.

b. $\frac{dy}{dx} = 0.2y,\ y(0) = 1$

$\frac{dy}{y} = 0.2\, dx$

$\ln y = 0.2x + C$

$y = e^{0.2x + C} = ce^{0.2x}$

Since $y(0) = 1$,

$1 = ce^{0.2(0)} = c$

Thus, $y(x) = e^{0.2x}$.

c. $y(2) = e^{0.2(2)} \approx 1.492$

15. **a.** Using a graphing calculator, we get $y(2) \approx 7.836$.

b. $y' + y = 2e^x,\ y(0) = 5$

$p(x) = 1,\ q(x) = 2e^x,\ I(x) = e^{\int dx} = e^x$

$e^x y' + e^x y = 2e^x \cdot e^x$

$\frac{d}{dx}(e^x y) = 2e^{2x}$

$e^x y = e^{2x} + C$

$y = e^x + Ce^{-x}$

Since $y(0) = 5$,

$5 = e^0 + Ce^{-0}$

$5 = 1 + C$

$C = 4$

Thus, $y(x) = e^x + 4e^{-x}$.

c. $y(2) = e^2 + 4e^{-2} \approx 7.930$

17. Using a graphing calculator, we get $y(3) \approx 1.733$.

19. Using a graphing calculator, we get $y(2.5) \approx 2.24$.

21. Using a graphing calculator, we get $y(3.8) \approx 2.828$.

23. **a.** With $n = 10$, $y(1) \approx 2.3096$.

b. With $n = 100$, $y(1) \approx 2.3309$.

c. With $n = 1000$, $y(1) = 2.3330$

25. $y' = 2e^{0.05y} + 2,\ y(0) = 0$. The initial point is (0, 0), and the slope is $g(x, y) = 2e^{0.05y} + 2$. The step size is $\frac{3-0}{3} = \frac{3}{3} = 1$

Segment 1

Slope $g(0, 0) = 2e^{0.05(0)} + 2 = 2e^0 + 2 = 4$

End point $x_1 = 1,\quad y_1 = 0 + 4(1) = 4$

Segment 2

Slope $g(1, 4) = 2e^{0.05(4)} + 2 = 2e^{0.2} + 2$

End point $x_2 = 2,\quad y_2 = 4 + (2e^{0.2} + 2)(1) \approx 8.44$

Segment 3

Slope $g(2, 8.44) = 2e^{0.05(8.44)} + 2 = 2e^{0.422} + 2$

End point $x_3 = 3,\quad y_3 = 8.44 + (2e^{0.422} + 2)(1) \approx 13.49$

Thus, about 13,490 were sold in the first 3 months.

27. $\frac{dy}{dt} = 1000 - 0.1y^{2/3}$, $y(0) = 0$. Using an Euler's method program, we get $y(2) \approx 1981$. Thus, the amount of sediment after 2 years is about 1981 tons.

29. $q' = -0.01q^2(1-q)$, $q(0) = 0.9$. Using an Euler's method program, we get $q(200) \approx 0.65$. Thus, the gene frequency after 200 generations is about 0.65.

31. Underestimate, since the slopes of the segments will underestimate the true slope on the interval, as in the diagram on p724.

33. $g(x, y) = \frac{f(x)}{g(y)}$ using $g(x, y)$ and $g(y)$ for two different functions.

35. $g(x, y) = q(x) - p(x) \cdot y$

37. Take the slope to be g evaluated at the *midpoint* of the segment. Alternatively, take the slope to be the *average* of g calculated at the two endpoints of the segment. (Other answers are possible.)

REVIEW EXERCISES AND CHAPTER TEST FOR CHAPTER 9

*****Please note that users of the Brief text should refer to the number in the parentheses: (Brief xx). Exercises #1–27 below are equivalent to the *Brief Applied Calculus* Chapter 6 Review Exercises #89–115.**

1. (Brief #89)

$$y^2 y' = x^2$$
$$y^2 \frac{dy}{dx} = x^2$$
$$y^2\, dy = x^2\, dx$$
$$\int y^2\, dy = \int x^2\, dx$$
$$\frac{y^3}{3} = \frac{x^3}{3} + C$$
$$y = \sqrt[3]{x^3 + C}$$

2. (Brief #90)

$$y' = x^2 y$$
$$\frac{dy}{dx} = x^2 y$$
$$\frac{dy}{y} = x^2\, dx$$
$$\int \frac{dy}{y} = \int x^2\, dx$$
$$\ln y = \frac{x^3}{3} + C$$
$$y = e^{\frac{x^3}{3} + C} = ce^{x^3/3}$$

3. (Brief #91)

$$y' = \frac{x^3}{x^4 + 1}$$
$$\frac{dy}{dx} = \frac{x^3}{x^4 + 1}$$
$$dy = \frac{x^3}{x^4 + 1} dx$$
$$\int dy = \int \frac{x^3}{x^4 + 1} dx$$
$$y = \frac{1}{4} \ln\left|x^4 + 1\right| + C$$

4. (Brief #92)

$$y' = xe^{-x^2}$$
$$\frac{dy}{dx} = xe^{-x^2}$$
$$\int dy = \int xe^{-x^2} dx$$
$$y = -\frac{1}{2} e^{-x^2} + C$$

5. (Brief #93)

$$y' = y^2$$
$$\frac{dy}{dx} = y^2$$
$$\frac{dy}{y^2} = dx$$
$$\int \frac{dy}{y^2} = \int dx$$
$$-\frac{1}{y} = x + C$$
$$y = \frac{1}{c - x}$$

6. (Brief #94)

$$y' = y^3$$
$$\frac{dy}{dx} = y^3$$
$$\int \frac{dy}{y^3} = \int dx$$
$$-\frac{y^{-2}}{2} = x + C$$
$$y^2 = \frac{1}{c - 2x}$$
$$y = \pm \frac{1}{\sqrt{c - 2x}}$$

7. (Brief #95)

$$y' = 1 - y$$
$$\frac{dy}{dx} = 1 - y$$
$$\frac{dy}{1-y} = dx$$
$$\int \frac{dy}{1-y} = \int dx$$
$$-\ln|1-y| = x + C$$
$$\ln|1-y| = c - x$$
$$1 - y = e^{c-x}$$
$$y = ke^{-x} + 1$$

8. (Brief #96)

$$y' = \frac{1}{y}$$
$$\frac{dy}{dx} = \frac{1}{y}$$
$$y\,dy = dx$$
$$\int y\,dy = \int dx$$
$$\frac{y^2}{2} = x + C$$
$$y = \pm\sqrt{2x + c}$$

9. (Brief #97)

$$y' = xy - y$$
$$\frac{dy}{dx} = y(x-1)$$
$$\frac{dy}{y} = (x-1)\,dx$$
$$\int \frac{dy}{y} = \int (x-1)\,dx$$
$$\ln|y| = \frac{x^2}{2} - x + C$$
$$y = e^{\frac{x^2}{2} - x + C}$$
$$y = ce^{\frac{x^2}{2} - x}$$

10. (Brief #98)

$$y' = x^2 + x^2 y$$
$$\frac{dx}{dy} = x^2(1+y)$$
$$\int \frac{dy}{1+y} = \int x^2\,dx$$
$$\ln|1+y| = \frac{x^3}{3} + C$$
$$1 + y = e^{\frac{x^3}{3} + C}$$
$$y = ce^{x^3/3} - 1$$

11. (Brief #99)

$$y^2 y' = 3x^2,\ y(0) = 1$$
$$y^2 \frac{dy}{dx} = 3x^2$$
$$y^2\,dy = 3x^2\,dx$$
$$\int y^2\,dy = \int 3x^2\,dx$$
$$\frac{y^3}{3} = x^3 + C$$
$$y^3 = 3x^3 + c$$
$$y = \sqrt[3]{3x^3 + c}$$

Since $y(0) = 1$

$$1 = \sqrt[3]{3(0)^3 + c}$$
$$c = 1$$

Thus, $y(x) = \sqrt[3]{3x^3 + 1}$.

12. (Brief #100)

$$y' = \frac{y}{x^2},\ y(1) = 1$$
$$\frac{dy}{dx} = \frac{y}{x^2}$$
$$\int \frac{dy}{y} = \int \frac{dx}{x^2}$$
$$\ln|y| = -\frac{1}{x} + C$$
$$y = e^{-\frac{1}{x} + C} = ce^{-1/x}$$

Since $y(1) = 1$,

$$1 = ce^{-1/1}$$
$$1 = ce^{-1}$$
$$e = c$$

Thus, $y(x) = ee^{-1/x} = e^{1-\frac{1}{x}}$.

13. (Brief #101)

$y' = \frac{y}{x^3}$, $y(1) = 1$

$$\frac{dy}{dx} = \frac{y}{x^3}$$

$$\frac{dy}{y} = \frac{dx}{x^3}$$

$$\int \frac{dy}{y} = \int \frac{dx}{x^3}$$

$$\ln|y| = -\frac{1}{2}x^{-2} + C$$

$$y = e^{\frac{-x^{-2}}{2}+C} = ce^{-x^{-2}/2}$$

Since, $y(1) = 1$,

$$1 = ce^{-(1)^{-2}/2}$$

$$1 = ce^{-1/2}$$

$$c = e^{1/2}$$

Thus, $y(x) = e^{1/2}e^{-x^{-2}/2}$.

14. (Brief #102)

$y' = \sqrt[3]{y}$, $y(1) = 0$

$$\frac{dy}{dx} = y^{1/3}$$

$$\frac{dy}{y^{1/3}} = dx$$

$$\int y^{-1/3}\,dy = \int dx$$

$$\frac{3}{2}y^{2/3} = x + C$$

$$y^{2/3} = \frac{2}{3}x + c$$

$$y = \left(\frac{2}{3} + c\right)^{3/2}$$

Since $y(1) = 0$,

$$0 = \left[\frac{2}{3}(1) + c\right]^{3/2}$$

$$c = -\frac{2}{3}$$

Thus, $y(x) = \left(\frac{2}{3}x - \frac{2}{3}\right)^{3/2}$

15. (Brief #103)

a. Let $y(x)$ = bank balance (in thousands) after t years. Since the balance grows by 4 thousand plus 5% of the balance compounded continuously,

$y' = 4 + 0.05y \qquad y(0) = 10$

b. $y' = 4 + 0.05y$, $y(0) = 10$

$$y' - 0.05y = 4$$

$$p(t) = -0.05,\ q(t) = 4$$

$$I(t) = e^{\int -0.05dt} = e^{-0.05t}$$

$$e^{-0.05t}y' - 0.05e^{-0.05t}y = 4e^{-0.05t}$$

$$\frac{d}{dt}(e^{-0.05t}y) = 4e^{-0.05t}$$

$$e^{-0.05t}y = \frac{4}{-0.05}e^{-0.05t} + C$$

$$e^{-0.05t}y = -80e^{-0.05t} + C$$

$$y = -80 + Ce^{0.05t}$$

Since $y(0) = 10$,

$$10 = -80 + Ce^{0.05(0)}$$

$$90 = Ce^0 = C$$

Thus, $y(t) = 90e^{0.05t} - 80$.

c. After 10 years,

$y(10) = 90e^{0.05(10)} - 80 \approx 68.385$

Thus, the balance is \$68,385 after 10 years.

16. (Brief #104)

a. Let $y(t)$ = the amount of discharge into the lake. Each year the amount of pollution grows by 4 tons and decreases by 0.25 times y. Thus,

$y' = 4 - 0.25y$

We assume that at $t = 0$, the amount of discharge is 0; that is, $y(0) = 0$.

b. $y' = 4 - 0.25y$, $y(0) = 0$

$$y' + 0.25y = 4$$

$$p(t) = 0.25,\ q(t) = 4,\ I(t) = e^{\int 0.25\,dt} = e^{0.25t}$$

$$e^{0.25t}y' + 0.25e^{0.25t}y = 4e^{0.25t}$$

$$\frac{d}{dt}(e^{0.25t}y) = 4e^{0.25t}$$

$$e^{0.25t}y = 16e^{0.25t} + C$$

$$y = 16 + Ce^{-0.25t}$$

Since $y(0) = 0$,

$$0 = 16 + Ce^{-0.25(0)}$$

$$-16 = Ce^0 = C$$

Thus, $y(t) = 16 - 16e^{-0.25t}$.

17. (Brief #105)

$$y' = 2.3(106 - y),\ y(0) = 70$$
$$\frac{dy}{dt} = 2.3(106 - y)$$
$$\frac{dy}{106 - y} = 2.3\,dt$$
$$\int \frac{dy}{106 - y} = \int 2.3\,dt$$
$$-\ln|106 - y| = 2.3t + C$$
$$106 - y = e^{-2.3t + C}$$
$$y = 106 - ce^{-2.3t}$$

$$70 = 106 - Ce^{-2.3(0)}$$
$$70 = 106 - C$$
$$36 = C$$
$$y = 106 - 36e^{-2.3t}$$

18. (Brief #106)

$y(1) = 106 - 36e^{-2.3(1)} \approx 102.4$ degrees
$y(2) = 106 - 36e^{-2.3(2)} \approx 105.6$ degrees
$y(3) = 106 - 36e^{-2.3(3)} \approx 105.96$ degrees
At $t = 3$, $y(t)$ is very close to the actual temperature.

19. (Brief #107)

a.

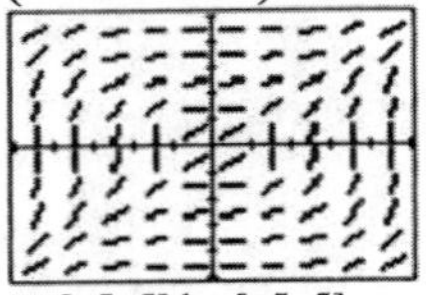
on [–5, 5] by [–5, 5]

b.

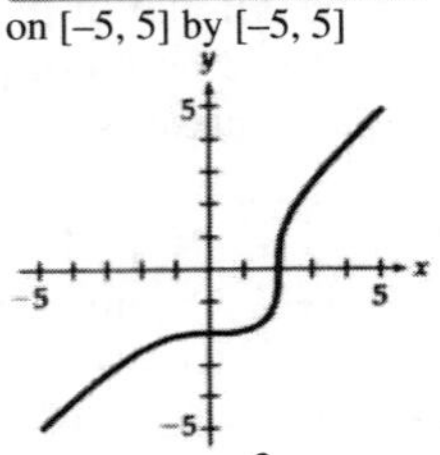

c.

$$\frac{dy}{dx} = \frac{x^2}{y^2}$$
$$y^2\,dy = x^2\,dx$$
$$\int y^2\,dy = \int x^2\,dx$$
$$\frac{1}{3}y^3 = \frac{1}{3}x^3 + C$$
$$y^3 = x^3 + C$$
$$y = x\sqrt[3]{x^3 + C}$$
$$-2 = \sqrt[3]{0 + C}$$
$$-8 = C$$
$$y = \sqrt[3]{x^3 - 8}$$

d.

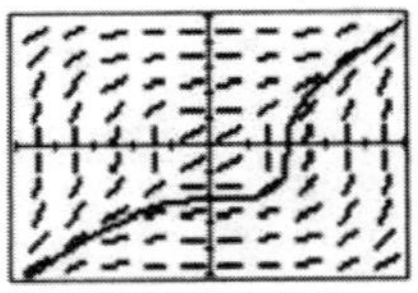
on [–5, 5] by [–5, 5]

20. (Brief #108)

a.

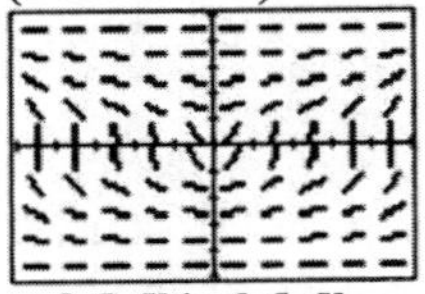
on [–5, 5] by [–5, 5]

b.

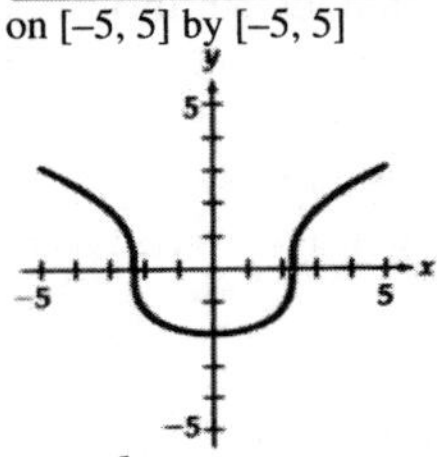

c.

$$\frac{dy}{dx} = \frac{x}{y^2}$$
$$y^2\,dy = x\,dx$$
$$\int y^2\,dy = \int x\,dx$$
$$\frac{y^3}{3} = \frac{1}{2}x^2 + C$$
$$y^3 = \frac{3}{2}x^2 + C$$
$$y = \sqrt[3]{\frac{3}{2}x^2 + C}$$
$$-2 = \sqrt[3]{\frac{3}{2}\cdot 0 + C}$$
$$-8 = C$$
$$y = \sqrt[3]{\frac{3}{2}x^2 - 8}$$

d.

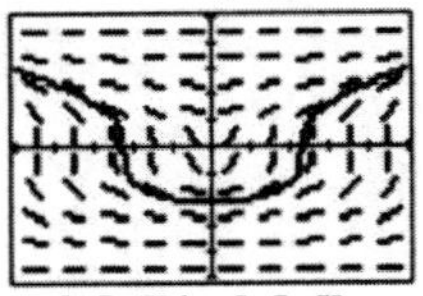
on [–5, 5] by [–5, 5]

21. **(Brief #109)**
Since the increase in the price of a movie ticket is proportional to its price, growth is unlimited and
$y' = ay,\ y(0) = \$7.85$
where $y(t)$ = the price of a movie ticket t years after 2010 and $a = 0.04$. Thus, the solution is
$y = ce^{0.04t}$.
Since $y(0) = 7.85$,
$7.85 = c^{0.04(0)} = c$
Thus, $y(t) = 7.85e^{0.04t}$
In 2018, $t = 8$,
$y(8) = 7.85e^{0.04(8)} \approx \$10.81$

22. **(Brief #110)**
Since the increase in spending is proportional to spending, growth is unlimited and
$y' = ay,\ y(0) = 132$ billion
where $y(t)$ = the amount spent (in billions of dollars) t years after 2008 and $a = 0.12$. Thus, the solution is $y = ce^{0.12t}$.
Since $y(0) = 132$,
$132 = ce^{0.12(0)} = c$
Thus, $y(t) = 132e^{0.12t}$
In 2018, $t = 10$.
$y(10) = 132e^{0.12(10)} \approx \$438$ billion

23. **(Brief #111)**
Since the growth in the number of infected people is proportional to both the number infected and the number uninfected, use the logistic growth model. Let $y(t)$ = the number of infected people. Then
$y' = ay(M - y)$
for $M = 8000$ and $y(0) = 10$. The solution is
$$y(t) = \frac{8000}{1 + ce^{-a(8000)t}}$$
Since $y(0) = 10$,
$$10 = \frac{8000}{1 + ce^{-8000a(0)}}$$
$$10 = \frac{8000}{1 + ce^{0}}$$
$$1 + c = 800$$
$$c = 799$$
Thus, $y(t) = \frac{8000}{1+799e^{-8000at}}$.
To find a, use the fact that $y(1) = 150$.
$$150 = \frac{8000}{1 + 799e^{-8000a(1)}}$$
$$1 + 799e^{-8000a} = \frac{8000}{150}$$
$$799e^{-8000a} = \frac{160}{3} - 1$$
$$e^{-8000a} = \frac{1}{799}\left(\frac{157}{3}\right)$$
$$-8000a = \ln\left(\frac{157}{2397}\right)$$
$$a = -\frac{1}{8000}\ln\left(\frac{157}{2397}\right) \approx 0.00034$$
Thus, since $8000(0.00034) \approx 2.73$, we get
$$y(t) = \frac{8000}{1 + 799e^{-2.73t}}$$
After 2 weeks, there will be
$$y(2) = \frac{8000}{1 + 799e^{-2.73(2)}} \approx 1818 \text{ cases.}$$

24. **(Brief #112)**
Since growth is proportional to both those who know and those who don't know, use the logistic growth model. Let $y(t)$ = the number of people who have heard the rumor after t days. Then
$y' = ay(M - y)$
for $M = 500$ and $y(0) = 2$. The solution is
$$y(t) = \frac{500}{1 + ce^{-500at}}$$
Since $y(0) = 2$,
$$2 = \frac{500}{1 + ce^{-500a(0)}}$$
$$2 = \frac{500}{1 + ce^{0}}$$
$$1 + c = 250$$
$$c = 249$$
Thus, $y = \frac{500}{1+249e^{-500at}}$.
To find a, use the fact that $y(1) = 75$.
$$75 = \frac{500}{1 + 249e^{-500a(1)}}$$
$$1 + 249e^{-500a} = \frac{20}{3}$$
$$e^{-500a} = \frac{1}{249}\left(\frac{17}{3}\right)$$
$$a = -\frac{1}{500}\ln\left(\frac{17}{747}\right) \approx 0.0076$$
Thus, $500(0.0076) \approx 3.78$ and
$$y(t) = \frac{500}{1 + 249e^{-3.78t}}$$
In 2 days, the number of students who will have heard the rumor is $y(2) = \frac{500}{1+249e^{-3.78(2)}} \approx 443$.

25. **(Brief #113)**
Let $y(t)$ = total sales after t months.
Since sales growth is proportional to the distance between the maximum and the sales, use the limited growth model. Thus
$y' = a(M - y), \ \ y(0) = 0$
for $M = 10{,}000$. The solution is
$y(t) = 10{,}000(1 - e^{-at})$
Since $y(7) = 3000$,

$$3000 = 10{,}000(1 - e^{-a(7)})$$
$$0.3 = 1 - e^{-7a}$$
$$-0.7 = -e^{-7a}$$
$$e^{-7a} = 0.7$$
$$-7a = \ln 0.7$$
$$a = \frac{1}{7}\ln 0.7 \approx 0.051$$

Thus, $y(t) = 10{,}000(1 - e^{-0.051t})$.
The sales after the first year ($t = 12$) are
$y(12) = 10{,}000(1 - e^{-0.051(12)}) \approx 4577$

26. **(Brief #114)**
Let $y(t)$ = the rate at which the mail carrier sorts letters after t weeks on the route. Since $y(0) = 0$ and the increase is proportional to the distance from the maximum, use the limited growth model. Thus,
$y' = a(M - y)$
for $M = 60$. The solution is
$y(t) = 60(1 - e^{-at})$
Since $y(2) = 25$,

$$25 = 60(1 - e^{-a(2)})$$
$$\frac{5}{12} = 1 - e^{-2a}$$
$$-\frac{7}{12} = -e^{-2a}$$
$$-2a = \ln\left(\frac{7}{12}\right)$$
$$a = -\frac{1}{2}\ln\left(\frac{7}{12}\right) \approx 0.269$$

Thus, $y(t) = 60(1 - e^{-0.269t})$. To find the value of t such that $y(t) = 50$, solve the equation

$$50 = 60(1 - e^{-0.269t})$$
$$\frac{5}{6} = 1 - e^{-0.269t}$$
$$-\frac{1}{6} = -e^{-0.269t}$$
$$t = -\frac{1}{0.269}\ln\left(\frac{1}{6}\right) \approx 6.7 \text{ weeks}$$

27. **(Brief #115)**

Let $y(t)$ = the number of people (in thousands) who have seen the ad in t weeks. Since $y(0) = 0$ and the growth is proportional to the distance from the maximum (that is, the people who have not seen the ad), use the limited growth model. Thus

$$y' = a(M - y)$$

for $M = 500$. The solution is

$$y(t) = 500(1 - e^{-at})$$

Since $y(2) = 200$,

$$200 = 500(1 - e^{-2a})$$

$$\frac{2}{5} = 1 - e^{-2a}$$

$$-2a = \ln\left(\frac{3}{5}\right)$$

$$a = -\frac{1}{2}\ln\left(\frac{3}{5}\right) = 0.255$$

Thus, $y(t) = 500(1 - e^{-0.255t})$.

The number of weeks it will take to reach 400,000 people is

$$400 = 500(1 - e^{-0.255t})$$

$$0.8 = 1 - e^{-0.255t}$$

$$e^{-0.255t} = 0.2$$

$$t = -\frac{1}{0.255}\ln 0.2 \approx 6.3 \text{ weeks}$$

28. Let $y(t)$ = sales in thousands. Since sales are growing in proportion to both the number sold and the size of the remaining market, use the logistic growth model. Thus

$$y' = ay(M - y)$$

for $M = 40$ and $y(0) = 1$. The solution is

$$y(t) = \frac{40}{1 + ce^{-a(40)t}}$$

Since $y(0) = 1$,

$$1 = \frac{40}{1 + ce^{-40a(0)}}$$

$$1 + ce^{0} = 40$$

$$c = 39$$

Thus, $y(t) = \frac{40}{1 + 39e^{-at}}$. Since $y(1) = 4$,

$$4 = \frac{40}{1 + 39e^{-a(1)}}$$

$$1 + 39e^{-a} = 10$$

$$e^{-a} = \frac{9}{39}$$

$$a = -\ln\frac{9}{39} \approx 1.47$$

Thus, $y(t) = \frac{40}{1 + 39e^{-1.47t}}$.

The company will sell 20,000 fax machines when $y(t)$ = 20 thousand. Thus,

$$20 = \frac{40}{1 + 39e^{-1.47t}}$$

$$1 + 39e^{-1.47t} = 2$$

$$e^{-1.47t} = \frac{1}{39}$$

$$t = -\frac{1}{1.47}\ln\left(\frac{1}{39}\right) \approx 2.5 \text{ years}$$

29. $xy - 5y = 4x^7$

$$y' - \frac{5}{x}y = 4x^6$$

$$p(x) = -\frac{5}{x},\ q(x) = 4x^6,$$

$$I(x) = e^{\int -\frac{5}{x}dx} = e^{-5\ln x} = e^{\ln x^{-5}} = x^{-5}$$

$$x^{-5}y' - 5x^{-6}y = 4x^6 \cdot x^{-5}$$

$$\frac{d}{dx}(x^{-5}y) = 4x$$

$$x^{-5}y = \frac{4x^2}{2} + C$$

$$y = 2x^7 + Cx^5$$

30. $xy' + \frac{1}{2}y = 3x$

$$y' + \frac{1}{2x}y = 3$$

$$p(x) = \frac{1}{2x},\ q(x) = 3,$$

$$I(x) = e^{\int \frac{1}{2x}dx} = e^{\frac{1}{2}\ln x} = e^{\ln x^{1/2}} = x^{1/2}$$

$$x^{1/2}y' + x^{1/2}\left(\frac{1}{2x}y\right) = 3x^{1/2}$$

$$\frac{d}{dx}(x^{1/2}y) = 3x^{1/2}$$

$$x^{1/2}y = 2x^{3/2} + C$$

$$y = 2x + Cx^{-1/2}$$

31. $y' + xy = x$

$p(x) = x,\ q(x) = x,\ I(x) = e^{\int x\,dx} = e^{x^2/2}$

$e^{x^2/2}y' + xe^{x^2/2}y = xe^{x^2/2}$

$\frac{d}{dx}(e^{x^2/2}y) = xe^{x^2/2}$

$e^{x^2/2}y = e^{x^2/2} + C$

$y = 1 + Ce^{-x^2/2}$

32. $y' - 2xy = -4x$

$p(x) = -2x,\ q(x) = -4x,\ I(x) = e^{\int -2x\,dx} = e^{-x^2}$

$e^{-x^2}y' - 2xe^{-x^2}y = -4xe^{-x^2}$

$\frac{d}{dx}(e^{-x^2}y) = -4xe^{-x^2}$

$e^{-x^2}y = 2e^{-x^2} + C$

$y = 2 + Ce^{x^2}$

33. $xy' + y = xe^x$ Note that $\frac{dx}{x}(x) = 1$, which is the coefficient of y.

$\frac{d}{dx}(xy) = xe^x$

$xy = xe^x - \int x^0 e^x\,dx = xe^x - e^x + C$

$y = e^x - x^{-1}e^x + Cx^{-1}$

34. $xy' + y = x\ln x$ Note that $\frac{d}{dx}(x) = 1$, which is the coefficient of y.

$\frac{d}{dx}(xy) = x\ln x$

$xy = \frac{1}{1+1}x^{1+1}\ln x - \frac{1}{(1+1)^2}x^{1+1} + C = \frac{1}{2}x^2\ln x - \frac{1}{4}x^2 + C$

$y = \frac{1}{2}x\ln x - \frac{1}{4}x + Cx^{-1}$

35. $y' - xy = 0,\ y(0) = 3$

$p(x) = -x,\ q(x) = 0,\ I(x) = e^{\int -x\,dx} = e^{-x^2/2}$

$y'e^{-x^2/2} - xe^{-x^2/2}y = 0$

$\frac{d}{dx}(e^{-x^2/2}y) = 0$

$e^{-x^2/2}y = C$

$y = Ce^{x^2/2}$

Since $y(0) = 3$,

$3 = Ce^{0^2/2} = C$

Thus, $y(t) = 3e^{x^2/2}$.

36. $y' + 4xy = 0,\ y(0) = 2$

$p(x) = 4x,\ q(x) = 0,\ I(x) = e^{\int 4x\,dx} = e^{2x^2}$

$e^{2x^2}y' + 4xe^{2x^2}y = 0$

$\frac{d}{dx}(e^{2x^2}y) = 0$

$e^{2x^2}y = C$

$y = Ce^{-2x^2}$

Since $y(0) = 2$,

$2 = Ce^{-2(0)^2} = C$

Thus, $y(t) = 2e^{-2x^2}$.

37. $xy' + 2x = 6x,\ y(1) = 0$

$x^2y' + 2xy = 6x^2$

$\frac{d}{dx}(x^2y) = 6x^2$

$x^2y = 2x^3 + C$

$y = 2x + Cx^{-2}$

Since $y(1) = 0$,

$0 = 2(1) + C(1)^{-2}$

$C = -2$

Thus, $y(t) = 2x - 2x^{-2}$.

38. $xy' - y = 4,\ y(1) = 6$

$y' - \frac{1}{x}y = \frac{4}{x}$

$p(x) = -\frac{1}{x},\ q(x) = \frac{4}{x},$

$I(x) = e^{\int -\frac{1}{x}dx} = e^{-\ln x} = x^{-1}$

$x^{-1}y' - x^{-2}y = 4x^{-2}$

$\frac{d}{dx}(x^{-1}y) = 4x^{-2}$

$x^{-1}y = -4x^{-1} + C$

$y = -4 + Cx$

Since $y(1) = 6$,

$6 = -4 + C(1)$

$C = 10$

Thus, $y(t) = -4 + 10x$.

39. **a.** $y' = 0.05y + 10e^{0.05t},\ y(0) = 100$

$y' - 0.05y = 10e^{0.05t}$

$p(t) = -0.05,\ q(t) = 10e^{0.05t},$

$I(t) = e^{\int -0.05 dt} = e^{-0.05t}$

$e^{-0.05t}y' - 0.05e^{-0.05t}y = 10e^{0.05t}(e^{-0.05t})$

$\frac{d}{dx}(e^{-0.05t}y) = 10$

$e^{-0.05t}y = 10t + C$

$y = 10te^{0.05t} + Ce^{0.05t}$

Since $y(0) = 100$,

$100 = 10(0)e^{0.05(0)} + Ce^{0.05(0)}$

$100 = 0 + Ce^0 = C$

Thus, $y(t) = 10te^{0.05t} + 100e^{0.05t}$.

b. The value of the fund after 5 years is

$y(5) = 10(5)e^{0.05(5)} + 100e^{0.05(5)} \approx 192.604$

The value after 5 years is about \$192,604.

c. 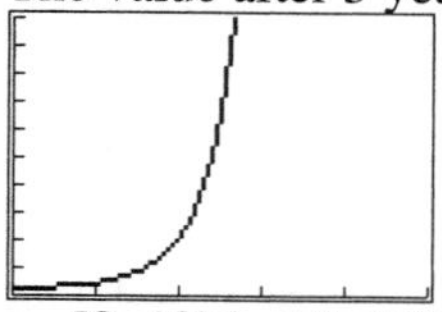

on [0, 10] by [0, 250]

The fund will reach \$250,000 in about 7.33 years

40. **a.** $p' = 0.02p + 0.1e^{0.01t},\ p(0) = 100$

$p' - 0.02p = 0.1e^{0.01t}$

$p(t) = -0.02,\ q(t) = 0.1e^{0.01t}$

$I(t) = e^{\int -0.02 dt} = e^{-0.02t}$

$e^{-0.02t}p' - 0.02e^{-0.02t}p = 0.1e^{0.01t}(e^{-0.02t})$

$\frac{d}{dt}(e^{-0.02t}p) = 0.1e^{-0.01t}$

$e^{-0.02t}p = \frac{0.1}{-0.01}e^{-0.01t} + C$

$p = -10e^{-0.01t}(e^{0.02t}) + Ce^{0.02t}$

$= -10e^{0.01t} + Ce^{0.02t}$

Since $P(0) = 100$,

$100 = -10e^{0.01(0)} + Ce^{0.02(0)}$

$110 = Ce^0 = C$

Thus, $y(t) = -10e^{0.01t} + 110e^{0.02t}$.

b. At $t = 10$,

$y(10) = -10e^{0.01(10)} + 110e^{0.02(10)} \approx 123$

In 10 years, the population will be about 123 million.

c. 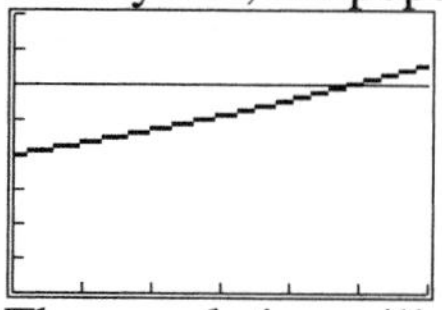

The population will reach 150 million in about 19.4 years.

41. **a.** Let $y(t)$ = amount of oxygen in the room after t minutes.

Incoming oxygen = 500 cubic feet × 0.20 = 100 cubic feet

Outgoing oxygen = $y\left(\frac{500}{10,000}\right) = 0.05y$

Thus, the rate of change of oxygen is $y' = 100 - 0.05y$ and $y(0) = 10,000(0.1) = 1000$.

b. $y' + 0.05y = 100,\ y(0) = 1000$

$p(t) = 0.05,\ q(t) = 100,$

$$I(t) = e^{\int 0.05\,dt} = e^{0.05t}$$

$$e^{0.05t}y' + 0.05e^{0.05t}y = 100e^{0.05t}$$

$$\frac{d}{dt}(e^{0.05t}y) = 100e^{0.05t}$$

$$e^{0.05t}y = \frac{100}{0.05}e^{0.05t} + C$$

$$y = 2000 + Ce^{-0.05t}$$

Since $y(0) = 1000$,

$1000 = 2000 + Ce^{-0.05(0)}$

$-1000 = Ce^0 = C$

Thus, $y(t) = 2000 - 1000e^{-0.05t}$.

c. After 15 minutes,

$y(15) = 2000 - 1000e^{-0.05(15)} \approx 1528$ cubic feet

d. In the long run, $y(t) \to 2000 + 0 = 2000$ because $\lim_{t\to\infty} Ce^{-0.05t} = 0$. Thus, the long-run oxygen content is 2000 cubic feet.

e.

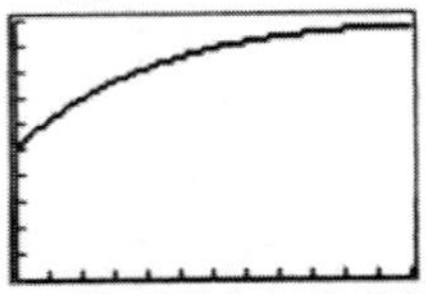

on [0, 60] by [0, 2000]

The oxygen will reach 19% (which is $0.19 \cdot 10,000 = 1900$ cubic feet) in about 46 minutes.

42. **a.** $p' = 0.2p + 2,\ p(0) = 5$

$p' - 0.2p = 2$

$p(t) = -0.2,\ q(t) = 2$

$$I(t) = e^{\int -0.2\,dt} = e^{-0.2t}$$

$$e^{-0.2t}p' - 0.2e^{-0.2t}p = 2e^{-0.2t}$$

$$\frac{d}{dt}(e^{-0.2t}p) = 2e^{-0.2t}$$

$$e^{-0.2t}p = \frac{2}{-0.2}e^{-0.2t} + C$$

$$p = -10 + Ce^{0.2t}$$

Since $p(0) = 5$,

$5 = -10 + Ce^{0.2(0)}$

$15 = Ce^0 = C$

Thus $p = -10 + 15e^{0.2t}$

b. After 2 years,

$p(2) = -10 + 15e^{0.2(2)} \approx 12$ eagles

c.

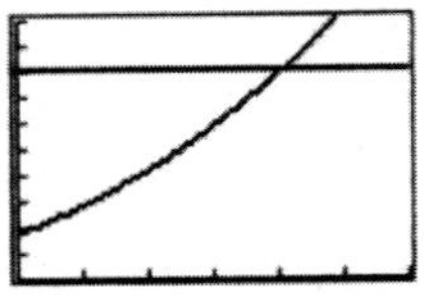

on [0, 6] by [0, 30]

The eagle population will reach 24 in about 4 years.

43. **a.** $y' = 2 - 2y$, $y(0) = 2$. The initial point is (0, 2), and the slope is $g(x, y) = 2 - 2y$. The step size is $\frac{1-0}{4} = 0.25$

Segment 1

Slope $g(0, 2) = 2 - 2(2) = -2$

End point $x_1 = 0.25$, $y_1 = 2 + (-2)(0.25) = 1.5$

Segment 2

Slope $g(0.25, 1.5) = 2 - 2(1.5) = -1$

End point $x_2 = 0.5$, $y_2 = 1.5 + (-1)(0.25) = 1.25$

Segment 3

Slope $g(0.5, 1.25) = 2 - 2(1.25) = -0.5$

End point $x_3 = 0.75$, $y_3 = 1.25 + (-0.5)(0.25) \approx 1.13$

Segment 4

Slope $g(0.75, 1.13) = 2 - 2(1.13) = -0.26$

End point $x_4 = 1$, $y_4 = 1.13 + (-0.26)(0.25) \approx 1.07$

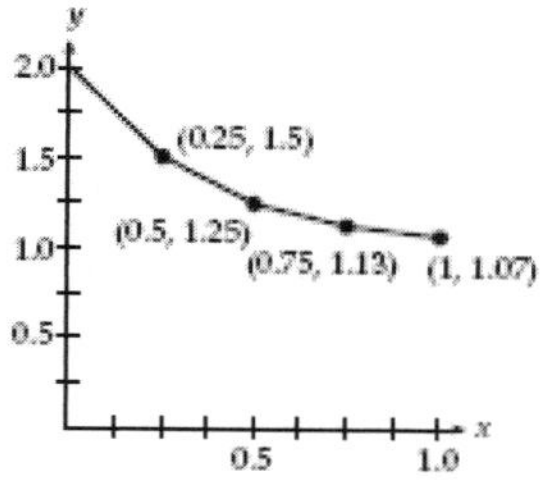

b. $y' = 2 - 2y$ or $y' + 2y = 2$, $y(0) = 2$

$p(x) = 2$, $q(x) = 2$, $I(x) = e^{\int 2\,dx} = e^{2x}$

$e^{2x}y' + 2e^{2x} = 2e^{2x}$

$\frac{d}{dx}(e^{2x}y) = 2e^{2x}$

$e^{2x}y = e^{2x} + C$

$y = 1 + Ce^{-2x}$

Since $y(0) = 2$,

$2 = 1 + Ce^{-2(0)}$

$1 = Ce^0 = C$

Thus, $y(x) = e^{-2x} + 1$.

c. $y(1) = e^{-2(1)} + 1 \approx 1.14$

44. **a.** $y' = \frac{4x}{x}$, $y(0) = 1$. The initial point is (0, 1), and the slope is $g(x, y) = \frac{4x}{y}$. The step size is $\frac{1-0}{4} = 0.25$

Segment 1

Slope $g(0, 1) = \frac{4(0)}{1} = 0$

End point $x_1 = 0.25, \quad y_1 = 1 + 0.(0.25) = 1$

Segment 2

Slope $g(0.25, 1) = \frac{4(0.25)}{1} = 1$

End point $x_2 = 0.5, \quad y_2 = 1 + 1(0.25) = 1.25$

Segment 3

Slope $g(0.5, 1.25) = \frac{4(0.5)}{1.25} = 1.6$

End point $x_3 = 0.75, \quad y_3 = 1.25 + 1.6(0.25) = 1.65$

Segment 4

Slope $g(0.75, 1.65) = \frac{4(0.75)}{1.65} = \frac{3}{1.65}$

End point $x_4 = 1, \quad y_4 = 1.65 + \frac{3}{1.65}(0.25) \approx 2.1$

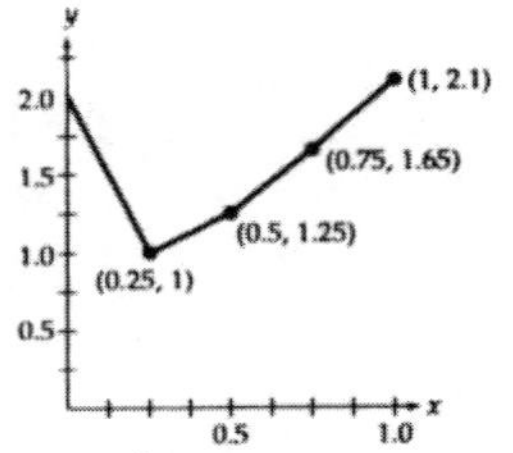

b. $y' = \frac{4x}{y}$

$$\frac{dy}{dx} = \frac{4x}{y}$$
$$y\,dy = 4x\,dx$$
$$\int y\,dy = \int 4x\,dx$$
$$\frac{y^2}{2} = 2x^2 + C$$
$$y^2 = 4x^2 + c$$

Since $y(0) = 1$,

$$1^2 = 4(0)^2 + c$$
$$c = 1$$

Thus, the particular solution is $y = \sqrt{4x^2 + 1}$.

c. At $x = 1$,

$$y^2 = 4(1)^2 + 1 = 5$$
$$y = \sqrt{5} \approx 2.24$$

45. $y' = \frac{1}{y^2 + 1}$, $y(0) = 1$. Using an Euler's method program, we get $y(2) \approx 1.70$.

46. $y' = ye^{-x}$, $y(1) = 5$. Using an Euler's method program, we get $y(2.5) \approx 6.66$.

47. **a.**
$$\begin{aligned} p' &= 0.1(1-p) + 0.1p(1-p) \\ &= (0.1 + 0.1p)(1-p) \\ &= 0.1(1+p)(1-p) \\ &= 0.1(1-p^2) \end{aligned}$$

b. $p' = 0.1(1-p^2)$, $p(0) = 0.3$. Using an Euler's method program, $p(4) \approx 0.61$. Thus, about 61% have heard of the product after 4 weeks.

48. $y' = 0.2y - 0.1\sqrt[3]{y}$, $y(0) = 100$. Using an Euler's method program, we get $y(2) \approx 148$. Thus, the population after 2 weeks is about 148.

Chapter 10: Sequences and Series

EXERCISES 10.1

1. $\sum_{i=1}^{5}\frac{1}{i+1}=\frac{1}{1+1}+\frac{1}{2+1}+\frac{1}{3+1}+\frac{1}{4+1}+\frac{1}{5+1}$

$=\frac{1}{2}+\frac{1}{3}+\frac{1}{4}+\frac{1}{5}+\frac{1}{6}$

3. $\sum_{k=1}^{4}\left(-\frac{1}{3}\right)^k=\left(-\frac{1}{3}\right)^1+\left(-\frac{1}{3}\right)^2+\left(-\frac{1}{3}\right)^3+\left(-\frac{1}{3}\right)^4$

$=-\frac{1}{3}+\frac{1}{9}-\frac{1}{27}+\frac{1}{81}$

5. $\sum_{n=1}^{6}\frac{1-(-1)^n}{2}=\frac{1-(-1)^1}{2}+\frac{1-(-1)^2}{2}+\frac{1-(-1)^3}{2}+\frac{1-(-1)^4}{2}+\frac{1-(-1)^5}{2}+\frac{1-(-1)^6}{2}$

$=\frac{1+1}{2}+\frac{1-1}{2}+\frac{1+1}{2}+\frac{1-1}{2}+\frac{1+1}{2}+\frac{1-1}{2}=1+0+1+0+1+0$

7. $\frac{1}{3}+\frac{1}{9}+\frac{1}{27}+\frac{1}{81}+\cdots=\frac{1}{3}+\frac{1}{3^2}+\frac{1}{3^3}+\frac{1}{3^4}+\cdots=\sum_{i=1}^{\infty}\frac{1}{3^i}$

9. $-2+4-8+16-\cdots=(-2)^1+(-2)^2+(-2)^3+(-2)^4+\cdots=\sum_{i=1}^{\infty}(-2)^i$

11. $-1+2-3+4-\cdots=(-1)^1(1)+(-1)^2(2)+(-1)^3(3)+(-1)^4(4)+\cdots=\sum_{i=1}^{\infty}(-1)^i i$

13. $1+2+2^2+2^3+\cdots+2^9$

$a=1,\ r=2,\ n=10$

$S_n=a\frac{1-r^n}{1-r}$

$S_{10}=(1)\frac{1-2^{10}}{1-2}=2^{10}-1=1023$

15. $3+3\cdot 4+3\cdot 4^2+\cdots+3\cdot 4^5$

$a=3,\ r=4,\ n=6$

$S_n=a\frac{1-r^n}{1-r}$

$S_6=3\left(\frac{1-4^6}{1-4}\right)=4^6-1=4095$

17. $3-3\cdot 2+3\cdot 2^2-3\cdot 2^3+\cdots+3\cdot 2^6$

$a=3,\ r=-2,\ n=7$

$S_n=a\frac{1-r^n}{1-r}$

$S_7=3\left(\frac{1-(-2)^7}{1-(-2)}\right)=2^7+1=129$

19. $4+\frac{4}{5}+\frac{4}{25}+\frac{4}{125}+\cdots$

$a=4,\ r=\frac{1}{5}$

$S=\frac{a}{1-r}=\frac{4}{1-\frac{1}{5}}=\frac{4}{\frac{4}{5}}=5$

21. $2-\frac{2}{3}+\frac{2}{9}-\frac{2}{27}+\frac{2}{81}-\cdots$

$a=2,\ r=\frac{-\frac{2}{3}}{2}=-\frac{2}{2\cdot 3}=-\frac{1}{3}$

$S=\frac{a}{1-r}=\frac{2}{1-\left(-\frac{1}{3}\right)}=\frac{2}{\frac{4}{3}}=\frac{3}{2}$

23. $\frac{1}{3}+\frac{2}{3}+\frac{4}{3}+\frac{8}{3}+\frac{16}{3}+\cdots$

$a=\frac{1}{3},\ r=\frac{\frac{2}{3}}{\frac{1}{3}}=\frac{2}{3}\cdot\frac{3}{1}=2$

Since $|r|>1$, the series diverges.

25. $\frac{3}{2}+\frac{3^2}{2^3}+\frac{3^3}{2^5}+\frac{3^4}{2^7}+\frac{3^5}{2^9}+\cdots$

$a=\frac{3}{2},\ r=\frac{\frac{3^2}{2^3}}{\frac{3}{2}}=\frac{3^2}{2^3}\cdot\frac{2}{3}=\frac{3}{2^2}=\frac{3}{4}$

$S=\frac{a}{1-r}=\frac{\frac{3}{2}}{1-\frac{3}{4}}=\frac{\frac{3}{2}}{\frac{1}{4}}=\frac{3}{2}\cdot 4=6$

27. $8+6+\frac{9}{2}+\frac{27}{8}+\cdots$

$a=8,\ r=\frac{6}{8}=\frac{3}{4}$

$S=\frac{a}{1-r}=\frac{8}{1-\frac{3}{4}}=\frac{8}{\frac{1}{4}}=32$

29. $\sum_{i=1}^{\infty}\frac{100}{5^i},\ a=\frac{100}{5^1}=20,\ r=\frac{\frac{100}{5^2}}{\frac{100}{5^1}}=\frac{5}{25}=\frac{1}{5}$

$S=\frac{a}{1-r}=\frac{20}{1-\frac{1}{5}}=\frac{20}{\frac{4}{5}}=25$

31. $\sum_{j=1}^{\infty}\frac{3^j}{7\cdot 2^j},\ a=\frac{3}{7\cdot 2}=\frac{3}{14},$

$r=\frac{\frac{3^2}{7\cdot 2^2}}{\frac{3}{7\cdot 2}}=\frac{3^2\cdot 2}{3\cdot 2^2}=\frac{3}{2}$

Since $|r|>1$, the series diverges.

33. $\sum_{j=0}^{\infty}\left(-\frac{1}{2}\right)^j,\ a=\left(-\frac{1}{2}\right)^0=1,\ r=\frac{\left(-\frac{1}{2}\right)^2}{-\frac{1}{2}}=-\frac{1}{2}$

$S=\frac{a}{1-r}=\frac{1}{1-\left(-\frac{1}{2}\right)}=\frac{1}{\frac{3}{2}}=\frac{2}{3}$

35. $\sum_{k=1}^{\infty}(1.01)^k,\ a=1.01,\ r=\frac{(1.01)^2}{1.01}=1.01$

Since $|r|>1$, the series diverges.

37. $\sum_{k=0}^{\infty}(0.99)^k,\ a=(0.99)^0=1,\ r=\frac{(0.99)^2}{0.99}=0.99$

$S=\frac{a}{1-r}=\frac{1}{1-0.99}=\frac{1}{0.01}=100$

39. $\sum_{j=0}^{\infty}\frac{3^j}{4^{j+1}},\ a=\frac{3^0}{4^{0+1}}=\frac{1}{4},\ r=\frac{\frac{3}{4^2}}{\frac{3^0}{4^1}}=\frac{3\cdot 4}{1\cdot 4^2}=\frac{3}{4}$

$S=\frac{a}{1-r}=\frac{\frac{1}{4}}{1-\frac{3}{4}}=\frac{\frac{1}{4}}{\frac{1}{4}}=1$

41. $0.\overline{36}=\frac{36}{100}+\frac{36}{100^2}+\frac{36}{100^3}+\cdots$

$a=\frac{36}{100},\ r=\frac{\frac{36}{100^2}}{\frac{36}{100}}=\frac{36}{100^2}\cdot\frac{100}{36}=\frac{1}{100}$

$S=\frac{a}{1-r}=\frac{\frac{36}{100}}{1-\frac{1}{100}}=\frac{\frac{36}{100}}{\frac{99}{100}}=\frac{36}{99}=\frac{4}{11}$

43. $2.\overline{54}=2+\left(\frac{54}{100}+\frac{54}{100^2}+\frac{54}{100^3}+\cdots\right)$

$a=\frac{54}{100},\ r=\frac{1}{100}$

$S=\frac{a}{1-r}=\frac{\frac{54}{100}}{1-\frac{1}{100}}=\frac{\frac{54}{100}}{\frac{99}{100}}=\frac{54}{99}$

Thus, $2.\overline{54}=2\frac{54}{99}=2\frac{6}{11}$.

45. $0.\overline{027} = \frac{27}{1000} + \frac{27}{1000^2} + \frac{27}{1000^3} + \cdots$

$a = \frac{27}{1000},\ r = \frac{\frac{27}{1000^2}}{\frac{27}{1000}} = \frac{27}{1000^2} \cdot \frac{1000}{27} = \frac{1}{1000}$

$S = \frac{a}{1-r} = \frac{\frac{27}{1000}}{1-\frac{1}{1000}} = \frac{\frac{27}{1000}}{\frac{999}{1000}} = \frac{27}{999} = \frac{1}{37}$

47. $0.\overline{012345679} = \frac{12345679}{10^9} + \frac{12345679}{10^{18}} + \frac{12345679}{10^{27}} + \cdots$

$a = \frac{12345679}{10^9},\ r = \frac{1}{10^9}$

$S = \frac{\frac{12345679}{10^9}}{1-\frac{1}{10^9}} = \frac{\frac{12345679}{10^{10}}}{\frac{99999999}{10^9}} = \frac{12345679}{999999999} = \frac{1}{81}$

49. **a.** For the Yonker's fine, $a = 100$ and $r = 2$. For 18 days,

$S_{18} = 100\frac{1-2^{18}}{1-2} = 100(2^{18}-1) = \$26,214,300$

b. In 24 days, the fine would have reached over $1 billion.

51. For 6% compounded monthly, the monthly interest rate is $\frac{0.06}{12} = 0.005$, so the deposits are multiplied by 1.005. Monthly deposits are $300 for 40 years = 40 · 12 = 480 months. The amount of annuity is

$300\frac{1-1.005^{480}}{1-1.005} \approx \$597,447$

53. (Present value) $= 4167 + \frac{4167}{1.005} + \frac{4167}{(1.005)^2} + \cdots + \frac{4167}{(1.005)^{239}}$

$a = 4167,\ r = \frac{1}{1.005}$, and

(Present value) $4167\frac{1-\left(\frac{1}{1.005}\right)^{240}}{1-\left(\frac{1}{1.005}\right)} \approx \$584,541$

55. **a.** Amount saved $1+2+4+8+\cdots+2^{30}$

Thus, $a = 1,\ r = 2$, and

$S_{31} = 1\frac{1-2^{31}}{1-2} = 2^{31} - 1 = 2,147,483,647$ pennies

$= \$21,474,836.47$

b. In 27 days, the savings will be more than $1 million.

57. You travel

$10 + 10(1.1) + 10(1.1)^2 + 10(1.1)^3 + \cdots$

$a = 10,\ r = 1.1$, and we need

$3000 = 10\frac{1-1.1^n}{1-1.1}$

Using a graphing calculator, this sum is over 3000 for $n = 37$. Thus, it will take 37 days to reach the opposite coast.

59. **a.** Maximum amount

$= 12 + 12(0.2) + 12(0.2)^2 + \cdots$

$a = 12,\ r = 0.02$, and

Maximum amount $= \frac{12}{1-0.2} = 15$ units

Minimum amount $=$ maximum $\cdot\, 0.2$

$= 15 \cdot 0.2 = 3$ units

b. Using a graphing calculator, the sum reaches 14.9 in the fourth term. Thus, on the fourth day, the maximum amount reaches 14.9 units.

61. Total effect $= \dfrac{50 \text{ billion}}{1-0.9 \text{ billion}} = \$500$ billion
Multiplier = 10

63. Total effect $= \dfrac{300 \text{ million}}{1-0.95 \text{ million}} = \$6000$ million
=\$6 billion
Multiplier = 20

65. Multiplier $= \dfrac{1}{1-0.75} = 4$

67. $(\text{Capital value}) = 300 + \dfrac{300}{1+0.015} + \dfrac{300}{(1+0.015)^2} + \dfrac{300}{(1+0.015)^3} + \cdots$

$a = 300$ and $r = \frac{1}{1.015}$. Thus

$(\text{Capital value}) = \dfrac{300}{1-\frac{1}{1.015}} = \$20,300$

69. **a.** $(\text{Vertical distance}) = 6 + \frac{2}{3}(6\cdot 2) + \left(\frac{2}{3}\right)^2(6\cdot 2) + \left(\frac{2}{3}\right)^3(6\cdot 2) + \cdots$

$= -6 + (6\cdot 2) + \frac{2}{3}(6\cdot 2) + \left(\frac{2}{3}\right)^2(6\cdot 2) + \left(\frac{2}{3}\right)^3(6\cdot 2) + \cdots$

$a = 12,\ r = \dfrac{2}{3}$

$(\text{Vertical distance}) = -6 + \dfrac{12}{1-\frac{2}{3}} = -6 + \dfrac{12}{\frac{1}{3}} = 30$ feet

b. Using a graphing calculator, we get that the vertical distance exceeds 29 feet at the ninth bounce.

71. After an immigration (using $a = 800$ and $r = 0.95$):
Population $= 800 + 800(0.95) + 800(0.95)^2 + \cdots$
$= \dfrac{800}{1-0.95} \approx 16,000$
Before an immigration:
Population = 16,000 – 800 = 15,200

73. 5281 terms

75. True

77. $r = \dfrac{1}{e^2} < 1$; the series converges

79. Converges, $r = \dfrac{1}{10}$

81. The sum formula does not apply since $r = 2$.

83. $150 + 150(0.3) + 150(0.3)^2 + \cdots + 150(0.3)^{n-1} = \sum_{i=1}^{n} 150(0.3)^{i-1}$

For $n = 1$, $S = 150$.
For $n = 2$, $S = 195$.
For $n = 3$, $S = 208.5$.
For $n = 4$, $S = 212.55$.
For $n = 5$, $S = 213.765$.
For $n = 6$, $S = 214.1295$.
For $n = 7$, $S = 214.23885$.

a. The third dose increases the cumulative amount to over 200 mg.
b. The fifth dose increases the cumulative amount to over 213 mg.

85. **a.** $a = 1, r = -1$

$$\frac{a}{1-r} = \frac{1}{1-(-1)} = \frac{1}{2}$$

b. No, since $r = -1$, the restriction of $|r| < 1$ is violated.

c. For $n = 1$, $S = 1$.
For $n = 2$, $S = 0$.
For $n = 3$, $S = 1$.
For $n = 4$, $S = 0$.
For $n = 5$, $S = 1$.

The partial sums are 1, 0, 1, 0, … which, in the long run average to $\frac{1}{2}$.

EXERCISES 10.2

1. $f(x) = e^{2x}$, $f'(x) = 2e^{2x}$, $f''(x) = 4e^{2x}$, $f'''(x) = 8e^{2x}$

$$p_3(x) = e^{2(0)} + \frac{2e^{2(0)}}{1!}x + \frac{4e^{2(0)}}{2!}x^2 + \frac{8e^{2(0)}}{3!}x^3$$
$$= 1 + 2x + 2x^2 + \frac{4}{3}x^3$$

3. $f(x) = \sqrt{x+1}$, $f'(x) = \frac{1}{2}(x+1)^{-1/2}$, $f''(x) = -\frac{1}{4}(x+1)^{-3/2}$, $f'''(x) = \frac{3}{8}(x+1)^{-5/2}$

$$p_3(x) = \sqrt{0+1} + \frac{\frac{1}{2}(0+1)^{-1/2}}{1!}x + \frac{-\frac{1}{4}(0+1)^{-3/2}}{2!}x^2 + \frac{\frac{3}{8}(0+1)^{-5/2}}{3!}x^3$$
$$= 1 + \frac{1}{2}x - \frac{1}{8}x^2 + \frac{1}{16}x^3$$

5. $f(x) = 7 + x - 3x^2$, $f'(x) = 1 - 6x$, $f''(x) = -6$, $f'''(x) = 0$

$$p_3(x) = \left[7 + 0 - 3(0)^2\right] + \frac{1-6(0)}{1!}x + \frac{-6}{2!}x^2 + \frac{0}{3!}x^3$$
$$= 7 + x - 3x^2$$

7. $f(x) = e^{x^2}$, $f'(x) = 2xe^{x^2}$, $f''(x) = 2e^{x^2} + 2x(2xe^{x^2}) = 2e^{x^2} + 4x^2e^{x^2}$,
$f'''(x) = 2(2xe^{x^2}) + 8xe^{x^2} + 4x^2(2xe^{x^2}) = 12xe^{x^2} + 8x^3e^{x^2}$

$$p_3(x) = e^{(0)^2} + \frac{2(0)e^{(0)^2}}{1!}x + \frac{2e^{(0)^2} + 4(0)^2e^{(0)^2}}{2!}x^2 + \frac{12(0)e^{(0)^2} + 8(0)^3e^{(0)^2}}{3!}x^3$$
$$= 1 + 0 \cdot x + \frac{2}{2!}x^2 + \frac{0+0}{3!}x^3 = 1 + x^2$$

9. **a.** $f(x)=\ln(x+1),\ f'(x)=\frac{1}{x+1},\ f''(x)=-\frac{1}{(x+1)^2},\ f'''=\frac{2}{(x+1)^3},\ f^{(4)}(x)=-\frac{6}{(x+1)^4}$

$$p_4(x)=\ln(0+1)+\left(\frac{1}{0+1}\right)\frac{x}{1!}+\left[-\frac{1}{(0+1)^2}\right]\frac{x^2}{2!}+\left[\frac{2}{(0+1)^3}\right]\frac{x^3}{3!}+\left[-\frac{6}{(0+1)^4}\right]\frac{x^4}{4!}$$
$$=0+\frac{1}{1}x-\left(\frac{1}{1}\right)\frac{x^2}{2}+\left(\frac{2}{1}\right)\frac{x^3}{6}+\left(-\frac{6}{1}\right)\frac{x^4}{24}$$
$$=x-\frac{x^2}{2}+\frac{x^3}{3}-\frac{x^4}{4}$$

b.

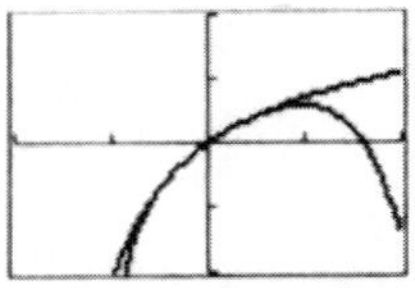

on [–2, 2] by [–2, 2]

11. **a.** $f(x)=\cos x,\ f'(x)=-\sin x,\ f''(x)=-\cos x,\ f'''(x)=\sin x,\ f^{(4)}(x)=\cos x$

$$p_4(x)=\cos 0+(-\sin 0)\frac{x}{1!}+(-\cos 0)\frac{x^2}{2!}+(\sin 0)\frac{x^3}{3!}+(\cos 0)\frac{x^4}{4!}$$
$$=1+(0)\frac{x}{1}+(-1)\frac{x^2}{2}+(0)\frac{x^3}{6}+(1)\frac{x^4}{24}$$
$$=1-\frac{x^2}{2}+\frac{x^4}{24}$$

b.

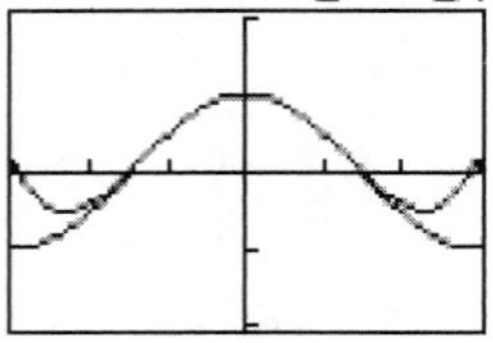

on [–π, π] by [–2, 2]

13. The second Taylor polynomial for e^x at $x=0$ is $p_2(x)=1+x+\frac{1}{2!}x^2$. Replacing x by x^2, we get that the fourth Taylor polynomial for e^{x^2} at $x=0$ is

$$p_4(x)=1+x^2+\frac{1}{2!}(x^2)^2=1+x^2+\frac{1}{2!}x^4$$

15. The fifth Taylor polynomial for $\sin x$ at $x=0$ is $p_5(x)=x-\frac{x^3}{3!}+\frac{x^5}{5!}$. Replacing x by $2x$, we get that the fifth Taylor polynomial for $\sin 2x$ at $x=0$ is

$$p_5(x)=2x-\frac{(2x)^3}{3!}+\frac{(2x)^5}{5!}=2x-\frac{8x^3}{3!}+\frac{32x^5}{5!}$$

17. **a.** $f(x)=e^x,\ f'(x)=e^x,\ f''(x)=e^x,\ f'''(x)=e^x,\ f^{(4)}(x)=e^x$

$$p_4(x)=e^0+e^0\frac{x}{1!}+e^0\frac{x^2}{2!}+e^0\frac{x^3}{3!}+e^0\frac{x^4}{4!}=1+x+\frac{x^2}{2!}+\frac{x^3}{3!}+\frac{x^4}{4!}$$

b. $$e^{-1/2}\approx p_4\left(-\frac{1}{2}\right)=1-\frac{1}{2}+\frac{1}{2!}\left(-\frac{1}{2}\right)^2+\frac{1}{3!}\left(-\frac{1}{2}\right)^3+\left(-\frac{1}{2}\right)^4\left(\frac{1}{4!}\right)=1-\frac{1}{2}+\frac{1}{2^3}+\frac{1}{6}\left(-\frac{1}{2^3}\right)+\frac{1}{24}\approx 0.60677$$

c. First, calculate $f^{(5)}(x)$.

$f^{(5)}(x)=e^x$

Since $f^{(5)}(x)=e^x$ is an increasing function, we consider the right-hand end point as where f is maximized.

$f^{(5)}(0)=e^0=1$

Thus, choose $M = 1$.

$$\left|R_4\left(-\frac{1}{2}\right)\right|\le\frac{1}{(4+1)!}\left|-\frac{1}{2}\right|^{4+1}=\frac{1}{5!}\cdot\frac{1}{2^5}\approx 0.0003 \text{ (rounded up)}$$

d. $e^{-1/2}\approx 0.6068$ with error less than 0.0003.

e.

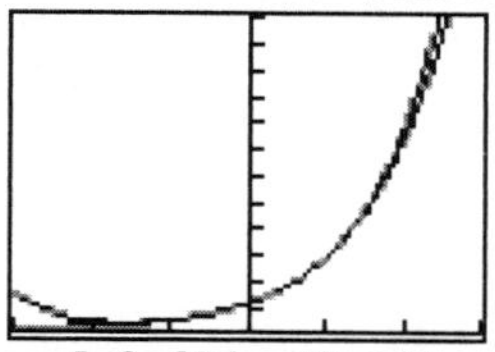

on [–3, 3] by [0, 12]

19. **a.** All derivatives of cos x are ±sin x or ±cos x. Therefore $\left|f^{(n)}(x)\right|\le|\pm1|=1$ because the maximum of sin x and cos x is 1 and the minimum of sin x and cos x is –1. Thus, choose $M = 1$.

$$|R_n(x)|\le\frac{1}{(n+1)!}|1|^{n+1}<0.0002 \quad\text{and}\quad |R_n(x)|\le\frac{1}{(n+1)!}|-1|^{n+1}<0.0002$$

Thus, $p_n(x)$ has error less than 0.0002 on the interval [–1, 1] when

$$\frac{1}{(n+1)!}<0.0002$$

$$\frac{1}{0.0002}<(n+1)!$$

Thus, $n + 1 = 7$ or $n = 6$.

Now, find the derivatives of $f(x)$.

$f(x)=\cos x,\ f'(x)=-\sin x,\ f''(x)=-\cos x,\ f'''(x)=\sin x,\ f^{(4)}(x)=\cos x,$
$f^{(5)}(x)=-\sin x,\ f^{(6)}(x)=-\cos x$

$$p_6(x)=\cos 0+(-\sin 0)\left(\frac{x}{1!}\right)+(-\cos 0)\frac{x^2}{2!}+(\sin 0)\frac{x^3}{3!}+(\cos 0)\frac{x^4}{4!}+(-\sin 0)\frac{x^5}{5!}+(-\cos 0)\frac{x^6}{6!}$$

$$=1-0x+(-1)\frac{x^2}{2!}+0\frac{x^3}{3!}+1\frac{x^4}{4!}+(-0)\frac{x^5}{5!}-1\frac{x^6}{6!}$$

$$=1-\frac{x^2}{2!}+\frac{x^4}{4!}-\frac{x^6}{6!}$$

b.

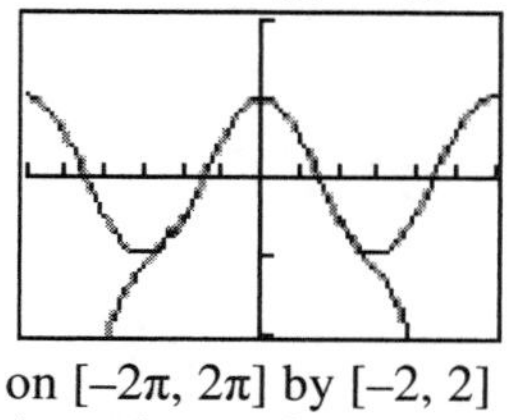

on [–2π, 2π] by [–2, 2]

Actual error for $-1\le x\le 1$ is less than 0.00003.

21. **a.** Find $p_1(x)$ for $f(x)=\sqrt[3]{1+x}$ at $x=0$.

$$f'(x)=\frac{1}{3}(1+x)^{-2/3}$$
$$p_1(x)=\sqrt[3]{1+0}+\frac{1}{3}(1+0)^{-2/3}\frac{x}{1!}=1+\frac{x}{3}$$

b. $f''(x)=-\frac{2}{9}(1+x)^{-5/3}$ Since $f''(x)$ is decreasing function, the maximum value on the interval occurs at the left endpoint. We need

$$|R_1(x)|\le\frac{M}{(1+1)!}|x|^{1+1}\le 0.01 \quad\text{or}\quad \frac{M}{2}x^2\le 0.01$$

with $\left|-\frac{2}{9}(1+t)^{-5/3}\right|\le M$ for $0\le t\le x$.

Since $\frac{2}{9}(1+t)^{-5/3}$ is decreasing on [0, x], it reaches a maximum at the left endpoint.

Let $M=\frac{2}{9}(1+0)^{-5/3}=\frac{2}{9}$.

Then

$$\frac{\frac{2}{9}}{2}x^2\le 0.01$$
$$\frac{1}{9}x^2\le 0.01$$
$$x^2\le 0.09$$
$$x\le 0.3$$

Thus, the error is less than 0.01 on the interval [0, 0.3].

c.

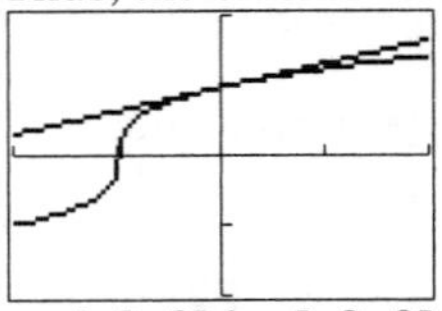

on [–2, 2] by [–2, 2]

23. **a.** $f(x)=\sqrt{x}$, $f'(x)=\frac{1}{2}x^{-1/2}$, $f''(x)=-\frac{1}{4}x^{-3/2}$, $f'''(x)=\frac{3}{8}x^{-5/2}$

$$p_3(x)=\sqrt{1}+\frac{1}{2}(1)^{-1/2}\frac{(x-1)}{1!}-\frac{1}{4}(1)^{-3/2}\frac{(x-1)^2}{2!}+\frac{3}{8}(1)^{-5/2}\frac{(x-1)^3}{3!}$$
$$=1+\frac{1}{2}(x-1)-\frac{1}{8}(x-1)^2+\frac{1}{16}(x-1)^3$$

b.

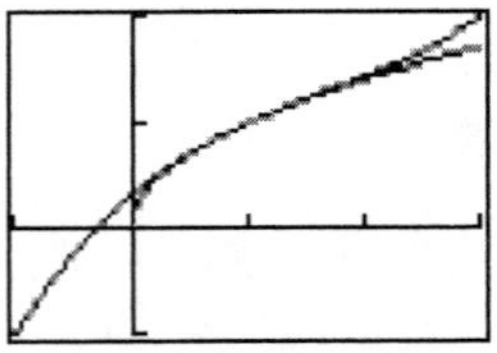

on [–1, 3] by [–1, 2]

25. **a.** $f(x)=\sin x$, $f'(x)=\cos x$, $f''(x)=-\sin x$, $f'''(x)=-\cos x$

$$p_3(x)=\sin\pi+\cos\pi\frac{(x-\pi)}{1!}+(-\sin\pi)\frac{(x-\pi)^2}{2!}+(-\cos\pi)\frac{(x-\pi)^3}{3!}$$
$$=0-1(x-\pi)-0\frac{(x-\pi)^2}{2!}-(-1)\frac{(x-\pi)^3}{3!}$$
$$=-(x-\pi)+\frac{1}{6}(x-\pi)^3$$

b. 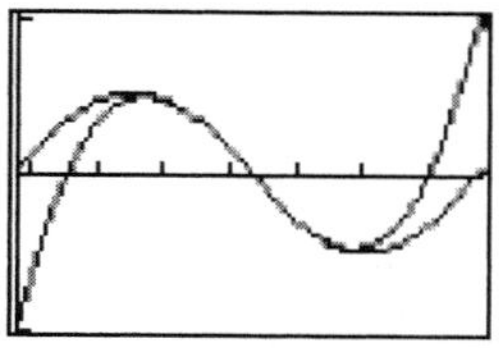

on $[0, 2\pi]$ by $[-2, 2]$

27. **a.** $f(x) = \sqrt[3]{x}$, $f'(x) = \frac{1}{3}x^{-2/3}$, $f''(x) = -\frac{2}{9}x^{-5/3}$

$$p_2(x) = \sqrt[3]{1} + \frac{1}{3}(1)^{-2/3}\frac{(x-1)}{1!} + \left[-\frac{2}{9}(1)^{-5/3}\right]\frac{(x-1)^2}{2!}$$

$$= 1 + \frac{1}{3}(x-1) - \frac{2}{9}\frac{(x-1)^2}{2!} = 1 + \frac{1}{3}(x-1) - \frac{1}{9}(x-1)^2$$

b. $\sqrt[3]{1.3} = f(1.3) \approx 1 + \frac{1}{3}(1.3-1) - \frac{1}{9}(1.3-1)^2 = 1.09$

c. $f'''(x) = \frac{10}{27}x^{-8/3}$ is a decreasing function, and so it has a maximum in $[1, 1.3]$ at $x = 1$. Thus, let $M = |f'''(1)| = \left|\frac{10}{27}(1)^{-8/3}\right| = \frac{10}{27}$. Thus,

$$|R_2(1.3)| \le \frac{\frac{10}{27}}{(2+1)!}|1.3-1|^{2+1} \approx 0.002 \text{ (rounded up)}$$

d. $\sqrt[3]{1.3} \approx 1.091$ with error less than 0.002.

29. **a.** $f(x) = \ln x$, $f'(x) = \frac{1}{x}$, $f''(x) = -\frac{1}{x^2}$, $f'''(x) = \frac{2}{x^3}$, $f^{(4)}(x) = -\frac{6}{x^4}$, $f^{(5)}(x) = \frac{24}{x^5}$, $f^{(6)}(x) = -\frac{120}{x^6}$

$$p_6(x) = \ln 1 + \frac{1}{1}\frac{(x-1)}{1!} + \left(-\frac{1}{1^2}\right)\frac{(x-1)^2}{2!} + \left(\frac{2}{1^3}\right)\frac{(x-1)^3}{3!} + \left(-\frac{6}{1^4}\right)\frac{(x-1)^4}{4!} + \frac{24}{1^5}\frac{(x-1)^5}{5!} + \left(-\frac{120}{1^6}\right)\frac{(x-1)^6}{6!}$$

$$= 0 + (x-1) - \frac{(x-1)^2}{2} + \frac{2}{6}(x-1)^3 - \frac{6}{24}(x-1)^4 + \frac{24}{120}(x-1)^5 - \frac{120}{720}(x-1)^6$$

$$= 0 + (x-1) - \frac{(x-1)^2}{2} + \frac{1}{3}(x-1)^3 - \frac{1}{4}(x-1)^4 + \frac{1}{5}(x-1)^5 - \frac{1}{6}(x-1)^6$$

b. $\ln\frac{3}{2} \approx p_6\left(\frac{3}{2}\right) = \left(\frac{3}{2}-1\right) - \frac{1}{2}\left(\frac{3}{2}-1\right)^2 + \frac{1}{3}\left(\frac{3}{2}-1\right)^3 - \frac{1}{4}\left(\frac{3}{2}-1\right)^4 + \frac{1}{5}\left(\frac{3}{2}-1\right)^5 - \frac{1}{6}\left(\frac{3}{2}-1\right)^6$

$= 0.4046875$

c. $f^{(7)}(x) = \frac{720}{x^7}$ is a decreasing function, and it has a maximum value in $[1, 1.5]$ at $x = 1$. Thus, let

$M = \left|f^{(7)}(1)\right| = \frac{720}{1^7} = 720$, and

$$\left|R_6\left(\frac{3}{2}\right)\right| \le \frac{720}{(6+1)!}\left|\frac{3}{2}-1\right|^{6+1} \approx 0.002 \text{ (rounded up)}$$

d. $\ln\frac{3}{2} \approx 0.405$ with error less than 0.002.

31. $f(x) = e^x$, $f'(x) = e^x$, $f''(x) = e^x$

$$p_2(x) = e^0 + e^0\frac{x}{1!} + e^0\frac{x^2}{2!} = 1 + x + \frac{x^2}{2}$$

$$e^{-0.25} = p_2(-0.25) = 1 - 0.25 + \frac{(-0.25)^2}{2} \approx 0.78$$

Check: Using a calculator, $e^{-0.25} \approx 0.7788$.

33. To solve 4 cos $x = x + 3$, first find the second Taylor polynomial at $x = 0$.

$f(x) = \cos x,\ f'(x) = -\sin x,\ f''(x) = -\cos x$

$p_2(x) = \cos 0 - (\sin 0)x - (\cos 0)\frac{x^2}{2!} = 1 - \frac{x^2}{2}$

Substituting $p_2(x)$ into the equation, we get

$$4\left(1 - \frac{x^2}{2}\right) = x + 3$$
$$4 - 2x^2 - x - 3 = 0$$
$$2x^2 + x - 1 = 0$$
$$(2x-1)(x+1) = 0$$
$$x = \frac{1}{2}, -1$$

Thus, supply will equal demand in about $\frac{1}{2}$ year.

35. x

37. $f(x) = \frac{1}{x}$ is undefined at $x = 0$, no linear approximation at $x = 0$.

39. 1 (or any number greater than 1)

41. left

43. $1 + x + x^2 + x + \ldots + x^n = \frac{1 - x^{n+1}}{1 - x}$

45. Let $u = f'(t)$ $\quad dv = dt$

$du = f''(t)\,dt$ $\quad v = t - x$ or $-(x - t)$

Using integration by parts,

$$\int_a^x f'(t)\,dt = -(x-t)\,f'(t)\Big|_a^x - \int_a^x [-(x-t)]\,f''(t)\,dt$$
$$= \underbrace{-(x-x)\,f'(x)}_{0} + (x-a)\,f'(a) + \int_a^x f''(t)(x-t)\,dt$$
$$= (x-a)\,f'(a) + \int_a^x f''(t)(x-t)\,dt$$

47. Let $u = f^{(k+1)}(t)$ $\quad dv = (x-t)^k\,dt$

$du = f^{(k+2)}(t)\,dt$ $\quad v = \frac{-1}{k+1}(x-t)^{k+1}$

Using integration by parts,

$$\int_a^x f^{(k+1)}(t)(x-t)^k\,dt = -\frac{1}{k+1}(x-t)^{k+1}\,f^{(k+1)}(t)\Big|_a^x - \int_a^x \frac{-1}{k+1}(x-t)^{k+1}\,f^{(k+2)}(t)\,dt$$
$$= \underbrace{-\frac{1}{k+1}(x-x)^{k+1}\,f^{(k+1)}(x)}_{0} + \frac{1}{k+1}(x-a)^{k+1}\,f^{(k+1)}(a) + \frac{1}{k+1}\int_a^x f^{(k+2)}(t)(x-t)^{t+1}\,dt$$
$$= \frac{1}{k+1}(x-a)^{k+1}\,f^{(k+1)}(a) + \frac{1}{k+1}\int_a^x f^{(k+2)}(t)(x-t)^{t+1}\,dt$$

49. When $\left|f^{(n+1)}(t)\right| \le M$, then

$$\left|R_n(x)\right| = \left|\frac{1}{n!}\int_a^x f^{(n+1)}(t)(x-t)^n\,dt\right| \le \left|\frac{1}{n!}M\int_a^x (x-t)^n\,dt\right|$$

$$\le \left|\frac{M}{n!}\cdot\frac{1}{n+1}(x-t)^{n+1}\Big|_a^x\right|$$

or

$$\left|R_n(x)\right| \le \frac{M}{(n+1)!}|x-a|^{n+1}$$

EXERCISES 10.3

1. $1-\frac{x}{2}+\frac{x^2}{2^2}-\frac{x^3}{2^3}+\cdots$

$$c_n = \frac{(-1)^n}{2^n}x^n,\ c_{n+1} = \frac{(-1)^{n+1}}{2^{n+1}}x^{n+1}$$

$$\frac{c_{n+1}}{c_n} = \frac{\frac{(-1)^{n+1}}{2^{n+1}}x^{n+1}}{\frac{(-1)^n}{2^n}x^n} = \frac{(-1)^{n+1}}{2^{n+1}}\frac{x^{n+1}}{x^n}\cdot\frac{2^n}{(-1)^n}$$

$$= -\frac{1}{2}x$$

Therefore,

$$r = \lim_{n\to\infty}\left|\frac{c_{n+1}}{c_n}\right| = \left|-\frac{1}{2}x\right| = \frac{1}{2}|x| < 1$$

Thus, $|x| < 2$ and $R = 2$.

3. $\frac{x}{1\cdot2}+\frac{x^2}{2\cdot3}+\frac{x^3}{3\cdot4}+\frac{x^4}{4\cdot5}+\cdots$

$$c_n = \frac{x^n}{n(n+1)},\ c_{n+1} = \frac{x^{n+1}}{(n+1)(n+2)}$$

$$\frac{c_{n+1}}{c_n} = \frac{\frac{x^{n+1}}{(n+1)(n+2)}}{\frac{x^n}{n(n+1)}} = \frac{x^{n+1}}{(n+1)(n+2)}\cdot\frac{n(n+1)}{x^n}$$

$$= \frac{n}{n+2}x$$

$r = \lim_{n\to\infty}\left|\frac{c_{n+1}}{c_n}\right| = \lim_{n\to\infty}\left|\frac{n}{n+2}x\right| = |x| < 1$ because $\frac{n}{n+2} \to 1$. Therefore, $R = 1$.

5. $\sum_{n=0}^{\infty}\frac{(-1)^n x^n}{n!},\ c_n = \frac{(-1)^n x^n}{n!},\ c_{n+1}\frac{(-1)^{n+1}x^{n+1}}{(n+1)!}$

$$\frac{c_{n+1}}{c_n} = \frac{\frac{(-1)^{n+1}x^{n+1}}{(n+1)!}}{\frac{(-1)^n x^n}{n!}} = \frac{(-1)^{n+1}x^{n+1}}{(n+1)!}\cdot\frac{n!}{(-1)^n x^n}$$

$$= -\frac{x}{n+1}$$

$$r = \lim_{n\to\infty}\left|\frac{c_{n+1}}{c_n}\right| = \lim_{n\to\infty}\left|-\frac{x}{n+1}\right| = 0|x| = 0$$

Since the limit is 0 for all x, $R = \infty$.

7. $\sum_{n=0}^{\infty}\frac{n!x^n}{5^n},\ c_n = \frac{n!x^n}{5^n},\ c_{n+1} = \frac{(n+1)!x^{n+1}}{5^{n+1}}$

$$\frac{c_{n+1}}{c_n} = \frac{\frac{(n+1)!x^{n+1}}{5^{n+1}}}{\frac{n!x^n}{5^n}} = \frac{(n+1)!x^{n+1}}{5^{n+1}}\cdot\frac{5^n}{n!x^n} = \frac{(n+1)}{5}x$$

$$\lim_{n\to\infty}\left|\frac{c_{n+1}}{c_n}\right| = \lim_{n\to\infty}\left|\frac{(n+1)}{5}x\right|$$

This limit is infinite for all $x \approx 0$. Thus, $R = 0$.

9. $\sum_{n=0}^{\infty}\frac{x^{2n}}{n!},\ c_n = \frac{x^{2n}}{n!},\ c_{n+1} = \frac{x^{2(n+1)}}{(n+1)!}$

$$\frac{c_{n+1}}{c_n} = \frac{\frac{x^{2(n+1)}}{(n+1)!}}{\frac{x^{2n}}{n!}} = \frac{x^{2(n+1)}}{(n+1)!}\cdot\frac{n!}{x^{2n}} = \frac{x^2}{n+1}$$

$$\lim_{n\to\infty}\left|\frac{c_{n+1}}{c_n}\right| = \lim_{n\to\infty}\left|\frac{x^2}{n+1}\right| = 0$$

Since the limit is 0 for all x, $R = \infty$.

11. $f(x) = \ln(1+x)$, $f'(x) = \frac{1}{1+x}$, $f''(x) = -\frac{1}{(1+x)^2}$, $f'''(x) = \frac{2}{(1+x)^3}$

The Taylor series is

$$\ln(1+0) + \frac{\frac{1}{1+0}}{1!}x + \frac{-\frac{1}{(1+0)^2}}{2!}x^2 + \frac{\frac{2}{(1+0)^3}}{3!}x^3 + \cdots = 0 + 1\cdot x - \frac{1}{2}x^2 + \frac{2}{6}x^3 - \cdots$$
$$= x - \frac{1}{2}x^2 + \frac{1}{3}x^3 - \cdots$$

13. $f(x) = \cos 2x$, $f'(x) = -2\sin 2x$, $f''(x) = -4\cos 2x$, $f'''(x) = 8\sin 2x$

The Taylor series

$$\cos 2(0) + \frac{-2\sin 2(0)}{1!}x + \frac{-4\cos 2(0)}{2!}x^2 + \frac{8\sin 2(0)}{3!}x^3 + \cdots = 1 + \frac{0}{1!}x - \frac{4}{2!}x^2 + \frac{0}{3!}x^3 + \cdots$$
$$= 1 - 2x^2 + \cdots$$

15. $f(x) = \sqrt{2x+1}$, $f'(x) = (2x+1)^{-1/2}$, $f''(x) = -(2x+1)^{-3/2}$, $f'''(x) = 3(2x+1)^{-5/2}$

The Taylor series is

$$\sqrt{2(0)+1} + [2(0)+1]^{-1/2}\frac{x}{1!} + [-2(0)+1]^{-3/2}\frac{x^2}{2!} + 3[2(0)+1]^{-5/2}\frac{x^3}{3!} + \cdots = 1 + x - \frac{x^2}{2!} + \frac{3x^3}{3!} - \cdots$$

17. **a.** $f(x) = e^{x/5}$, $f'(x) = \frac{1}{5}e^{x/5}$, $f''(x) = \frac{1}{25}e^{x/5}$, $f'''(x) = \frac{1}{125}e^{x/5}$

The Taylor series is

$$e^{0/5} + \frac{1}{5}e^{0/5}\frac{x}{1!} + \frac{1}{25}e^{0/5}\frac{x^2}{2!} + \frac{1}{125}e^{0/5}\frac{x^3}{3!} + \cdots = 1 + \frac{x}{5} + \frac{x^2}{5^2 2!} + \frac{x^3}{5^3 3!} + \cdots$$

b. The Taylor series at $x = 0$ for e^x is

$$1 + x + \frac{x^2}{2!} + \frac{x^3}{3!} + \cdots$$

Substituting $\frac{x}{5}$ for x, we get that the Taylor series for $e^{x/5}$ at $x = 0$ is

$$1 + \frac{x}{5} + \left(\frac{x}{5}\right)^2\frac{1}{2!} + \frac{1}{3!}\left(\frac{x}{5}\right)^3 + \cdots = 1 + \frac{x}{5} + \frac{x^2}{5^2 2!} + \frac{x^3}{5^3 3!} + \cdots$$

19. $f(x) = \sin x$, $f'(x) = \cos x$, $f''(x) = -\sin x$, $f'''(x) = -\cos x$, $f^{(4)}(x) = \sin x$

The Taylor series for $\sin x$ at $x = \frac{\pi}{2}$ is

$$\sin\frac{\pi}{2} + \left(\cos\frac{\pi}{2}\right)\frac{\left(x-\frac{\pi}{2}\right)}{1!} - \left(\sin\frac{\pi}{2}\right)\frac{\left(x-\frac{\pi}{2}\right)^2}{2!} - \left(\cos\frac{\pi}{2}\right)\frac{\left(x-\frac{\pi}{2}\right)^3}{3!} + \left(\sin\frac{\pi}{2}\right)\frac{\left(x-\frac{\pi}{2}\right)^4}{4!} - \ldots$$
$$= 1 + 0\frac{\left(x-\frac{\pi}{2}\right)}{1!} - 1\frac{\left(x-\frac{\pi}{2}\right)^2}{2!} - 0\frac{\left(x-\frac{\pi}{2}\right)^3}{3!} + 1\frac{\left(x-\frac{\pi}{2}\right)^4}{4!} - \ldots$$
$$= 1 - \frac{\left(x-\frac{\pi}{2}\right)^2}{2!} + \frac{\left(x-\frac{\pi}{2}\right)^4}{4!} - \ldots$$

21. Multiply the Taylor series for $\frac{1}{1-x}$ by x^2. The Taylor series for $\frac{x^2}{1-x}$ is

$$x^2 + x^3 + x^4 + x^5 + \cdots$$

23. Substitute x^2 for x in the Taylor series for $\sin x$. The Taylor series for $\sin x^2$ is

$$x^2 - \frac{(x^2)^3}{3!} + \frac{(x^2)^5}{5!} - \frac{(x^2)^7}{7!} + \cdots$$
$$= x^2 - \frac{x^6}{3!} + \frac{x^{10}}{5!} - \frac{x^{14}}{7!} + \cdots$$

25. The Taylor series for e^x is

$$e^x = 1 + \frac{x}{x!} + \frac{x^2}{2!} + \frac{x^3}{3!} + \frac{x^4}{4!} + \cdots$$

Thus $e^x - 1 = \frac{x}{1!} + \frac{x^2}{2!} + \frac{x^3}{3!} + \frac{x^4}{4!} + \cdots$

and $\frac{e^x - 1}{x} = 1 + \frac{x}{2!} + \frac{x^2}{3!} + \frac{x^3}{4!} + \cdots$

27.

$$\sin x = x - \frac{x^3}{3!} + \frac{x^5}{5!} - \frac{x^7}{7!} + \cdots$$

$$-\sin x = -x + \frac{x^3}{3!} - \frac{x^5}{5!} + \frac{x^7}{7!} - \cdots$$

$$x - \sin x = \frac{x^3}{3!} - \frac{x^5}{5!} + \frac{x^7}{7!} + \cdots$$

$$\frac{x - \sin x}{x^3} = \frac{1}{3!} - \frac{x^2}{5!} + \frac{x^4}{7!} - \cdots$$

29. $\frac{1}{1+x} = 1 - x + x^2 - x^3 + \cdots$ for $|x| < 1$

$$\int \frac{1}{1+x}\,dx = \int (1 - x + x^2 - x^3 + \cdots)\,dx$$

$$\ln(1+x) = x - \frac{x^2}{2} + \frac{x^3}{3} - \frac{x^4}{4} + \cdots$$

31. **a.** Substitute $-x^2$ for x in the Taylor series for $\frac{1}{1-x}$.

$$\frac{1}{1-(-x)^2} = 1 + (-x^2) + (-x^2)^2 + (-x^2)^3 + \cdots$$

$$\frac{1}{1+x^2} = 1 - x^2 + x^4 - x^6 + \cdots$$

b. $c_n = (-1)^{n+1} x^{2(n-1)}$, $c_{n+1} = (-1)^{n+2} x^{2n}$

$$\frac{c_{n+1}}{c_n} = \frac{(-1)^{n+2} x^{2n}}{(-1)^{n+1} x^{2(n-1)}} = (-1)(x^{2n} \cdot x^{-(2n-2)}) = -x^2$$

$$\lim_{n\to\infty} \left|\frac{c_{n+1}}{c_n}\right| = \lim_{n\to\infty} \left|-x^2\right| = x^2$$

For convergence, $x^2 < 1$ or $-1 < x < 1$. Thus, $R = 1$.

c.

$$\int_0^1 \frac{1}{1+x^2}\,dx = \int_0^1 \left(1 - x^2 + x^4 - x^6 + \cdots\right) dx$$

$$= \left[x - \frac{x^3}{3} + \frac{x^5}{5} - \frac{x^7}{7} + \cdots\right]_0^1$$

$$= 1 - \frac{1}{3} + \frac{1}{5} - \frac{1}{7} + \cdots$$

d. From Section 8.5, #78

$$\int_0^1 \frac{1}{1+x^2}\,dx = \frac{\pi}{4}$$

From part (c),

$$\int_0^1 \frac{1}{1+x^2}\,dx = 1 - \frac{1}{3} + \frac{1}{5} - \frac{1}{7} + \cdots$$

Therefore,

$$1 - \frac{1}{3} + \frac{1}{5} - \frac{1}{7} + \cdots = \frac{\pi}{4}$$

33.

$$e^x = 1 + \frac{x}{1!} + \frac{x^2}{2!} + \frac{x^3}{3!} + \frac{x^4}{4!} + \frac{x^5}{5!} + \cdots$$

$$e^{-x} = 1 - \frac{x}{1!} + \frac{x^2}{2!} - \frac{x^3}{3!} + \frac{x^4}{4!} - \frac{x^5}{5!} + \cdots$$

$$e^x + e^{-x} = 2 + 2\frac{x^2}{2!} + 2\frac{x^4}{4!} + \cdots$$

$$\frac{e^x + e^{-x}}{2} = 1 + \frac{x^2}{2!} + \frac{x^4}{4!} + \cdots$$

35. **a.** $e^{0.5} \approx 1.64872$

b. $e^{0.5} \approx 1 + 0.5 + \frac{0.5^2}{2!} + \frac{0.5^3}{3!} \approx 1.64583$

c. 7 terms will produce the estimate in (*a*).

37. $\cos x = 1 - \frac{x^2}{2!} + \frac{x^4}{4!} - \frac{x^6}{6!} + \cdots$

$$\frac{d}{dx}\cos x = \frac{d}{dx}\left(1 - \frac{x^2}{2!} + \frac{x^4}{4!} - \frac{x^6}{6!} + \cdots\right)$$
$$= 0 - 2\frac{x}{2!} + 4\frac{x^3}{4!} - \frac{6x^5}{6!} + \cdots$$
$$= -\frac{x}{1!} + \frac{x^3}{3!} - \frac{x^5}{5!} + \cdots$$
$$= -\left(x - \frac{x^3}{3!} + \frac{x^5}{5!} - \cdots\right) = -\sin x$$

39. **a.** $e^{-t^2/2} = 1 + \frac{-t^2/2}{1!} + \frac{(-t^2/2)^2}{2!} + \frac{(-t^2/2)^3}{3!} + \frac{(-t^2/2)^4}{4!} + \cdots$

$$= 1 - \frac{t^2}{2} + \frac{t^4}{2^2 2!} - \frac{t^6}{2^3 3!} + \frac{t^8}{2^4 4!} - \cdots$$

b. $\int_0^x e^{-t^2/2}\,dt = \int_0^x \left(1 - \frac{t^2}{2} + \frac{t^4}{2^2 2!} - \frac{t^6}{2^3 3!} + \frac{t^8}{2^4 4!} - \cdots\right) dt$

$$= \left(t - \frac{t^3}{6} + \frac{t^5}{5\cdot 2^2 2!} - \frac{t^7}{7\cdot 2^3 3!} + \frac{x^9}{9\cdot 2^4 4!} - \cdots\right)\Bigg|_0^x$$
$$= x - \frac{x^3}{6} + \frac{x^5}{5\cdot 2^2 2!} + \frac{x^7}{7\cdot 2^3 3!} + \frac{x^9}{9\cdot 2^4 4!} - \cdots$$

$$0.4\int_0^1 e^{-t^2/2}\,dt = 0.4x - \frac{0.4x^3}{6} + \frac{0.4x^5}{5\cdot 2^2\cdot 2!} - \frac{0.4x^7}{7\cdot 2^3 3!} + \frac{0.4x^9}{9\cdot 2^4\cdot 4!} - \cdots$$

c. $\frac{1}{\sqrt{2\pi}}\int_0^1 e^{-t^2/2}\,dt \approx 0.4(1) - \frac{0.4(1)^3}{6} + \frac{0.4(1)^5}{5\cdot 2^2\cdot 2!} \approx 0.34$

So about 34% have IQs in the specified range.

41. **a.** $\cos t = 1 - \frac{t^2}{2!} + \frac{t^4}{4!} - \frac{t^6}{6!} + \cdots$

$\cos t^2 = 1 - \frac{t^4}{2!} + \frac{t^8}{4!} - \frac{t^{12}}{6!} + \cdots$

b. $\int_0^x \cos t^2\,dt = \int_0^x \left(1 - \frac{t^4}{2!} + \frac{t^8}{4!} - \frac{t^{12}}{6!} + \cdots\right) dt$

$$= \left(t - \frac{t^5}{5\cdot 2!} + \frac{t^9}{9\cdot 4!} - \frac{t^{13}}{13\cdot 6!} + \cdots\right)\Bigg|_0^x$$
$$= x - \frac{x^5}{5\cdot 2!} + \frac{x^9}{9\cdot 4!} - \frac{x^{13}}{13\cdot 6!} + \cdots$$

c. $\int_0^1 \cos t^2 \approx 1 - \frac{(1)^5}{5\cdot 2!} + \frac{(1)^9}{9\cdot 4!} \approx 0.905$ thousand.

So there are about 905 sales.

43. No (If the radius of convergence from 0 does not extend beyond 3, it cannot reach 4.)

45. 17

47. $e^{-1} = \frac{1}{e}$

49. –1 (using the series for $\cos\pi$)

51. $a = 10$ (the midpoint of the interval)

53. $\sin x = \sum_{n=0}^{\infty} \frac{(-1)^n x^{2n+1}}{(2n+1)!}$

$$\frac{c_{n+1}}{c_n} = \frac{(-1)^{n+1} x^{2(n+1)+1}}{[2(n+1)+1]!} \div \frac{(-1)^n x^{2n+1}}{(2n+1)!}$$
$$= \frac{(-1)^{n+1} x^{2n+3}}{(2n+3)!} \cdot \frac{(2n+1)!}{(-1)^n x^{2n+1}}$$
$$= (-1)\frac{x^2}{(2n+2)(2n+3)} = -\frac{x^2}{(2n+2)(2n+3)}$$

$$\lim_{n\to\infty}\left|\frac{c_{n+1}}{c_n}\right| = \lim_{n\to\infty}\left|-\frac{x^2}{(2n+2)(2n+3)}\right| = 0$$

Since the limit is 0 for all x, the series for $\sin x$ converges for all x.

55. a. $S_n - S_{n-1} = (c_1 + c_2 + c_3 + \cdots + c_n) - (c_1 + c_2 + c_3 + \cdots + c_{n-1})$
$= (c_1 - c_1) + (c_2 - c_2) + (c_3 - c_3) + \cdots + (c_{n-1} - c_{n-1}) + c_n$
$= c_n$

b. Since the series converges, say $\lim_{n\to\infty} S_n = a$.

$\lim_{n\to\infty} c_n = \lim_{n\to\infty}(S_n - S_{n-1}) = \lim_{n\to\infty} S_n - \lim_{n\to\infty} S_{n-1}$
$= a - a = 0$

Therefore $\lim_{n\to\infty} c_n = 0$.

57. a. $\frac{c_{n+1}}{c_n} = \frac{\frac{|x|^{n+1}}{(n+1)!}}{\frac{|x|^n}{n!}} = \frac{|x|^{n+1}}{(n+1)!} \cdot \frac{n!}{|x|^n} = \frac{|x|}{n+1}$

Thus, $\lim_{n\to\infty}\left|\frac{c_{n+1}}{c_n}\right| = \lim_{n\to\infty}\frac{|x|}{n+1} = 0$

The series converges for all x.

b. Since the series converges, by the nth term test,

$\lim_{n\to\infty}\frac{|x|^n}{n!} = \lim_{n\to\infty} c_n = 0$

$\lim_{n\to\infty}\frac{|x|^{2n+1}}{(2n+1)!} = \lim_{n\to\infty} c_{2n+1} = 0$

c. First, use the error formula to show that $|R_{2n}(x)| \le \frac{1}{(2n+1)!}|x|^{2n+1}$.

Since $f(x) = \sin x$, all derivatives are $\pm\cos x$ or $\pm\sin x$. Since $\left|f^{(n+1)}(t)\right| \le 1$ for all t, we may let $M = 1$.

Thus, $|R_{2n}(x)| \le \frac{M}{(2n+1)!}|x-0|^{2n+1}$

$= \frac{1}{(2n+1)!}|x|^{2n+1}$.

Since this expression converges to 0 by part (b), $|R_{2n}(x)| \to 0$ as $n \to \infty$.

59. a. Suppose $\lim_{n\to\infty}\left|\frac{c_{n+1}}{c_n}\right| = r > 1$. Then $r - 1 > 0$, so (by the definition of limit) there is an integer N such that $\left|\frac{c_{n+1}}{c_n}\right|$ is within $r - 1$ units of r for $n > N$, that is,

$\left|\left|\frac{c_{n+1}}{c_n}\right| - r\right| \le r - 1$ for all $n > N$.

Since $r - \left|\frac{c_{n+1}}{c_n}\right| \le \left|\left|\frac{c_{n+1}}{c_n}\right| - r\right|$, we conclude that, for $n > N$,

$r - \left|\frac{c_{n+1}}{c_n}\right| \le r - 1$, and so $\left|\frac{c_{n+1}}{c_n}\right| \ge 1$.

b. $\left|\frac{c_{n+1}}{c_n}\right| \ge 1$ for $n \ge N$. Thus $\left|\frac{c_{N+1}}{c_N}\right| \ge 1$ and $|c_{N+1}| \ge |c_N|$.

$\left|\frac{c_{N+2}}{c_{N+1}}\right| \ge 1$ or $|c_{N+2}| \ge |c_{N+1}| \ge |c_N|$

$\left|\frac{c_{N+3}}{c_{N+2}}\right| \ge 1$ or $|c_{N+3}| \ge |c_{N+2}| \ge |c_{N+1}| \ge |c_N|$

c. Since $0 < |c_n| \le |c_{N+1}| \le |c_{N+2}| \le |c_{N+3}| \le \cdots$, c_n does not approach 0 as $n \to \infty$. Thus, by the nth term test, the series diverges.

61. **a.** Let $x = 2$,

$$(x-1)-\frac{(x-1)^2}{2}+\frac{(x-1)^3}{3}-\frac{(x-1)^4}{4}+\cdots$$

$$(2-1)-\frac{(2-1)^2}{2}+\frac{(2-1)^3}{3}-\frac{(2-1)^4}{4}+\cdots$$

$$=1-\frac{1}{2}+\frac{1}{3}-\frac{1}{4}+\frac{1}{5}-\cdots$$

b. $1-\frac{1}{2}+\frac{1}{3}-\frac{1}{4}+\frac{1}{5}-\cdots$

$$=\left(1-\frac{1}{2}\right)+\left(\frac{1}{3}-\frac{1}{4}\right)+\left(\frac{1}{5}-\frac{1}{6}\right)+\cdots$$

Since the terms in each set of parenthesis are positive, so it is a sum of positive terms.

c. $1-\frac{1}{2}+\frac{1}{3}-\frac{1}{4}+\frac{1}{5}-\cdots$

$$=1-\left(\frac{1}{2}-\frac{1}{3}\right)-\left(\frac{1}{4}-\frac{1}{5}\right)-\left(\frac{1}{6}-\frac{1}{7}\right)+\cdots$$

The series begins with 1, with positive quantities subtracted, and so it must be less than 1.

63. **a.** $e^x = 1+x+\frac{x^2}{2}+\frac{x^3}{3!}+\cdots$. Thus,

$$e^x > \frac{x^3}{3!} = \frac{x^3}{6}.$$

b. Replacing x by ax, we get $e^{ax} > \frac{(ax)^3}{6} = \frac{a^3x^3}{6}$.

c. Taking the reciprocal, we get $e^{-ax} < \frac{6}{a^3x^3}$.

d. $x^2e^{-ax} < \frac{6}{a^3x}$

e. $\lim_{n\to\infty} x^2e^{-ax} \le \lim_{n\to\infty} \frac{6}{a^3x} = 0$. Since $x^2e^{-ax} > 0$ for $x > 0$, this implies that $\lim_{n\to\infty} x^2e^{-ax} = 0$.

EXERCISES 10.4

1. $f(x) = x^3 + x - 4$

$f'(x) = 3x^2 + 1$

$x_0 = 1$

$$x_1 = x_0 - \frac{f(x_0)}{f'(x_0)} = 1 - \frac{f(1)}{f'(1)} = 1 - \frac{-2}{4} = \frac{3}{2}$$

$$x_2 = x_1 - \frac{f(x_1)}{f'(x_1)} = \frac{3}{2} - \frac{f\left(\frac{3}{2}\right)}{f'\left(\frac{3}{2}\right)} = \frac{3}{2} - \frac{0.875}{7.75} \approx 1.39$$

3. $f(x) = e^x - 3x$

$f'(x) = e^x - 3$

$x_0 = 0$

$$x_1 = 0 - \frac{f(0)}{f'(0)} = -\frac{1}{-2} = \frac{1}{2}$$

$$x_2 = \frac{1}{2} - \frac{f\left(\frac{1}{2}\right)}{f'\left(\frac{1}{2}\right)} = \frac{1}{2} - \frac{e^{1/2} - \frac{3}{2}}{e^{1/2} - 3} \approx 0.61$$

5. To approximate $\sqrt{5}$, consider $x^2 = 5$ or $x^2 - 5 = 0$.

$f(x) = x^2 - 5$

$f'(x) = 2x$

$x_0 = 2$

$$x_1 = 2 - \frac{f(2)}{f'(2)} = 2 - \frac{-1}{4} = 2.25$$

$$x_2 = 2.25 - \frac{f(2.25)}{f'(2.25)} = 2.25 - \frac{2.25^2 - 5}{2(2.25)} \approx 2.236$$

$$x_3 = 2.236 - \frac{f(2.236)}{f'(2.236)} = 2.236 - \frac{2.236^2 - 5}{2(2.236)} \approx 2.236$$

7. $f(x) = x^3 + 3x - 8$, $f'(x) = 3x^2 + 3$

$x_0 = 2$

$$x_1 = 2 - \frac{f(2)}{f'(2)} = 2 - \frac{6}{15} = 1.6$$

$$x_2 = 1.6 - \frac{f(1.6)}{f'(1.6)} \approx 1.516$$

$$x_3 = 1.516 - \frac{f(1.516)}{f'(1.516)} \approx 1.513$$

$$x_4 = 1.513 - \frac{f(1.513)}{f'(1.513)} \approx 1.513$$

9. $f'(x) = e^x + 3x + 2,\ f'(x) = e^x + 3$

$x_0 = -1$

$x_1 = -1 - \dfrac{f(-1)}{f'(-1)} \approx -0.812$

$x_2 = -0.812 - \dfrac{f(-0.812)}{f'(-0.812)} \approx -0.814$

$x_3 = -0.814 - \dfrac{f(-0.814)}{f'(-0.814)} \approx -0.814$

11. $x = \sqrt[3]{130}$

$x^3 = 130$

$x^3 - 130 = 0$

$x = 5.065797019$

13. $x = 0.486389036$

15. $x = 3.693441359$

17. $x = 0.869647810$

19. $x = 1.106060158$

21.

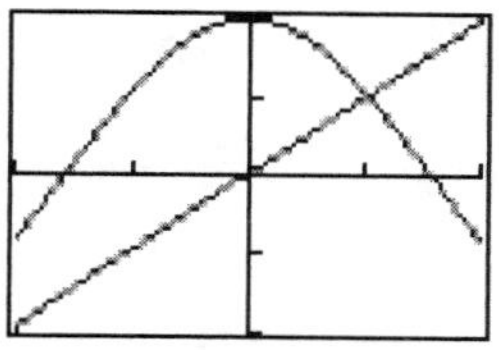

on [–2, 2] by [–2, 2]

Choose $x_0 = 1$. Then $x \approx 1.029866529$.

23.

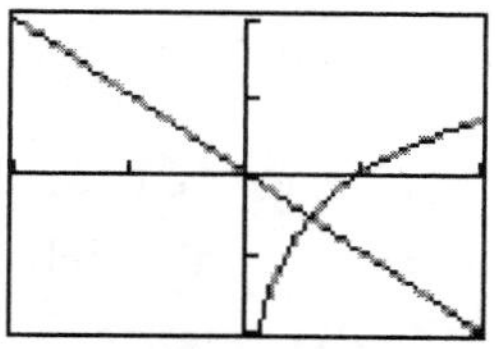

on [–2, 2] by [–2, 2]

Choose $x_0 = 0.5$. Then $x = 0.567143290$.

25. $x_0 = 1$

$x_1 = -1$

$x_2 = 1$

$x_3 = -1$

$x_4 = 1$

Newton's method does not work because $x^2 + 3 = 0$ has no real number solutions.

27. It converges in one step to 2.

29. $x_1 = 0.9000012$

$x_2 = 0.81000241$

$x_3 = 0.72900365$

$x_4 = 0.65610493$

$x_5 = 0.59049627$

$x_6 = 0.53144867$

$x_7 = 0.47830607$

$x_8 = 0.43047797$

$x_9 = 0.387432958$

$x_{10} = 0.348692760$

$x_{11} = 0.313826925$

$x_{12} = 0.28244806$

$x_{13} = 0.25420750$

$x_{14} = 0.22879147$

$x_{15} = 0.20591757$

$x_{16} = 0.18533163$

$x_{17} = 0.16680494$

$x_{18} = 0.15013164$

$x_{19} = 0.13512647$

$x_{20} = 0.12162270$

It takes 102 iterations to reach a number that rounds to 0.001.

31. $S(p) = p,\ D(p) = 10e^{-0.1p}$. Since we need the root of $S(p) = D(p)$, let $f(p) = S(p) - D(p)$. Then

$f(p) = p - 10e^{-0.1p}$

$f'(p) = 1 + e^{-0.1p}$

$x_0 = 4$

$x_1 = 4 - \dfrac{f(4)}{f'(4)} \approx 5.62$

$x_2 = 5.62 - \dfrac{f(5.62)}{f'(5.62)} \approx 5.67$

Thus, the equilibrium price is \$5.67.

33. $S(p)=2e^{0.5p}$, $D(p)=10-p$. To find the root of $S(p)=D(p)$, let $f(p)=S(p)-D(p)=2e^{0.5p}-(10-p)$. Choosing $x_0=5$, we get the root $p\approx 2.61$. The equilibrium price is about \$2.61.

35. Since the amount of the loan = 15,000 and each installment payment is 5000, then

$$15,000=\frac{5000}{(1+r)}+\frac{5000}{(1+r)^2}+\frac{5000}{(1+r)^3}+\frac{5000}{(1+r)^4}$$

for r = the internal rate of return. Let $x = 1 + r$. Then, dividing by 5000, we get

$$3=\frac{1}{x}+\frac{1}{x^2}+\frac{1}{x^3}+\frac{1}{x^4}$$

$$3x^4-x^3-x^2-x-1=0$$

Choosing $x_0=1$, we get $x\approx 1.126$. Since $r=x-1$, then $r\approx 0.126$. Thus, the internal rate of return is 12.6%.

37. $f(t)=5t-30e^{-0.2t}$. Choosing $t_0=1$, we get $t\approx 3.18$. The two effects will be in equilibrium in about 3.18 weeks.

39. $0=mx+b$

$$x=-\frac{b}{m}$$

41. $f(x+17)=0$, $x=-4$, when $f(13)=0$

43. It converges in one step.

45. To the right of the zero

47. The iterations will be exactly the same, since changing the sign of the function will not change the f/f' in the iteration formula.

REVIEW EXERCISES AND CHAPTER TEST FOR CHAPTER 10

1. $\frac{3}{10}+\frac{9}{100}+\frac{27}{1000}+\cdots$ $\quad r=\frac{\frac{9}{100}}{\frac{3}{10}}=\frac{9}{100}\cdot\frac{10}{3}=\frac{3}{10}$

$$S=\frac{\frac{3}{10}}{1-\frac{3}{10}}=\frac{\frac{3}{10}}{\frac{7}{10}}=\frac{3}{7}$$

2. $\frac{2}{5}-\frac{4}{25}+\frac{8}{125}+\cdots$ $\quad r=\frac{-\frac{4}{25}}{\frac{4}{5}}=-\frac{4}{25}\cdot\frac{5}{2}=-\frac{2}{5}$

$$S=\frac{\frac{2}{5}}{1-\left(-\frac{2}{5}\right)}=\frac{\frac{2}{5}}{\frac{7}{5}}=\frac{2}{7}$$

3. $\frac{3}{100}-\frac{9}{100}+\frac{27}{100}-\cdots$ $\quad r=\frac{-\frac{9}{100}}{\frac{3}{100}}=-3$

Since $|r|=3>1$, the series diverges.

4. $\frac{2}{3}-\frac{4}{9}+\frac{8}{27}-\cdots$ $\quad r=\frac{-\frac{4}{9}}{\frac{2}{3}}=-\frac{4}{9}\cdot\frac{3}{2}=-\frac{2}{3}$

$$S=\frac{\frac{2}{3}}{1-\left(-\frac{2}{3}\right)}=\frac{\frac{2}{3}}{\frac{5}{3}}=\frac{2}{5}$$

5. $10+8+\frac{32}{5}+\frac{128}{25}+\cdots$ $\quad r=\frac{8}{10}=\frac{4}{5}$

$$S=\frac{10}{1-\frac{4}{5}}=\frac{10}{\frac{1}{5}}=50$$

6. $\sum_{i=0}^{\infty}\left(-\frac{3}{4}\right)^i$ $\quad r=\frac{\left(-\frac{3}{4}\right)^{i+1}}{\left(-\frac{3}{4}\right)^i}=\left(-\frac{3}{4}\right)^{i+1-i}=-\frac{3}{4}$

$$S=\frac{\left(-\frac{3}{4}\right)^0}{1-\left(-\frac{3}{4}\right)}=\frac{1}{\frac{7}{4}}=\frac{4}{7}$$

7. $\sum_{k=0}^{\infty} 6(-0.2)^k \qquad r = \frac{6(-0.2)^{k+1}}{6(-0.2)^k} = -0.2$

$S = \frac{6(-0.2)^0}{1-(-0.2)} = \frac{6}{1.2} = 5$

8. $\sum_{n=0}^{\infty} \frac{2^n}{3} \qquad r = \frac{\frac{2^{n+1}}{3}}{\frac{2^n}{3}} = 2$

Since $|r| = 2 > 1$, the series diverges.

9. $0.\overline{54} = \frac{54}{100} + \frac{54}{100^2} + \frac{54}{100^3} + \cdots \qquad r = \frac{\frac{54}{100^2}}{\frac{54}{100}} = \frac{1}{100}$

$0.\overline{54} = S = \frac{\frac{54}{100}}{1-\frac{1}{100}} = \frac{\frac{54}{100}}{\frac{99}{100}} = \frac{54}{99} = \frac{6}{11}$

10. $0.\overline{72} = \frac{72}{100} + \frac{72}{100^2} + \frac{72}{100^3} + \cdots \qquad r = \frac{\frac{72}{100^2}}{\frac{72}{100}} = \frac{1}{100}$

$0.\overline{72} = S = \frac{\frac{72}{100}}{1-\frac{1}{100}} = \frac{\frac{72}{100}}{\frac{99}{100}} = \frac{8}{11}$

11. The deposit is \$6000 per month for 8 years = 96 months and $r = \frac{0.06}{12} = 0.005$.

$(\text{Value of annuity}) = a\frac{1-r^n}{1-r}$

For $a = 6000$, $r = 1.005$, and $n = 96$,

$(\text{Value of annuity}) = 6000\frac{1-1.005^{96}}{1-1.005} \approx \$736{,}971$

12. We know that the amount of an annuity from n payments of D dollars with a compound interest rate i per period is

$$D+D(1+i)+D(1+i)^2+\cdots+D(1+i)^{n-1} = D\left[\frac{1-(1+i)^n}{1-(1+i)}\right] = D\left[\frac{1-(1+i)^n}{1-1-i}\right] = D\left[\frac{1-(1+i)^n}{-i}\right] = D\left[\frac{(1+i)^n-1}{i}\right]$$

13. **a.** Maximum amount of digoxin $= 0.26 + 0.26(0.35) + 0.26(0.35)^2 + \cdots$.
$a = 0.26$, $r = 0.35$.

Maximum $= \frac{0.26}{1-0.35} = 0.4$ mg

Minimum = maximum · 0.35 = 0.14 mg

b. Using a graphing calculator, we get that the cumulative S_n reaches 0.39 mg just after the fourth dose. It reaches 0.399 mg just after the sixth dose.

14. **a.** $1000 + 1000(0.90) + 1000(0.90)^2 + \cdots$.
$a = 1000$, $r = 0.9$, $S = \frac{1000}{1-0.9} = 10{,}000$

The multiplier is $\frac{10{,}000}{1000} = 10$.

b. The amount surpasses \$9000 on the 22nd term. The amount surpasses \$9900 on the 44th term.

15. **a.** $f(x) = \sqrt{x+9}$, $f'(x) = \frac{1}{2}(x+9)^{-1/2}$, $f''(x) = -\frac{1}{4}(x+9)^{-3/2}$, $f'''(x) = \frac{3}{8}(x+9)^{-5/2}$

$$p_3(x) = \sqrt{0+9} + \frac{1}{2}(0+9)^{-1/2}\frac{x}{1!} - \frac{1}{4}(0+9)^{-3/2}\frac{x^2}{2!} + \frac{3}{8}(0+9)^{-5/2}\frac{x^3}{3!}$$
$$= 3 + \frac{1}{2}\left(\frac{1}{3}\right)x - \frac{1}{4}\left(\frac{1}{3\cdot 9}\right)\frac{x^2}{2!} + \frac{3}{8}\left(\frac{1}{9^2\cdot 3}\right)\frac{x^3}{3!}$$
$$= 3 + \frac{1}{6}x - \frac{1}{216}x^2 + \frac{1}{3888}x^3$$

b.

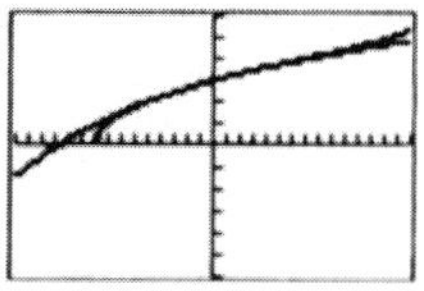

on [–15, 15] by [–6, 6]

16. **a.** $f(x) = e^{3x}$, $f'(x) = 3e^{3x}$, $f''(x) = 9e^{3x}$, $f'''(x) = 27e^{3x}$

$$p_3(x) = e^{3(0)} + 3e^{3(0)}\frac{x}{1!} + 9e^{3(0)}\frac{x^2}{2!} + 27e^{3(0)}\frac{x^3}{3!}$$
$$= 1 + 3x + \frac{9}{2}x^2 + \frac{9}{2}x^3$$

b.

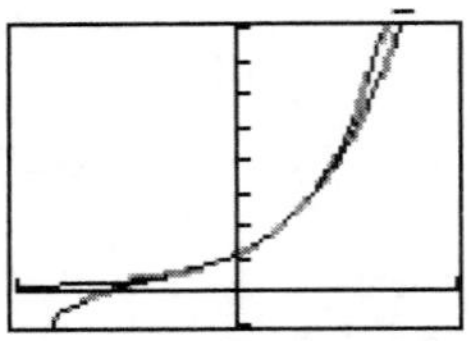

on [–1, 1] by [–1, 8]

17. **a.** $f(x) = \ln(2x+1)$, $f'(x) = \frac{2}{(2x+1)}$, $f''(x) = \frac{-4}{(2x+1)^2}$, $f'''(x) = \frac{16}{(2x+1)^3}$

$$p_3(x) = \ln[2(0)+1] + \frac{2}{[2(0)+1]}\frac{x}{1!} - \frac{4}{[2(0)+1]^2}\frac{x^2}{2!} + \frac{16}{[2(0)+1]^3}\frac{x^3}{3!}$$
$$= 0 + \frac{2}{1}x - \frac{4}{2}x^2 + \frac{16}{6}x^3$$
$$= 2x - 2x^2 + \frac{8}{3}x^3$$

b.

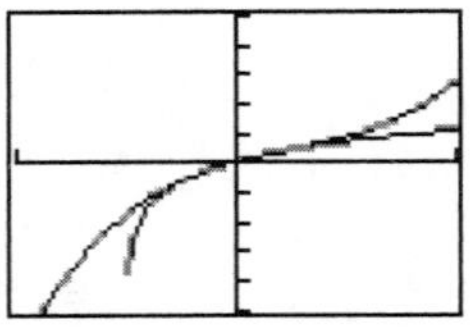

on [–1, 1] by [–5, 5]

18. **a.** $f(x) = \sin^2 x$, $f'(x) = 2\sin x \cos x$, $f''(x) = 2\cos^2 x - 2\sin^2 x$, $f'''(x) = 4\cos x \sin x - 4\sin x \cos x$

$$p_3(x) = \sin^2 0 + 2\sin 0 \cos 0 \frac{x}{1!} + (2\cos^2 0 - 2\sin^2 0)\frac{x^2}{2!} - 4(\cos 0 \sin 0 + \sin 0 \cos 0)\frac{x^3}{3!}$$
$$= 0^2 + (2\cdot 0\cdot 1)\frac{x}{1!} + (2\cdot 1 - 2\cdot 0)\frac{x^2}{2!} - 4(1\cdot 0 + 0\cdot 1)\frac{x^3}{3!}$$
$$= \frac{2x^2}{2!} = x^2$$

b.

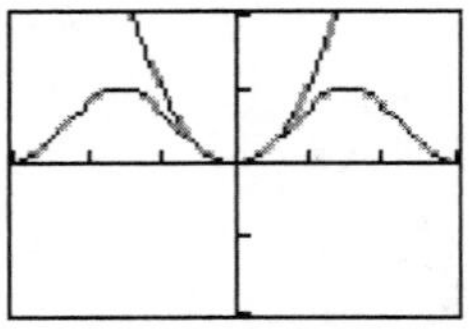

on [–3, 3] by [–2, 2]

19. **a.** $f(x)=\sqrt{x+1},\ f'(x)=\frac{1}{2}(x+1)^{-1/2},\ f''(x)=-\frac{1}{4}(x+1)^{-3/2},\ f'''(x)=\frac{3}{8}(x+1)^{-5/2}$

$$p_3(x)=\sqrt{0+1}+\frac{1}{2}(0+1)^{-1/2}\frac{x}{1!}-\frac{1}{4}(0+1)^{-3/2}\frac{x^2}{2!}+\frac{3}{8}(0+1)^{-5/2}\frac{x^3}{3!}$$
$$=1+\frac{1}{2}x-\frac{1}{8}x^2+\frac{1}{16}x^3$$

b. $\sqrt{2}\approx p_3(1)=1+\frac{1}{2}(1)-\frac{1}{8}(1)^2+\frac{1}{16}(1)^3=1.4375$

c. $f^{(4)}(x)=-\frac{15}{16}(x+1)^{-7/2}$. Since $\left|f^{(4)}(x)\right|$ is a decreasing function on [0, 1], the maximum occurs at $x=0$.

Let $M=\left|f^{(4)}(0)\right|=\frac{15}{16}$. Then

$$\left|R_3(x)\right|\le\frac{M}{(3+1)!}|x|^{(3+1)}=\frac{\frac{15}{16}x^4}{4!}$$
$$=\frac{15}{384}x^4$$
$$\left|R_3(1)\right|\le\frac{15}{384}(1)^4\approx 0.04 \text{ (rounded up)}$$

d. $\sqrt{2}\approx 1.44$ with error less than 0.04.

20. **a.** $f(x)=\cos x,\ f'(x)=-\sin x,\ f''(x)=-\cos x,\ f'''(x)=\sin x,\ f^{(4)}(x)=\cos x$

$$p_4(x)=\cos 0-\sin 0\frac{x}{1!}-\cos 0\frac{x^2}{2!}+\sin 0\frac{x^3}{3!}+\cos 0\frac{x^4}{4!}$$
$$=1-0-\frac{x^2}{2!}+0+\frac{x^4}{4!}=1-\frac{x^2}{2!}+\frac{x^4}{4!}$$

b. $\cos 1\approx p_4(1)=1-\frac{1^2}{2!}+\frac{1^4}{4!}\approx 0.54167$

c. $f^{(5)}(x)=-\sin x$. Since $|-\sin x|\le 1$, let $M=1$.

$$\left|R_4(x)\right|\le\frac{1}{(n+1)!}|x|^{n+1}$$
$$\left|R_4(x)\right|\le\frac{1}{5!}(1)^5=\frac{1}{120}\approx 0.009 \text{ (rounded up)}$$

d. $\cos 1\approx 0.542$ with error less than 0.009.

21. $f(x)=(1+x)^{1/n},\ f'(x)=\frac{1}{n}(1+x)^{1/n-1}$

$$p_1(x)=(1+0)^{1/n}+\frac{1}{n}(1+0)^{1/n-1}\frac{x}{1!}=1+\frac{1}{n}x$$

22. We know that the nth Taylor polynomial for e^x is

$$p_n(x) = 1 + \frac{x}{1!} + \frac{x^2}{2!} + \frac{x^3}{3!} + \cdots \frac{x^n}{n!}$$

Since $f^{(n)}(x) = e^x$ and $f^{(n)}$ is increasing on [–1, 1], the maximum occurs at $e^1 = e$. Thus $M = e$.

$$|R_n(x)| \le \frac{e}{(n+1)!}|1|^{n+1} = \frac{e}{(n+1)!}$$

n	$\frac{e}{(n+1)!}$
3	0.11326
4	0.02265
5	0.00378
6	0.00054
7	0.00007

Thus, $p_7(x) = 1 + \frac{x}{1!} + \frac{x^2}{2!} + \frac{x^3}{3!} + \frac{x^4}{4!} + \frac{x^5}{5!} + \frac{x^6}{6!} + \frac{x^7}{7!}$ has error less than 0.0001.

23. **a.** $f(x) = \sin x,\ f'(x) = \cos x,\ f''(x) = -\sin x,\ f'''(x) = -\cos x,\ f^{(4)}(x) = \sin x$

$$p_4(x) = \sin\frac{\pi}{2} + \left(\cos\frac{\pi}{2}\right)\left(x - \frac{\pi}{2}\right) - \left(\sin\frac{\pi}{2}\right)\frac{(x-\frac{\pi}{2})^2}{2!} - \left(\cos\frac{\pi}{2}\right)\frac{(x-\frac{\pi}{2})^3}{3!} + \left(\sin\frac{\pi}{2}\right)\frac{(x-\frac{\pi}{2})^4}{4!}$$

$$= 1 + 0\left(x - \frac{\pi}{2}\right) - 1\frac{(x-\frac{\pi}{2})^2}{2!} - 0\frac{(x-\frac{\pi}{2})^3}{3!} + 1\frac{(x-\frac{\pi}{2})^4}{4!}$$

$$= 1 - \frac{(x-\frac{\pi}{2})^2}{2!} + \frac{(x-\frac{\pi}{2})^4}{4!}$$

b.

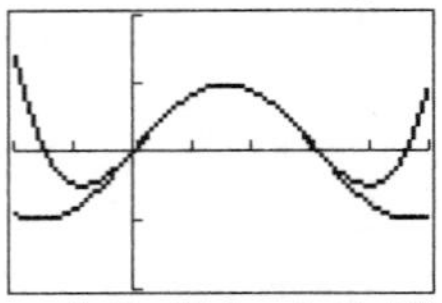

on [–2, 5] by [–2, 2]

24. **a.** $f(x) = x^{3/2},\ f'(x) = \frac{3}{2}x^{1/2},\ f''(x) = \frac{3}{4}x^{-1/2},\ f'''(x) = -\frac{3}{8}x^{-3/2}$

$$p_3(x) = 4^{3/2} + \frac{3}{2}(4)^{1/2}(x-4) + \frac{3}{4}(4)^{-1/2}\frac{(x-4)^2}{2!} - \frac{3}{8}(4)^{-3/2}\frac{(x-4)^3}{3!}$$

$$= 8 + \frac{3}{2}(2)(x-4) + \frac{3}{4}\left(\frac{1}{2}\right)\frac{(x-4)^2}{2!} - \frac{3}{8}\left(\frac{1}{8}\right)\frac{(x-4)^3}{3!}$$

$$= 8 + 3(x-4) + \frac{3}{16}(x-4)^2 - \frac{1}{128}(x-4)^3$$

b.

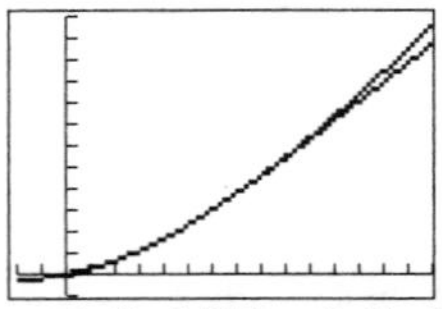

on [–2, 15] by [–5, 60]

25. a. $f(x)=x^{2/3},\ f'(x)=\frac{2}{3}x^{-1/3},\ f''(x)=-\frac{2}{9}x^{-4/3}$

$p_2(x)=(1)^{2/3}+\frac{2}{3}(1)^{-1/3}(x-1)-\frac{2}{9}(1)^{-4/3}\frac{(x-1)^2}{2!}=1+\frac{2}{3}(x-1)-\frac{1}{9}(x-1)^2$

b. $(1.2)^{2/3}=f(1.2)\approx p_2(1.2)=1+\frac{2}{3}(1.2-1)-\frac{1}{9}(1.2-1)^2$
≈ 1.12889

c. $f'''(x)=\frac{8}{27}x^{-7/3}$ is decreasing function on [1, 1.2], so the maximum is at $x = 1$. Let $M=f'''(1)=\frac{8}{27}$.

d. $|R_2(x)|\le\frac{\frac{8}{27}}{(2+1)!}|x-1|^{2+1}=\frac{8}{27\cdot 3!}|x-1|^3$

$|R_2(1.2)|\le\frac{8}{27\cdot 3!}(0.2)^3\approx 0.0004$ (rounded up)

e. $(1.2)^{2/3}\approx 1.1289$ with error less than 0.0004.

26. $p_2(x)=1+x+\frac{x^2}{2}$

Thus,

$e^{-0.22}\approx p_2(-0.22)=1-0.22+\frac{(-0.22)^2}{2}\approx 0.8$

27. $p_2(x)=1-\frac{x^2}{2!}+\frac{x^4}{4!}$

$\cos 0.45\approx p_2(0.45)=1-\frac{(0.45)^2}{2!}+\frac{(0.45)^4}{4!}\approx 0.9$

Thus, $55-24\cos 0.45\approx 55-24(0.9)\approx 33°$

28. a. $e^x\approx 1+x+\frac{x^2}{2!}$

$e^{0.1t^2}\approx 1+0.1t^2+\frac{(0.1t^2)^2}{2!}$

$200e^{0.1t^2}\approx 200+20t^2+2\frac{t^4}{2}=200+20t^2+t^4$

b. For $t=2$,

$200e^{0.1(2)^2}=200e^{0.4}\approx 200+20(2)^2+(2)^4$
$=296$ units

c. Using a calculator, we get $200e^{0.4}\approx 298$. The estimate is approximately 2 less.

29. $\sum_{n=0}^{\infty}\frac{nx^n}{3^n},\ c_n=\frac{nx^n}{3^n},\ c_{n+1}=\frac{(n+1)x^{n+1}}{3^{n+1}}$

$\frac{c_{n+1}}{c_n}=\frac{\frac{(n+1)x^{n+1}}{3^{n+1}}}{\frac{nx^n}{3^n}}=\frac{(n+1)x^{n+1}}{3^{n+1}}\cdot\frac{3^n}{nx^n}=\frac{(n+1)}{3n}x$

$\lim_{n\to\infty}\left|\frac{c_{n+1}}{c_n}\right|=\lim_{n\to\infty}\left|\frac{(n+1)x}{3n}\right|$
$=\lim_{n\to\infty}\left|\left(\frac{nx}{3n}+\frac{x}{3n}\right)\right|$
$=\left|\frac{x}{3}\right|$

Thus, $\frac{|x|}{3}<1$ or $|x|<3$, so $R=3$.

30. $\frac{x}{\sqrt{1}}+\frac{x^2}{\sqrt{2}}+\frac{x^3}{\sqrt{3}}+\frac{x^4}{\sqrt{4}}+\cdots=\sum_{n=1}^{\infty}\frac{x^n}{\sqrt{n}}$

$\frac{c_{n+1}}{c_n}=\frac{\frac{x^{n+1}}{\sqrt{n+1}}}{\frac{x^n}{\sqrt{n}}}=\frac{\sqrt{n}}{\sqrt{n+1}}x$

$\lim_{n\to\infty}\left|\frac{c_{n+1}}{c_n}\right|=\lim_{n\to\infty}\left(\frac{\sqrt{n}}{\sqrt{n+1}}\cdot\frac{\frac{1}{\sqrt{n}}}{\frac{1}{\sqrt{n}}}\right)|x|$
$=\lim_{n\to\infty}\frac{|x|}{\sqrt{1+\frac{1}{n}}}=|x|$

Thus, $|x|<1$ or $R=1$.

31. $\sum_{n=0}^{\infty}\frac{n!x^n}{5^n},$

$\frac{c_{n+1}}{c_n}=\frac{\frac{(n+1)!x^{n+1}}{5^{n+1}}}{\frac{n!x^n}{5^n}}=\frac{(n+1)!x^{n+1}}{5^{n+1}}\cdot\frac{5^n}{n!x^n}$
$=\frac{(n+1)x}{5}$

$\lim_{n\to\infty}\left|\frac{c_{n+1}}{c_n}\right|=\lim_{n\to\infty}\left|\frac{(n+1)x}{5}\right|$

The limit is infinite for $x\ge 0$, so $R=0$.

32. $\displaystyle\sum_{n=1}^{\infty}\frac{(-1)^n x^n}{(n-1)!}$

$$\frac{c_{n+1}}{c_n}=\frac{\frac{(-1)^{n+1}x^{n+1}}{(n)!}}{\frac{(-1)^n x^n}{(n-1)!}}=\frac{(-1)^{n+1}x^{n+1}}{n!}\cdot\frac{(n-1)!}{(-1)^n x^n}$$
$$=\frac{(-1)x}{n}$$

$\displaystyle\lim_{n\to\infty}\left|\frac{c_{n+1}}{c_n}\right|=\lim_{n\to\infty}\left|\frac{(-1)x}{n}\right|=0.$ Thus $R=\infty$.

33. First, replace x by x^2 in the series for $\frac{1}{1-x}$.

$$\frac{1}{1-x^2}=1+x^2+(x^2)^2+(x^2)^3+\cdots$$
$$=1+x^2+x^4+x^6+\cdots$$

Multiply by x^3.

$$\frac{x^3}{1-x^2}=x^3+x^5+x^7+x^9+\cdots$$

34. Replace x by $2x$ in the series $\cos x$.

$$\cos 2x=1-\frac{(2x)^2}{2!}+\frac{(2x)^4}{4!}-\frac{(2x)^6}{6!}+\cdots$$
$$=1-\frac{2^2x^2}{2!}+\frac{2^4x^4}{4!}-\frac{2^6x^6}{6!}+\cdots$$
$$x\cos 2x=x-\frac{2^2x^3}{2!}+\frac{2^4x^5}{4!}-\frac{2^6x^7}{6!}+\cdots$$

35. Replace x by x^2 in the series for e^x.

$$e^{x^2}=1+x^2+\frac{(x^2)^2}{2!}+\frac{(x^2)^3}{3!}+\frac{(x^2)^4}{4!}+\cdots$$
$$=1+x^2+\frac{x^4}{2!}+\frac{x^6}{3!}+\frac{x^8}{4!}+\cdots$$
$$xe^{x^2}=x+x^3+\frac{x^5}{2!}+\frac{x^7}{3!}+\frac{x^9}{4!}+\cdots$$

36. Replace x by $-x$ in the series for e^x.

$$e^{-x}=1-x+\frac{x^2}{2!}-\frac{x^3}{3!}+\frac{x^4}{4!}-\cdots$$
$$xe^{-x}=x-x^2+\frac{x^3}{2!}-\frac{x^4}{3!}+\frac{x^5}{4!}-\cdots$$

37. **a.** $f(x)=\frac{1}{1-2x},\ f'(x)=\frac{2}{(1-2x)^2},\ f''(x)=\frac{8}{(1-2x)^3},\ f'''(x)=\frac{48}{(1-2x)^4}$

$f(x)=\frac{1}{1-2x},\ f'(x)=\frac{2}{(1-2x)^2},\ f''(x)=\frac{8}{(1-2x)^3},\ f'''(x)=\frac{48}{(1-2x)^4}$

$$\frac{1}{1-2x}=\frac{1}{1-2(0)}+\frac{2}{[1-2(0)]^2}x+\frac{8}{[1-2(0)]^3}\frac{x^2}{2!}+\frac{48}{[1-2(0)]^4}\frac{x^3}{3!}+\cdots$$
$$=1+2x+2^2x^2+2^3x^3+\cdots$$

b. Replace x by $2x$ in the series for $\frac{1}{1-x}$.

$$\frac{1}{1-2x}=1+2x+(2x)^2+(2x)^3+\cdots=1+2x+2^2x^2+2^3x^3+\cdots$$

38. **a.** $f(x)=e^{-\frac{1}{2}x},\ f'(x)=-\frac{1}{2}e^{-\frac{1}{2}x},\ f''(x)=\frac{1}{4}e^{-\frac{1}{2}x},\ f'''(x)=-\frac{1}{8}e^{-\frac{1}{2}x}$

$$e^{-\frac{1}{2}x}=e^{-\frac{1}{2}(0)}-\frac{1}{2}e^{-\frac{1}{2}(0)}x+\frac{1}{4}e^{-\frac{1}{2}(0)}\frac{x^2}{2!}-\frac{1}{8}e^{-\frac{1}{2}(0)}\frac{x^3}{3!}+\cdots$$
$$=1-\frac{1}{2}x+\frac{x^2}{2^2 2!}-\frac{x^3}{2^3 3!}+\cdots$$

b. Replace x by $-\frac{1}{2}x$ in the series for e^x.

$$e^{-\frac{1}{2}x}=1-\frac{1}{2}x+\frac{\left(-\frac{1}{2}x\right)^2}{2!}+\frac{\left(-\frac{1}{2}x\right)^3}{3!}+\cdots$$
$$=1-\frac{1}{2}x+\frac{x^2}{2^2 2!}-\frac{x^3}{2^3 3!}+\cdots$$

39. **a.** $f(x)=\cos x,\ f'(x)=-\sin x,\ f''(x)=-\cos x,\ f'''(x)=\sin x,\ f^{(4)}(x)=\cos x$

$$\cos x=\cos\frac{\pi}{2}-\left(\sin\frac{\pi}{2}\right)\left(x-\frac{\pi}{2}\right)-\left(\cos\frac{\pi}{2}\right)\frac{\left(x-\frac{\pi}{2}\right)^2}{2!}+\left(\sin\frac{\pi}{2}\right)\frac{\left(x-\frac{\pi}{2}\right)^3}{3!}+\left(\cos\frac{\pi}{2}\right)\frac{\left(x-\frac{\pi}{2}\right)^4}{4!}$$
$$=0-1\left(x-\frac{\pi}{2}\right)-0\frac{\left(x-\frac{\pi}{2}\right)^2}{2!}+1\frac{\left(x-\frac{\pi}{2}\right)^3}{3!}+0\frac{\left(x-\frac{\pi}{2}\right)^4}{4!}$$
$$=-\left(x-\frac{\pi}{2}\right)+\frac{\left(x-\frac{\pi}{2}\right)^3}{3!}-\cdots$$

b.

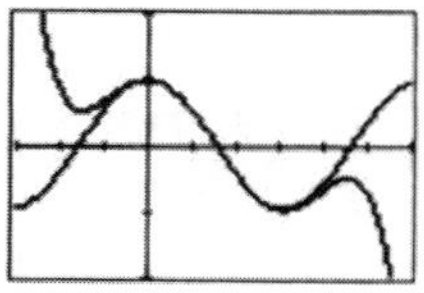

on [–3, 6] by [–2, 2]

40.
$$\sin(-x)=-x-\frac{(-x)^3}{3!}+\frac{(-x)^5}{5!}-\frac{(-x)^7}{7!}+\cdots$$
$$=-x+\frac{x^3}{3!}-\frac{x^5}{5!}+\frac{x^7}{7!}-\cdots$$
$$=-\left(x-\frac{x^3}{3!}+\frac{x^5}{5!}-\frac{x^7}{7!}+\cdots\right)=-\sin x$$
$$\cos(-x)=1-\frac{(-x)^2}{2!}+\frac{(-x)^4}{4!}-\frac{(-x)^6}{6!}+\cdots$$
$$=1-\frac{x^2}{2!}+\frac{x^4}{4!}-\frac{x^6}{6!}+\cdots=\cos x$$

41. **a.**
$$\frac{1}{1+t^2}=1-t^2+(-t^2)^2+(-t^2)^3+\cdots$$
$$=1-t^2+t^4-t^6+\cdots$$

b.
$$\int_0^x\frac{1}{1+t^2}\,dt=\int_0^x(1-t^2+t^4-t^6+\cdots)\,dx$$
$$=\left(t-\frac{t^3}{3}+\frac{t^5}{5}-\frac{t^7}{7}+\cdots\right)\Bigg|_0^x$$
$$=x-\frac{x^3}{3}+\frac{x^5}{5}-\frac{x^7}{7}+\cdots$$

c.
$$\int_0^{1/2}\frac{1}{1+t^2}\,dt=\frac{1}{2}-\frac{\left(\frac{1}{2}\right)^3}{3}+\frac{\left(\frac{1}{2}\right)^5}{5}$$
≈ 0.465 thousand

So about 465 cars.

d.
$$\int_0^{1/2}\frac{1}{1+t^2}\,dt\approx 0.464$$

42. **a.**
$$\frac{1}{x^2(1-x)}=\frac{1}{x^2}(1+x+x^2+x^3+x^4+\cdots)$$
$$=\frac{1}{x^2}+\frac{1}{x}+1+x+x^2+\cdots$$

b.
$$\int\frac{1}{x^2(1-x)}\,dx=-\frac{1}{x}+\ln x+x+\frac{1}{2}x^2+\frac{1}{3}x^3+\ldots$$

c.
$$\int_{0.2}^{0.5}\frac{1}{x^2(1-x)}\,dx$$
$$=\left(-\frac{1}{x}+\ln x+x+\frac{1}{2}x^2+\frac{1}{3}x^3\right)\Bigg|_{0.2}^{0.5}$$
$$\approx 4.36$$

d. (2.5)(4.36) = 10.9 so 11 generations are necessary. The result agrees with the solution of Example 3 in Section 2 of Chapter 6.

43.
$$x=\sqrt{26}$$
$$x^2=26$$
$x^2-26=0$ Choose $x_0=5$.
$$f'(x)=2x$$
$$x_0=5$$
$$x_1=5-\frac{f(5)}{f'(5)}=5.1$$
$$x_2=5.1-\frac{f(5.1)}{f'(5.1)}\approx 5.099$$
$$x_3=5.099-\frac{f(5.099)}{f'(5.099)}\approx 5.099$$

44. $f(x)=x^3-30$

$f'(x)=3x^2$ Choose $x_0=3$.

$x_0=3$

$x_1=3-\frac{f(3)}{f'(3)}\approx 3.111$

$x_2=3.111-\frac{f(3.111)}{f'(3.111)}\approx 3.107$

$x_3=3.107-\frac{f(3.107)}{f'(3.107)}\approx 3.107$

45. $f(x)=x^3+3x-3$

$f'(x)=3x^2+3$

$x_0=1$

$x_1=1-\frac{f(1)}{f'(1)}=\frac{5}{6}$

$x_2=\frac{5}{6}-\frac{f\left(\frac{5}{6}\right)}{f'\left(\frac{5}{6}\right)}\approx 0.818$

$x_3=0.818-\frac{f(0.818)}{f'(0.818)}\approx 0.818$

46. $f(x)=e^x+10x-4$

$f'(x)=e^x+10$

$x_0=0$

$x_1=0-\frac{f(0)}{f'(0)}=\frac{3}{11}\approx 0.273$

$x_2=0.273-\frac{f(0.273)}{f'(0.273)}\approx 0.269$

$x_3=0.269-\frac{f(0.269)}{f'(0.269)}\approx 0.269$

47. $x^3-100=0$. Choose $x_0=4$. Then $x\approx 4.641588834$.

48. $x^2-10=0$. Choose $x_0=3$. Then $x\approx 3.162277660$.

49. $e^x-5x-3=0$. Choose $x_0=1$. Then $x\approx -0.475711445$.

50. $e^x+4x-100=0$, $x_0=3$. Then $x\approx 4.411048937$.

51. $x^3=5\ln x+10$

$x^3-5\ln x-10=0$, $x_0=2$

Then, $x\approx 2.435766883$.

52. $x^4=2+\sin x$

$x^4-\sin x-2=0$, $x_0=0$

Then $x\approx -1.033503908$.

53. Since the amount of the investment is \$20,000 and each payment is \$9000,

$20{,}000=\frac{9000}{(1+r)}+\frac{9000}{(1+r)^2}+\frac{9000}{(1+r)^3}$

where r = the internal rate of return.
Let $x = 1 + r$.

$$20{,}000=\frac{9000}{x}+\frac{9000}{x^2}+\frac{9000}{x^3}$$

$$20=\frac{9}{x}+\frac{9}{x^2}+\frac{9}{x^3}$$

$$20x^3=9x^2+9x+9$$

$20x^3-9x^2-9x-9=0$

Using Newton's method, we get $x\approx 1.166$.
Since $r = x - 1$, $r = 0.166 = 16.6\%$. The internal rate of return is 16.6%.

54. $S(p)=20+3p$, $D(p)=50-8\sqrt{p}$. To find the value of p at which $S(p) = D(p)$, let $f(p) = S(p) - D(p) = 20 + 3p - (50-8\sqrt{p})=0$.
Using Newton's method, we get $p\approx 4.40$.
The market equilibrium price is about \$4.40.

55. $C(x)=3x+100$, $R(x)=40\sqrt{x}$

To find the break-even quantity, let

$f(x)=C(x)-R(x)=0$

$3x+100-40\sqrt{x}=0$

Choose $x_0=10$. Newton's method gives $x\approx 11.11$ (so about 11,110) units.

Chapter 11: Probability

EXERCISES 11.1

1. **a.** The event "rolling at least a 3" is made up of the events "rolling a 3, 4, 5, or 6." Since each event has probability $\frac{1}{6}$,

$$\begin{aligned} P(\text{rolling at least a 3}) &= P(3)+P(4)+P(5)+P(6) \\ &= \frac{1}{6}+\frac{1}{6}+\frac{1}{6}+\frac{1}{6}=\frac{4}{6}=\frac{2}{3} \end{aligned}$$

b. The event "rolling an odd number" is made up of the events "rolling a 1, 3, or 5." Since each event has probability $\frac{1}{6}$,

$$\begin{aligned} P(\text{rolling an odd number}) &= P(1)+P(3)+P(5) \\ &= \frac{1}{6}+\frac{1}{6}+\frac{1}{6}=\frac{3}{6}=\frac{1}{2} \end{aligned}$$

3. **a.** The event of tossing a coin three times has 8 possible elementary events: HHH, THH, HTH, HHT, TTH, THT, HTT, TTT. Each elementary event has probability $\frac{1}{8}$. Thus,

$$\begin{aligned} P(\text{tossing exactly one head}) &= P(\text{HTT})+P(\text{THT})+P(\text{TTH}) \\ &= \frac{1}{8}+\frac{1}{8}+\frac{1}{8}=\frac{3}{8} \end{aligned}$$

b.

$$\begin{aligned} P(\text{tossing exactly two heads}) &= P(\text{HHT})+P(\text{HTH})+P(\text{THH}) \\ &= \frac{1}{8}+\frac{1}{8}+\frac{1}{8}=\frac{3}{8} \end{aligned}$$

5. **a.** $\mu = E(X) = x_1 \cdot p_1 + x_2 \cdot p_2 = 9\cdot\frac{1}{2}+11\cdot\frac{1}{2}=\frac{9}{2}+\frac{11}{2}=\frac{20}{2}=10$

$$\begin{aligned} \text{Var}(X) &= (x_1-\mu)^2\cdot p_1+(x_2-\mu)^2\cdot p_2 \\ &= (9-10)^2\cdot\frac{1}{2}+(11-10)^2\cdot\frac{1}{2}=\frac{1}{2}+\frac{1}{2}=1 \end{aligned}$$

$\sigma(X)=\sqrt{\text{Var}(X)}=\sqrt{1}=1$

b. $\mu = E(Y) = y_1\cdot p_1 + y_2\cdot p_2 = 0\cdot\frac{1}{2}+20\cdot\frac{1}{2}=10$

$$\begin{aligned} \text{Var}(Y) &= (y_1-\mu)^2\cdot p_1+(y_2-\mu)^2\cdot p_2 \\ &= (0-10)^2\cdot\frac{1}{2}+(20-10)^2\cdot\frac{1}{2}=100\cdot\frac{1}{2}+100\cdot\frac{1}{2}=100 \end{aligned}$$

$\sigma(Y)=\sqrt{\text{Var}(Y)}=\sqrt{100}=10$

7. **a.** The events in X are 4, 7, and 10, and each event has probability $\frac{1}{3}$. Thus, the probability distribution is

$P(X=4)=\frac{1}{3},\ P(X=7)=\frac{1}{3},\ P(X=10)=\frac{1}{3}$

b.

$$\begin{aligned} \mu = E(X) &= x_1\cdot p_1+x_2\cdot p_2+x_3\cdot p_3 \\ &= 4\cdot\frac{1}{3}+7\cdot\frac{1}{3}+10\cdot\frac{1}{3}=\frac{4}{3}+\frac{7}{3}+\frac{10}{3}=\frac{21}{3}=7 \end{aligned}$$

c.

$$\begin{aligned} \text{Var}(X) &= (x_1-\mu)^2\cdot p_1+(x_2-\mu)^2\cdot p_2+(x_3-\mu)^2\cdot p_3 \\ &= (4-7)^2\cdot\frac{1}{3}+(7-7)^2\cdot\frac{1}{3}+(10-7)^2\cdot\frac{1}{3} \\ &= \frac{9}{3}+\frac{9}{3}+\frac{18}{3}=6 \end{aligned}$$

9. **a.** The events in X are 2, 4, and 8, and

$$P(X=2)=\frac{1}{2},\ P(X=4)=\frac{1}{4},\ P(X=8)=\frac{1}{4}$$

b. $\mu = E(X) = x_1 \cdot p_1 + x_2 \cdot p_2 + x_3 \cdot p_3$

$$= 2\cdot\frac{1}{2}+4\cdot\frac{1}{4}+8\cdot\frac{1}{4}=1+1+2=4$$

c. $\text{Var}(X) = (x_1-\mu)^2 \cdot p_1 + (x_2-\mu)^2 \cdot p_2 + (x_3-\mu)^2 \cdot p_3$

$$= (2-4)^2\cdot\frac{1}{2}+(4-4)^2\cdot\frac{1}{4}+(8-4)^2\cdot\frac{1}{4}$$

$$= 2+0+4=6$$

11. **a.** The probability of ending up in the top 30° of the circle is proportional to the area of the sector. Since there are 30° in a circle, the area of the sector is $\frac{30}{360}=\frac{1}{12}$ of the total area.

Thus,

$P(\text{ending up in the top } 30°) = \frac{1}{12}$

b. The area of the top 10° is $\frac{10}{360}=\frac{1}{36}$ the area of the circle. Thus,

$P(\text{ending up in the top } 10°) = \frac{1}{36}$

c. If the point is exactly upwards, the area of the sector is 0. Thus,

P(ending up exactly upwards) = 0

13. $$\mu = E(X) = \sum_{k=0}^{\infty} x_k p_k = \sum_{k=0}^{\infty} k\frac{e^{-a}a^k}{k!} = \sum_{k=1}^{\infty}\frac{e^{-a}a^k}{(k-1)!}$$

$$= e^{-a}\cdot a\sum_{k=1}^{\infty}\frac{a^{k-1}}{(k-1)!} = e^{-a}\cdot a\left(1+\frac{a}{1}+\frac{a^2}{2!}+\frac{a^3}{3!}+\cdots\right)$$

$$= e^{-a}\cdot a\cdot e^{a} = a$$

15. Since the mean is 2, by Exercise 13, $a = 2$.

a. $P(X=0)=e^{-2}\frac{2^0}{0!}\approx 0.135$

b. $P(X=1)=e^{-2}\frac{2^1}{1!}=2e^{-2}\approx 0.271$

c. $P(X=2)=e^{-2}\frac{2^2}{2!}=e^{-2}\cdot\frac{4}{2}=2e^{-2}\approx 0.271$

d. $P(X=3)=e^{-2}\frac{2^3}{3!}=e^{-2}\cdot\frac{8}{6}=\frac{4}{3}e^{-2}\approx 0.180$

17. Since the mean is 2, by Exercise 13, $a = 2$. Use POISSONCDF.

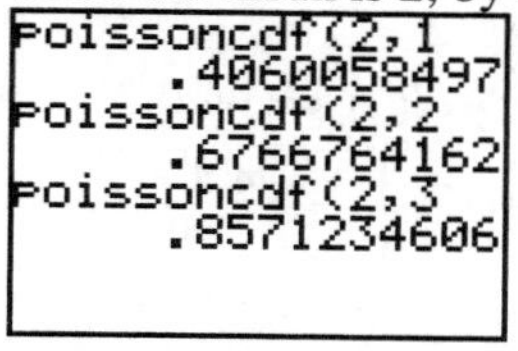

a. $P(X \le 1) \approx 0.406$

b. $P(X \le 2) \approx 0.677$

c. $P(X \le 3) \approx 0.857$

d. $P(X > 3) = 1 - P(X \le 3) \approx 1 - 0.857 = 0.143$

19. **a.** {BBBB, BBBG, BBGB, BBGG, BGBB, BGBG, BGGB, BGGG, GBBB, GBBG, GBGB, GBGG, GGBB, GGBG, GGGB, GGGG}

b. The probability of each event is $\frac{1}{16}$.

c. $P(X = 0) = P(\text{BBBB}) = \frac{1}{16}$

$$\begin{aligned} P(X = 1) &= P(\text{BBBG}) + P(\text{BBGB}) + P(\text{BGBB}) + P(\text{GBBB}) \\ &= \frac{1}{16} + \frac{1}{16} + \frac{1}{16} + \frac{1}{16} = \frac{4}{16} = \frac{1}{4} \end{aligned}$$

$$\begin{aligned} P(X = 2) &= P(\text{BBGG}) + P(\text{BGGB}) + P(\text{GGBB}) + P(\text{BGBG}) + P(\text{GBGB}) + P(\text{GBBG}) \\ &= \frac{1}{16} + \frac{1}{16} + \frac{1}{16} + \frac{1}{16} + \frac{1}{16} + \frac{1}{16} = \frac{6}{16} = \frac{3}{8} \end{aligned}$$

$$\begin{aligned} P(X = 3) &= P(\text{BGGG}) + P(\text{GBGG}) + P(\text{GGBG}) + P(\text{GGGB}) \\ &= \frac{1}{16} + \frac{1}{16} + \frac{1}{16} + \frac{1}{16} = \frac{4}{16} = \frac{1}{4} \end{aligned}$$

$$P(X = 4) = P(\text{GGGG}) = \frac{1}{16}$$

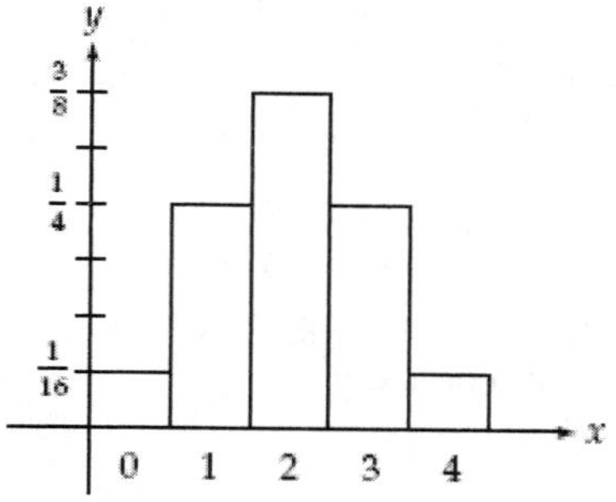

d. $P(X \ge 1) = 1 - P(X = 0) = 1 - \frac{1}{16} = \frac{15}{16}$

e.
$$\begin{aligned} \mu = E(X) &= x_1 p_1 + x_2 p_2 + x_3 p_3 + x_4 p_4 + x_5 p_5 \\ &= 0 \cdot \frac{1}{16} + 1 \cdot \frac{1}{4} + 2 \cdot \frac{3}{8} + 3 \cdot \frac{1}{4} + 4 \cdot \frac{1}{16} \\ &= \frac{1}{4} + \frac{3}{4} + \frac{3}{4} + \frac{1}{4} = 2 \end{aligned}$$

$$\begin{aligned} \text{Var}(X) &= (x_1 - \mu)^2 \cdot p_1 + (x_2 - \mu)^2 \cdot p_2 + (x_3 - \mu)^2 \cdot p_3 + (x_4 - \mu)^2 \cdot p_4 + (x_5 - \mu)^2 \cdot p_5 \\ &= (0-2)^2 \cdot \frac{1}{16} + (1-2)^2 \cdot \frac{1}{4} + (2-2)^2 \cdot \frac{3}{8} + (3-2)^2 \cdot \frac{1}{4} + (4-2)^2 \cdot \frac{1}{16} \\ &= \frac{1}{4} + \frac{1}{4} + 0 + \frac{1}{4} + \frac{1}{4} = 1 \end{aligned}$$

21. $\mu = E(X)$

$$= 10{,}000 \cdot \frac{1}{10} + 2000 \cdot \frac{1}{2} + 1000 \cdot \frac{3}{10} + 0 \cdot \frac{1}{10}$$

$$= 1000 + 1000 + 300 + 0 = 2300$$

Since the price is \$500 more than $E(X)$, the price should be \$2800.

23. To use the Poisson distribution, we note that the average is 4. Thus, $a = 4$.

$$P(X \le 2) = P(X = 2) + P(X = 1) + P(X = 0)$$

$$= e^{-4}\frac{4^2}{2!} + e^{-4}\frac{4^1}{1!} + e^{-4}\frac{4^0}{0!}$$

$$= 8e^{-4} + 4e^{-4} + e^{-4} = 13e^{-4} \approx 0.238$$

25. Since the mean is 1.93, use a Poisson distribution with $a = 1.93$.

$$P(X \le 2)$$

$$= P(X = 2) + P(X = 1) + P(X = 0)$$

$$= e^{-1.93}\frac{1.93^2}{2!} + e^{-1.93}\frac{1.93}{1!} + e^{-1.93}\frac{1.93^0}{0!}$$

$$= 1.86245e^{-1.93} + 1.93e^{-1.93} + e^{-1.93}$$

$$\approx 0.70$$

27. Since the mean is 1.7, use a Poisson distribution with $a = 1.7$.

$$P(X = 0) = e^{-1.7}\frac{1.7^0}{0!} \approx 0.18 \text{ or } 18\%$$

29. Since the mean is 2, use a Poisson distribution with $a = 2$.

$$P(X < 5) = P(X = 4) + P(X = 3) + P(X = 2) + P(X = 1) + P(X = 0)$$

$$= e^{-2}\frac{2^4}{4!} + e^{-2}\frac{2^3}{3!} + e^{-2}\frac{2^2}{2!} + e^{-2}\frac{2^1}{1!} + e^{-2}\frac{2^0}{0!}$$

$$\approx 0.947$$

To find the probability of getting 5 or more defects, we calculate

$P(X \ge 5) = 1 - P(X < 5) \approx 1 - 0.947 = 0.053$ or less than 6%

31. Since the mean is 2, we use a Poisson distribution with $a = 2$.

a. For no traffic jams,

$$P(X = 0) = e^{-2}\frac{2^0}{0!} \approx 0.135 \text{ or about } 14\%$$

b. For at least 1 traffic jam,

$$P(X > 1) = 1 - P(X = 7) \approx 1 - 0.14 = 0.86$$

or about 86%

c. For 3 or fewer traffic jams,

$$P(X \le 3) = \sum_{k=0}^{3} e^{-2}\frac{2^k}{k!}$$

$$\approx 0.135 + 0.271 + 0.271 + 0.180$$

$$= 0.857 \text{ or about } 86\%$$

33. Since the mean is 7, we use a Poisson distribution with $a = 7$.

a. $P(X \le 3) = \sum_{k=0}^{3} e^{-7}\frac{7^k}{k!}$

$$\approx 0.0009 + 0.0064 + 0.0223 + 0.0521$$

$$= 0.0817 \text{ or about } 8\%$$

b. $P(X \le 5) = \sum_{k=0}^{5} e^{-7}\frac{7^k}{k!}$

$$\approx 0.001 + 0.006 + 0.022 + 0.052 + 0.091 + 0.128$$

$$= 0.300 \text{ or about } 30\%$$

35. Since the mean is 2.4, we use a Poisson distribution with $a = 2.4$.

a. $P(X = 0) = e^{-2.4}\frac{2.4^0}{0!} \approx 0.091$ or about 9%

b. For no more than 5,

$$P(X \le 5) = \sum_{k=0}^{5} e^{-2.4}\frac{2.4^k}{k!}$$

$$\approx 0.091 + 0.218 + 0.261 + 0.209 + 0.125 + 0.060$$

$$= 0.964 \text{ or about } 96\%$$

37. $\frac{1}{275}$

39. $P(X = 5)$

41. False. (See Example 9.)

43. 105; 10

45. $X = 5$ (In general, if the variance is zero, then the random variable is a constant.)

47.
1. This is the definition of variance.
2. Expand $(x_i - \mu)^2 = x^2 - 2\mu x_i + \mu^2$.
3. Multiply by p_i and separate into 3 sums.
4. Take the constants 2μ and μ^2 outside the sums.
5. This results follows from the definition of $\sum_{i=1}^{n} x_i p_i = \mu$ and the fact that $\sum_{i=1}^{n} p_i = 1$
6. This follows because $-2\mu\mu + \mu^2 \cdot 1 = -2\mu^2 + \mu^2 = -\mu^2$.

EXERCISES 11.2

1. The function $ax^2(1-x)$ is nonnegative on [0, 1]. To find the area under the curve, integrate the function.

$$\int_0^1 ax^2(1-x)\,dx = \int_0^1 (ax^2 - ax^3)\,dx = \left(\frac{ax^3}{3} - \frac{ax^4}{4}\right)\Bigg|_0^1$$
$$= \frac{a(1)^3}{3} - \frac{a(1)^4}{4} - \left[\frac{a(0)^3}{3} - \frac{a(0)^4}{4}\right]$$
$$= \frac{a}{3} - \frac{a}{4} = \frac{a}{12}$$

Since $\frac{a}{12}$ must equal 1, $a = 12$.

3. The function ax^2 is nonnegative on [0, 3]. To find the area under the curve, integrate the function.

$$\int_0^3 ax^2\,dx = \frac{ax^3}{3}\Bigg|_0^3 = \frac{a(3)^3}{3} - \frac{a(0)^3}{3} = 9a$$

Since $9a$ must equal 1, $a = \frac{1}{9}$.

5. The function $\frac{a}{x}$ is nonnegative on [1, e]. To find the area under the curve, integrate the function.

$$\int_1^e \frac{a}{x}\,dx = a\ln x\Bigg|_1^e = a\ln e - a\ln 1 = a \cdot 1 - a \cdot 0 = a$$

Thus, $a = 1$.

7. The function axe^x is nonnegative on [0, 1]. To find the area under the curve, integrate the function.

$$\int_0^1 axe^x\,dx$$

Let $u = x$ and $dv = e^x\,dx$. Then $du = dx$ and $v = \int e^x\,dx = e^x$

$$\int_0^1 axe^x = a\left(xe^x\Big|_0^1 - \int_0^1 e^x\,dx\right) = a(xe^x - e^x)\Big|_0^1$$
$$= a[(1\cdot e^1 - e^1) - (0\cdot e^0 - e^0)]$$
$$= a[e - e - (0-1)] = a[-(-1)] = a$$

Thus, $a = 1$.

9. The function ae^{-x^2} is nonnegative on [–1, 1]. The area under the curve is found by integrating the function.

$$\int_{-1}^{1} ae^{-x^2}\,dx = a\int_{-1}^{1} e^{-x^2}\,dx \approx a(1.49)$$

using FnInt. Since 1.49a must equal 1, $a = \frac{1}{1.49} \approx 0.670$.

11. **a.** $\mu = E(X) = \int_0^1 x(3x^2)\,dx = \int_0^1 3x^3\,dx = \frac{3}{4}x^4\Big|_0^1$

$= \frac{3}{4}(1)^4 - \frac{3}{4}(0)^4 = \frac{3}{4}$

b. $\text{Var}(X) = \int_0^1 \left(x - \frac{3}{4}\right)^2 (3x^2)\,dx$

$= \int_0^1 \left(x^2 - \frac{3}{2}x + \frac{9}{16}\right)(3x^2)\,dx$

$= \int_0^1 \left(3x^4 - \frac{9}{2}x^3 + \frac{27}{16}x^2\right)dx$

$= \left(\frac{3}{5}x^5 - \frac{9}{8}x^4 + \frac{9}{16}x^3\right)\Big|_0^1$

$= \frac{3}{5}(1)^5 - \frac{9}{8}(1)^4 + \frac{9}{16}(1)^3 - \left[\frac{3}{5}(0)^5 - \frac{9}{8}(0)^4 + \frac{9}{16}(0)^3\right]$

$= \frac{3}{5} - \frac{9}{8} + \frac{9}{16} = \frac{48}{80} - \frac{90}{80} + \frac{45}{80} = \frac{3}{80} = 0.0375$

c. $\sigma(X) = \sqrt{\frac{3}{80}} \approx 0.194$

13. **a.** $\mu = E(X) = \int_0^1 x \cdot 12x^2(1-x)\,dx = \int_0^1 (12x^3 - 12x^4)\,dx$

$= \left(3x^4 - \frac{12}{5}x^5\right)\Big|_0^1 = 3(1)^4 - \frac{12}{5}(1)^5 - \left[3(0)^4 - \frac{12}{5}(0)^5\right]$

$= 3 - \frac{12}{5} = \frac{3}{5}$

b. $\text{Var}(X) = \int_0^1 \left(x - \frac{3}{5}\right)^2 \cdot 12x^2(1-x)\,dx$

$= \int_0^1 \left(x^2 - \frac{6}{5}x + \frac{9}{25}\right)(12x^2 - 12x^3)\,dx$

$= \int_0^1 \left[12x^4 - \frac{72}{5}x^3 + \frac{108}{25}x^2 - \left(12x^5 - \frac{72}{5}x^4 + \frac{108}{25}x^3\right)\right]dx$

$= \int_0^1 \left(-12x^5 + \frac{132}{5}x^4 - \frac{468}{25}x^3 + \frac{108}{25}x^2\right)dx$

$= \left(-2x^6 + \frac{132}{25}x^5 - \frac{117}{25}x^4 + \frac{36}{25}x^3\right)\Big|_0^1$

$= \left[-2(1)^6 + \frac{132}{25}(1)^5 - \frac{117}{25}(1)^4 + \frac{36}{25}(1)^3\right] - 0$

$= -2 + \frac{132}{25} - \frac{117}{25} + \frac{36}{25} = -2 + \frac{51}{25} = \frac{1}{25}$

c. $\sigma(X) = \sqrt{\frac{1}{25}} = \frac{1}{5}$

15. **a.** $\mu = E(X) = \int_0^1 \left(x \cdot \frac{4}{\pi} \frac{1}{1+x^2} \right) dx$

$= \int_0^1 \left(\frac{4}{\pi} \frac{x}{1+x^2} \right) dx \approx 0.441$

b. $\text{Var}(X) = \int_0^1 \left[(x-0.441)^2 \cdot \frac{4}{\pi} \frac{1}{1+x^2} \right] dx$

≈ 0.079

c. $\sigma(X) = \sqrt{0.079} \approx 0.280$

17. **a.** $F(x) = \int_A^x f(t)\,dt = \int_1^x \frac{1}{12} t\,dt = \left. \frac{t^2}{24} \right|_1^x = \frac{x^2}{24} - \frac{1}{24}$

b. $P(3 \le X \le 5) = F(5) - F(3)$

$= \frac{5^2}{24} - \frac{1}{24} - \left(\frac{3^2}{24} - \frac{1}{24} \right) = \frac{25}{24} - \frac{9}{24} = \frac{16}{24} = \frac{2}{3}$

19. $P\left(X \le \frac{1}{3} \right) = F\left(\frac{1}{3} \right) = \int_0^{1/3} \frac{2}{(1+x)^2} dx$

Let $u = 1 + x$. For $x = 0$, $u = 1$

$du = dx$ For $x = \frac{1}{3}$, $u = \frac{4}{3}$.

$P\left(X \le \frac{1}{3} \right) = F\left(\frac{1}{3} \right) = 2 \int_1^{4/3} \frac{1}{u^2} du$

$= 2 \int_1^{4/3} u^{-2} du$

$= 2 \left(\frac{u^{-1}}{-1} \right) \Bigg|_1^{4/3} = -\frac{2}{u} \Bigg|_1^{4/3}$

$= -\frac{2}{\frac{4}{3}} - \left(-\frac{2}{1} \right) = -\frac{6}{4} + 2 = \frac{1}{2}$

21. To find $P(2 \le X \le 3)$, just evaluate $F(3) - F(2)$.

$P(2 \le X \le 3) = F(3) - F(2)$

$= \frac{3}{3} - \frac{2}{3} = \frac{1}{3}$

23. **a.** $P(0.5 \le X \le 1) = F(1) - F(0.5)$

$= 1^2 - 0.5^2 = 0.75$

b. $f(x) = \frac{d}{dx} F(x) = \frac{d}{dx} x^2 = 2x$ on $[0, 1]$

25. **a.** $F(A) = \int_A^A f(t)\,dt = 0$

b. $F(B) = P(A \le X \le B) = 1$ because the probability of all the possible values of $[A, B]$ is 1.

27. $P\left(-\frac{1}{2} \le X \le \frac{1}{2} \right) = F\left(\frac{1}{2} \right) - F\left(-\frac{1}{2} \right)$

$= \int_{-1/2}^{1/2} \frac{2}{\pi} \frac{1}{1+x^2} dx - \int_{-1/2}^{-1/2} \frac{2}{\pi} \frac{1}{1+x^2} dx$

$= \int_{-1/2}^{1/2} \frac{2}{\pi} \frac{1}{1+x^2} dx \approx 0.590$

29. To find $P(0.4 \le X \le 0.6)$, first find $F(x)$.

$$F(x) = \int_0^x 6t(1-t)\,dt = \int_0^x (6t - 6t^2)\,dt = (3t^2 - 2t^3)\Big|_0^x$$
$$= 3x^2 - 2x^3 - 3(0)^2 + 2(0)^3 = 3x^2 - 2x^3$$
$$P(0.4 \le X \le 0.6) = F(0.6) - F(0.4)$$
$$= 3(0.6)^2 - 2(0.6)^3 - [3(0.4)^2 - 2(0.4)^3]$$
$$= 1.08 - 0.432 - (0.48 - 0.128)$$
$$= 1.08 - 0.432 - 0.352 = 0.296$$

31. **a.** $E(X) = \int_0^4 x \cdot f(x)\,dx = \int_0^4 x \cdot \frac{3}{64}x^2\,dx = \int_0^4 \frac{3}{64}x^3\,dx$

$$= \frac{3}{256}x^4\Big|_0^4 = \frac{3}{256}\cdot 4^4 - \frac{3}{256}\cdot 0^4 = 3$$

b. To find $P(2 \le X \le 4)$, first find $F(x)$.

$$F(x) = \int_0^x \frac{3}{64}t^2\,dt = \frac{1}{64}t^4\Big|_0^x = \frac{x^3}{64}$$

$$P(2 \le X \le 4) = F(4) - F(2) = \frac{4^3}{64} - \frac{2^3}{64} = 1 - \frac{1}{8} = \frac{7}{8}$$

33. **a.** Since $F(x) = P(X \le x)$, the probability of hitting the area inside the circle of radius x is $\frac{\pi x^2}{\pi \cdot 9^2} = \frac{x^2}{81}$.

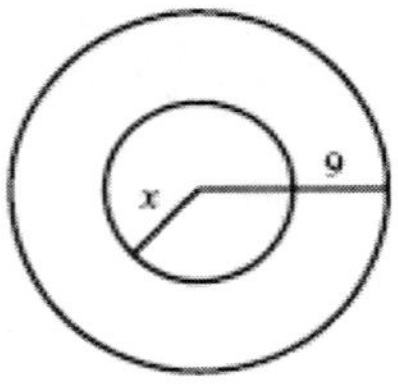

Thus, $F(x) = \frac{x^2}{81}$ on [0, 9].

b. $f(x) = \frac{d}{dx}F(x) = \frac{d}{dx}\left(\frac{x^2}{81}\right) = \frac{2}{81}x$ on [0, 9]

c. $E(X) = \int_0^9 x \cdot \frac{2}{81}x\,dx = \int_0^9 \frac{2}{81}x^2\,dx = \frac{2x^3}{243}\Big|_0^9$

$$= \frac{2\cdot 9^3}{243} - \frac{2\cdot 0^3}{243} = 2\cdot 3 = 6$$

d. $P(3 \le X \le 6) = \int_3^6 f(x)\,dx = \int_3^6 \frac{2}{81}x\,dx$

$$= \frac{x^2}{81}\Big|_3^6 = \frac{6^2}{81} - \frac{3^2}{81}$$
$$= \frac{36-9}{81} = \frac{27}{81} = \frac{1}{3}$$

e. $P(3 \le X \le 6) = F(6) - F(3) = \frac{6^2}{81} - \frac{3^2}{81}$

$$= \frac{36-9}{81} = \frac{27}{81} = \frac{1}{3}$$

35. **a.** $E(X) = \int_0^2 x \cdot f(x)\,dx = \int_0^2 x \cdot 0.75x(2-x)\,dx$

$$= \int_0^2 (1.5x^2 - 0.75x^3)\,dx$$
$$= \left(0.5x^3 - \frac{0.75x^4}{4}\right)\Big|_0^2$$
$$= 0.5\cdot 2^3 - \frac{0.75}{4}\cdot 2^4 - \left(0.5\cdot 0^3 - \frac{0.75}{4}\cdot 0^4\right)$$
$$= 4 - 3 = 1$$

b. $\text{Var}(X) = \int_0^2 (x-1)^2[0.75x(2-x)]\,dx = 0.2$

$\sigma(X) = \sqrt{0.2} \approx 0.447$

c. $P(X \ge 1.5) = 1 - P(X < 1.5) = 1 - F(1.5)$

$$= 1 - \int_0^{1.5} 0.75x(2-x)\,dx$$
$\approx 1 - 0.844 = 0.156$ or about 16%.

37. a.

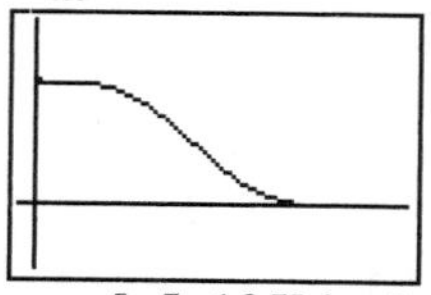

on [–5, 105] by [–0.5, 1.5]

b. Using EVALUATE with $x = 30$, $S = 0.8308$

39. To find $P(1 \le X \le 3)$, first find $F(x)$.

$$F(x) = \int_0^x 0.55e^{-0.37t}dt$$
$$= \frac{0.55}{-0.37}\left(e^{-0.37t}\right)\Big|_0^x$$
$$= \frac{0.55}{-0.37}e^{-0.37x}$$
$$P(1 \le X \le 3) = F(3) - F(1)$$
$$= \frac{0.55}{-0.37}\left[e^{-0.37(3)} - e^{-0.37(1)}\right]$$
$$\approx 0.537 \text{ or about } 54\%$$

41. To find $P(1000 \le X \le 5000)$, first find $F(x)$.

$$F(x) = \int_0^x 1250t^{-2.1}dt$$
$$= \frac{1250}{-1.1}\left(t^{-1.1}\right)\Big|_0^x$$
$$= \frac{1250}{-1.1}x^{-1.1}$$
$$P(1000 \le X \le 5000)$$
$$= F(5000) - F(1000)$$
$$= \frac{1250}{-1.1}\left[(5000)^{-1.1} - (1000)^{-1.1}\right]$$
$$\approx 0.473 \text{ or about } 47\%$$

43. In *discrete* probability it means that the event *cannot* occur, while in *continuous* probability it means that the event is *overwhelmingly unlikely* (like guessing one number from infinitely many).

45. $F(x+7)$ (because the graph is shifted 7 units to the left)

47. There is more area in the *outer* 6 inches of the target than in the *inner* 6 inches, so it is more likely that the dart will land *beyond* 6 inches, making the mean *greater* than 6.

49. True

51. 20; 10

EXERCISES 11.3

1. a. The probability density function on [0, 10] is $f(x) = \frac{1}{10}$.

b. $E(X) = \frac{B}{2} = \frac{10}{2} = 5$

c. $\text{Var}(X) = \frac{10^2}{12} = \frac{100}{12} = \frac{25}{3}$

d. $\sigma(X) = \sqrt{\frac{25}{3}} \approx 2.89$

e. $P(8 \le X \le 10) = \int_8^{10} f(x)dx = \int_8^{10} \frac{1}{10}dx$

$$= \frac{x}{10}\Big|_8^{10} = \frac{10}{10} - \frac{8}{10} = \frac{1}{5}$$

3. a. The probability density function for [0, B] $= \frac{1}{B}$.

Thus,

$$F(x) = \int_0^x \frac{1}{B}dt = \frac{t}{B}\Big|_0^x = \frac{x}{B} - \frac{0}{B} = \frac{x}{B} \text{ [on 0, } B]$$

b. Since each point is equally likely, the number of points above the midpoint should be the same as the number of points below the midpoint. Thus, the median is $\frac{B}{2}$.

5. **a.** $E(X) = \frac{30}{2} = 15$

b. $\text{Var}(X) = \frac{30^2}{12} = 75$

c. $\sigma(X) = \sqrt{75} = 5\sqrt{3} \approx 8.7$

d. $P(X > 24) = 1 - P(X \le 24) = 1 - \int_0^{24} \frac{1}{30}\,dx$

$$= 1 - \frac{x}{30}\Big|_0^{24} = 1 - \left(\frac{24}{30} + \frac{0}{30}\right)$$
$$= \frac{6}{30} = \frac{1}{5}$$

7. **a.** $F(x) = \int_A^x f(t)\,dt = \int_A^x \frac{1}{B-A}\,dt$

$$= \frac{t}{B-A}\Big|_A^x = \frac{x}{B-A} - \frac{A}{B-A}$$
$$= \frac{x-A}{B-A} \text{ on } [A, B]$$

b. $E(X) = \int_A^B x \cdot f(x)\,dx = \int_A^B x \cdot \frac{1}{B-A}\,dx$

$$= \frac{1}{B-A} \cdot \int_A^B x\,dx = \frac{1}{B-A}\left(\frac{x^2}{2}\right)\Big|_A^B$$
$$= \frac{1}{B-A}\left(\frac{B^2}{2} - \frac{A^2}{2}\right)$$
$$= \frac{1}{2}\left(\frac{B^2 - A^2}{B-A}\right) = \frac{B+A}{2}$$

c. $\text{Var}(X) = \int_A^B x^2 \cdot \frac{1}{B-A}\,dx - \left(\frac{B+A}{2}\right)^2$

$$= \frac{1}{B-A}\int_A^B x^2\,dx - \left(\frac{B+A}{2}\right)^2$$
$$= \frac{1}{B-A}\left(\frac{x^3}{3}\right)\Big|_A^B - \left(\frac{B+A}{2}\right)^2$$
$$= \frac{1}{B-A}\left(\frac{B^3}{3} - \frac{A^3}{3}\right) - \left(\frac{B+A}{2}\right)^2$$
$$= \frac{B-A}{B-A} \cdot \frac{B^2 + AB + A^2}{3} - \frac{B^2 + 2AB + A^2}{4}$$
$$= \frac{4B^2 + 4AB + 4A^2}{12} - \frac{3B^2 + 6AB + 3A^2}{12}$$
$$= \frac{B^2 - 2AB + A^2}{12} = \frac{(B-A)^2}{12}$$

9. **a.** $E(X) = \frac{1}{a} = \frac{1}{5}$

b. $\text{Var}(X) = \frac{1}{a^2} = \frac{1}{5^2} = \frac{1}{25}$

c. $\sigma(X) = \sqrt{\frac{1}{25}} = \frac{1}{5}$

11. **a.** $E(X) = \frac{1}{a} = \frac{1}{\frac{1}{2}} = 2$

b. $\text{Var}(X) = \frac{1}{a^2} = \frac{1}{\left(\frac{1}{2}\right)^2} = 4$

c. $\sigma(X) = \sqrt{4} = 2$

13. If the mean is 3, then $\frac{1}{a} = 3$ or $a = \frac{1}{3}$. Thus, $f(x) = \frac{1}{3}e^{-\frac{1}{3}x}$ on $[0, \infty)$.

15. If $\text{Var}(X) = \frac{1}{4}$, then $\frac{1}{a^2} = \frac{1}{4}$. Thus, $a^2 = 4$ or $a = 2$. $f(x) = 2e^{-2x}$ on $[0, \infty)$.

17. **a.** $F(x) = \int_0^x f(t)\,dt = \int_0^x ae^{-at}\,dt$

$= \left(a\dfrac{e^{-at}}{-a} \right) \Bigg|_0^x$

$= -e^{-ax} + e^{-a(0)}$

$= -e^{-ax} + 1$ on $[0, \infty)$.

b. To find the median, we must find x such that $P(X \le 0.5)$. Since $P(X \le 5) = F(0.5)$, then

$$\begin{aligned} 0.5 &= -e^{-ax} + 1 \\ e^{-ax} &= 0.5 \\ e^{ax} &= 2 \\ ax &= \ln 2 \\ x &= \frac{\ln 2}{a} \approx \frac{0.693}{a} \end{aligned}$$

19. Since the mean time is 20 days, then $\frac{1}{a} = 20$ or $a = \frac{1}{20}$. Thus, $f(x) = \frac{1}{20}e^{-\frac{1}{20}x}$.

$P(X \le 30) = F(30) = \int_0^{30} \frac{1}{20}e^{-\frac{1}{20}x}\,dx \approx 0.777$

21. Since the mean is 10, $\frac{1}{a} = 10$ or $a = \frac{1}{10}$. Thus

$f(x) = \frac{1}{10}e^{-\frac{1}{10}x}$.

$P(X \le 2) = F(2) = \int_0^2 \frac{1}{10}e^{-\frac{1}{10}x}\,dx \approx 0.181$

23. **a.** $f(x) = 0.125e^{-0.125x}$ on $[0, \infty)$.

$P(X \le 2) \int_0^2 0.125e^{-0.125x}\,dx = -e^{-0.125x}\Big|_0^2 \approx 0.22$ or 22%

b. $P(X \le 12) \int_0^{12} 0.125e^{-0.125x}\,dx = -e^{-0.125x}\Big|_0^{12} \approx 0.78$ or 78%

25. The median is the point x such that $P(X \le x) = 0.5$.

$P(X \le x) = F(x) = \int_0^x f(t)\,dt$

$= \int_0^x 2t\,dt = t^2\Big|_0^x = x^2 - 0 = x^2$

Thus, the median is the point x such that

$x^2 = 0.5$

$x \approx 0.707$

27. $P(X \le x) = F(x) = \int_0^x \frac{1}{2}t\,dt$

$= \dfrac{t^2}{4}\Bigg|_0^x = \dfrac{x^2}{4} - \dfrac{0^2}{4} = \dfrac{x^2}{4}$

Thus, the median is the point x such that

$\dfrac{x^2}{4} = 0.5$

$x^2 = 2$

$x \approx 1.41$

29. By the "Alternative Formula for Variance" on page 767, and $E(x)=\frac{1}{a}$, from page 777,

$$\text{Var}(X)=\int_0^\infty x^2\cdot ae^{-ax}dx-\left(\frac{1}{a}\right)^2.$$

Using "integration by parts" with $u=x^2$ and $dv=ae^{-ax}dx$,

$$\int_0^\infty x^2\cdot ae^{-ax}dx-\left(\frac{1}{a}\right)^2=-x^2\cdot e^{-ax}-\int_0^\infty -e^{-ax}\cdot 2xdx$$

so

$$\int_0^\infty x^2\cdot ae^{-ax}dx=\lim_{B\to\infty}\left(-x^2e^{-ax}\right)\Big|_0^B+\frac{2}{a}\int_0^\infty xae^{-ax}dx$$

By Exercise 63 on page 730 and the calculation of $E(x)$ on page 777, this is

$$=0+\frac{2}{a}\cdot\frac{1}{a}=\frac{2}{a^2}$$

Thus

$$\text{Var}(X)=\frac{2}{a^2}-\frac{1}{a^2}=\frac{1}{a^2}.$$

31. Since the mean is 7, $\frac{1}{a}=7$ or $a=\frac{1}{7}$. Thus,

$f(x)=\frac{1}{7}e^{-x/7}$.

$P(X\le 3)=F(3)=\int_0^3\frac{1}{7}e^{-x/7}dx\approx 0.35$ or 35%

33. Since the mean is 700, $\frac{1}{a}=700$ or $a=\frac{1}{700}$. Thus,

$f(x)=\frac{1}{700}e^{-x/700}$.

$P(X\le 100)=F(100)=\int_0^{100}\frac{1}{100}e^{-x/700}dx$

≈ 0.13 or 13%

35. Since the mean is 1000, $\frac{1}{a}=1000$ or

$a=\frac{1}{1000}=0.001$. Thus, $f(x)=0.0001e^{-0.001x}$

$P(800\le X\le 1200)=\int_{800}^{1200}0.001e^{-0.001x}dx$

≈ 0.15 or 15%

37. The probability density function is $f(x)=ae^{-ax}$

$S(x)=P(X\ge x)=1-P(X\le x)=1-\int_0^x ae^{-at}dt$

$=1-F(x)$

$=e^{-ax}$

39. $f(x)=\frac{1}{20}$ on $[0,\ 20]$

41. $f(x)=\frac{1}{6}$ on $[0,\ 6]$

42. $f(x)=5e^{-5x}$ for $0\le x<\infty$

43. $f(x)=2e^{-2x}$ for $0\le x<\infty$

43. $f(x)=2e^{-2x}$ for $0\le x<\infty$

45. A Poisson random variable with parameter 3

47. The median (because there are some very large families).

49. The error is the word "median." The reporter should have said "mean."

EXERCISES 11.4

1. $P(0\le Z\le 1.95)=0.4744$

3. $P(-2.55\le Z\le 0.48)=0.4946+0.1844=0.6790$

5. $P(1.4\le Z\le 2.8)=0.4974-0.4192=0.0782$

7. $P(Z<-3.1)=1-0.499-0.5=0.001$

9. $P(Z \le 3.45) = 0.4997 + 0.5 = 0.9997$

11. $P(10 \le X \le 11) = \int_{10}^{11} \frac{1}{2\sqrt{2\pi}} e^{-\frac{1}{2}\left(\frac{x-9}{2}\right)^2} dx$
≈ 0.15

13. $P(X \le -6) = P(X \le -5) - \int_{-6}^{-5} \frac{1}{2\sqrt{2\pi}} e^{-\frac{1}{2}\left[\frac{x-(-5)}{2}\right]^2} dx$

Since $\mu = -5$, $P(X \le -5) = 0.5$. Thus,

$$P(X \le -6) = 0.5 - \int_{-6}^{-5} \frac{1}{2\sqrt{2\pi}} e^{-\frac{1}{2}\left[\frac{x-(-5)}{2}\right]^2} dx$$
$$\approx 0.5 - 0.1915 = 0.3085$$

15. Since the mean is 210 and standard deviation is 30, the probability density function is

$$f(x) = \frac{1}{30\sqrt{2\pi}} e^{-\frac{1}{2}\left(\frac{x-210}{30}\right)^2}.$$

$$P(X \le 250) = P(X \le 210) + P(210 \le X \le 250)$$
$$= 0.5 + \int_{210}^{250} \frac{1}{30\sqrt{2\pi}} e^{-\frac{1}{2}\left(\frac{x-210}{30}\right)^2} dx$$
$$\approx 0.5 + 0.4088 = 0.9088 \text{ or } 91\%$$

17. $\mu = 22$, $\sigma = 3$, $f(x) = \frac{1}{3\sqrt{2\pi}} e^{-\frac{1}{2}\left(\frac{x-22}{3}\right)^2}$

$$P(X > 20) = P(20 < X < 22) + P(X \ge 22)$$
$$= \int_{20}^{22} \frac{1}{3\sqrt{2\pi}} e^{-\frac{1}{2}\left(\frac{x-22}{3}\right)^2} dx + 0.5$$
$$\approx 0.2475 + 0.5 = 0.7475$$

Thus, $P(X \le 20) = 1 - 0.7475 = 0.2525$

19. $\mu = 2000$, variance = 160,000. Since $\sigma = \sqrt{\text{variance}}$, $\sigma = \sqrt{160{,}000} = 400$.

$$P(X > 2500) = 1 - P(X \le 2500)$$
$$= 1 - P(X \le 2400) - P(2400 \le X \le 2500)$$
$$= 1 - \left(0.5 + \frac{0.682}{2}\right) - \int_{2400}^{2500} \frac{1}{400\sqrt{2\pi}} e^{-\frac{1}{2}\left(\frac{x-2000}{400}\right)^2}$$
$$\approx 1 - 0.841 - 0.053 = 0.106$$

21. $\mu = 190$, $\sigma = 40$

$$P(\text{a man qualifies}) = \int_{117}^{238} \frac{1}{40\sqrt{2\pi}} e^{-\frac{1}{2}\left(\frac{x-190}{40}\right)^2} dx$$
$$\approx 0.851 \text{ or about } 85\%$$

23. $\mu = 100$, $\sigma = 15$. Using the graphing calculator program that finds x such that $P(X < x) = 0.98$ we get $x = 131$.

25. Let x = amount of soft drink. Since X is a random variable with $\sigma = 0.25$, we must find σ such that $P(X \ge 32) = 0.98$. Since

$0.98 = P(X \ge 32) = 1 - P(X < 32)$

we must find x such that

$P(X < 32) = 0.02$

Using the normal distribution table, we get $P(X < 32) = 0.02$ at the value $\mu - 2.055\sigma$.

Since $\sigma = 0.25$, we get

$$\mu - 2.055(0.25) = 32$$
$$\mu - 0.51 = 32$$
$$\mu = 32.51$$

27. Mean 0 and standard deviation 2

29. Mean 5 and standard deviation 2

31. $P(Y > 4)$

33. Y has the greater mean and the greater standard deviation.

35. It has mean 47.8.

37. Since $f(x) = \frac{1}{\sqrt{2x}} e^{-\frac{1}{2}x^2}$ is the normal standard probability density function with mean 0 and standard deviation 1, the maximum of the function occurs at the mean, that is, $x = 0$.
At $x = 0$,

$$f(0) = \frac{1}{\sqrt{2\pi}} e^{-\frac{1}{2}0^2} = \frac{1}{\sqrt{2\pi}} \cdot 1 = \frac{1}{\sqrt{2\pi}}$$

39.
$$f(x) = \frac{1}{\sqrt{2\pi}} e^{-\frac{1}{2}x^2}$$
$$f'(x) = -\frac{1}{2} \cdot 2x \cdot \frac{1}{\sqrt{2\pi}} e^{-\frac{1}{2}x^2} = -\frac{x}{\sqrt{2\pi}} e^{-\frac{1}{2}x^2}$$
$$f''(x) = -\frac{1}{\sqrt{2\pi}} e^{-\frac{1}{2}x^2} - \frac{x}{\sqrt{2\pi}}\left(-xe^{-\frac{1}{2}x^2}\right)$$
$$= -\frac{1}{\sqrt{2\pi}} e^{-\frac{1}{2}x^2} + x^2\left(\frac{1}{\sqrt{2\pi}} e^{-\frac{1}{2}x^2}\right) = 0$$
$$-1 + x^2 = 0$$
$$x^2 = 1$$
$$x = \pm 1$$

$$f(1) = \frac{1}{\sqrt{2\pi}} e^{-\frac{1}{2}(1)^2} = \frac{1}{\sqrt{2\pi}} e^{-\frac{1}{2}}$$
$$f(-1) = \frac{1}{\sqrt{2\pi}} e^{-\frac{1}{2}(-1)^2} = \frac{1}{\sqrt{2\pi}} e^{-\frac{1}{2}}$$

Thus, the points of inflection are $\left(1, \frac{1}{\sqrt{2\pi}} e^{-\frac{1}{2}}\right)$ and $\left(-1, \frac{1}{\sqrt{2\pi}} e^{-\frac{1}{2}}\right)$.

41. The steps follow from the given hints.

REVIEW EXERCISES AND CHAPTER TEST FOR CHAPTER 11

1. The events in X are 2, 8, and 12, and $P(2) = \frac{1}{2}$, $P(8) = \frac{1}{4}$, $P(12) = \frac{1}{4}$.

a.
$$E(X) = x_1 \cdot p_1 + x_2 \cdot p_2 + x_3 \cdot p_3$$
$$= 2 \cdot \frac{1}{2} + 8 \cdot \frac{1}{4} + 12 \cdot \frac{1}{4} = 1 + 2 + 3 = 6$$

b.
$$\text{Var}(X) = (x_1 - \mu)^2 \cdot p_1 + (x_2 - \mu)^2 \cdot p_2 + (x_3 - \mu)^2 \cdot p_3$$
$$= (2-6)^2 \cdot \frac{1}{2} + (8-6)^2 \cdot \frac{1}{4} + (12-6)^2 \cdot \frac{1}{4}$$
$$= 8 + 1 + 9 = 18$$

c. $\sigma(X) = \sqrt{\text{Var}(X)} = \sqrt{18} = 3\sqrt{2} \approx 4.24$

2. The events in X are 0, 2, 4, and 6, and each has probability $\frac{1}{4}$.

a. $$\begin{aligned} E(X) &= x_1 \cdot p_1 + x_2 \cdot p_2 + x_3 \cdot p_3 + x_4 \cdot p_4 \\ &= 0 \cdot \frac{1}{4} + 2 \cdot \frac{1}{4} + 4 \cdot \frac{1}{4} + 6 \cdot \frac{1}{4} \\ &= 0 + \frac{1}{2} + 1 + \frac{3}{2} = 3 \end{aligned}$$

b. $$\begin{aligned} \text{Var}(X) &= (x_1 - \mu)^2 \cdot p_1 + (x_2 - \mu)^2 \cdot p_2 + (x_3 - \mu)^2 \cdot p_3 + (x_4 - \mu)^2 \cdot p_4 \\ &= (0-3)^2 \cdot \frac{1}{4} + (2-3)^2 \cdot \frac{1}{4} + (4-3)^2 \cdot \frac{1}{4} + (6-3)^2 \cdot \frac{1}{4} \\ &= \frac{9}{4} + \frac{1}{4} + \frac{1}{4} + \frac{9}{4} = \frac{20}{4} = 5 \end{aligned}$$

c. $\sigma(X) = \sqrt{\text{Var}(X)} = \sqrt{5} \approx 2.24$

3. The events are 2,000,000 and 50,000, and $P(2{,}000{,}000) = 0.4$ and $P(50{,}000) = 0.6$.
$E(X) = x_1 \cdot p_1 + x_2 \cdot p_2 = 2{,}000{,}000(0.4) + 50{,}000(0.6) = \$830{,}000$

4. $E(A) = x_1 \cdot p_1 + x_2 \cdot p_2 = 1{,}200{,}000(0.1) + 2000(0.9) = \$121{,}800$
$E(B) = x_1 \cdot p_1 + x_2 \cdot p_2 = 1{,}000{,}000(0.15) + 1000(0.85) = \$150{,}850$
The speculator should buy property B.

5. Since the mean is 2, use the Poisson distribution with $a = 2$.
$$\begin{aligned} P(X \le 2) &= P(X = 2) + P(X = 1) + P(X = 0) \\ &= e^{-2} \frac{2^2}{2!} + e^{-2} \frac{2^1}{1!} + e^{-2} \frac{2^0}{0!} \approx 0.677 \end{aligned}$$

6. Since the mean is 2.63, use the Poisson distribution with $a = 2.63$.
$$\begin{aligned} P(2 \le X \le 4) &= P(X = 4) + P(X = 3) + P(X = 2) \\ &= e^{-2.63} \frac{2.63^4}{4!} + e^{-2.63} \frac{2.63^3}{3!} + e^{-2.63} \frac{2.63^2}{2!} \\ &\approx 0.612 \end{aligned}$$

7. Since the mean is 4, use the Poisson distribution with $a = 4$.
$P(X = 0) = e^{-4} \frac{4^0}{0!} \approx 0.018$ or about 2%

8. Since there are 500 chips for 100 cookies, each cookie averages 5 chips. Use a Poisson distribution with $a = 5$.
$P(X = 0) = e^{-5} \frac{5^0}{0!} \approx 0.0067$

9. Since the mean is 6.46, use a Poisson distribution with $a = 6.46$.

a. $P(X \le 4) = \sum_{k=0}^{4} e^{-6.46} \frac{6.46^k}{k!} \approx 0.228$ or about 23%

b. $P(X \le 6) = \sum_{k=0}^{6} e^{-6.46} \frac{6.46^k}{k!} \approx 0.533$ or about 53%

c. $P(X \le 8) = \sum_{k=0}^{8} e^{-6.46} \frac{6.46^k}{k!} \approx 0.796$ or about 80%

d. $P(X > 9) = 1 - P(X \le 8) = 1 - 0.796 \approx 0.204$ or about 20%

10. Since the mean is 2, use a Poisson distribution with $a = 2$.

$$P(X > 4) = 1 - P(X \le 4) = 1 - \sum_{k=0}^{4} e^{-2}\frac{2^k}{k!}$$
$$\approx 1 - 0.95 = 0.05 \text{ or } 5\%$$

11. The function $ax^2(2-x)$ is nonnegative on [0, 2]. To find the area under the curve, integrate the function.

$$\int_0^2 ax^2(2-x)\,dx = \int_0^2 2ax^2 - ax^3\,dx = \left(\frac{2ax^3}{3} - \frac{ax^4}{4}\right)\Bigg|_0^2$$
$$= \frac{2a(2)^3}{3} - \frac{a(2)^4}{4} - \left[\frac{2a(0)^3}{3} - \frac{a(0)^4}{4}\right]$$
$$= \frac{16a}{3} - 4a = \frac{4}{3}a$$

Since $\frac{4}{3}a$ must equal 1, $a = \frac{3}{4}$.

12. The function $a\sqrt{9-x}$ is nonnegative on [0, 9]. To find the area under the curve, integrate the function.

$$\int_0^9 a\sqrt{9-x}\,dx$$

Let $u = 9 - x$. Then $du = -dx$. For $x = 0$, $u = 9$. For $x = 9$, $u = 0$.

$$\int_0^9 a\sqrt{9-x}\,dx = \int_9^0 a\sqrt{u}(-du)$$
$$= -a\int_9^0 \sqrt{u}\,du = -a\left(\frac{2}{3}u^{3/2}\right)\Bigg|_9^0$$
$$= -a\left(\frac{2}{3}\cdot 0^{3/2} - \frac{2}{3}\cdot 9^{3/2}\right)$$
$$= -a(-18) = 18a$$

Since $18a$ must equal 1, $a = \frac{1}{18}$.

13. The function $\frac{a}{(1+x)^2}$ is nonnegative on [0, 1]. To find the area under the curve, integrate the function.

$$\int_0^1 \frac{a}{(1+x)^2} = a\left[\frac{(1+x)^{-1}}{-1}\right]\Bigg|_0^1 = -a\left[\frac{1}{1+x}\right]\Bigg|_0^1$$
$$= -a\left(\frac{1}{1+1} - \frac{1}{1+0}\right) = \frac{1}{2}a$$

Since $\frac{1}{2}a$ must equal 1, $a = 2$.

14. The function $a \sin x$ is nonnegative on $[0, \pi]$. To find the area under the curve, integrate the function.

$$\int_0^\pi a\sin x\,dx = a(-\cos x)\Big|_0^\pi$$
$$= a[-\cos\pi - (-\cos 0)]$$
$$= a[-(-1) - (-1)] = 2a$$

Since $2a$ must equal 1, $a = \frac{1}{2}$.

15. The function $a\sqrt{1-x^2}$ is nonnegative on [–1, 1]. To find the area under the curve, integrate the function.

$$\int_{-1}^1 a\sqrt{1-x^2}\,dx = \int_{-1}^1 \sqrt{1-x^2}\,dx \approx a(1.57)$$

Since $1.57a$ must equal 1, $a = \frac{1}{1.57} \approx 0.637$.

16. The function $ae^{-|x|}$ is nonnegative on [–2, 2]. To find the area under the curve, integrate the function.

$$\int_{-2}^2 ae^{-|x|}\,dx = a\int_{-2}^2 e^{-|x|}\,dx \approx 1.729a$$

Since $1.729a$ must equal 1, $a = \frac{1}{1.729} \approx 0.578$.

17. **a.** $P(1 \le X \le 2) = \int_1^2 \frac{1}{9}x^2 dx = \frac{x^3}{27}\Big|_1^2 = \frac{2^3}{27} - \frac{1^3}{27} = \frac{7}{27}$

b. $F(x) = \int_0^x \frac{1}{9}t^2\, dt = \frac{t^3}{27}\Big|_0^x = \frac{x^3}{27} - 0 = \frac{x^3}{27}$ on [0, 3]

c. $P(1 \le X \le 2) = F(2) - F(1) = \frac{2^3}{27} - \frac{1^3}{27} = \frac{7}{27}$

18. **a.** $P(4 \le X \le 9) = F(9) - F(4) = \frac{1}{4}\sqrt{9} - \frac{1}{4} - \left(\frac{1}{4}\sqrt{4} - \frac{1}{4}\right)$

$= \frac{3}{4} - \frac{1}{4} - \left(\frac{1}{2} - \frac{1}{4}\right) = \frac{1}{4}$

b. $f(x) = \frac{d}{dx}F(x) = \frac{d}{dx}\left(\frac{1}{4}\sqrt{x} - \frac{1}{4}\right) = \frac{1}{4}\left(\frac{1}{2}x^{-1/2}\right) = \frac{1}{8}x^{-1/2}$ on [1, 25]

c. $P(4 \le X \le 9) = \int_4^9 f(x)\,dx = \int_4^9 \frac{1}{8}x^{-1/2}\,dx = \frac{1}{4}x^{1/2}\Big|_4^9$

$= \frac{1}{4}\sqrt{9} - \frac{1}{4}\sqrt{4} = \frac{3}{4} - \frac{1}{2} = \frac{1}{4}$

19. **a.** $E(X) = \int_0^1 x \cdot f(x)\,dx = \int_0^1 x \cdot \frac{3}{2}(1 - x^2)\,dx = \int_0^1 \left(\frac{3}{2}x - \frac{3}{2}x^3\right)dx$

$= \left(\frac{3}{4}x^2 - \frac{3}{8}x^4\right)\Big|_0^1 = \frac{3}{4}(1)^2 - \frac{3}{8}(1)^4 - \left[\frac{3}{4}(0)^2 - \frac{3}{8}(0)^4\right] = \frac{3}{8}$

b.

$\text{Var}(X) = \int_0^1 x^2 f(x)\,dx - \mu^2 = \int_0^1 x^2 \cdot \frac{3}{2}(1 - x^2)\,dx - \left(\frac{3}{8}\right)^2$

$= \int_0^1 \left(\frac{3}{2}x^2 - \frac{3}{2}x^4\right)dx - \frac{9}{64} = \left(\frac{1}{2}x^3 - \frac{3}{10}x^5\right)\Big|_0^1 - \frac{9}{64}$

$= \left\{\frac{1}{2}(1)^3 - \frac{3}{10}(1)^5 - \left[\frac{1}{2}(0)^3 - \frac{3}{10}(0)^5\right]\right\} - \frac{9}{64}$

$= \frac{1}{2} - \frac{3}{10} - \frac{9}{64} = \frac{38}{640} = \frac{19}{320}$

c. $\sigma(X) = \sqrt{\frac{19}{320}} \approx 0.244$

20. **a.** $E(X) = \int_1^{16} x \cdot f(x)\,dx = \int_1^{16} x \cdot \frac{2}{3}x^{-3/2}\,dx = \int_1^{16} \frac{2}{3}x^{-1/2}\,dx$

$= \left(\frac{2}{3} \cdot 2x^{1/2}\right)\Big|_1^{16} = \frac{4}{3}(16)^{1/2} - \frac{4}{3}(1)^{1/2} = \frac{16}{3} - \frac{4}{3} = 4$

b. $\text{Var}(X) = \int_1^{16} x^2 f(x)\,dx - \mu^2 = \int_1^{16} x^2 \cdot \frac{2}{3}x^{-3/2}\,dx - 4^2$

$= \int_1^{16} \frac{2}{3}x^{1/2}\,dx - 16 = \left[\frac{2}{3}\left(\frac{2}{3}x^{3/2}\right)\right]\Big|_1^{16} - 16$

$= \frac{4}{9}(16^{3/2} - 1^{3/2}) - 16 = \frac{4}{9}(64 - 1) - 16 = 12$

c. $\sigma(X) = \sqrt{12} \approx 3.46$

21. **a.** $E(X)=\int_0^8 x\cdot f(x)\,dx=\int_0^8 x\cdot\frac{1}{32}x\,dx=\int_0^8\frac{x^2}{32}dx=\left.\frac{x^3}{96}\right|_0^8$

$=\frac{8^3}{96}-\frac{0^3}{96}\approx 5.33$

b. To find the value x such that $P(X\le x)=0.95$, first find $F(x)$.

$F(x)=\int_0^x\frac{1}{32}t\,dt=\left.\frac{t^2}{64}\right|_0^x=\frac{x^2}{64}-0=\frac{x^2}{64}$

Since we want x such that $F(x)=0.95$, we have

$\frac{x^2}{64}=0.95$

$x^2=60.8$

$x=\sqrt{60.8}\approx 7.8$ million kilowatt hours

22. **a.** First find $F(x)$

$F(x)=\int_0^x 12t(1-t)^2\,dt=\int_0^x(12t-24t^2+12t^3)\,dt$

$=(6t^2-8t^3+3t^4)\Big|_0^x=6x^2-8x^3+3x^4$

$P\left(X\le\frac{1}{2}\right)=F\left(\frac{1}{2}\right)=6\left(\frac{1}{2}\right)^2-8\left(\frac{1}{2}\right)^3+3\left(\frac{1}{2}\right)^4$

$=\frac{3}{2}-1+\frac{3}{16}=\frac{11}{16}$

b. $E(X)=\int_0^1 x\cdot 12x(1-x)^2\,dx=\int_0^1(12x^2-24x^3+12x^4)\,dx$

$=\left(4x^3-6x^4+\frac{12}{5}x^5\right)\Big|_0^1=4(1)^3-6(1)^4+\frac{12}{5}(1)^5-\left[4(0)^3-6(0)^4+\frac{12}{5}(0)^5\right]$

$=4-6+\frac{12}{5}=\frac{2}{5}$

23. **a.** $E(X)=\int_{-1}^1 x\cdot\frac{2}{\pi}\sqrt{1-x^2}\,dx=\frac{2}{\pi}\int_{-1}^1 x\sqrt{1-x^2}\,dx=\frac{2}{\pi}(0)=0$

b. $\text{Var}(X)=\int_{-1}^1 x^2 f(x)\,dx-\mu^2=\int_{-1}^1 x^2\cdot\frac{2}{\pi}\sqrt{1-x^2}\,dx-0^2$

$=\frac{2}{\pi}\int_{-1}^1 x^2\sqrt{1-x^2}\,dx\approx 0.25$

c. $\sigma(X)=\sqrt{\text{Var}(X)}=\sqrt{0.25}=0.5$

24. **a.** $E(X)=\int_0^1 x\cdot 504x^3(1-x)^5\,dx=\int_0^1 504x^4(1-x)^5\,dx$

≈ 0.4

b. $\text{Var}(X)=\int_0^1 x^2\cdot f(x)\,dx-\mu^2=\int_0^1 x^2\cdot 504x^3(1-x)^5\,dx-0.4^2$

$=\int_0^1 504x^5(1-x)^5\,dx-0.4^2\approx 0.0218$

c. $\sigma(X)=\sqrt{\text{Var}(X)}=\sqrt{0.0218}\approx 0.1477$

25. $P(-1 \le X \le 1) = \int_{-1}^{1} \frac{1}{2\pi}\sqrt{4-x^2}\,dx \approx 0.609$

26. x = number of minutes a person takes to complete a task

$f(x) = 105x^2(1-x)^4$ on [0, 1]

$$P(0.25 < X < 0.5) = \int_{0.25}^{0.5} 105x^2(1-x)^4\,dx \approx 0.53$$

27. **a.** The uniform probability density function on [0, 50] is $f(x) = \frac{1}{50}$.

b. $E(X) = \int_0^{50} x \cdot f(x)\,dx = \int_0^{50} x \cdot \frac{1}{50}\,dx = \frac{x^2}{100}\Big|_0^{50} = \frac{50^2}{100} - \frac{0}{100} = 25$

c. $\text{Var}(X) = \int_0^{50} x^2 \cdot f(x)\,dx - \mu^2 = \int_0^{50} \frac{x^2}{50}\,dx - 25^2$

$$= \frac{x^3}{150}\Big|_0^{50} - 625 = \frac{50^3}{150} - \frac{0^3}{150} - 625 = \frac{625}{3}$$

d. $\sigma(X) = \sqrt{\text{Var}(X)} = \sqrt{\frac{625}{3}} \approx 14.4$

e. $P(15 \le X \le 45) = \int_{15}^{45} \frac{1}{50}\,dx = \frac{x}{50}\Big|_{15}^{45} = \frac{45}{50} - \frac{15}{50} = \frac{3}{5}$

28. The uniform probability density function on [0, 2] is $f(x) = \frac{1}{2}$.

a. $E(X) = \int_0^2 x \cdot f(x)\,dx = \int_0^2 x \cdot \frac{1}{2}\,dx = \frac{x^2}{4}\Big|_0^2 = \frac{2^2}{4} - \frac{0^2}{4} = 1$

b. $\text{Var}(X) = \int_0^2 x^2 \cdot f(x)\,dx - \mu^2 = \int_0^2 x^2 \cdot \frac{1}{2}\,dx - 1^2$

$$= \frac{x^3}{6}\Big|_0^2 - 1 = \frac{2^3}{6} - \frac{0^3}{6} - 1 = \frac{4}{3} - 1 = \frac{1}{3}$$

c. $\sigma(X) = \sqrt{\text{Var}(X)} = \sqrt{\frac{1}{3}} = 0.577$

d. $P(X < 0.5) = F(0.5) = \int_0^{0.5} \frac{1}{2}\,dx = \frac{x}{2}\Big|_0^{0.5} = \frac{0.5}{2} - \frac{0}{2} = 0.25$

29. **a.** For the exponential probability density function $f(x) = 0.01e^{-0.01x}$, $E(X) = \frac{1}{0.01} = 100$.

b. $\text{Var}(X) = \frac{1}{(0.01)^2} = \frac{1}{0.0001} = 10{,}000$

30. Since the mean is 4, $E(X) = \frac{1}{a} = 4$. Thus, $a = \frac{1}{4}$ and $f(x) = \frac{1}{4}e^{-\frac{1}{4}x}$ on $[0, \infty)$.

31. Since the mean is 20, $\frac{1}{a}=20$ or $a=\frac{1}{20}$. Thus, the exponential probability density function is $f(x)=\frac{1}{20}e^{-\frac{1}{20}x}$.

a. $P(X\geq 60)=1-P(X<60)$

$$=1-\int_0^{60}\frac{1}{20}e^{-\frac{1}{20}x}\,dx=1-\left(\frac{1}{20}\cdot\frac{e^{-\frac{1}{20}x}}{-\frac{1}{20}}\right)\Bigg|_0^{60}=1+e^{-\frac{1}{20}x}\Bigg|_0^{60}$$
$$=1+(e^{-\frac{1}{20}(60)}-e^{-\frac{1}{20}(0)})$$
$$=1+e^{-3}-1=e^{-3}\approx 0.0498 \text{ or about } 5\%$$

b. $P(X\geq 180)=1-P(X<180)$

$$=1-\int_0^{180}\frac{1}{20}e^{-\frac{1}{20}x}\,dx=1-\left(\frac{1}{20}\cdot\frac{e^{-\frac{1}{20}x}}{-\frac{1}{20}}\right)\Bigg|_0^{180}=1+e^{-\frac{1}{20}x}\Bigg|_0^{180}$$
$$=1+(e^{-\frac{1}{20}(180)}-e^{-\frac{1}{20}(0)})$$
$$=1+e^{-9}-1=e^{-9}\approx 0.00012 \text{ or about } 0.012\%$$

[*Note*: Although this is a very small proportion, it predicts that of the 7.5 million flights annually, more than 900 will have such runway delays, which agrees with flight records.]

32. Since the mean is 1000, $\frac{1}{a}=1000$ or $a=\frac{1}{1000}$. Thus, the exponential probability density function is $f(x)=\frac{1}{1000}e^{-\frac{1}{1000}x}$.

$$P(X\leq 500)=\int_0^{500}\frac{1}{1000}e^{-\frac{1}{1000}x}\,dx=-e^{-\frac{1}{1000}x}\Bigg|_0^{500}$$
$$=-e^{-\frac{500}{1000}}-(-e^{-\frac{1}{1000}(0)})$$
$$=-e^{-\frac{1}{2}}+e^0\approx 0.393 \text{ or about } 39\%$$

33. Since the median is the number that is the middle point of the distribution, we are looking for the number a such that $F(a)=0.5$.

$$0.5=F(a)=\int_0^a\frac{3}{8}x^2\,dx$$
$$=\frac{1}{8}x^3\Bigg|_0^a=\frac{1}{8}a^3-\frac{1}{8}(0)^3=\frac{a^3}{8}$$

Thus,

$$\frac{a^3}{8}=0.5$$
$$a^3=4$$
$$a=\sqrt[3]{4}\approx 1.59$$

34. $$0.5=F(a)=\int_0^a 3e^{-3x}\,dx=-e^{-3x}\Bigg|_0^a$$
$$=-e^{-3a}-(-e^{-3(0)})=1-e^{-3a}$$

Thus,

$$e^{-3a}=0.5$$
$$a=\frac{\ln 0.5}{-3}\approx 0.231$$

35. x = the time between calls to a fire department is an exponential random variable with mean 4. Thus, $\frac{1}{a}=4$ or $a=\frac{1}{4}$. The exponential probability density function is $f(x)=\frac{1}{4}e^{-\frac{1}{4}x}$.

$$P(X<1)=\int_0^1\frac{1}{4}e^{-\frac{1}{4}x}\,dx\approx 0.221 \text{ or about } 22\%$$

36. x = the time between arrivals of ships is an exponential random variable with mean 5. Thus, $\frac{1}{a}=5$ or $a=\frac{1}{5}$. The exponential probability density function is $f(x)=\frac{1}{5}e^{-\frac{1}{5}x}$.

$$P(X<1)=\int_0^1\frac{1}{5}e^{-\frac{1}{5}x}\,dx\approx 0.181 \text{ or about } 18\%$$

37. $P(0.95\leq Z\leq 2.54)=0.4945-0.3289=0.1655$

38. $P(-1.11\leq Z\leq 1.44)=0.4251+0.3665=0.7916$

39. $P(Z<2.5)=0.4938+0.5=0.9938$

40. $P(Z > 1) = 1 - P(Z \le 1)$
$= 1 - (0.5 + 0.3413) = 0.1587$

41. $P(10 < X \le 15) = \int_{10}^{15} \frac{1}{2\sqrt{2\pi}} e^{-\frac{1}{2}\left(\frac{x-12}{2}\right)^2} dx$
$= 0.7745$

42. $P(0 < X \le 1.8) = \int_{0}^{1.8} \frac{1}{0.4\sqrt{2\pi}} e^{-\frac{1}{2}\left(\frac{x-1}{0.4}\right)^2} dx \approx 0.971$

43. Let x = the lifetime of miles of a tire. Then x is a normal random variable with $\mu = 50,000$ and $\sigma = 5000$.

$P(X > 40,000) = 1 - P(X \le 40,000)$

$$= 1 - \left[P(X \le 50,000) - \int_{40,000}^{50,000} \frac{1}{5000\sqrt{2\pi}} e^{-\frac{1}{2}\left(\frac{x-50,000}{5000}\right)^2} dx \right]$$

$= 1 - (0.5 - 0.4772) = 0.9772$ or about 98%

44. Let x = the number of cigarettes smoked. Then x is a normal random variable with $\mu = 28$ and $\sigma = 10$.

$P(X \ge 40) = 1 - P(X < 40)$

$= 1 - [P(X \le 28) + P(28 < X < 40)]$

$$= 1 - \left(0.5 + \int_{28}^{40} \frac{1}{10\sqrt{2\pi}} e^{-\frac{1}{2}\left(\frac{x-28}{10}\right)^2} dx \right)$$

$= 1 - 0.5 - 0.3849 = 0.115$ or about 12%

45. Let x = the height of a woman. Then x is a normal random variable with $\mu = 64.6$ and $\sigma = 2.9$

$$P(60 \le X \le 72) = \int_{60}^{72} \frac{1}{2.9\sqrt{2\pi}} e^{-\frac{1}{2}\left(\frac{x-64.6}{2.9}\right)^2} dx$$

≈ 0.938 or about 94%

46. $\mu = 496$, $\sigma = 115$. Using the graphing calculator program to find x such that $P(X > x) = 0.7$ we get $x = 556.3$, so a score of 556 would put you in the top 30%.

CUMULATIVE REVIEW—CHAPTERS 1–11

1. **a.** The horizontal line through (–2, 5) is $y = 5$.
b. The vertical line through (–2, 5) is $x = -2$.

2. $\left(\frac{1}{4}\right)^{-3/2} = 4^{3/2} = (\sqrt{4})^3 = 2^3 = 8$

3.
$$f'(x) = \lim_{h\to 0} \frac{f(x+h) - f(x)}{h} = \lim_{h\to 0} \frac{3(x+h)^2 - 7(x+h) + 1 - (3x^2 - 7x + 1)}{h}$$
$$= \lim_{h\to 0} \frac{3(x^2 + 2hx + h^2) - 7(x+h) + 1 - 3x^2 + 7x - 1}{h}$$
$$= \lim_{h\to 0} \frac{3x^2 + 6hx + 3h^2 - 7x - 7h + 1 - 3x^2 + 7x - 1}{h}$$
$$= \lim_{h\to 0} \frac{6hx + 3h^2 - 7h}{h} = \lim_{h\to 0} 6x + 3h - 7 = 6x - 7$$

4. $T(t) = 12\sqrt{t^3} + 225 = 12t^{3/2} + 225$

$T'(t) = \frac{3}{2} \cdot 12t^{1/2} = 18t^{1/2}$

$T'(4) = 18(4)^{1/2} = 36$

After 4 hours, the temperature is rising at the rate of 36 degrees per hour.

$T''(t) = \frac{1}{2} \cdot 18t^{-1/2} = 9t^{-1/2}$

$T''(4) = 9(4)^{-1/2} = 9 \cdot \frac{1}{2} = \frac{9}{2}$

After 4 hours, the temperature rate of increase is rising by 4.5 degrees per hour per hour.

5. $\frac{d}{dx}\sqrt[3]{x^3+8} = \frac{d}{dx}(x^3+8)^{1/3} = \frac{1}{3}(x^3+8)^{-2/3} \cdot 3x^2$

$= x^2(x^3+8)^{-2/3}$

6. $f(x) = (2x+3)^3(3x+2)^4$

$f'(x) = 3(2x+3)^2(2)(3x+2)^4 + (2x+3)^3 \cdot 4(3x+2)^3 \cdot 3$

$= 6(2x+3)^2(3x+2)^4 + 12(2x+3)^3(3x+2)^3$

7. $f(x) = x^3 - 12x^2 - 60x + 400$

$f'(x) = 3x^2 - 24x - 60$

$f''(x) = 6x - 24$

Set $f'(x) = 0$ and solve.

$3x^2 - 24x - 60 = 0$

$x^2 - 8x - 20 = 0$

$(x-10)(x+2) = 0$

$x = 10, -2$

There are critical values at $x = -2$ and $x = 10$.

$f' > 0$ | $f' = 0$ | $f' < 0$ | $f' = 0$ | $f' > 0$

$x = -2$ | $x = 10$

↗ → ↘ → ↗

rel max | rel min

Set $f''(x) = 0$ and solve.

$6x - 24 = 0$

$x = 4$

$f'' < 0$ | $f'' = 0$ | $f'' > 0$

con dn | $x = 4$ | con up

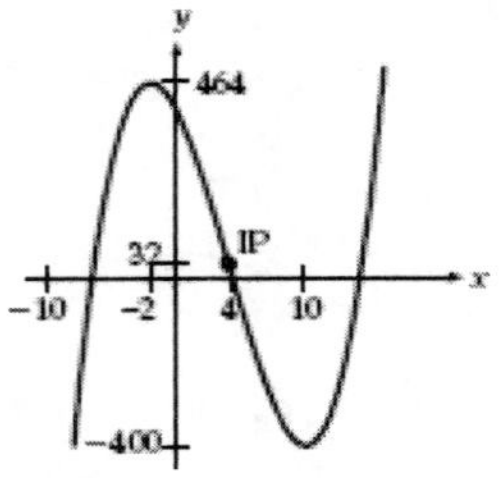

8. $f(x) = \sqrt[3]{x} + 1 = x^{1/3} + 1$

$f'(x) = \frac{1}{3}x^{-2/3}$

$f''(x) = -\frac{2}{3} \cdot \frac{1}{3}x^{-5/3} = -\frac{2}{9}x^{-5/3}$

Set $f'(x) = 0$ and solve.

$\frac{1}{3}x^{-2/3} = 0$

$\frac{1}{3x^{2/3}} = 0$

There are no solutions. Set the denominator equal to 0 to find a critical point.

$3x^{2/3} = 0$

$x = 0$

$f' > 0$ | $f' > 0$

↗ | $x = 0$ | ↗

$f''(x) = -\frac{2}{9}x^{-5/3}$ is also undefined at $x = 0$.

$f'' > 0$ | $f'' < 0$

con up | $x = 0$ | con dn

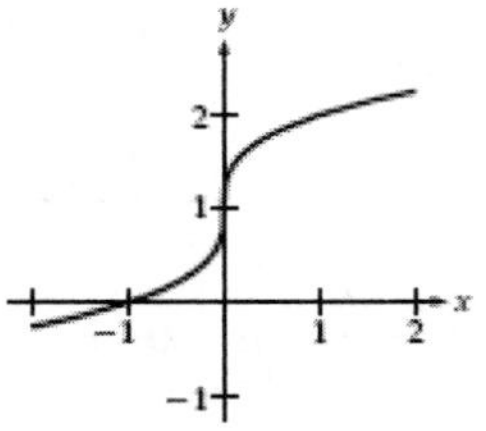

9. $f(x)=\frac{1}{x^2-4}=(x^2-4)^{-1}$

$f'(x)=-(x^2-4)^{-2}(2x)=-2x(x^2-4)^{-2}$

$f''(x)=-2(x^2-4)^{-2}+(-2x)(-2)(x^2-4)^{-3}(2x)$
$\quad=-2(x^2-4)^{-2}+8x^2(x^2-4)^{-3}$

Set $f'(x)=0$ and solve.

$\frac{-2x}{(x^2-4)^2}=0$

$x=0$

Note that $f'(x)$ is undefined at $x = -2, 2$. Thus, there are critical points at $x = 0, -2, 2$.

$f'>0$		$f'>0$		$f'<0$		$f'<0$
↗	$x=-2$	↗	$x=0$	↘	$x=2$	↘

Set $f''(x)=0$ and solve.

$\frac{-2}{(x^2-4)^2}+\frac{8x^2}{(x^2-4)^3}=0$

$-2(x^2-4)+8x^2=0$

$6x^2+8=0$

This has no solutions, but f'' is also undefined at $x = -2, 2$.

$f''>0$		$f''<0$		$f''>0$
con up	$x=-2$	con dn	$x=2$	con up

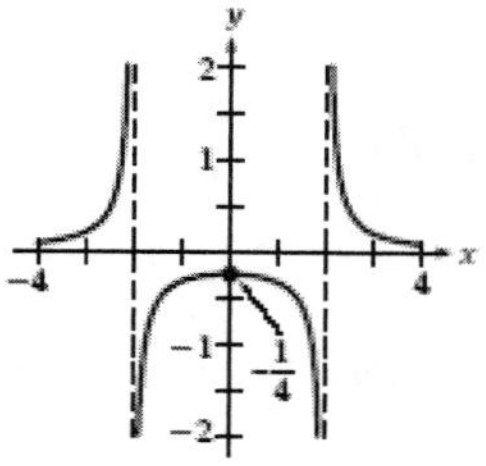

10. Let x = the length of a side perpendicular to the wall. Let y = the length of the side parallel to the wall. Since the total length of fence is 300, then

$$3x + y = 300$$
$$y = 300 - 3x$$

We wish to maximize the area.

$$A = xy = x(300 - 3x) = 300x - 3x^2$$
$$A' = 300 - 6x = 0$$
$$6x = 300$$
$$x = 50$$

Thus, $y = 300 - 3(50) = 300 - 150 = 150$, and the largest area is $50 \cdot 150 = 7500$ square feet.

11. Let x = the length of one side of the square to be cut. Then the length of each side of the bottom of the box is $12 - 2x$.

$$V = l \cdot w \cdot h$$
$$= (12 - 2x)(12 - 2x)x$$
$$= (144 - 48x + 4x^2)x$$
$$= 4x^3 - 48x^2 + 144x$$
$$V' = 12x^2 - 96x + 144 = 0$$
$$x^2 - 8x + 12 = 0$$
$$(x - 6)(x - 2) = 0$$
$$x = 6, 2$$

For $x = 2$, $V = (12 - 4)(12 - 4)(2) = 128$ cubic inches.
For $x = 6$, $V = (12 - 12)(12 - 12)(6) = 0$.
Thus, the largest box has a volume of 128 cubic inches.

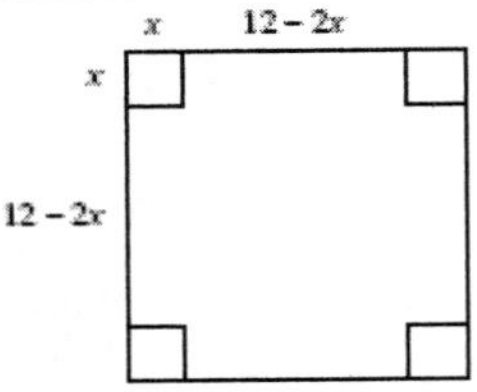

12. $$xy - 3y = 5$$
$$1 \cdot y + x \cdot \frac{dy}{dx} - 3\frac{dy}{dx} = 0$$
$$(x - 3)\frac{dy}{dx} = -y$$
$$\frac{dy}{dx} = \frac{-y}{x - 3}$$

At (4, 5), $\frac{dy}{dx} = \frac{-5}{4 - 3} = -5$

13. $$A = \pi r^2$$
$$\frac{dA}{dt} = 2\pi r \frac{dr}{dt}$$

Since $\frac{dA}{dt} = 3$ and $r = 2$, then

$$3 = 2\pi(2)\frac{dr}{dt}$$
$$\frac{dr}{dt} = \frac{3}{4\pi} \approx 0.24 \text{ mile per day}$$

14. **a.** For semiannual compounding, $r = \frac{0.06}{2} = 0.03$ and $n = 4 \cdot 2 = 8$.

$$P(1 + r)^n = 5000(1 + 0.03)^8 \approx \$6333.85$$

b. For continuous compounding, $r = 0.06$ and $n = 4$.

$$Pe^{rn} = 5000e^{0.06(4)} \approx \$6356.25$$

15. For continuous compounding, $r = 0.09$. We wish to find when P reaches $1.75P$.

$$Pe^{0.09n} = 1.75P$$
$$e^{0.09n} = 1.75$$
$$n = \frac{\ln 1.75}{0.09} \approx 6.22 \text{ years}$$

16. For quarterly compounding, $r = \frac{0.08}{4} = 0.02$ and $n = 4 \cdot 25 = 100$. To find how much to deposit, solve

$$P(1 + 0.02)^{100} = 100{,}000$$
$$P = \frac{100{,}000}{1.02^{100}} \approx \$13{,}803.30$$

17. **a.** $$\frac{d}{dx}\left(\frac{e^{2x}}{x}\right) = \frac{xe^{2x} \cdot 2 - 1 \cdot e^{2x}}{x^2} = \frac{2xe^{2x} - e^{2x}}{x^2}$$

b. $$\frac{d}{dx}\ln(x^3 + 1) = \frac{1}{x^3 + 1} \cdot 3x^2 = \frac{3x^2}{x^3 + 1}$$

18. $E(p) = p \cdot D(p) = p \cdot 12{,}000e^{-0.04p} = 12{,}000pe^{-0.04p}$

$$\begin{aligned} E'(p) &= 12{,}000e^{-0.04p} + 12{,}000p(-0.04)e^{-0.04p} \\ &= 12{,}000e^{-0.04p} - 480pe^{-0.04p} = 0 \\ 12{,}000 - 480p &= 0 \\ p &= \$25 \end{aligned}$$

19. $G(t) = 3 + 5e^{0.06t}$

$G'(t) = 0.3e^{0.06t}$

The relative rate of change is

$$\frac{G'(t)}{G(t)} = \frac{0.3e^{0.06t}}{3 + 5e^{0.06t}}$$

At $t = 2$,

$$\frac{G'(2)}{G(2)} = \frac{0.3e^{0.06(2)}}{3 + 5e^{0.06(2)}} \approx 0.039 \text{ or } 3.9\%$$

20. **a.** $$\begin{aligned} \int (12x^3 - 4x + 5)\,dx &= 12 \cdot \frac{x^4}{4} - 4 \cdot \frac{x^2}{2} + 5x + C \\ &= 3x^4 - 2x^2 + 5x + C \end{aligned}$$

b. $$\int 6x^{-2x}\,dx = 6 \cdot \frac{e^{-2x}}{-2} + C = -3e^{-2x} + C$$

21. $$\int_1^e \frac{2}{x}\,dx = 2\int_1^e \frac{1}{x}\,dx = 2\ln x \Big|_1^e = 2(\ln e - \ln 1) = 2(1 - 0) = 2$$

22. $$\begin{aligned} C(x) &= \int MC(x)\,dx \\ &= \int_{25}^{100} 3x^{-1/2}\,dx = 3\frac{x^{1/2}}{\frac{1}{2}}\Bigg|_{25}^{100} = 6x^{1/2}\Big|_{25}^{100} \\ &= 6(100)^{1/2} - 6(25)^{1/2} = 60 - 30 \\ &= 30 \text{ thousand dollars} \end{aligned}$$

23. $$\begin{aligned} \int_0^t 120e^{0.15x}\,dx &= 120 \cdot \frac{e^{0.15x}}{0.15}\Bigg|_0^t = 800e^{0.15x}\Big|_0^t \\ &= 800e^{0.15t} - 800e^{0.15(0)} \\ &= 800e^{0.15t} - 800 \end{aligned}$$

24. First, find the points of intersection

$$\begin{aligned} x^2 + 4 &= 6x + 4 \\ x^2 - 6x &= 0 \\ x(x - 6) &= 0 \\ x &= 0, 6 \end{aligned}$$

At $x = 1$, $y = (1)^2 + 4 = 5$ and $y = 6(1) + 4 = 10$.

Thus, $y = 6x + 4$ is the upper curve

$$\begin{aligned} \text{Area} &= \int_0^6 [(6x + 4) - (x^2 + 4)]\,dx \\ &= \int_0^6 (6x - x^2)\,dx = \left(3x^2 - \frac{x^3}{3}\right)\Bigg|_0^6 \\ &= 3(6)^2 - \frac{6^3}{3} - \left[3(0)^2 - \frac{0^3}{3}\right] = 36 \text{ square units} \end{aligned}$$

25. $$\begin{aligned} (\text{Average population}) &= \frac{1}{b - a}\int_a^b P(t)\,dt \\ &= \frac{1}{10 - 0}\int_0^{10} 60e^{0.04t}\,dt = \frac{1}{10}\left(60 \cdot \frac{e^{0.04t}}{0.04}\right)\Bigg|_0^{10} \\ &= 150e^{0.04t}\Big|_0^{10} = 150e^{0.04(10)} - 150e^{0.04(0)} \approx 73.8 \end{aligned}$$

26. **a.** $\int \frac{x\,dx}{\sqrt{x^2+1}}$ Let $u = x^2+1$; then $du = 2x\,dx$.

$$\int \frac{\frac{1}{2}}{\sqrt{u}}du = \frac{1}{2}\left(\frac{u^{1/2}}{\frac{1}{2}}+C\right) = u^{1/2}+C = \sqrt{x^2+1}+C$$

b. $\int_0^{-1} xe^{-x^2}dx$ Let $u = -x^2$. Then $du = -2x\,dx$.
For $x = 0$, $u = -0^2 = 0$.
For $x = 1$, $u = -(1)^2 = -1$

$$\int_0^{-1} e^u\left(\frac{du}{-2}\right) = -\frac{1}{2}e^u\Big|_0^{-1} = -\frac{1}{2}e^{-1} - \left(-\frac{1}{2}e^0\right)$$
$$= -\frac{1}{2}e^{-1}+\frac{1}{2}$$

27. $\int x^2 \ln x\,dx$ Let $u = \ln x$, $dv = x^2\,dx$ Then $du = \frac{1}{x}dx$, $v = \frac{x^3}{3}$.

$$\int x^2 \ln x\,dx = \ln x \cdot \frac{x^3}{3} - \int \frac{x^3}{3}\cdot\frac{1}{x}dx = \frac{x^3}{3}\ln x - \int \frac{x^2}{3}dx$$
$$= \frac{x^3}{3}\ln x - \frac{x^3}{9}+C$$

28. $\int \frac{x}{\sqrt{x-1}}dx$ Use Formula 13 with $a = 1$, $b = -1$.

$$\int \frac{x}{\sqrt{x-1}}dx = \frac{2(1)x-4(-1)}{3(1)^2}\sqrt{x-1}+C$$
$$= \frac{2x+4}{3}\sqrt{x-1}+C$$

29.

$$\int_0^{\infty} e^{-2x}dx = \lim_{a\to\infty}\int_0^a e^{-2x}dx = \lim_{a\to\infty}\left(\frac{e^{-2x}}{-2}\right)\Big|_0^a$$
$$= \lim_{a\to\infty}\left[-\frac{e^{-2a}}{2} - \left(-\frac{e^0}{2}\right)\right]$$
$$= \left(\lim_{a\to\infty} -\frac{e^{-2a}}{2}\right)+\frac{1}{2} = 0+\frac{1}{2} = \frac{1}{2}$$

30. $\int_0^1 e^{\sqrt{x}}dx$, $n = 4$ $\Delta x = \frac{1-0}{4} = \frac{1}{4}$

x	$f(x) = e^{\sqrt{x}}$
0	$1 \to 0.5$
$\frac{1}{4}$	1.649
$\frac{1}{2}$	2.028
$\frac{3}{4}$	2.377
1	$2.718 \to 1.359$

$0.5+1.649+2.028+2.377+1.359 = 7.913$

$$\int_0^1 e^{\sqrt{x}}dx \approx 7.913\left(\frac{1}{4}\right) \approx 1.98$$

31. $\int_0^1 e^{\sqrt{x}}\,dx,\ n=4 \qquad \Delta=\frac{1-0}{4}=\frac{1}{4}$

x	$e^{\sqrt{x}}$	Weight	$e^{\sqrt{x}}\cdot$ weight
0	1	1	1
$\frac{1}{4}$	1.649	4	6.596
$\frac{1}{2}$	2.028	2	4.056
$\frac{3}{4}$	2.377	4	9.508
1	2.718	1	2.718
			23.878

$\int e^{\sqrt{x}}\,dx \approx 23.878\left(\frac{0.25}{3}\right)\approx 1.99$

FnInt produces 2.000.

32. $f(x,y)=xe^y-y\ln(x^2+1)$

$f_x(x,y)=e^y-y\left(\frac{1}{x^2+1}\right)\cdot 2x=e^y-y\frac{2x}{x^2+1}$

$f_y(x,y)=xe^y-\ln(x^2+1)$

33. $f(x,y)=3x^2+2y^2-2xy-8x-4y+15$

$f_x(x,y)=6x-2y-8=0$

$f_y(x,y)=4y-2x-4=0$

Solving these systems simultaneously gives $x=2,\ y=2$.

$f_{xx}(x,y)=6,\ f_{yy}(x,y)=4,$ and $f_{xy}(x,y)=-2.$ Thus,

$D=(6)(4)-(-2)^2=24-4=20>0$

and $f_{xx}>0$. Thus, f has a relative minimum at (2, 2) and

$f(2,2)=3(2)^2+2(2)^2-2(2)(2)-8(2)-4(2)+15=3$

f has no maximum.

34.

x	y	xy	x^2
2	4	8	4
4	3	12	16
6	0	0	36
8	−1	−8	64
$\Sigma x=20$	$\Sigma y=6$	$\Sigma xy=12$	$\Sigma x^2=120$

$a=\frac{n\Sigma xy-(\Sigma x)(\Sigma y)}{n\Sigma x^2-(\Sigma x)^2}=\frac{4(12)-(20)(6)}{4(120)-(20)^2}$

$=\frac{48-120}{480-400}=-0.9$

$b=\frac{1}{n}(\Sigma y-a\Sigma x)=\frac{1}{4}[6+0.9(20)]=6$

The least squares line is $y=-0.9x+6$.

35. Maximize $f(x, y) = 100 - 2x^2 - 3y^2 + 2xy$ subject to $2x + y = 18$.

$F(x, y, \lambda) = 100 - 2x^2 - 3y^2 + 2xy + \lambda(2x + y - 18)$

$F_x = -4x + 2y + 2\lambda = 0$ or $\lambda = 2x - y$

$F_y = -6y + 2x + \lambda = 0$ or $\lambda = -2x + 6y$

$F_\lambda = 2x + y - 18 = 0$

Equating the equations for λ, we have

$$\begin{aligned} 2x - y &= -2x + 6y \\ 4x &= 7y \\ x &= \frac{7}{4}y \end{aligned}$$

Substituting into the equation for F_λ, we get

$$\begin{aligned} 2\left(\frac{7}{4}y\right) + y - 18 &= 0 \\ \frac{7}{2}y + y &= 18 \\ \frac{9}{2}y &= 18 \\ y &= 4 \\ x &= \frac{7}{4}(4) = 7 \end{aligned}$$

The maximum constrained value of f occurs at (7, 4) and

$$\begin{aligned} f(7, 4) &= 100 - 2(7)^2 - 3(4)^2 + 2(7)(4) \\ &= 100 - 98 - 48 + 56 = 10 \end{aligned}$$

36. $f(x, y) = 5x^2 - 2xy + 2y^3 - 12$

$f_x = 10x - 2y$

$f_y = -2x + 6y^2$

Total differential $df = (10x - 2y)\,dx + (-2x + 6y^2)\,dy$

37. $$\begin{aligned} \text{Volume} &= \int_0^3 \int_0^3 (x^2 + y^2)\,dy\,dx = \int_0^3 \left(x^2 y + \frac{y^3}{3}\right)\Bigg|_0^3 dx \\ &= \int_0^3 \left(3x^2 + \frac{3^3}{3}\right) dx = \int_0^3 (3x^2 + 9)\,dx = (x^3 + 9x)\Big|_0^3 \\ &= 3^3 + 9(3) - [0^3 + 9(0)] = 54 \text{ cubic units} \end{aligned}$$

38. **a.** $135° = 135° \cdot \frac{\pi}{180°} = \frac{3\pi}{4}$

b. $\frac{5\pi}{6} = \frac{5\pi}{6} \cdot \frac{180°}{\pi} = 150°$

39. **a.** $\sin\frac{3\pi}{4}=\frac{y}{r}=\frac{1}{\sqrt{2}}=\frac{\sqrt{2}}{2}$

b. $\cos\frac{3\pi}{4}=\frac{x}{r}=\frac{-1}{\sqrt{2}}=-\frac{\sqrt{2}}{2}$

c. $\sin\frac{5\pi}{6}=\frac{1}{2}$

d. $\cos\frac{5\pi}{6}=\frac{-\sqrt{3}}{2}=-\frac{\sqrt{3}}{2}$

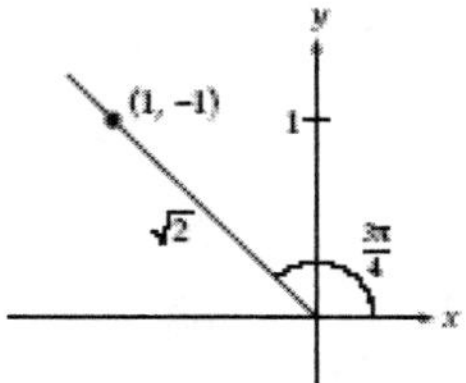

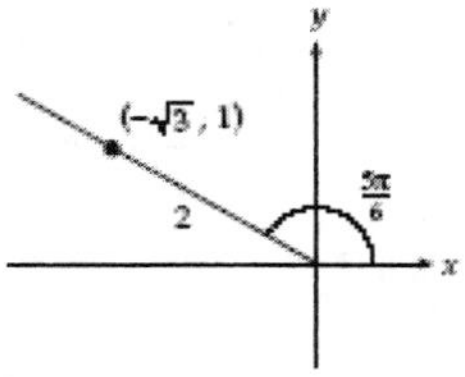

40.

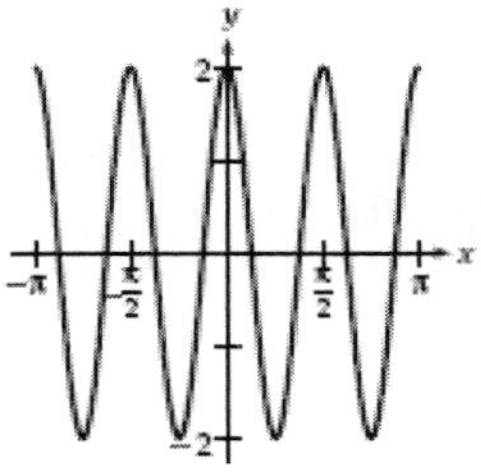

41. $\frac{d}{dx}(\sin x\cos x)=\cos x\cdot\cos x+\sin x\cdot(-\sin x)$
$=\cos^2 x-\sin^2 x$

42. $\int e^{\cos x}\sin x\,dx$ Let $u=\cos x$, $du=-\sin x\,dx$.
$\int e^{\cos x}\sin x\,dx=\int e^{u}(-du)=-e^{u}+C$
$=-e^{\cos x}+C$

43. $\text{Area}=\int_0^{\pi/2}\cos x\,dx=\sin x\Big|_0^{\pi/2}$
$=\sin\frac{\pi}{2}-\sin 0=1$ square unit

44. **a.** $\tan\frac{5\pi}{6}=\frac{y}{x}=\frac{1}{-\sqrt{3}}=-\frac{\sqrt{3}}{3}$

b. $\cot\frac{5\pi}{6}=\frac{x}{y}=\frac{-\sqrt{3}}{1}=-\sqrt{3}$

c. $\sec\frac{3\pi}{4}=\frac{r}{x}=\frac{\sqrt{2}}{-1}=-\sqrt{2}$

d. $\csc\frac{3\pi}{4}=\frac{r}{y}=\frac{\sqrt{2}}{1}=\sqrt{2}$

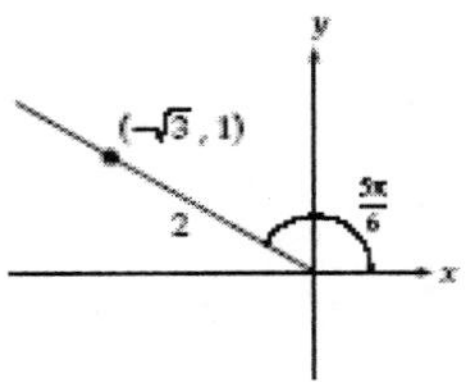

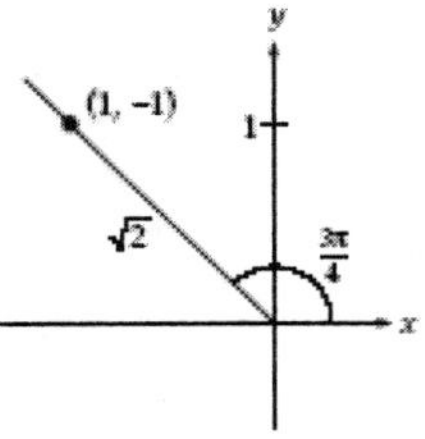

45. **a.** $\frac{d}{dx}\tan(x^2+1) = \sec^2(x^2+1)\cdot 2x = 2x\sec^2(x^2+1)$

b. $\int \sec(2t+1)\tan(2t+1)\,dt$ Let $u = 2t+1;\ du = 2\,dt.$

$$\int \sec u \tan u\left(\frac{du}{2}\right) = \frac{1}{2}\sec u + C = \frac{1}{2}\sec(2t+1) + C$$

46. **a.** $y' = \frac{dy}{dx} = x^3 y$

$$\frac{dy}{y} = x^3 dx$$

$$\int \frac{dy}{y} = \int x^3 dx$$

$$\ln y = \frac{x^4}{4} + C$$

$$y = e^{x^4/4+C} = Ce^{x^4/4}$$

b. Since $y(0) = 2$,

$2 = Ce^{(0)^4/4} = C\cdot e^0 = C$

Thus, the particular solution is $y = 2e^{x^4/4}$.

47. $y' + 6xy = 0, \quad y(0) = 4$

$$\frac{dy}{dx} + 6xy = 0$$

$$\frac{dy}{dx} = -6xy$$

$$\frac{dy}{y} = -6x\,dx$$

$$\int \frac{dy}{y} = \int -6x\,dx$$

$$\ln y = -3x^2 + C$$

$$y = e^{-3x^2+C} = Ce^{-3x^2}$$

Since $y(0) = 4$,

$4 = Ce^{-3(0)^2} = C\cdot e^0 = c$

Thus, $y = 4e^{-3x^2}$.

48. $y' = \frac{12x}{y}$, $y(0) = 2$. The initial point is (0, 2), and the slope is $g(x, y) = \frac{12x}{y}$. The step size is

$$\frac{1-0}{4} = 0.25$$

Segment 1

Slope $g(0, 2) = \frac{12 \cdot 0}{2} = 0$

Endpoint $x_1 = 0.25$, $y_1 = 2 + (0)(0.25) = 2$

Segment 2

Slope $g(0.25, 2) = \frac{12 \cdot 0.25}{2} = \frac{3}{2}$

Endpoint $x_2 = 0.5$, $y_2 = 2 + \left(\frac{3}{2}\right)(0.25) = 2.375$

Segment 3

Slope $g(0.5, 2.375) = \frac{12 \cdot 0.5}{2.375} \approx 2.53$

Endpoint $x_3 = 0.75$, $y_3 = 2.375 + 2.53(0.25) \approx 3.008$

Segment 4

Slope $g(0.75, 3.008) = \frac{12 \cdot 0.75}{3.008} \approx 2.99$

Endpoint $x_4 = 1$, $y_4 = 3.008 + 2.99(0.25) \approx 3.756$

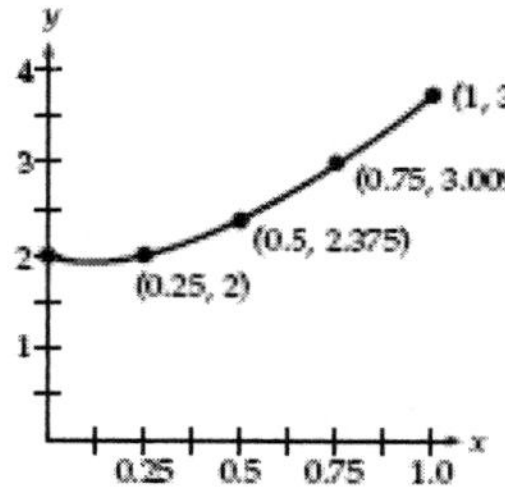

49. $\frac{\frac{16}{25}}{\frac{4}{5}} = \frac{16}{25} \cdot \frac{5}{4} = \frac{4}{5}$

Since the ratio is $\frac{4}{5} < 1$, the series converges.

$$S = \frac{a}{1-r} = \frac{\frac{4}{5}}{1 - \frac{4}{5}} = \frac{\frac{4}{5}}{\frac{1}{5}} = 4$$

50. $f(x) = e^{2x}$, $f'(x) = 2e^{2x}$, $f''(x) = 4e^{2x}$, $f'''(x) = 8e^{2x}$

$$p_3(x) = e^{2(0)} + 2e^{2(0)} \frac{x}{1!} + 4e^{2(0)} \frac{x^2}{2!} + 8e^{2(0)} \frac{x^3}{3!}$$
$$= 1 + 2x + \frac{4}{2}x^2 + \frac{8}{6}x^3 = 1 + 2x + 2x^2 + \frac{4}{3}x^3$$

51. Substitute x^2 for x in the Taylor series for $\sin x$.

$$\sin x^2 \approx x^2 - \frac{(x^2)^3}{3!} + \frac{(x^2)^5}{5!} - \frac{(x^2)^7}{7!} + \cdots$$
$$= x^2 - \frac{x^6}{3!} + \frac{x^{10}}{5!} - \frac{x^{14}}{7!} + \cdots$$

$$c_n = (-1)^{n+1} \frac{x^{2(2n-1)}}{(2n-1)!}, \quad c_{n+1} = (-1)^{n+2} \frac{x^{2(2n+1)}}{(2n+1)!}$$

$$\lim_{n\to\infty} \left| \frac{c_{n+1}}{c_n} \right| = \lim_{n\to\infty} \left| \frac{(-1)^{n+2} \frac{x^{2(2n+1)}}{(2n+1)!}}{(-1)^{n+1} \frac{x^{2(2n-1)}}{(2n-1)!}} \right|$$
$$= \lim_{n\to\infty} \left| (-1) \frac{x^{4n+2}}{x^{4n-2}} \cdot \frac{(2n-1)!}{(2n+1)!} \right|$$
$$= \lim_{n\to\infty} \left| (-1) \frac{x^4}{(2n+1)(2n)} \right| = 0$$

Thus, since the limit approaches 0 for all x, the radius of convergence is $R = \infty$.

52. To estimate $\sqrt{105}$, consider the function $f(x) = x^2 - 105$. Thus, $f'(x) = 2x$.

x_n	$x_n - \frac{f(x_n)}{f'(x_n)}$
10	10.25
10.25	10.2469512
10.2469512	10.2469507

53. Since the mean is 4, use a Poisson distribution with $a = 4$.

$$P(X \le 3) = \sum_{k=0}^{3} e^{-4}\frac{4^k}{k!}$$
$$= e^{-4}\frac{4^0}{0!} + e^{-4}\frac{4^1}{1!} + e^{-4}\frac{4^2}{2!} + e^{-4}\frac{4^3}{3!} \approx 0.433$$

54. The uniform probability density function on [0, 18] is $f(x) = \frac{1}{18}$.

a. $$E(X) = \int_0^{18} x \cdot f(x)\,dx = \int_0^{18} \frac{x}{18}\,dx = \frac{x^2}{36}\bigg|_0^{18}$$
$$= \frac{18^2}{36} - \frac{0^2}{36} = 9$$

b. $$\text{Var}(X) = \int_0^{18} x^2 f(x)\,dx - \mu^2$$
$$= \int_0^{18} \frac{x^2}{18}\,dx - (9)^2$$
$$= \frac{x^3}{54}\bigg|_0^{18} - 81 = \left(\frac{18^3}{54} - \frac{0^3}{54}\right) - 81 = 27$$

55. Since the mean is 1, $\frac{1}{a} = 1$ or $a = 1$. Thus, the exponential probability density function is $f(x) = 1$.

$$P\left(X \le \frac{3}{2}\right) = \int_0^{3/2} e^{-x}\,dx = \frac{e^{-x}}{-1}\bigg|_0^{3/2}$$
$$= -e^{-3/2} - (-e^0) \approx 0.777$$

56. $$P(11 \le X \le 13) = \int_{11}^{13} \frac{1}{2\sqrt{2\pi}} e^{-\frac{1}{2}\left(\frac{x-12}{2}\right)^2}\,dx$$
$$\approx 0.383$$